T0371610

Introduction to Optical and Optoelectronic Properties of Nanostructures

Get to grips with the fundamental optical and optoelectronic properties of nanostructures. This comprehensive guide makes a wide variety of modern topics accessible, including up-to-date material on the optical properties of monolayer crystals, plasmonics, nanophotonics, ultraviolet quantum well lasers, and wide-bandgap materials and heterostructures. The unified, multidisciplinary approach makes it ideal for those in disciplines spanning nanoscience, physics, materials science, and optical, electrical, and mechanical engineering.

Building on work first presented in *Quantum Heterostructures* (Cambridge 1999), this volume draws on years of research and teaching experience. Rigorous coverage of basic principles makes it an excellent resource for senior undergraduates, and detailed mathematical derivations illuminate concepts for graduate students, researchers, and professional engineers. The examples with solutions included in the text and end-of-chapter problems allow students to use this text to enhance their understanding.

Vladimir V. Mitin is a SUNY Distinguished Professor in the Department of Electrical Engineering and in the Department of Physics at the University at Buffalo, State University of New York.

Viacheslav A. Kochelap is Professor of Theoretical Physics and Head of the Theoretical Physics Department at the Institute of Semiconductor Physics, National Academy of Sciences of Ukraine.

Mitra Dutta is the Vice Chancellor for Research and a University Distinguished Professor at the University of Illinois at Chicago and also holds appointments in the Departments of Electronic and Computer Engineering and Physics.

Michael A. Stroscio is a University Distinguished Professor and Richard and Loan Hill Professor at the University of Illinois at Chicago and also holds appointments in the Departments of Electrical and Computer Engineering, Bioengineering, and Physics.

Introduction to Optical and Optoelectronic Properties of Nanostructures

VLADIMIR V. MITIN
University at Buffalo, State University of New York

VIACHESLAV A. KOCHELAP
Lashkaryov Institute of Semiconductor Physics, Ukraine

MITRA DUTTA
University of Illinois at Chicago

MICHAEL A. STROSCIO
University of Illinois at Chicago

CAMBRIDGE
UNIVERSITY PRESS

University Printing House, Cambridge CB2 8BS, United Kingdom

One Liberty Plaza, 20th Floor, New York, NY 10006, USA

477 Williamstown Road, Port Melbourne, VIC 3207, Australia

314–321, 3rd Floor, Plot 3, Splendor Forum, Jasola District Centre, New Delhi – 110025, India

79 Anson Road, #06-04/06, Singapore 079906

Cambridge University Press is part of the University of Cambridge.

It furthers the University's mission by disseminating knowledge in the pursuit of education, learning, and research at the highest international levels of excellence.

www.cambridge.org
Information on this title: www.cambridge.org/9781108428149
DOI: 10.1017/9781108674522

First published 2019

Printed in the United Kingdom by TJ International Ltd, Padstow Cornwall

A catalogue record for this publication is available from the British Library.

Library of Congress Cataloging-in-Publication Data

ISBN 978-1-108-42814-9 Hardback

Contents

Preface

There are many monographs and textbooks devoted to nanostructures and nanoelectronics. It is generally accepted that microstructure science and technology deal with microstructures with sizes down to 0.1 micrometer, while sizes below 100 nm (0.1 micrometer) are considered in nanoscience and nanoelectronics. Nanoelectronics was and is predominantly driven by applications in processors and memories of computers and their follow-on systems. This is why the overwhelming majority of textbooks deal with the progress in the technology, covering so-called top-down technology which produces nanostructures and nanostructure devices whose critical dimensions are reduced to less than 100 nm. Textbooks have devoted less attention to the optical properties and optoelectronics of nanostructures, the subject of this book.

A new burst of basic research and applications of nanostructures was stimulated in 2004 by a "scotch tape" technology in which a single layer of graphene was obtained by peeling it from a graphite surface. The simplicity of "scotch tape" technology and the very unusual properties of graphene initiated interest in structures with a thickness of one atomic layer. Many materials have weak bonding between the layers, and monolayers or bilayers can easily be produced with "scotch tape" technology. In spite of the fact that the "scotch tape" technology is not applicable for industrial production, it is convenient and inexpensive for wide use in research. These monolayer structures demonstrated interesting electrical, optical, and optoelectronic properties. Today different methods of fabrication of graphene have been developed, including epitaxial technologies. A number of other one-atom-thick material structures have been discovered and studied. These monolayer materials constitute a new class of two-dimensional crystals which complements conventional two-dimensional heterostructures. Industrially applicable technologies for these structures are currently under active development.

Modern educational programs need to address the rapidly evolving facets of nanoscience and nanotechnology. A new generation of researchers, technologists, and engineers has to be trained in the emerging nano-disciplines. With the purpose of contributing to education in the nano-fields, we present this textbook providing a unifying framework for the basic ideas needed to understand recent developments underlying the optical and optoelectronic properties of nanostructures. This book grew out of the authors' research and teaching experience in these subjects. We have found that many of the ideas and achievements can be explained in a relatively simple setting. We have designed this textbook primarily for senior undergraduate students and first-year graduate students, who will have been trained in different fields, including nanoscience, physics and optical engineering, materials science and engineering, and mechanical engineering. To reach such a broad audience, materials are presented in such a way that an instructor can choose the level of presentation depending on the backgrounds of the students. We include details of derivations and the mathematical justification of concepts in some sections. These detailed sections may be omitted in an undergraduate curriculum. The book contains some solved examples in the text and control questions for student self-evaluation at the end of each chapter.

Although this book has been written basically for the student, the material presented in Chapters 4–8 should be useful for a wide audience of researchers working on different subjects within the multidisciplinary area of nanoscience. The core of this textbook is in Chapters 4–8 but, in order to make the book self-contained, we give in Chapters 1–3 the relevant background material on nanostructures and their electronic properties. Appendix A discusses the Schrödinger wave equation for particles and contains major relations and equations from quantum mechanics. Appendix B includes tables of units as well as major physical constants.

In Chapter 1, we present in concise form the main subject of the book, namely the recent and diverse trends in nanotechnologies; novel concepts of nanodevices are also reviewed briefly. These trends make it clear why understanding the fundamentals of the optics and optoelectronics of nanostructures is of great importance. Chapters 2 and 3 have been written for students who have not taken courses in nanoelectronics.

In Chapter 2, we present an overview of the basic materials that are exploited in nanoelectronics. We introduce the major properties of semiconductors and materials that demonstrate great potential for application in nanoelectronics and optoelectronics. Special attention is paid to carbon nanotubes, graphene, transition-metal dichalcogenides and other single-atomic-layer materials.

In Chapter 3, we discuss the basic physical concepts related to the behavior of particles in the nanoworld. Keeping in mind the diverse variants of nanostructures, we analyze a number of particular examples which highlight important quantum properties of particles in heterostructures. Recently, substantial efforts in research have been made in the area of reconfigurable materials. The last section in Chapter 3 briefly describes three examples of reconfigurable materials based on nanostructures, namely on graphene, quantum wells, and quantum dots. Many of the examples analyzed can serve as the simplest underlying models of nanostructures that are exploited in later chapters.

For supplemental information on the fundamentals presented in Chapters 2 and 3, students can refer to the following recent books: *Quantum Mechanics for Nanostructures*, V. Mitin, D. Sementsov, and N. Vagidov, Cambridge University Press, 2010, *Introduction to Nanoelectronics: Nanotechnology, Engineering, Science, and Applications*, V. Mitin, V. Kochelap, and M. Stroscio, Cambridge University

Press, 2008, and *Quantum Heterostructures: Microelectronics and Optoelectronics*, V. Mitin, V. Kochelap, and M. Stroscio, Cambridge University Press, 1999.

Chapter 4 covers the properties of light and light–semiconductor interactions. We start with a brief review of the basic concepts of electromagnetic fields. We define the classical characteristics of electromagnetic fields, such as the energy, intensity, and density of states, and we introduce the concept of the quanta of these fields – the photons. By analyzing electromagnetic fields in free space and in optical resonators, we show that resonators drastically change the structure and the behavior of electromagnetic fields. Next we study the interaction of light with matter and define three principal optical processes: spontaneous emission, stimulated emission, and stimulated absorption. We review the optical properties of bulk semiconductors and define the major optical characteristics of semiconductors, with specific emphasis on the III–V compounds.

Chapter 5 emphasizes that the electrodynamics of heterostructures differs from that of bulk materials. The spatial nonuniformity causes specific characteristics of the interaction of light with matter, including light propagation, absorption, etc. We introduce several parameters which characterize the interaction of light with matter for different cases of light propagation. The parameters for light absorption are calculated for type-I heterostructure quantum wells. The optical properties of monolayer materials are discussed. Various factors affecting the optical properties of low-dimensional electrons, such as the broadening of spectra due to intraband scattering processes, excitonic effects, etc., are analyzed in this chapter.

Chapter 6 discusses the influence of an external electric field on the refractive index and the absorption coefficient, i.e., the effect of the electric field on the propagation of light through a material or on the reflection of the light, which is known as the electro-optical effect. We study the electro-optical effect for quantum heterostructures including quantum wells, double- and multiple-quantum-well structures, and superlattices. We consider the quantum-confined Stark effect, the Burstein–Moss effect, and the effect of destroying excitons in gated heterostructures. We show that electro-optical effects have great potential for optoelectronic applications using the electric field to control light and for realizing the tunable generation of microwave emission. We consider nonlinear optical effects in quantum heterostructures that occur at high intensities of light. Also in Chapter 6 we present a new emerging branch of the subject – plasmonics for optoelectronics. We discuss subwavelength light phenomena, which involve surface plasmons and plasmon excitation in low-dimensional electron gases.

Chapters 7 and 8 are devoted to the applications of nanostructures. We analyze the effect of optical confinement. Our discussion focuses on heterostructures which facilitate the simultaneous formation of optical modes and the realization of laser oscillations which convert electric power into laser emission. A variety of different laser designs are analyzed, such as: quantum well and quantum wire lasers, including devices operating in the terahertz (THz), infrared, visible, and ultraviolet spectral regions; the quantum cascade laser; and single-monolayer photodetectors and emitters. In particular, we consider ultraviolet-emitting devices based on the group-III-nitride materials. We discuss the use of plasmonic effects for optoelectronic devices of subwavelength scale. Other emerging optoelectronic devices based on nanophotonic effects are highlighted in Chapter 8. For example, special attention is paid to silicon photonics. Chapter 8 concludes with a brief discussion of adaptive photodetectors.

The authors would like to acknowledge help from their professional colleagues, friends, and students in useful discussions and encouragement, especially Dr. Andrei Sergeev, Professor Taiichi Otsuji, and Victor Ryzhii and their students Yifan Du and Xiang Zhang, who helped with some typing and formatting, and Trevor McDonough and Sinan Janabi, who helped in preparing and obtaining permissions for figures. Without their contributions and sacrifices, this work would not have been completed. In addition, VM would like to thank the many devoted and hardworking PhD students and research associates who have worked in his group over the years and helped to create and support an atmosphere of professional development and inspiration. VK acknowledges his colleagues from the Institute of Semiconductor Physics of the National Academy of Sciences, Ukraine, and in particular, Professor Alexander Belyaev, Dr. Valery Sokolov, Dr. Boris Glavin, and Dr. Vadim Koroteev for the numerous discussions of problems of semiconductor physics and devices. MD and MAS would like to thank each other as well as their coauthors VM and VK. MD and MAS would also like to thank their graduate students from the North Carolina State University, Duke University, and the University of Illinois at Chicago. Finally, the authors wish to thank their loved ones for their support and for forgiving us for not devoting more time to them while working on this book. In particular, VM is grateful to his granddaughter Christina and grandson Anthony, VK is deeply thankful to his granddaughter Anastasia, and MD and MAS thank their children Gautam Stroscio, Elizabeth Stroscio and Marshall Stroscio and granddaughter Lili Belle Stroscio, who have all been attentive during this period of writing.

1 Some Trends in Optoelectronics

This book is intended to provide the foundations of the physics and engineering of the optical properties of quantum heterostructures and the basics of optoelectronics devices.

A wide variety of quantum heterostructures and devices has become possible due to dramatic improvements in semiconductor materials and technology as well as to a deeper understanding of the underlying physics and new device concepts. This variety of heterostructures and devices is enhanced by the recently discovered single- and few-monolayer crystals (true two-dimensional materials). Progress in each of these areas has been stimulated, in part, by the enormous demands for information and communication technologies as well as by numerous special applications.

In this chapter we analyze briefly trends in the optics and optoelectronics of quantum heterostrucures and two-dimensional crystals and discuss the use of these nanostructured materials to realize devices with greatly enhanced performance.

In order to trace how the dominant trends are evolving, Fig. 1.1 illustrates the relationships between end-use technologies, the physics of the materials and devices, and new device concepts. The upper level of this chart presents end-use technologies. As is well known, these information and communication technologies are essential to the present-day functioning and progress of society. There are also other special applications based on optoelectronics; these applications underlie many high-technology industries, including those supporting aeronautics, space, and the military. These end-use technologies are based on supporting technical capabilities.

The next level of Fig. 1.1 focuses on the general demands for technical capabilities and systems. Modern information technology depends heavily on systems that are highly integrated, with great numbers of devices per unit area or on a single chip; moreover, there are increasing demands for high-speed operation and low power consumption. Communication technology relies on microwave and optical-fiber transmission and is based on systems operating with high-frequency electrical and optical signals. Special applications result in additional demands such as high-temperature operation and the handling of high-power signals. The technical systems consist of active devices, passive elements, a number of inter-device connections, etc. The next level of Fig. 1.1 highlights the demands on single devices. For high-performance systems, single devices have to be as small as possible and it is highly desirable to minimize the number of interconnections on a chip. Clearly, efficient conversion from electrical to optical signals and vice versa is necessary.

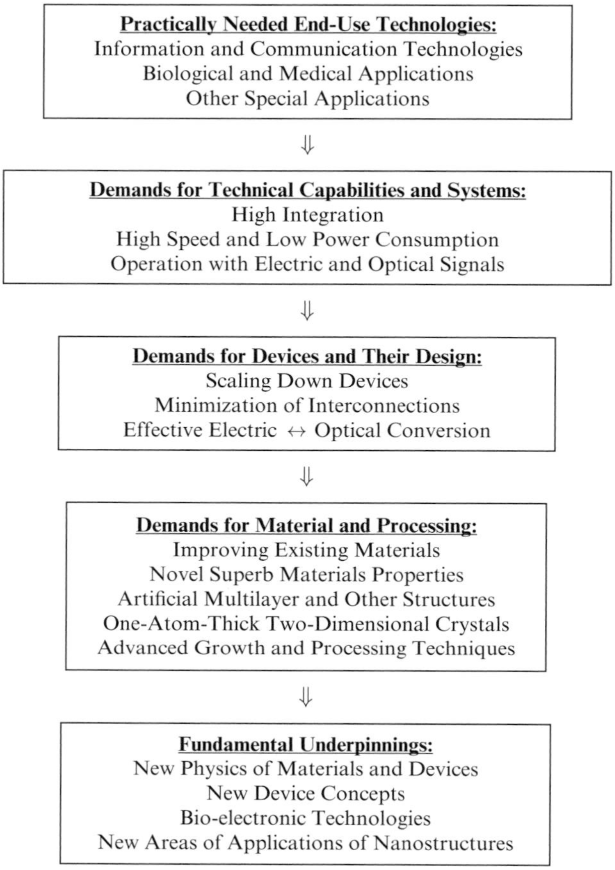

Practically Needed End-Use Technologies:
Information and Communication Technologies
Biological and Medical Applications
Other Special Applications

Demands for Technical Capabilities and Systems:
High Integration
High Speed and Low Power Consumption
Operation with Electric and Optical Signals

Demands for Devices and Their Design:
Scaling Down Devices
Minimization of Interconnections
Effective Electric $\leftrightarrow$ Optical Conversion

⇓

Demands for Material and Processing:
Improving Existing Materials
Novel Superb Materials Properties
Artificial Multilayer and Other Structures
One-Atom-Thick Two-Dimensional Crystals
Advanced Growth and Processing Techniques

Fundamental Underpinnings:
New Physics of Materials and Devices
New Device Concepts
Bio-electronic Technologies
New Areas of Applications of Nanostructures

Figure 1.1 The relationships between end-use technologies, technical systems, single devices, material science and engineering, physics and new device concepts, and systems.

It is possible to achieve many of these goals through advances in materials and processing, as presented in the next level of Fig. 1.1. This includes improving existing semiconductor and optical materials, developing new materials with superb properties and "perfectness", and fabricating artificial structures such as multilayered structures and other semiconductor heterostructures. Techniques for the processing of materials and heterostructures are essential for these advances. Currently, processes such as patterning, etching, implantation, metallization and others are carried out with nanoscale control. Finally, the basis for future progress in all these technologies is the physics of the new materials, structures, and devices. A principal task for researchers in these fields is to establish the fundamental properties of the materials, to model processes in devices, and to find ultimate regimes and operation limits for devices. It is equally important that scientists and engineers generate new conceptual ideas of devices.

In the processes of achieving minimum device sizes and ultra-high levels of integration it is necessary to identify the limiting and critical parameters for improved performance. In reality, these parameters depend on the integrated elements of each individual material system. It is known now that it is possible to achieve improvements in the parameters through materials engineering. Examples include growing alloys and fabricating high-quality artificial multilayered nanostructures composed of single- and few-monolayer crystals.

The simplest multilayered structure has a single heterojunction; a single heterojunction structure is made of two different materials. At the interface of such a heterojunction, the electronic properties can be changed to improve selected physical characteristics. In particular, electrons can be confined in a thin layer near the interface and spatially separated from their parent impurities; this so-called modulation doping greatly enhances the electron mobility. In fact, the layers with confined electrons are so thin that electrons become *quantized*; that is, they obey the laws of quantum physics. The same is valid for different multilayered structures, which can be grown with high quality. Two- and one-dimensional electron channels including quantum wells and quantum wires, as well as "cells" for electrons known as quantum boxes or quantum dots, are currently being fabricated on large wafers and used in electronic and optoelectronic devices. Such structures are known as *quantum semiconductor heterostructures*. The progress in heterostructure technology has been made possible largely as a result of new advances in fabrication techniques. In Table 1.1, we give a very brief summary of some of the important steps now used in the growth, characterization, and processing of heterostructures.

In the 1960s and 1970s, molecular-beam epitaxy was invented, developed, and employed to fabricate high-quality and ultra-thin layers and superlattices. Qualitative electron-beam and X-ray microscope technologies were used to characterize the perfectness of structures, including interface disorder. During this period, lithographic and etching methods suitable for microscale devices were proposed and realized. In the 1980s and later, new epitaxial techniques were developed; these included metal–organic vapor-phase epitaxy, metal–organic molecular-beam epitaxy, and others. These innovations made possible the fabrication of layers with atomic-level accuracy. Desirable spatial-modulation doping by impurities has become possible, including δ-doping – the doping of one or a few atomic monolayers. Thin-layer fabrication techniques facilitated atomic-scale control and the use of materials with quite different lattice parameters. Such layers are strained, but in many cases they can be almost perfect. New tools – scanning tunneling microscopy and atomic-force microscopy – emerged, which portend numerous applications in high-precision fabrication. Lithography and etching methods were improved to the point that they can be used for nanoscale structuring. Finally, femtosecond spectroscopy progressed substantially, moving into ottosecond spectroscopy, and became heavily used to characterize heterostructures.

Heterostructures based on Si/Ge have significant potential since these structures are compatible with Si technology. Many properties of these structures portend

Table 1.1 Advances in growth, characterization, and processing of quantum heterostructures

Period	Methods
1970s–1980s	***Growth and fabrication methods***
	Molecular-beam epitaxy
	Ultrathin-layer fabrication
	Superlattice fabrication
	Characterization methods
	Lithographic microstructuring
	Qualitative electron-beam and X-ray microscopies
1990s–2000s	***Growth and fabrication methods***
	Metal–organic vapor-phase epitaxy
	Metal–organic molecular-beam epitaxy
	Atomic-layer-accuracy fabrication
	δ-doping
	Controlled strained layers
	Fabrication methods based on chemistry and biology
	Assembling inorganic nanoblocks with biomolecules
	Characterization methods
	Lithography and etching for nanostructuring
	Dip-pen nanolithography
	Quantitative electron-beam and X-ray microscopies
	Scanning tunneling microscopy (STM)
	Atomic-force microscopy (AFM)
	Picosecond and femtosecond spectroscopy
	Terahertz time-domain spectroscopy
2000s–2018	***Growth and fabrication methods***
	Methods of fabrication of quantum wires and quantum dots of controllable compositions and sizes
	Micromechanical cleavage techniques for extraction of two-dimensional crystals
	Extreme ultraviolet interference lithography for tens of nanometers resolution
	Fabrication of structures for subwavelength optics
	Methods of generation and detection of THz emission
	Characterization methods
	Angle-resolved photoemission spectroscopy
	Near-field optical spectroscopy which uses plasmonic effects
	Sub-femtosecond spectroscopy
	Time-delay spectroscopy

devices with advantages over Si devices. In particular, Si/Ge heterostructure bipolar transistors can operate at frequencies up to the sub-terahertz range.

Recently, another type of heterostructure – silicon on insulator – has received a great deal of attention and has significant promise. The term silicon-on-insulator technology refers to the exploitation of a layered silicon–insulator–silicon substrate instead of the conventional silicon substrates widely used in the semiconductor industry. In silicon-on-insulator systems, the silicon junction is located above an electrical insulator, typically silicon dioxide or sapphire. The choice of insulator depends largely on the intended application. For example, silicon on sapphire is used for high-performance radio frequency applications, while silicon on silicon dioxide is used in nanoelectronic devices and in silicon optoelectronics. Indeed, the growing demand for instant and reliable communication requires the integration of microelectronic and optical devices in optoelectronic circuits. Silicon-on-insulator structures facilitate the fabrication of optical waveguides and other optical devices. For example, the buried insulator enables the propagation of infrared light in the silicon layer, on the basis of total internal reflection.

The very first two-dimensional crystal – graphene – was discovered in 2004. After that, various other two-dimensional crystals were discovered. Two-dimensional crystals are atomically thin materials with atoms strongly bound in one crystal plane. Owing to the reduced dimensionality, charge carriers in these materials are strongly confined to one crystalline plane. This leads to a significant modification of the electronic band structure and, in particular, radical changes in optical behavior, giving rise to exciting new physical effects. These effects indicate the great potential for applications of two-dimensional materials in optoelectronics over a very wide spectral range (from terahertz to ultraviolet electromagnetic spectra).

Very recently novel ways to confine and control light (electromagnetic emission in general) on dimensional scales smaller than the light wavelength have been proposed. In such cases, one can exploit the interaction of light and conduction electrons at an interface between a dielectric and a conducting material (metal or semiconductor). In many instances such interaction is associated with surface localized plasmons – collective oscillations of the electrons against a fixed background of positive-ion cores. As the result of the interaction of light waves and plasmons, coupled excitations – plasmon-polaritons – are formed. The corresponding research branch is known as plasmonics. The most intriguing plasmonic effects are an enhancement of the optical fields in the subwavelength domain and the possibility of light control on subwavelength scales. Plasmonics promises a number of new optoelectronic devices and applications.

In this book, we will study the conditions associated with the transition between the classical and quantum regimes of operation, as well as the quantum physics of new microelectronic devices and concepts.

Now, we consider briefly optoelectronics, which complements microelectronics in many applications and systems. First of all, optoelectronics provides means to make electronic systems compatible with lightwave communication technologies. Furthermore, optoelectronics can be used to accomplish the tasks of the acquisition, storage, and processing of information. Advances in optoelectronics have made significant contributions to the transmission of information via optical fibers (including communication between processing machines as well as within them), to the high-capacity mass storage of information in laser disks, and to a number of other specific applications.

The principal components of optoelectronic systems are light sources, sensitive optical detectors, and properly designed light waveguides, for example, optical fibers. These devices and passive optical elements are fabricated with optically active semiconductor materials. The III–V, IV–IV, and II–VI compounds belong to this group; most of these compounds have a direct bandgap, which makes them suitable materials for optoelectronic devices. Using direct-bandgap semiconductor materials, two main types of light sources have been developed: light-emitting diodes, which produce spontaneous incoherent emission; and lasers, which emit stimulated coherent light. In both cases, electrical energy is converted into light energy. The general goals for these devices include electric control, high-speed optical tuning, and achieving operation in the desired optical spectral range.

Optoelectronic devices and systems employ a variety of different optical and electro-optical effects. Quantum heterostructures provide a means to enhance many of the effects known in bulk-like materials, such as excitonic effects and optical nonlinearities near the fundamental edge of optical absorption. Quantum heterostructures also exhibit new optical effects.

The original semiconductor light-emitting diodes were homojunctions, i.e., they were made of one material, usually GaAs, doped to form a p–n junction. For these light-emitting diodes, the injection of electrons and holes from both sides of the junction into the active region provides the population inversion necessary for light emission. It is necessary to have a very high current density in order to achieve stimulated emission due to the radiative recombination of highly nonequilibrium electrons and holes in the active region. Semiconductor heterojunction lasers are quantum devices. They have superior properties to homojunction light-emitting diodes and are preferable for many technologies. These lasers employ two heterojunctions, and they are quite compact and are highly compatible with semiconductor electronic circuits. Double-heterojunction structures confine electrons and holes in a precisely defined active region and provide a waveguide for the stimulated emitted light. Such lasers have been designed successfully for different spectral ranges. For example, AlGaAs/GaAs double-heterostructure lasers operate in the 0.75–0.9 μm range while GaInAsP/InP lasers cover the 1.2–1.6 μm range, which is ideally suited for low-attenuation optical-fiber transmission.

Some optoelectronic applications require efficient short-wavelength emitting devices, which may be realized with the use of wide-bandgap semiconductor materials and their heterostructures. The direct-bandgap group-III nitrides present a suitable class of materials for such emitting devices. Recent progress in the fabrication of high-quality single-crystal GaN and ternary and quaternary alloys such as AlGaN, InGaN, and AlGaInN and the successful development of p-type doping technologies for these wide-bandgap materials have facilitated the realization of short-wavelength emitting devices.

There are several different critical parameters of semiconductor lasers: the threshold current, temperature sensitivity, modulation bandwidth, speed of modulation, coherence, etc. All these demands can be met if nonequilibrium electrons and holes are squeezed together in a sufficiently narrow active region; accordingly, quantum effects in electron transport become significant. The demand for advanced semiconductor lasers promotes the reliance on various heterostructure materials. Devices with low costs and long life are required as well. There are several types of heterostructure laser: quantum well injection lasers, surface emitting lasers, quantum wire and quantum dot lasers, quantum-cascade lasers, and short-wavelength injection lasers.

The trends in optoelectronics are toward scaling down the sizes of these devices and achieving high levels of integration in systems such as arrays of light-emitting diodes, laser arrays, and integrated systems with other electronic elements on the same chip. Of particular importance is that there is a fundamental limit to the size scaling of optical devices: light cannot be spatially confined below λ/n_{ref}, where λ is the wavelength of light in vacuum and n_{ref} is the refractive index of the optical material. Light confinement on scales of the order of λ/n_{ref} is possible in waveguides or specially designed optical microcavities such as a Fabry–Pérot resonator with highly reflective multilayered mirrors.

Generally, there are two approaches to device operation with optical signals. The first, currently the most widely used, is optical-to-electrical signal conversion and subsequent processing by electronic means; these systems are referred to as hybrid optoelectronic systems. To achieve this goal, one needs optical detectors; in addition, optical modulators, optical gates, and other electrically controlled devices are used. The essential performance requirements are fast response, high sensitivity, and high quantum efficiency.

Special techniques for the growth of optically active semiconductor materials and their processing are being developed to fabricate these optoelectronic devices with sizes close to the previously mentioned fundamental optical limit and with sizes that lead to the confinement of electrons and holes in the quantum limit. Large arrays of emitting diodes or lasers, nonlinear elements, and optical detectors have been fabricated for this purpose. Their fabrication is based on the heterostructure manufacturing and processing techniques presented in Table 1.1 and tends to provide devices with long lifetimes for low costs. Thus, we can conclude that optoelectronics benefits substantially through the use of quantum heterostructures and becomes competitive with its microelectronic counterpart.

These recent and diverse trends in semiconductor heterostructures and device technologies as well as in novel device concepts are driving the establishment of new subdisciplines of optoelectronics based on quantum structures. These subdisciplines and their foundations will be studied in the following chapters.

The rest of this book is organized as follows. In Chapter 2, we consider the different materials used in optoelectronic applications. We start with the classification of dielectrics, semiconductors, and metals and define electron energy spectra, which determine the basic optical properties of crystals. For optoelectronic and optical applications, a critical issue is the engineering of electron spectra, which can be realized in heterostructures. Thus, we analyze the principles of such engineering of the spectra in semiconductor heterostructures. In Chapter 3 we present the basic electronic properties of quantum heterostructures: i.e., quantum wells, quantum wires, and quantum dots, as well as two-dimensional crystals. Key concepts of the quantum physics necessary for understanding the properties of quantum heterostructures are given in Appendix A. In Chapter 4, we discuss the properties of light and light–semiconductor interactions, review the optical properties of bulk semiconductors, and define the major optical characteristics of semiconductors with an emphasis on the specifics of direct- and indirect-bandgap semiconductors. The optical properties of quantum structures are studied in Chapter 5, where we analyze stimulated emission and other optical effects in quantum structures and one- and few-monolayer crystals. In Chapter 6, we study electro-optical and nonlinear optical effects for quantum heterostructures, including quantum wells, double- and multiple-quantum-well structures, and superlattices. We show that these effects have a great potential for optoelectronic applications. In Chapter 7, we analyze the applications of quantum heterostructures to devices emitting near-infrared, visible, and ultraviolet light; these devices exploit phototransitions between the valence and conduction bands, i.e., interband phototransitions. Finally, in Chapter 8 optoelectronic devices which exploit intraband phototransitions are presented. These include unipolar cascade lasers operating in the mid-infrared and terahertz ranges, and quantum-structure photodetectors. We also present the topic of silicon optoelectronics, which is important for communication technologies. This chapter concludes with a discussion of the prospective applications of two-dimensional crystals in optoelectronics.

2 Materials for Optoelectronic Applications

2.1 Introduction

We shall now overview the basic materials which are exploited in the optoelectronic applications considered in this book. A simple and intuitive classification of solids includes nonconducting materials, semiconductor materials, and metals, i.e., good conducting materials. Semiconductors occupy a place in between nonconducting materials and metals: semiconductor materials are conductive and optically active materials with electrical and optical properties varying over a wide range. Semiconductors are the principal candidates for use in optical and optoelectronic structures because they exhibit great flexibility, allowing good control of the electrical and optical properties and functions of optoelectronic devices.

The semiconductors exploited in optoelectronic applications are, in general, crystalline materials. Through proper regimes of growth, subsequent modifications and processing, doping by impurities, etc., one can fabricate nanostructures and nanodevices starting from these "bulk-like" materials.

Recently, other physical objects that demonstrate promising properties for optoelectronic applications have been discovered. These include two-dimensional monolayer crystals, which we will discuss in Section 2.10.

In this chapter, we consider different materials for optoelectronic applications. We start with the classification of dielectrics, semiconductors, and metals. Then, we define electron energy spectra, which determine the basic properties of the electrons in crystals. For optoelectronic applications, a critical issue is the engineering of electronic energy spectra, which can be realized in heterostructures. Thus, we analyze basic types of semiconductor heterostructures and their components, including two-dimensional and so-called wide-bandgap semiconductors, which emit photons with energies equal to or greater than that of visible photons. Finally, we briefly describe two-dimensional monolayer crystals: graphene, silicene, germanene, tinene, black phosphorus, monochalcogenides, dichalcogenides, trichalcogenides, and boron nitrides.

2.2 Semiconductors

In every solid, electrons can be characterized in terms of their energy levels. In crystals, the allowed electron energies typically have an energy band structure that

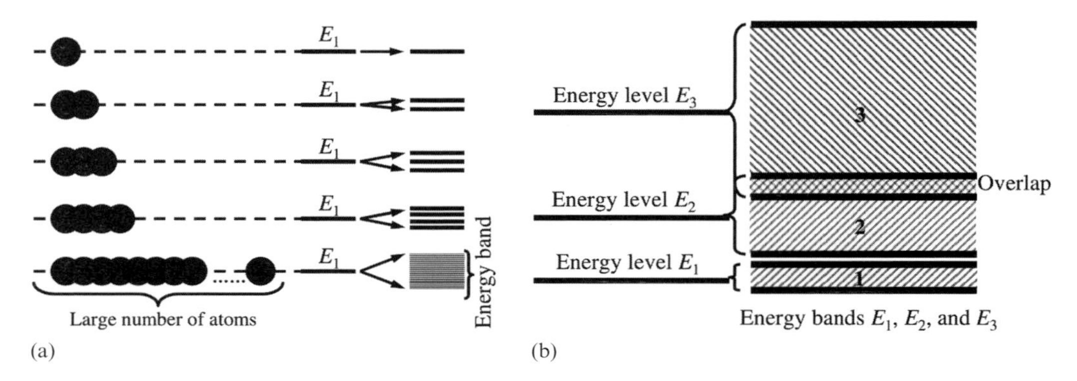

Figure 2.1 (a) The formation of an energy band. The ground state energy level E_1 of a single atom evolves into an energy band when a large number of single atoms are interacting with each other. (b) A schematic representation of energy bands 1, 2, and 3 that correspond to single-atom energy levels E_1, E_2, and E_3, respectively.

may be understood as follows. When two atoms, each with the same energy, come close to each other, the composite two-atom system is characterized by two close energy levels. Similarly, for a system of N atoms, every energy level of the isolated atom is split into N closely spaced levels. This assembly of close levels may be considered as an *energy band*. Figure 2.1(a) illustrates schematically the formation of such an energy band from a single atomic level. Since a single atom has a series of energy levels, the electron energies in a crystal constitute a series of energy bands that may be separated by energy gaps, or may overlap as illustrated by Fig. 2.1(b). As soon as the energy bands are formed, the electrons should be thought of as collectivized: they can no longer be attributed to certain atoms, since the energy bands characterize the whole system of N atoms.

A crucial point is the filling of the bands by electrons. We shall use the Pauli exclusion principle (see Appendix A). That is, two electrons cannot be in the same state. It is possible, as an example, for two electrons to be in the same energy state, but these electrons must be in different spin states; thus the electrons are in fact in different overall states. Under equilibrium conditions and at low ambient temperature, the lowest energy levels should be populated. As we will see later, the most important electrons in defining the optical properties of a semiconductor are those in the upper populated bands. Then, in principle, we obtain two possible cases.

First, all bands are completely filled and the filled bands are separated from the upper (empty) bands by an energy gap. This is the case illustrated by Figs. 2.2(a) and (b), for *dielectrics* (*insulators*) with bandgaps $E_g > 5$ eV and for semiconductors with bandgaps $E_g < 5$ eV, respectively. Actually, there is no difference between filling up energy bands for a dielectric (insulator) and for a semiconductor. The difference is in the energy gap between the upper filled band and the next empty band: for semiconductors this energy gap is much smaller than for dielectrics, as is illustrated in Figs. 2.2(a) and (b). This upper band is empty, at least at low temperature. Later, we will study the electronic band structure of semiconductor materials in more detail.

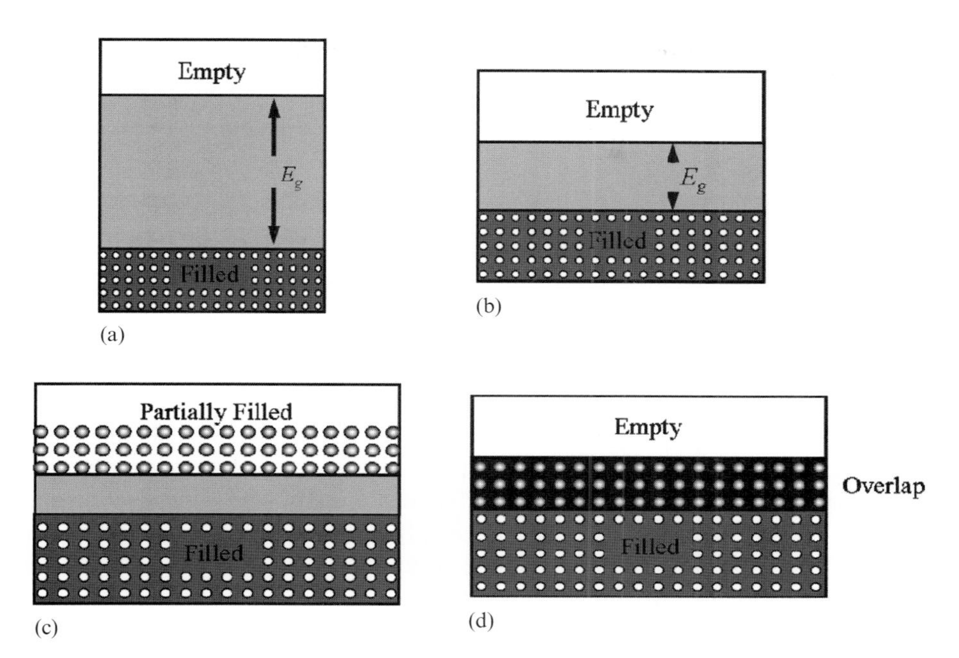

Figure 2.2 (a) The case of a dielectric: filled valence band and empty conduction band; $E_g > 5$ eV. (b) The case of a semiconductor: filled valence band and empty conduction band (at low temperature); $E_g < 5$ eV. (c) Electrons in a partially filled band may gain energy from an electric field and, as a result, transfer to empty ("free") levels of higher energy, thus exhibiting electrical conductivity. Accordingly, a structure with a partially filled band corresponds to a metal. (d) For overlapping bands the available electrons fill states in both bands.

Second, the upper bands that contain electrons are not completely filled, as in Figs. 2.2(c) and (d); this case corresponds to a metal. Indeed, to exhibit conductivity under an applied electric field, an electron should experience acceleration and gain energy. That is, an electron should be able to gain a small amount of energy and be transferred to close, but higher, energy levels. If all of the energy levels (a whole band) are already filled up, the electron cannot participate in the conduction process. This is the case of a *dielectric* (*insulator*). In contrast, if the energy band is not completely filled and there are empty energy levels available for the electrons, in an electric field the electron can move to upper levels and gain energy in a process that corresponds to electrical conductivity. This is the case of a *metal*.

Many important semiconductors are formed from elements from the central portion of the Periodic Table of Elements – groups II to VI, as shown in Table 2.1. In the center of Table 2.1 is silicon, Si, the backbone material of modern electronics. Silicon plays the central role in electronics just as steel plays a dominant role in metallurgy. Below Si is germanium, Ge. Nowadays, Ge is rarely used by itself; however, Ge–Si alloys play an increasingly important role in today's electronics technology. Besides the elemental materials, contemporary electronics also uses combinations of elements from group III and group V, and combinations of elements of groups II

Table 2.1 The central portion of the Periodic Table of Elements

Group II	Group III	Group IV	Group V	Group VI
Be	B	C	N	O
Mg	Al	Si	P	S
Zn	Ga	Ge	As	Se
Cd	In	Sn	Sb	Te
Hg	Tl	Pb	Bi	Po

and VI, as well as some more complicated combinations. These combinations are called *compound semiconductors*. By combining each element from group III with N, P, As, Sb, and Bi from group V, 25 different III–V compounds can be formed. The most widely used compound semiconductor is GaAs (gallium arsenide), and all III–V semiconductors are used to fabricate so-called *heterostructures*. A heterostructure is made of two different materials with a heterojunction boundary between them. The specific choice of heterostructure depends on the application.

Two or more compounds may be combined to form *alloys*. A common example is aluminum gallium arsenide, $Al_xGa_{1-x}As$, where x is the fraction of group III sites in the crystal occupied by Al atoms, and $1 - x$ is the fraction of group III sites occupied by Ga atoms, where x may take a value from 0 to 1. Hence, now we have not just 25 discrete compounds, but a continuous range of materials. Similarly to the III–V compounds, every element shown in the column for group II may be used together with every element in the column for group VI to create II–VI compounds, and again, by combining more than two of these elements, it is possible to create a continuous range of materials. As a result, it is possible to make compositionally different IV–IV, III–V, II–VI, and II–IV compounds.

2.3 Crystal Lattices: Bonding in Crystals

We start with the definition of a crystal. A *crystal* is a solid where the constituent atoms are arranged in a certain *periodic* fashion. That is, one can introduce a basic arrangement of atoms that is repeated throughout the entire solid. In other words, a crystal is characterized by a strictly periodic internal structure. Not all solids are crystals. In Fig. 2.3, for comparison, we present a crystalline solid (a), a solid without any periodicity (a so-called *amorphous* solid) (b), and a solid where only small regions are of a single-crystal material (a so-called *polycrystalline* solid) (c). As might be expected, crystalline materials can be the most perfect and controllable materials. Before studying periodic arrangements of atoms in crystals, we shall discuss different types of bonding in crystals.

2.3.1 Ionic Crystals

Ionic crystals are made up of positive and negative ions. The ionic bond results primarily from the attractive electrostatic interaction of neighboring ions with opposite

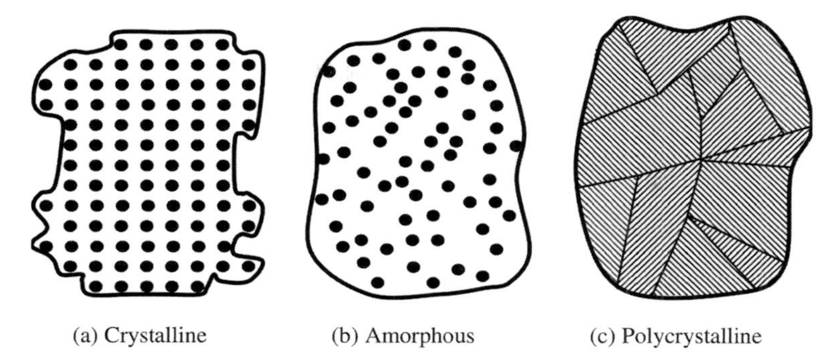

Figure 2.3 Three types of solids: (a) ordered crystalline and (b) amorphous materials are illustrated by microscopic views of the atoms, whereas (c) a polycrystalline structure is illustrated by a more macroscopic view of adjacent single-crystalline regions, each of which has a crystalline structure as in (a).

charges. There is also a repulsive interaction with other neighbors of the same charge. Attraction and repulsion together result in a balancing of forces that leads to the atoms being in stable equilibrium positions in such an ionic crystal. As for the electronic configuration in a crystal, it corresponds to a closed (completely filled) outer electronic shell. A good example of an ionic crystal is NaCl (salt). Neutral sodium, Na, and chlorine, Cl, atoms have the configurations $Na^{11}(1s^22s^22p^63s^1)$ and $Cl^{17}(1s^22s^22p^63s^23p^5)$, respectively. That is, the Na atom has only one valence electron, while one electron is necessary to complete the shell in the Cl atom. It turns out that a stable electronic configuration develops when the Na atom gives one valence electron to the Cl atom. Both of them become ions with opposite charges and the pair has a closed outer-shell configuration (like inert gases such as helium, He, and neon, Ne). The inner shells are, of course, completely filled both before and after binding of the two atoms. In general, for all elements with almost closed shells, there is a tendency to form ionic bonds and ionic crystals. These crystals are usually dielectrics (insulators).

2.3.2 Covalent Crystals

Covalent bonding is typical for atoms with a low level of outer-shell filling. An excellent example is provided by a Si crystal. The electron configuration of Si can be represented as core + $3s^23p^2$. To complete the outer $3s^23p^2$ shell, a silicon atom in a crystal forms four bonds with four other neighboring silicon atoms. The symmetry of the hybrid sp^3 orbitals dictates that these neighboring atoms should be situated in the corners of a tetrahedron as shown in Fig. 2.4. Then, the central Si atom and each of its nearest-neighbor Si atoms share two electrons. This provides so-called *covalent* bonds (symmetric combinations of the sp^3 orbitals) in the Si crystal. The four bonding sp^3 orbitals form an energy band that is completely *filled* by the valence electrons. This band is called the *valence* band. The antisymmetric combination of

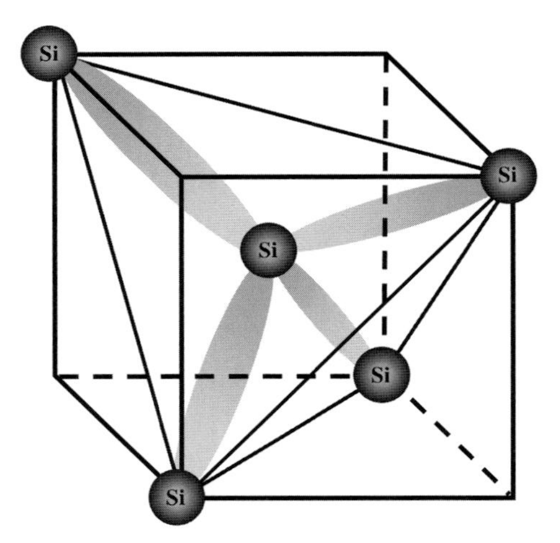

Figure 2.4 Four sp^3 hybrid bonding orbitals in Si crystal.

the sp^3 orbitals forms a band that is called the *conduction* band of the Si crystal. Not surprisingly, a covalent bond of this type plays a main role in Ge, which is also from group IV of the Periodic Table of Elements.

In fact, both of the previously discussed types of atomic bonding may exist simultaneously in a crystal, for example, in III–V compounds. Indeed, the electron configurations in Ga and As are core + $4s^2 4p^1$ (Ga) and core + $4s^2 4p^3$ (As).

When a GaAs crystal is formed, the As atom gives one valence electron to the Ga atom, which makes them both ions. The Coulomb interaction of these ions contributes to the ionic bonding in III–V compounds. Now each ion has only two 2s electrons and two 2p electrons, which are not enough to completely fill the shell. Therefore, the rest of the bonding goes through the formation of the sp^3 hybridized orbitals. This is the covalent contribution to the crystal bonding. We can conclude that the III–V compounds are materials with mixed bonding – partly ionic and partly covalent.

2.3.3 Metals

Metals such as Na, K, Ca, etc. consist of ions regularly situated in space. Each atom contributes an electron to an electron "sea" in which the ions are embedded. The system as a whole is neutral and stable. The electrons contribute significantly to the binding energy of metals. Binding energies for different types of crystals are given in Table 2.2.

2.3.4 Crystal Lattices

Now we return to the discussion of crystal periodicity. The periodic arrangement of atoms (ions) in a crystal forms the *lattice*. The positions of atoms in the lattice

Table 2.2 Binding energies for different types of crystals

Type of crystal coupling	Crystal	Binding energy per atom (eV)
Ionic	NaCl	7.9
	LiF	10.4
Covalent	Diamond, C	7.4
	Si	3.7
	Ge	3.7
Metallic	Na	1.1
	Fe	4.1
	Al	2.4
Molecular and inert-gas crystals	CH_4	0.1
	Ar	0.8

are defined as the *sites*. In principle, atoms always perform small-amplitude oscillations around the sites. However, in many cases we can neglect these small-amplitude oscillations and think of a crystal as a system of regularly distributed atoms (ions). In such a perfect and periodic crystal lattice, we can identify a region called a *unit cell*. Such a unit cell is representative of the entire lattice, since the crystal can be built by regular repeats in space of this element. The smallest unit cell is called the *primitive cell* of the lattice. The importance of the unit cell lies in the fact that by studying this representative element one can analyze a number of properties of the entire crystal. The primitive cell determines the fundamental characteristics of the crystal, including the basic electronic properties.

One of the most important properties of a perfect crystalline lattice is its *translational symmetry*. Translational symmetry is the property that the crystal is "carried" into itself under parallel translation in certain directions and for certain distances. For any three-dimensional lattice it is possible to define three non-coplanar fundamental *primitive translation vectors* (*basis vectors*) $\vec{a}_1$, $\vec{a}_2$, and $\vec{a}_3$, such that the position of any lattice site can be defined by the vector $\vec{R} = n_1\vec{a}_1 + n_2\vec{a}_2 + n_3\vec{a}_3$, where n_1, n_2, and n_3 are arbitrary integers. If we construct a parallelepiped using the basis vectors $\vec{a}_1$, $\vec{a}_2$, and $\vec{a}_3$, we obtain just the primitive cell.

Translational symmetry is illustrated in Fig. 2.5, where for simplicity we present a two-dimensional lattice where several different unit cells (A, B, and C) are shown. The unit cell may be taken as the core corresponding to the smallest magnitudes of the vectors $\vec{a}_1$ and $\vec{a}_2$. That is, the unit cell may be taken as cell A. Thus, the basis vectors of the primitive cell are the vectors $\vec{a}_1$ and $\vec{a}_2$ from the unit cell A. An arbitrary lattice site is at the point $\vec{R} = n_1\vec{a}_1 + n_2\vec{a}_2$ with integers n_1 and n_2. To visualize the translational symmetry of this lattice one can start from any point of the lattice and find all other equivalent positions in space by just applying translations that are integer multiples of the basis vectors.

Since many of the crystals used in electronics have so-called *cubic symmetry*, here we consider briefly such cubic lattices. For them, the unit cell may be selected in the form of a cube. There are three different types of cubic lattice. The *simple cubic*

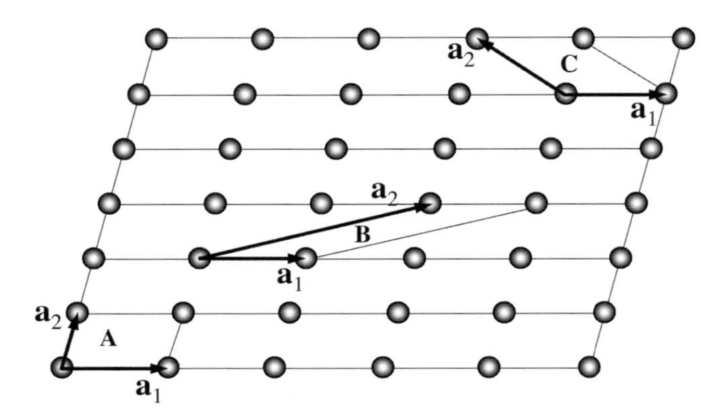

Figure 2.5 Two-dimensional lattice. Three unit cells are illustrated by A, B, and C. Two basis vectors are illustrated by $\vec{a}_1$ and $\vec{a}_2$.

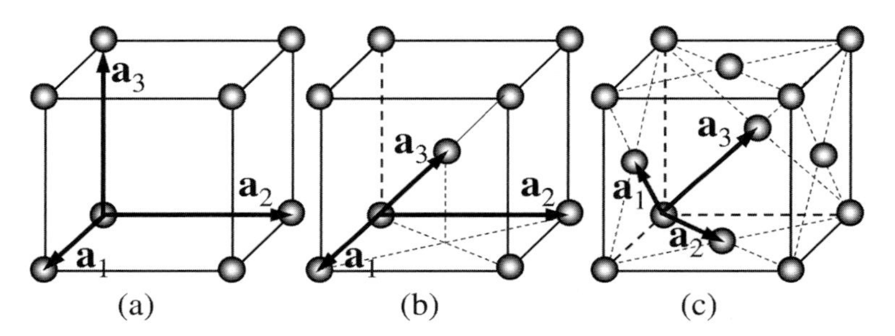

Figure 2.6 Three types of cubic lattice: (a) simple, (b) body-centered, and (c) face-centered. The basis vectors of the three different primitive cells are $\vec{a}_1$, $\vec{a}_2$, and $\vec{a}_3$.

lattice has atoms located at each corner of the cube, as shown in Fig. 2.6(a). The *body-centered* cubic lattice has an additional atom at the center of the cube, as shown in Fig. 2.6(b). The third type is the *face-centered* cubic lattice, which has atoms at the corners and also at the centers of the six faces as depicted in Fig. 2.6(c).

Specifically, the basic lattice structure for diamond, C, silicon, Si, and germanium, Ge, is the so-called *diamond lattice*. The diamond lattice consists of two face-centered cubic structures with the second structure being shifted by a quarter of a diagonal of the first structure, or by a distance $\vec{a}_1/4 + \vec{a}_2/4 + \vec{a}_3/4$ from each point of the first structure; here the vectors $\vec{a}_1$, $\vec{a}_2$, and $\vec{a}_3$ are the vectors defined in Fig. 2.6(a). In Fig. 2.7 the atoms of the first face-centered structure are shown in black and some atoms of the second one are shown in gray. Thus, a diamond lattice contains twice as many atoms per unit volume as a single face-centered cubic lattice. The four nearest-neighbor atoms to each atom are shown by complementary gray-scale for easier visualization. The parameter a, which characterizes the cubic lattice, is the so-called the *lattice constant*. The unit cell of volume $\mathcal{V} = a^3$ consists of eight atoms.

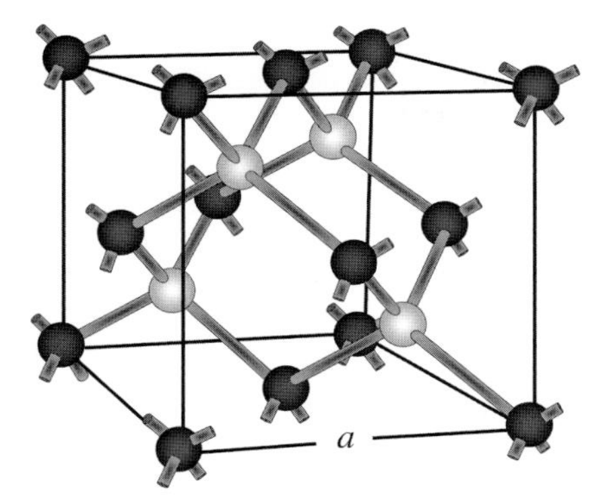

Figure 2.7 The diamond lattice is a member of the face-centered type of cubic lattice. The tetrahedral bonding arrangement of neighboring atoms is clear.

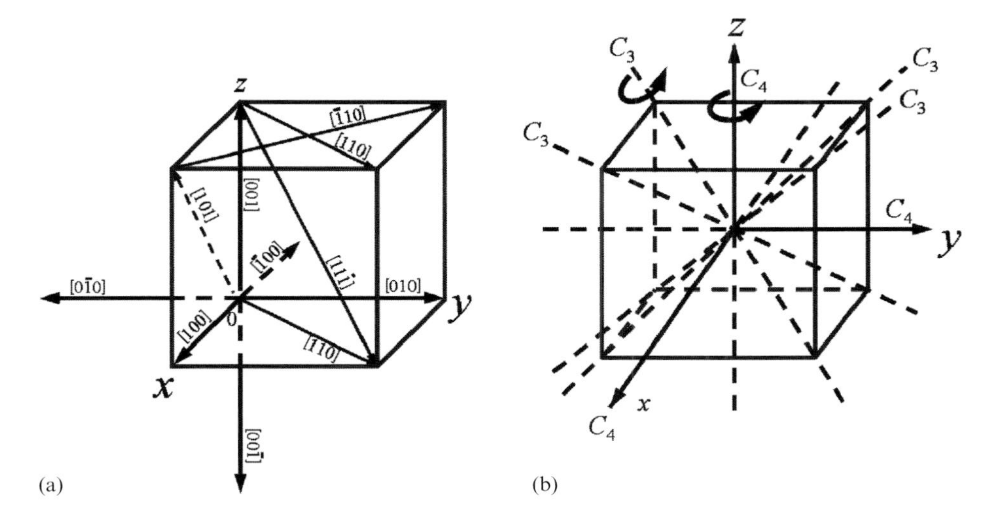

Figure 2.8 (a) Symmetry directions in cubic crystals. (b) Rotation symmetry in cubic crystals.

Besides translations, the crystal symmetry contains other symmetry elements, for example, specific rotations around high-symmetry axes. In cubic crystals, the axes directed along the basis vectors are equivalent and they are the symmetry axes. It is convenient to use a system of coordinates built on the basis vectors. The three symmetry axes may be denoted as [100], [010], and [001]. (Here we have used notation that is common in crystallography: [100] is a unit vector in the x-direction, [010] in the y-direction, and [001] in the z-direction.) The important symmetry directions of a cubic crystal are shown in Fig. 2.8(a). Directions of the type [110] and [111] are also important crystal directions. It is evident that these directions are equivalent to the opposite ones $[\bar{1}\bar{1}0]$ and $[\bar{1}\bar{1}\bar{1}]$, as well as to analogous ones. If the crystal is carried into itself on rotation through an angle $2\pi/n$ about some axis that passes through the crystal, then this axis is said to be an n-fold axis C_n. For example, in a cubic crystal there are three four-fold axes C_4 and four three-fold axes C_3, as shown in Fig. 2.8(b). The symmetry elements of the lattice make the analysis of crystal properties much simpler.

2.4 Electron Energy Bands

Summarizing the above analysis, we conclude that a crystal consists of nuclei and electrons. The valence electrons are collectivized by all of the nuclei and we can expect them to be relatively weakly coupled to atoms. The latter situation allows one to think of a crystal as a system with two relatively independent subsystems: the atom (ion) subsystem and the electron subsystem. In this section, we will consider the electron subsystem of crystals. Actually, in a crystal an electron moves in the electrostatic potential created by positively charged ions and all other electrons. This potential is frequently referred to as the *crystalline potential*, $W(\vec{r})$. We will consider one-particle states for electrons in an ideal crystal, and we will present a classification of these states and find a general form for the wave functions and energies.

2.4.1 An Electron in a Crystalline Potential

For an ideal crystal the crystalline potential is periodic, with the period of the crystalline lattice. Let $\vec{a}_j$, with $j = 1, 2$, and 3, be the three basis vectors of the lattice that define the three primitive translations. The periodicity of the crystalline potential, $W(\vec{r})$, implies that

$$W\left(\vec{r} + \sum_{j=1}^{3} n_j \vec{a}_j\right) = W(\vec{r}), \tag{2.1}$$

where $\vec{r}$ is an arbitrary point of the crystal and n_j are some integers. The one-particle wave function should satisfy the time-independent Schrödinger equation

$$\mathcal{H}\psi(\vec{r}) = \left(-\frac{\hbar^2}{2m}\nabla^2 + W(\vec{r})\right)\psi(\vec{r}) = E\psi(\vec{r}), \tag{2.2}$$

where m is the free-electron mass and $\mathcal{H}$ is the crystalline Hamiltonian. Equation (2.2) neglects interactions between electrons, and that is why the wave function, $\psi(\vec{r})$, is often called the *one-particle wave function* (see also the discussion of envelope functions in Appendix A).

Because of the potential periodicity of Eq. (2.1), the wave functions $\psi(\vec{r})$ may be classified and presented in a special form. To find this form we introduce the translation operator $\mathcal{T}_{\vec{d}}$ that acts on the coordinate vector $\vec{r}$ as

$$\mathcal{T}_{\vec{d}}\vec{r} = \vec{r} + \vec{d}, \quad \vec{d} = \sum_{j=1}^{3} n_j \vec{a}_j. \tag{2.3}$$

Upon applying this operator to the wave function, we find that the function $\mathcal{T}_{\vec{d}}\psi(\vec{r}) \equiv \psi(\vec{r}+\vec{d})$ is also a solution of Eq. (2.2) for the same energy E. Let us assume that the electron state with energy E is not degenerate. Then, we conclude that the two wave functions $\psi(\vec{r})$ and $\psi(\vec{r}+\vec{d})$ can differ only by some factor $C_{\vec{d}}$:

$$\psi(\vec{r}+\vec{d}) = C_{\vec{d}}\psi(\vec{r}). \tag{2.4}$$

From the normalization condition,

$$\int |\psi(\vec{r}+\vec{d})|^2 d\vec{r} = 1, \tag{2.5}$$

we obtain $|C_{\vec{d}}|^2 = 1$. Two different translations $\vec{d}_1$ and $\vec{d}_2$ should lead to the same result as the single translation $\vec{d} = \vec{d}_1 + \vec{d}_2$, i.e., $C_{\vec{d}_1} C_{\vec{d}_2} = C_{\vec{d}_1+\vec{d}_2}$. From this last result it follows that $C_{\vec{d}}$ may be represented in an exponential form:

$$C_{\vec{d}} = e^{i\vec{k}\vec{d}} = \exp\left(i\vec{k}\sum_{j=1}^{3} n_j\vec{a}_j\right), \tag{2.6}$$

where $\vec{k}$ is a constant vector. Thus, from Eq. (2.4) we get the wave function in *Bloch form*,

$$\psi(\vec{r}) = e^{-i\vec{k}\vec{d}}\psi(\vec{r}+\vec{d}) = e^{i\vec{k}\vec{r}}u_{\vec{k}}(\vec{r}), \tag{2.7}$$

where

$$u_{\vec{k}}(\vec{r}) = e^{-i\vec{k}(\vec{r}+\vec{d})}\psi(\vec{r}+\vec{d}). \tag{2.8}$$

One can check that the so-called *Bloch function* $u_{\vec{k}}(\vec{r})$ is a periodic function:

$$u_{\vec{k}}(\vec{r}+\vec{d}') = u_{\vec{k}}(\vec{r}), \quad \vec{d}' = \sum_{j=1}^{3} n_j\vec{a}_j.$$

Therefore, the stationary one-particle wave function in a crystalline potential has the form of a plane wave modulated by a Bloch function with the lattice periodicity. The vector $\vec{k}$ is called the *wavevector of the electron* in the crystal. This wavevector is one of the quantum numbers of electron states in crystals.

By applying the so-called *cyclic boundary conditions* to the crystal with a number of periods N_j along the direction $\vec{a}_j$,

$$\psi(\vec{r}+N_j\vec{a}_j) = \psi(\vec{r}), \quad N_j \to \infty, \tag{2.9}$$

we find for $\vec{k}$

$$\vec{k}\vec{a}_j N_j = 2\pi n_j, \quad n_j = 1, 2, 3, \ldots, N_j. \tag{2.10}$$

These allowed quasi-continuum values of $\vec{k}$ form the so-called (*first*) *Brillouin zone* in $\vec{k}$-space of the crystal. They are just those energy bands which we discussed above using simple qualitative considerations. It is important that the symmetry of the Brillouin zone in $\vec{k}$-space is determined by the crystal symmetry.

Let the one-particle energy corresponding to the wavevector $\vec{k}$ be $E = E(\vec{k})$. If the wavevector changes within the Brillouin zone, one gets a continuum energy band; i.e., an *electron energy band*. At fixed $\vec{k}$, the Schrödinger equation (2.2) has a number of solutions in the Bloch form:

$$\psi_{\alpha,\vec{k}}(\vec{r}) = \frac{1}{\sqrt{\mathcal{V}}} e^{i\vec{k}\vec{r}} u_{\alpha,\vec{k}}, \tag{2.11}$$

where α enumerates these solutions and, thus, the energy bands. Owing to the crystal periodicity, the Bloch function can be calculated within a single primitive cell. In Eq. (2.11) we have normalized the wave function $\psi_{\alpha,\vec{k}}$ for the crystal volume $\mathcal{V} = N\mathcal{V}_0$; $N = N_1 \times N_2 \times N_3$ and $\mathcal{V}_0$ are the number and volume of the primitive crystal cell, respectively. From the normalization of the wave function $\psi_{\alpha,\vec{k}}$ one obtains

$$\frac{1}{\mathcal{V}_0} \int_{\mathcal{V}_0} |u_{\alpha,\vec{k}}|^2 d\vec{r} = 1, \tag{2.12}$$

where the integral is calculated over the primitive cell. The latter formula allows one to estimate the order of the value of $u_{\alpha,\vec{k}} : |u_{\alpha,\vec{k}}| \approx 1$.

Thus, through this analysis we have established an extremely important property of the electrons in crystalline solids: despite the interaction of an electron with atoms and other electrons, in a perfect lattice the electron behaves much like a free electron. The electron can be characterized by a wavevector $\vec{k}$ and, thus, it possesses *momentum* $\hbar\vec{k}$. By considering phenomena that have spatial scales much greater than the distances between atoms (ions) in the primitive cell, we may omit the Bloch function $u_{\alpha,\vec{k}}$ and describe the electrons by a wave function in the form of a plane wave, $\psi_{\vec{k}}(\vec{r}) = A\ \exp(i\overrightarrow{kr})$, just as for a free particle. However, the wavevector changes inside the Brillouin zone in a manner that is specific for a given crystal and, in general, the *energy dispersion* $E = E_\alpha(\vec{k})$ can differ from that of the free electron considerably.

2.4.2 The Holes

According to the discussion given in Section 2.2, some of the energy bands are completely filled, while the others are almost empty. For our purposes, two of the bands are of great importance: the upper filled band and the lowest empty band. They are called the *valence* band, $E_v(\vec{k})$, and the *conduction* band, $E_c(\vec{k})$, respectively.

One of the ways to get an electron into the conduction band is to transfer an electron from the valence band to the conduction band. Thus, for analysis of the valence band, it is useful to adopt the *concept of a hole* as a new quasi-particle; i.e., a hole refers to an electron missing from the valence band. These quasi-particles can be introduced and described in terms of simple considerations. If the valence band is full, the total wavevector of all the electrons in the valence band is zero:

$$\vec{k}_{\mathrm{v}} = \sum_i \vec{k}_i = 0, \tag{2.13}$$

where the sum accounts for all occupied valence states. Let us assume that one of the electrons with wavevector $\vec{k}_{\mathrm{e}}$ is removed from the valence band. The total wavevector of the valence electrons becomes

$$\vec{k}_{\mathrm{v}} = \sum_i \vec{k}_i = -\vec{k}_{\mathrm{e}}. \tag{2.14}$$

On the other hand, removing this electron is identical to the creation of a hole in the valence band. One can attribute the wavevector of Eq. (2.14) to this hole: $\vec{k}_{\mathrm{h}} = -\vec{k}_{\mathrm{e}}$. Then, the energy of the valence electrons decreases by the factor $E_{\mathrm{v}}(\vec{k}_{\mathrm{e}})$, and, thus, one can also attribute the energy $E_{\mathrm{h}}(\vec{k}_{\mathrm{h}}) = -E_{\mathrm{v}}(\vec{k}_{\mathrm{e}})$ to this hole. If the energy band is symmetric, i.e., $E_{\mathrm{v}}(\vec{k}) = E_{\mathrm{v}}(-\vec{k})$, we can write for the hole energy

$$E_{\mathrm{h}}(\vec{k}_{\mathrm{h}}) = -E_{\mathrm{v}}(\vec{k}_{\mathrm{e}}) = -E_{\mathrm{v}}(-\vec{k}_{\mathrm{e}}) = -E_{\mathrm{v}}(\vec{k}_{\mathrm{h}}). \tag{2.15}$$

Thus, we can characterize the hole by a wavevector $\vec{k}_{\mathrm{h}}$ and energy $E_{\mathrm{h}}(\vec{k}_{\mathrm{h}})$, and consider the hole as a new quasi-particle created when the electron is removed from the valence band. In the conduction band, the electron energy, $E_{\mathrm{c}}(\vec{k})$, increases as the wavevector, $\vec{k}$, increases. In contrast, in a valence band, near the maximum energy of the band the electron energy, $E_{\mathrm{v}}(\vec{k})$, decreases as $\vec{k}$ increases. According to Eq. (2.15), the hole energy increases with the hole wavevector, $\vec{k}_{\mathrm{h}}$. That is, the hole behaves as a normal particle. Thus, one can introduce the velocity of the hole, $\vec{v}_{\mathrm{h}} = \partial E_{\mathrm{h}}(\vec{k}_{\mathrm{h}})/\partial \vec{k}_{\mathrm{h}}$, and then employ Newton's laws, etc. The absence of a negative charge in the valence band, when an electron is removed, makes it possible to characterize a hole by a positive elementary charge; that is, *holes carry positive electric charge*.

It is worth emphasizing that the similarity between electrons and holes is not complete: holes exist as quasi-particles only in a crystal, while electrons exist also in other physical media as well as in vacuum.

2.4.3 Symmetry of Crystals and Properties of Electron Spectra

Usually, the energy dispersion relations, $E_\alpha(\vec{k})$, are very complex and can be obtained only numerically by approximate methods.

Fortunately, the Brillouin zone possesses a symmetry which directly reflects the symmetry of the unit cell of the crystal in coordinate space. If a crystal is mapped into itself due to transformations in the form of certain rotations around the crystalline axes and of mirror reflections, one can speak about the *point symmetry* of directions in the crystal. In the Brillouin zone, this symmetry generates several points with high symmetry with respect to the transformations of the zone in $\vec{k}$-space. The extrema of the energy dispersion $E_\alpha(\vec{k})$ always coincide with these high-symmetry points. In particular, this fact allows one to simplify and solve the problem of obtaining the electron spectra. Near extrema, the energy spectra can

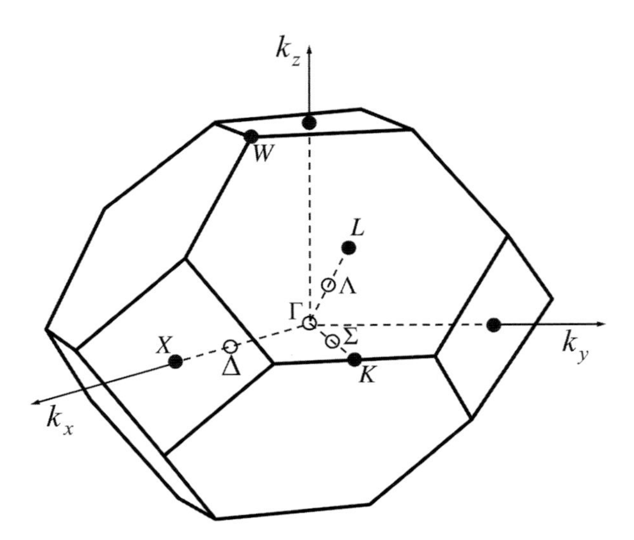

Figure 2.9 The first Brillouin zone of cubic crystals. The points of high symmetry shown here are Γ, L, Δ, and X (see also Table 2.3).

be approximated by expansion of $E_\alpha(\vec{k})$ in series with respect to deviations from the symmetry points. Such an expansion can be characterized by several constants which define the reciprocal effective-mass tensor:

$$\left(\frac{1}{m^*}\right)_{ij} = \begin{pmatrix} m_{xx}^{-1} & m_{xy}^{-1} & m_{xz}^{-1} \\ m_{yx}^{-1} & m_{yy}^{-1} & m_{yz}^{-1} \\ m_{zx}^{-1} & m_{zy}^{-1} & m_{zz}^{-1} \end{pmatrix}, \tag{2.16}$$

where i and j denote the x, y, and z coordinates.

Thus, due to the crystal symmetry the problem of finding an electron energy spectrum $E_\alpha(\vec{k})$ is reduced to the following steps: (a) determination of the high-symmetry $\vec{k}$ points of the Brillouin zone; (b) calculation of the energy positions of these extrema; and (c) analysis of the effective masses or other parameters of an expansion of $E_\alpha(\vec{k})$ within the extrema.

The structure and symmetry of the Brillouin zone for cubic crystals of group IV semiconductors and III–V compounds are very similar. Figure 2.9 shows the Brillouin zone of these semiconductor materials. The symmetry points are shown in Fig. 2.9 and presented in Table 2.3. Evidently, as a result of crystal symmetry, several points have the same symmetry; indeed, they are mapped into themselves under proper symmetry transformations. Such a degeneracy of the symmetry points is indicated in Table 2.3. In particular, the Γ, L, X, and Δ points are of central importance. They give the positions of the extrema of the electron energy in III–V compounds, Ge, and Si.

The band structures for GaAs and Si are presented in Fig. 2.10. The energy dispersions along two symmetric directions of the wavevectors [111] (from Γ to L) and [100] (from Γ to X) are shown. In each case the energy, E, is taken to be zero at the

Table 2.3 Symmetry points in group IV semiconductors and III–V compounds

Symmetry point	Position of extremum in $\vec{k}$-space	Degeneracy
Γ	0	1
L	$\pm(\pi/a)[111], \pm(\pi/a)\,[\bar{1}11], \pm(\pi/a)\,[1\bar{1}1], \pm(\pi/a)\,[11\bar{1}]$	4
Δ	$\pm\gamma(2\pi/a)[100], \pm\gamma(2\pi/a)[010], \pm\gamma(2\pi/a)[001], \lvert\gamma\rvert < 1$	6
X	$\pm(2\pi/a)[100], \pm(2\pi/a)[010], \pm(2\pi/a)[001]$	3

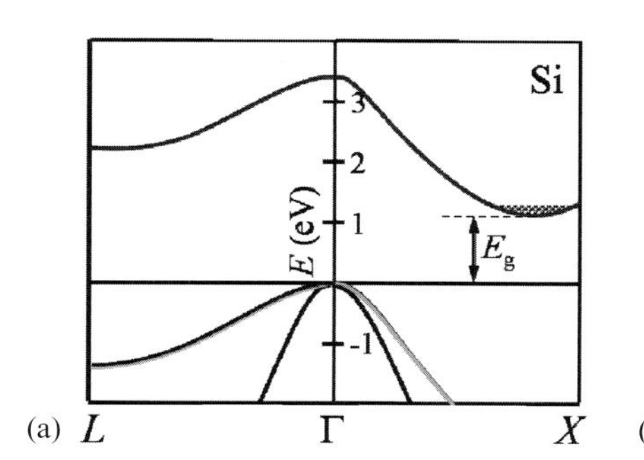

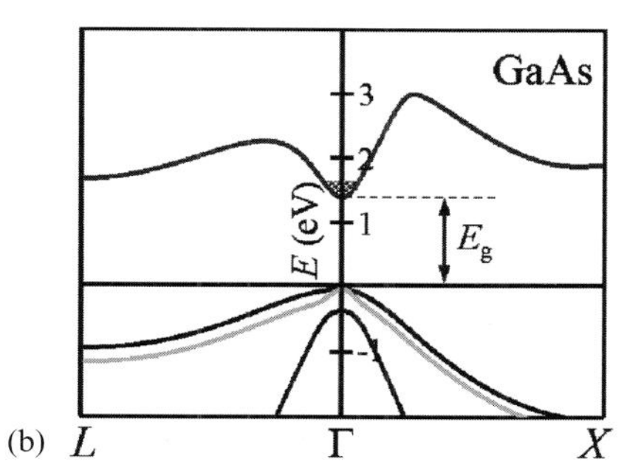

Figure 2.10 The band structure of Si and GaAs. The bandgap of Si is $E_g = 1.12$ eV; the bandgap of GaAs is $E_g = 1.42$ eV.

top of the highest valence band, which is located at Γ point for both of these materials. Since electrons tend to be near the energy minima, one can think of them as being located inside marked regions of $\vec{k}$-space. Frequently, these regions in $\vec{k}$-space are referred to as *energy valleys*, or simply *valleys*. Materials with several valleys are also called *many-valley semiconductors.* For III–V compounds there is only one valley, around the point $\vec{k} = 0$ (the Γ *valley*); however, in the case of Si, there exist six Δ *valleys* in accordance with the degeneracy of the Δ *points*. It is worth emphasizing that, in processes occurring far from equilibrium, other symmetry points also can play a considerable role.

The highest valence band and the lowest conduction band are separated by an *energy bandgap*, E_g. Let us postpone the analysis of valence bands for a while and focus on conduction bands. Conduction bands have different structures for the groups of materials considered in this section. The main difference is that for Si and Ge the lowest minima are located at the Δ and L points, respectively, while for most of the III–V compounds there is only one lowest minimum, which is at the Γ point. The difference is not simply quantitative; indeed, it is qualitative and leads to a series of important consequences in terms of the behavior of electrons.

As discussed previously, within the Γ, Δ, X and L points the electron dispersion curves, $E_\alpha(\vec{k})$, can be expanded in series with respect to deviations from the minima:

$$E_\alpha(k) = E(k_\beta) + \frac{1}{2}\left(\frac{\hbar^2}{m^*_\beta}\right)_{ij}(k_i - k_{i,\beta})(k_j - k_{j,\beta}). \tag{2.17}$$

Table 2.4 Energy band parameters for Si and Ge

	Si	Ge
Type of bandgap	Indirect	Indirect
Lowest minima	Δ points	L points
Degeneracy	6	4
E_g (eV)	1.12	0.664
Electrons		
m_l/m	0.98	1.64
m_t/m	0.19	0.082
Holes		
m_{hh}/m	0.50	0.44
m_{lh}/m	0.16	0.28
Δ_{so} (eV)	0.044	0.29

For the Γ point, we have the simplest case, that of the so-called *isotropic effective mass*, m^*:

$$\left(\frac{1}{m_\Gamma^*}\right)_{ij} = \frac{1}{m^*}\delta_{ij}. \tag{2.18}$$

For the Δ, X, and L points, i.e., for the Δ, X, and L valleys, the reciprocal effective-mass tensor has only two independent components, corresponding to the longitudinal, m_l, and transverse, m_t, masses. For example, in the case of Si, for a coordinate system with the z-axis along the axis where the Δ valley under consideration is located and the two other axes perpendicular to the first, one can obtain

$$\left(\frac{1}{m^*}\right)_{ij} = \begin{pmatrix} m_t^{-1} & 0 & 0 \\ 0 & m_t^{-1} & 0 \\ 0 & 0 & m_l^{-1} \end{pmatrix}, \tag{2.19}$$

For two group IV semiconductors, Si and Ge, the energy parameters are presented in Table 2.4. The degeneracy, given in Table 2.4, indicates the existence of six and four equivalent energy valleys for Si and Ge, respectively. According to Fig. 2.9, the L points are located at the edges of the Brillouin zone, i.e., only half of each energy valley lies inside the first Brillouin zone. This reduces the effective number of valleys to four. In Table 2.4 the bandgaps, E_g, and the electron longitudinal, m_l, and transverse, m_t, effective masses are also presented along with other quantities discussed below.

For III–V compounds, different situations occur for different materials: some of these compounds are direct-bandgap crystals and others are indirect materials (see Section 2.4.4). Thus, for the conduction bands of these materials the conduction band edge can be found either at the Γ point or at the L or X point. In Table 2.5, energy band parameters are shown for three III–V compounds; the direct or indirect nature of the bandgaps is indicated. Now we shall consider the valence bands. Their structure is more complicated. For group IV semiconductors and for III–V compounds, the top of the valence bands at $\vec{k}=0$ has high degeneracy, i.e., several

Table 2.5 Energy band parameters for some III–V compounds

	InAs	GaAs	AlAs
Type of bandgap	Direct	Direct	Indirect
Lowest minima	Γ point	Γ point	X points
E_g (eV)	0.354	1.42	2.16
Electrons			
m^*/m_0	0.025	0.067	0.124
Holes:			
m_{hh}/m_0	0.41	0.50	0.50
m_{lh}/m_0	0.26	0.07	0.26
Δ_{so} (eV)	0.38	0.34	0.28

valence bands have the same energy at this point. The degeneracy occurs because these bands originate from the bonding of three p-orbitals of the atoms composing the crystals. Thus, if one neglects the interaction between the spin of the electrons and their motion (the so-called *spin–orbit interaction*), one obtains three degenerate valence bands, each of which is also doubly degenerate as a result of electron spin. In fact, the spin–orbit interaction causes a splitting of these six-fold degenerate states. At $\vec{k} = 0$ they are split into (i) a quadruplet of states with degeneracy equal to 4 and (ii) a doublet of states with degeneracy equal to 2. This splitting of valence bands, Δ_{so}, at $\vec{k} = 0$ is shown in Fig. 2.10. One frequently refers to the lower valence band as the *split-off valence band*. At finite $\vec{k}$, the spin–orbit interaction leads to further splitting of the upper valence band into two branches: the *heavy-* and *light-hole bands*. The parameters characterizing these bands are the *heavy-hole*, m_{hh}, and *light-hole*, m_{lh}, masses. The effective masses of the light hole, m_{lh}, and heavy hole, m_{hh}, as well as the distance in energy to the split-off band, Δ_{so}, are presented for Si, Ge, and some of the III–V compounds in Tables 2.4 and 2.5.

It is worth mentioning that, despite the relative complexity of the picture presented for energy bands, the description of electron properties in terms of electron and hole quasi-particles using several $E_\alpha(\vec{k})$ dependences is immeasurably simpler than operating with the enormous number of valence electrons in the crystal.

2.4.4 Direct-Bandgap and Indirect-Bandgap Semiconductors

One of the important conclusions which can be drawn from the previously described energy band picture is related to the optical properties of crystals.

The absorption and emission of light can be interpreted as the absorption and emission of discrete "portions" of the light, or photons, with a specific energy and momentum. Let us apply this concept to consider light–crystal interactions. In a crystal, upon absorbing or emitting a photon of sufficient energy, an electron may be transferred between the valence and conduction bands. For a given frequency of light, ω, such a transition is possible if the energy and momentum conservation laws are satisfied:

$$E_c(\vec{k}_1) - E_v(\vec{k}_2) = \hbar\omega,$$
$$\vec{k}_1 - \vec{k}_2 = \pm\vec{q},$$

where $\vec{k}_1$ and $\vec{k}_2$ are the wavevectors of the electrons participating in the phototransition; here, $\vec{q}$ is the photon wavevector. The sign $+$ $(-)$ in the second set of equations stands for photon emission (absorption). Throughout the whole optical spectral region (from infrared to ultraviolet light), the wavelengths λ are much greater than the electron de Broglie wavelengths, as estimated in Appendix A. The photon wavevectors $q = 2\pi/\lambda$, in turn, are much smaller than the electron wavevectors $(|\vec{k}_1|, |\vec{k}_2| \gg |\vec{q}|)$. This property reduces the above equations to

$$\vec{k}_1 \approx \vec{k}_2 = \vec{k} \quad \text{and} \quad E_c(\vec{k}) - E_v(\vec{k}) = \hbar\omega.$$

In other words, under light absorption and emission the electrons transferred between the valence and conduction bands practically preserve their wavevectors: the electron wavevector changes very little. In an energy scheme like those presented in Fig. 2.10, the processes of light–crystal interaction can be interpreted as *vertical* electron interband transitions. Another conclusion following from this analysis is that interband phototransitions are possible only for light with energy quanta exceeding the bandgap, $\hbar\omega \geq E_g$. This finding implies that a pure semiconductor crystal is optically transparent for light with $\hbar\omega \leq E_g$ $(\lambda \geq ch/E_g)$, with c the velocity of light in vacuum. For different optoelectronic applications, the spectral range near the onset of light absorption/emission is critically important.

On combining this analysis with the previously described bandstructures of different materials we can see that phototransitions induced by light with energy near the bandgap are possible in semiconductors where the conduction and valence bands have a minimum and a maximum, respectively, at the same Γ point. For example, in GaAs one can transfer an electron from the valence band to the conduction band directly without a change in its momentum. Crystals of this type are called *direct-bandgap semiconductors*.

In contrast, in order to move an electron from the valence band to the conduction band in Si and Ge, one must not only add an amount of energy – greater than the minimum energy difference between the conduction and valence bands – to excite an electron, but also change its momentum by a large amount (comparable to the scale of the Brillouin zone). Such a semiconductor is called an *indirect-bandgap semiconductor*.

Summarizing, the band structure of a semiconductor material determines both its electrical and its optical properties. The manipulation of electrons using light, i.e., optoelectronic functions, is easier for a direct-bandgap semiconductor like GaAs, in contrast to silicon and other group IV materials. The situation is different in nanoscale Si and SiGe structures where the momentum conservation law is no longer rigorously obeyed.

Table 2.6 Bandgaps for III–V alloys

Alloy	E_g (eV)
$Al_xGa_{1-x}As$	$1.42 + 1.247x$
$Al_xIn_{1-x}As$	$0.360 + 2.012x + 0.698x^2$
$Ga_xIn_{1-x}As$	$0.360 + 1.064x$
$Ga_xIn_{1-x}Sb$	$0.172 + 0.139x + 0.415x^2$
$Al_xGa_{1-x}Sb$	$0.726 + 1.129x + 0.368x^2$
$Al_xIn_{1-x}Sb$	$0.172 + 1.621x + 0.430x^2$

2.4.5 Band Structures of Semiconductor Alloys

As emphasized in previous discussions, the energy band structure of a particular semiconductor determines its electrical and optical properties. For naturally existing semiconductor crystals like monoatomic Ge and Si, binary GaAs, etc., their fixed, and unalterable, energy band structures restrict their applications. One of the powerful tools for varying the band structure is based on alloying two or more semiconductor materials. Some alloys exhibit well-ordered crystal structures. Though an alloy always has some disorder of the constitutive atoms, contemporary technology facilitates partial control of this disorder and produces high-quality crystals. The properties of such materials can be interpreted in terms of nearly ideal periodic crystals.

Consider an alloy consisting of two components: A with a fraction x, and B with a fraction $(1 - x)$. If A and B have similar crystalline lattices, one can expect that the alloy A_xB_{1-x} has the same crystalline structure with the lattice constant a_c given by a combination of the lattice constants of materials A, a_A, and B, a_B. The simplest linear combination leads to the following equation (Vegard's law):

$$a_c = a_A x + a_B(1 - x). \tag{2.20}$$

Then, the symmetry analysis can be extended to these types of alloys. For SiGe alloys and III–V compounds, this leads us to the previously discussed symmetry properties of the energy bands. Since the band structures are similar, one can characterize different parameters of the alloy as a function of the fraction x. This approximation is often called the *virtual-crystal approximation*. For example, the bandgap of an alloy can be represented as $E_g^{alloy} = E_g(x)$. Such approximate dependences are given in Table 2.6 for III–V alloys. They correspond to the bandgaps, E_g, at the Γ points.

As the composition of an alloy varies, the internal structure of the energy bands changes significantly. For example, in the case of $Al_xGa_{1-x}As$ alloys, the lowest energy minimum of the Γ conduction band of GaAs is replaced by the six X minima of AlAs as the value of x is increased. Indeed, near the composition $x \approx 0.4$, the alloy transforms from a direct- to an indirect-bandgap material. The x-dependences of the effective masses for different electron energy minima as well as for heavy and

Table 2.7 Effective masses for the alloy $Al_xGa_{1-x}As$

Type of minimum	Effective mass, m_α^*/m
Γ point	$0.067 + 0.083x$
X minima	$0.32 - 0.06x$
L minima	$0.11 + 0.03x$
Heavy hole	$0.62 + 0.14x$
Light hole	$0.087 + 0.063x$

light holes are presented in Table 2.7 for $Al_xGa_{1-x}As$. Clearly, the established capability for fabricating a variety of high-quality materials provides an excellent tool for modifying the fundamental properties of materials.

2.5 Semiconductor Heterostructures

Further modification and engineering of material properties is possible with the use of *heterostructures*. Heterostructures are structures with two or more abrupt interfaces at the boundaries between the different material regions. With modern materials-growth techniques, it is possible to grow structures with transition regions between adjacent materials that have thicknesses of only one or two atomic monolayers.

2.5.1 Band-Offsets at Heterojunctions

Let us consider a junction between two different semiconductor materials, which generates an abrupt change in the energy gap as well as an abrupt change in the conduction and valence band energies. These abrupt changes result in band-offset steps.

To understand the principal novel features brought about by an abrupt energy change in the energy band structure, we need to deviate from the approach of the previous sections where, while considering the energy bands of different semiconductors, we analyzed energy structures in terms of the *relative positions* of the bands in each of the semiconductors. In this approach, the absolute values of the energies were not important and only the relative positions of the bands were taken into account. However, if two different materials are brought together, the absolute values of energies become critically important. There is a simple way to compare energy bands of different materials. Let us introduce the *vacuum level* of the electron energy, which coincides with the energy of an electron "outside" a material. It is obvious that the vacuum level may be taken to have the same value for any material. One can characterize the absolute energy position of the bottom of the conduction band with respect to this level, as shown in Fig. 2.11. The energy distance between the bottom of the conduction band and the vacuum level, χ, is called the *electron affinity*. In other words, the electron affinity is the energy required to remove an electron from the bottom of the conduction band to a position outside the material, i.e., to the

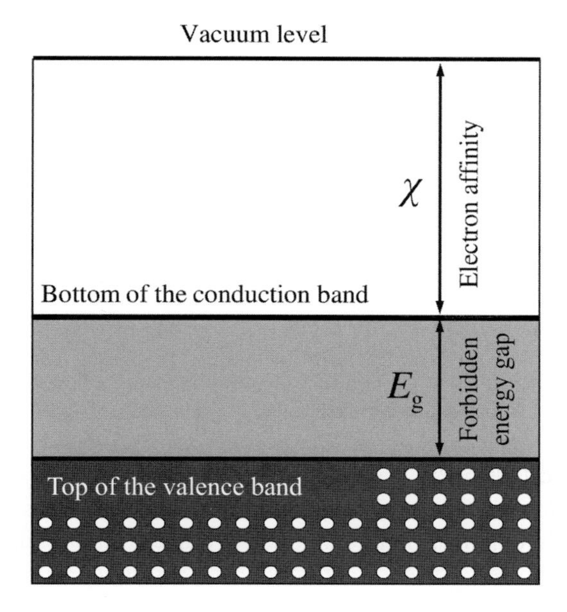

Figure 2.11 The electron affinity and vacuum level in a crystal.

so-called *vacuum level*. Thus, if we know the electron affinities for different materials, we know the values of the bottoms of the conduction bands with respect to each other.

With this definition of the electron affinity, one can calculate the discontinuity in the conduction band at an abrupt heterojunction of two materials, A and B:

$$\Delta E_c = E_{c,B} - E_{c,A} = \chi_A - \chi_B, \tag{2.21}$$

where $\chi_{A,B}$ are the electron affinities of materials A and B. Similarly, one can calculate the discontinuity of the valence band for the same heterojunction:

$$\Delta E_v = E_{v,B} - E_{v,A} = \chi_B - \chi_A + \Delta E_g, \tag{2.22}$$

where $\Delta E_g = E_{g,A} - E_{g,B}$ is the bandgap discontinuity for the heterojunction, with $E_{g,A}$ and $E_{g,B}$ the bandgaps of materials A and B, respectively. Thus, if this simple approach – the *electron affinity rule* – is applicable to a pair of semiconductor materials, one can calculate the band-offsets for an ideal heterojunction. Furthermore, if three materials, say A, B, and C, obey this rule, the following "transitivity" property is valid:

$$\Delta E_v(A/B) + \Delta E_v(B/C) + \Delta E_v(C/A) = 0,$$

where $\Delta E_v(A/B)$ is the valence band discontinuity at the A/B interface. Hence, it is possible to calculate the band-offset for one of three junctions, if the parameters for two of them are known.

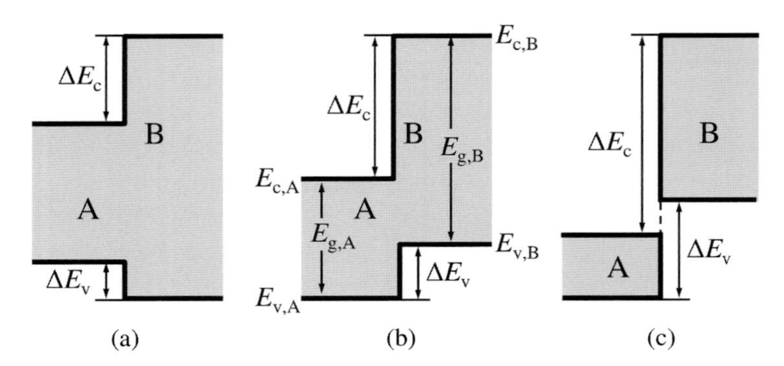

Figure 2.12 Three types of interface: type I (a); type II (b); and broken-gap line-up (c). The energy bandgaps of materials A and B are indicated in (b).

Unfortunately, this rule fails for many semiconductor pairs. One reason for this failure is the dissimilar character of chemical bonds in adjacent materials. The formation of new chemical bonds at such a heterojunction results in charge transfer across this junction and the consequent reconstruction of energy bands, which leads to the breakdown of the electron affinity rule. In real heterojunctions, band-offsets can depend on the quality of the interface, conditions of growth, etc.

On combining different values of the electron affinity and the energy bandgap, we can expect different band line-ups at the interface between two semiconductor materials. In Fig. 2.12, sketches of the three possible types of band discontinuity are presented. The most common line-up is of the "straddling" type presented in Fig. 2.12(a), with conduction and valence band-offsets of opposite signs and with the lowest conduction band states occurring in the same part of the structure as the highest valence band states. This case is referred to as a *type-I heterostructure*. The most widely studied heterojunction system, GaAs/$Al_xGa_{1-x}As$, is of this kind for $x < 0.4$. The next sketch, Fig. 2.12(b), depicts a heterostructure where the lowest conduction band minimum occurs on one side, and the highest valence band maximum on the other, with an energy separation between the two less than the lower of the two bulk bandgaps. This case represents a *type-II heterostructure*. The combination AlAs/$Al_xGa_{1-x}As$ for $x > 0.4$ and some Si/Si_xGe_{1-x} structures are of this kind. Figure 2.12(c) illustrates a broken-gap line-up, in which the bottom of the conduction band on one side drops below the top of the valence band on the other. An example of this band line-up is given by InAs/GaSb, with a break in the forbidden gap at the interface on the order of 150 meV.

2.5.2 Graded Semiconductors

Very often, instead of abrupt heterointerfaces, graded semiconductors are used. To illustrate the idea, consider first a homogeneous piece of a semiconductor, say, a piece of uniformly doped silicon, but with an electric field applied. Then, the band

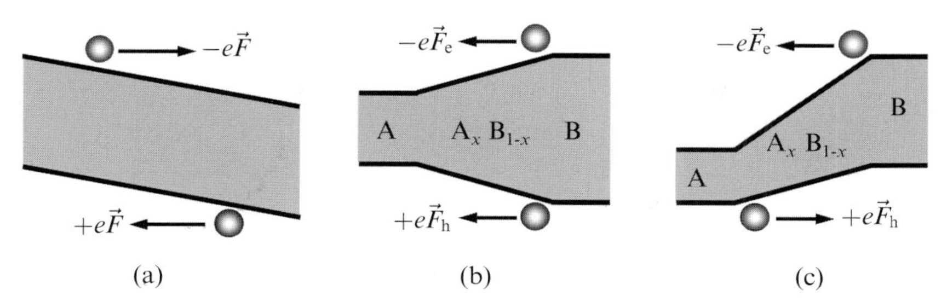

Figure 2.13 (a) An external electric field, $\vec{F}$, simply tilts the conduction and valence bands of a semiconductor. The forces $-e\vec{F}$ and $+e\vec{F}$ acting on the electron and the hole are equal in magnitude and opposite in direction. (b) The same direction of the forces $-e\vec{F}_e$ and $+e\vec{F}_h$ on the electron and the hole caused by a quasi-electric field in the conduction band, $\vec{F}_e$, and in the valence band, $\vec{F}_h$. (c) Forces $-e\vec{F}_e$ and $+e\vec{F}_h$ of opposite directions for electrons and holes.

diagram looks like that illustrated in Fig. 2.13(a) and is represented simply by two parallel tilted lines corresponding to the conduction and valence band edges. The separation between the two lines is the energy bandgap of the semiconductor; the slope of the two band edges is the elementary charge e multiplied by the electric field $\vec{F}$. When an electron or a hole is placed into this structure, a force $-e\vec{F}$ acts on the electron and $+e\vec{F}$ on the hole; the two forces are equal in magnitude and opposite in direction.

Slopes of the conduction and valence band edges arise in the case of a graded transition from one material to another. Graded transitions from a narrow-bandgap to a wide-bandgap semiconductor that correspond to the abrupt heterojunctions of Figs. 2.12(a) and (b) are shown in Figs. 2.13(b) and (c). As is obvious from Figs. 2.13(b) and (c), in the case of graded heterostructures, there is a built-in electric field that acts on electrons and holes. This field is called *quasi-electric*. The quasi-electric field does not exist in homogeneous crystals; that is why graded heterostructures can be used for new devices where the existence of a built-in electric field is required. Examples of materials used in graded nanostructure devices are Si_xGe_{1-x} and $Al_xGa_{1-x}As$, where x changes in the direction of growth of the structure. Graded structures and the accompanying quasi-electric forces introduce a new degree of freedom for the device designer and allow him or her to obtain effects that are basically impossible to obtain using only external (or *real*) electric fields.

2.6 Lattice-Matched and Lattice-Mismatched Materials

Now we shall consider some of the principal problems that arise in the fabrication of heterostructures. In general, one can grow any one layer on almost any other material. In practice, however, the interfacial quality of such artificially grown structures can vary enormously. Even when one fabricates a structure from two materials of the same group or from compounds of the same family, the artificially grown materials of the heterostructure may be very different from the corresponding bulk materials.

Table 2.8 Lattice constants for cubic semiconductor materials ($T = 300$ K)

Semiconductor	Lattice constant (Å)
SiC	3.0806
C	3.5668
Si	5.4309
GaP	5.4495
GaAs	5.6419
Ge	5.6461
AlAs	5.6611
InP	5.8687
InAs	6.0584

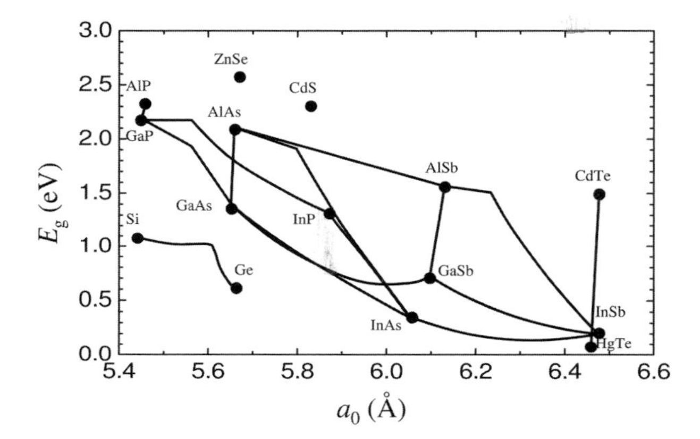

Figure 2.14 Room-temperature bandgaps E_g as functions of the lattice constant a_0 for selected III–V and II–VI compounds and selected group IV materials and their alloys.

First of all, the quality of the materials near heterointerfaces depends strongly on the ratio of lattice constants for the two materials.

In Table 2.8, lattice constants for several group IV semiconductors and III–V compound semiconductors are presented; all of the cases presented represent cubic crystals. The lattice constants for some other materials can be found from Fig. 2.14. Depending on the structural similarity and lattice constants of the constituent materials, there exist two principally different classes of heterointerfaces: lattice-matched and lattice-mismatched materials. Prior to an analysis of both classes, we highlight other factors affecting the quality and usefulness of heterointerfaces.

2.6.1 Valence Matching

Since there are still no rigorous rules for how one can realize a given level of quality for heterojunctions, we consider a few examples that illustrate some problems.

If lattice matching were the only obstacle, the Ge/GaAs system would be the ideal heterosystem, because, according to Table 2.8, it would allow one to realize

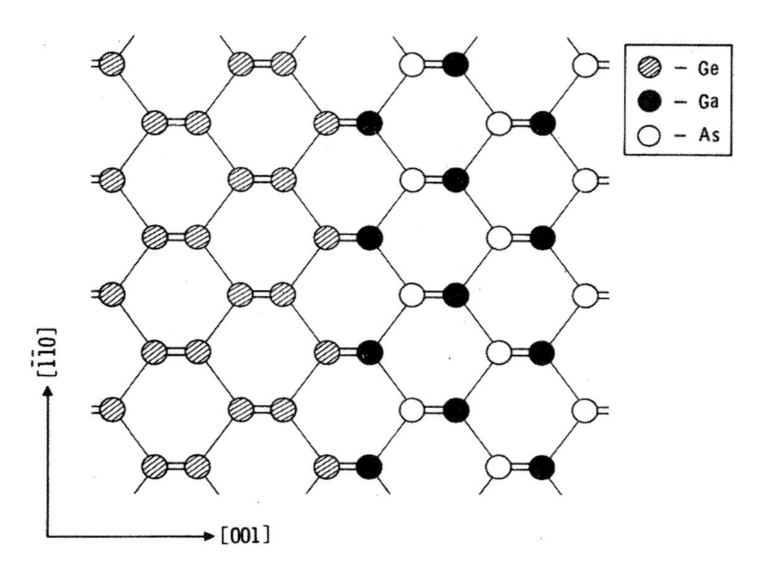

Figure 2.15 Departure from electrical neutrality at a "mathematically planar" (001)-oriented Ge/GaAs interface. The different atomic species – Ga or As atoms and Ge atoms – do not bring along the correct number of electrons to form electrically neutral Ga—Ge or As—Ge covalent bonds of two electrons per bond, and there will be an electric field of the order of 10^7 V cm^{-1} at the heterointerface. Reprinted with permission from Fig. 2 in W. A. Harrison, E. Kraut *et al.*, "Polar heterojunction interfaces" *Phys. Rev. B*, **18**, 4402 (1978). © 1978 by the American Physical Society.

the ideal combination of group IV semiconductors and III–V compounds. Indeed, based solely on lattice-constant matching, the Ge/GaAs system appears to be the most promising candidate. However, it turns out that there is the problem of chemical compatibility for this heterostructure. Covalent bonds between Ge on the one hand and Ga or As on the other are readily formed, but they are what could be called *valence-mismatched bonds*, meaning that the number of electrons provided by the atoms is not equal to the canonical number of exactly two electrons per covalent bond. Hence, the bonds themselves are not electrically neutral. Consider a hypothetical idealized (001)-oriented interface between Ge and GaAs, with Ge to the left of a "mathematical plane" and GaAs to the right, as shown in Fig. 2.15. In GaAs, an As atom brings along five electrons (resulting in 5/4 electrons per bond) and is surrounded by four Ga atoms, each of which brings along three electrons (3/4 electron per bond), adding up to the correct number of two (8/4) electrons per Ga–As covalent bond.

However, when, at an (001) interface, an Ga atom has two Ge atoms as bonding partners, each Ge atom brings along one electron per bond, which is half of an electron less than is required for bonding. Loosely speaking, the Ga atom does not "know" whether it is a constituent of GaAs or an acceptor in Ge. As a result, each Ge–Ga bond acts as an acceptor with a fractional charge. Analogously, each

Ge–As bond acts as a donor with the opposite fractional charge. To be electrically neutral, a Ge/GaAs interface would have to have equal numbers of these two charges, averaged not only over large distances but also locally. Given chemical bonding preferences, such an arrangement will not occur naturally during epitaxial growth. If only one kind of bond were present, as in Fig. 2.15, the interface charge would support a large electric field of 4×10^7 V cm^{-1}. Such a huge field would force atomic rearrangements, during the growth, trying to equalize the number of Ge–As and Ge–Ga bonds. However, these rearrangements will never go to completion but will leave behind locally fluctuating residual charges, with deleterious consequences for the electrical properties of the materials and any device applications.

Interfaces with perfect bond-charge cancellation may be readily drawn on paper, but in practice there will always remain some local deviations from perfect charge compensation, leading to performance-degrading random potential fluctuations along the interface. This argument applies also to other interfaces combining semiconductors from different groups of the periodic table.

The above discussion pertains to the most widely used (001)-oriented interface. The interface charge at a valence-mismatched interface actually depends on the crystallographic orientation. It has been shown that an ideal (112) interface between group IV and III–V compounds exhibits no interface charge. An important example is the GaP-on-Si interface, which has a sufficiently low defect density; as a result, it is used in different devices grown on Si. After these comments, we return to the discussion of the role and importance of lattice matching.

2.6.2 Lattice-Matched Materials

For lattice-matched structures, the lattice constants of the constituent materials *are nearly matched*; i.e., the lattice constants are within a small fraction of 1% of each other. There is no problem, in general, in growing high-quality heterostructures with such lattice-matched pairs of materials. By "high-quality" we mean that the interface structure is free of lattice imperfections such as interface defects, etc. Such imperfections result in poor electrical and optical properties and may lead to fast and widespread degradation of the structure. Figure 2.16(a) illustrates a lattice-matched layer B on a substrate A. One can expect that the layer can be grown on the substrate if both materials are from the same group and the binding energies and crystal structures are very similar.

According to the data of Fig. 2.14 and Table 2.8, the AlGaAs/GaAs system is an example of a lattice-matched material. The system has a very small mismatch in the lattice constants of only about 0.1% over the entire range of possible Al-to-Ga ratios in the AlGaAs system. As a result, such heterostructures can be grown free of mechanical strain and significant imperfections. Hence, these structures provide a practical way of tailoring band structures. In addition to these tailored electronic parameters, the elastic and other lattice properties can be different in layers composing such a lattice-matched heterostructure.

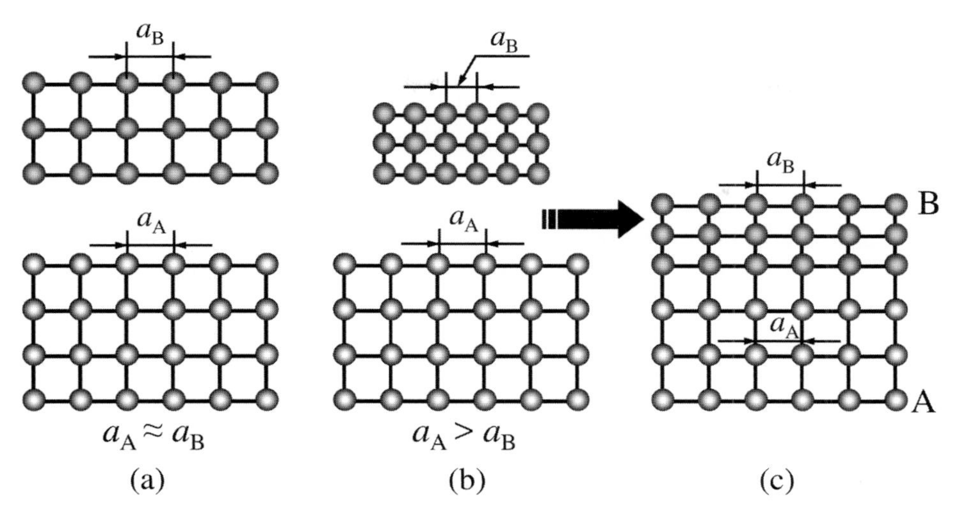

Figure 2.16 Lattice-matched materials (a) and lattice-mismatched materials (b); the resulting structure is strained (pseudomorphic) if the upper layer B adopts the lattice of substrate A (c).

2.6.3 Lattice-Mismatched Materials

The case of lattice-mismatched structures is characterized by a *finite lattice mismatch*. Figure 2.16(b) depicts this case. If one tries to match these lattices, strain in the plane of growth and a distortion along the growth axis arise. Thus, one obtains a strained layer with a lattice deformation. The lattice-mismatched structure also can be characterized by the *relative mismatch* of the lattice constants of the substrate, a_A, and the epilayer, a_B:

$$\epsilon = \frac{a_A - a_B}{a_A}. \tag{2.23}$$

Consider an elastic deformation of a lattice. It can be characterized by the vector of relative displacement $\vec{u}$. The displacement defines how any lattice point $\vec{r}$ moves to a new position, $\vec{r}' = \vec{r} + \vec{u}$, as a result of the deformation. Different regions of the crystal can be deformed differently. Thus, the displacement depends on the coordinates, $\vec{u} = \vec{u}(\vec{r})$. In fact, only relative displacements are important. They are determined by the *strain tensor*:

$$u_{ij} = \frac{1}{2}\left(\frac{\partial u_i}{\partial \zeta_j} + \frac{\partial u_j}{\partial \zeta_i}\right), \tag{2.24}$$

where the tensor u_{ij} has the following components:

$$u_{ij} = \begin{pmatrix} u_{xx} & u_{xy} & u_{xz} \\ u_{yx} & u_{yy} & u_{yz} \\ u_{zx} & u_{zy} & u_{zz} \end{pmatrix}. \tag{2.25}$$

Here i and j denote the x, y, and z components and $\zeta_i = x, y$, and z for $i = 1$, 2, and 3, respectively. In this discussion, we consider only the *diagonal components* of u_{ij}. They determine the change in the crystal volume, from $\mathcal{V}$ to $\mathcal{V}'$, produced by the strain:

$$\delta = \frac{\mathcal{V}' - \mathcal{V}}{\mathcal{V}} = u_{xx} + u_{yy} + u_{zz}. \tag{2.26}$$

The elastic energy density (elastic energy per unit volume) of a crystal may also be expressed in terms of the strain tensor. For cubic crystals, this energy is given by

$$U = \frac{1}{2} c_{11} \left(u_{xx}^2 + u_{yy}^2 + u_{zz}^2 \right) + c_{44} \left(u_{xy}^2 + u_{xz}^2 + u_{yz}^2 \right) + c_{12} \left(u_{xx} u_{yy} + u_{xx} u_{zz} + u_{yy} u_{zz} \right), \tag{2.27}$$

where c_{11}, c_{12}, and c_{44} are the *elastic constants or elastic moduli* of the crystal. Equation (2.5) represents the potential energy of an elastic isotropic medium. The elastic constants c_{12} and c_{44} are responsible for the anisotropy of crystals; this is why they equal zero for an elastic isotropic medium. As a result, Eq. (2.27) is recovered in this case with $\Lambda = c_{11}$.

The *stress tensor* is defined in terms of the derivatives of the elastic energy density with respect to the strain tensor components:

$$\sigma_{ij} = \frac{\partial U}{\partial u_{ij}}. \tag{2.28}$$

The boundary conditions at a surface or at an interface may be formulated in terms of the stress tensor:

$$\sigma_{ij} N_j = f_i, \tag{2.29}$$

where $\vec{N}$ is a vector perpendicular to the surface, and $\vec{f}$ is an external force applied to the surface.

These equations are sufficient for calculations of the strain of a layer A grown on a mismatched substrate B. Let the lattice constants of these two materials be a_A and a_B, respectively. In this discussion, both materials are assumed to be cubic crystals and the direction of growth is along the [001]-direction. If the layer A adopts the lattice periodicity of the substrate B, the in-plane strain of the layer is

$$u_{xx} = u_{yy} = u_{||} = 1 - \frac{a_B}{a_A}. \tag{2.30}$$

There should be no stress in the direction of growth. Thus, from Eq. (2.29) it follows that $\sigma_{zz} = 0$. Calculating σ_{zz} from Eqs. (2.27) and (2.28), from the obtained result, $\sigma_{zz} = c_{11} u_{zz} + c_{12} \left(u_{xx} + u_{yy} \right)$, we find the strain in the direction perpendicular to the layer:

$$u_{zz} = -\frac{2c_{12}}{c_{11}} u_{||}. \tag{2.31}$$

Thus, the strain can be found through the mismatch of the lattice constants.

The strain results in two types of effects: (1) the strain can generate different imperfections and defects; and (2) the strain in the layer leads to a change in the symmetry of the crystal lattice, for example, from cubic to tetragonal or to rhombohedral, etc. Of course, the latter effect can modify the energy band structure of the layer.

2.7 Lattice-Matched and Pseudomorphic Heterostructures, Si/Ge Heterostructures, and Lattice-Matched III–V Heterostructures

2.7.1 Lattice-Matched and Pseudomorphic Heterostructures

Here, we consider imperfections generated by the strain from a lattice mismatch. In order to understand the nature of the formation of imperfections in a layered structure, let us consider the characteristic energies of the structure. First of all, a layer grown on a substrate with a mismatched lattice should possess *extra elastic energy*, E_{el}, caused by the strain. This energy is a function of the thickness of the layer, d, and increases with increasing d. In the simplest case, of uniform strain, the elastic energy can be calculated through its density U: $E_{el} = U \times d \times S$, where S is the area of the layer. On the other hand, the generation of misfit defects requires some energy. Let us denote this energy by E_{im}. If the extra elastic energy exceeds the energy associated with the imperfection, i.e., if $E_{el}(d) > E_{im}$, the system will relax to a new state with lower energy and imperfections will be generated. That is, the extra strain energy is the main physical reason for the instability and degradation of heterostructures fabricated from materials with a large mismatch of lattice constants.

Since the value of E_{im} remains finite even in thin layers, for certain thicknesses we may get $E_{el}(d) < E_{im}$. Thus, there is not sufficient strain energy and imperfections will not be generated. Such strained heterostructures can be of high quality. Hence, in some approximations, for each pair of materials there exists a critical thickness of the layers, d_{cr}; if $d < d_{cr}$ the lattice mismatch is accommodated by the layer strain without the generation of defects. The corresponding layered systems are called *pseudomorphic heterostructures*. In general, a pseudomorphic layer of material possesses some characteristics similar to those of the substrate and may possibly have the same lattice structure as the substrate material. In our case, a crystalline semiconductor layer grown on another semiconductor takes on the in-plane lattice periodicity of the substrate semiconductor. Figure 2.16(c) illustrates the case when the deposited layer adopts the lattice periodicity of the substrate material. The $Ga_{1-x}Al_xAs/Ga_{1-x}In_xAs$ and $GaAs/Ga_{1-x}In_xAs$ structures are examples of such systems. In fact, these heterostructures are used to improve the characteristics of the so-called *heterojunction-field-effect transistors*, which will be considered in Chapter 8. In spite of significant mismatches of lattice constants, these structures are virtually free of interface defects due to the small, nanometer-scale, thicknesses of the pseudomorphic layers used in the fabrication of functioning heterojunction-field-effect transistors.

It is sometimes possible to grow defect-free systems with layer thicknesses exceeding the critical thickness. However, such systems are metastable, and this may lead to device degradation as a result of the generation of misfit defects driven

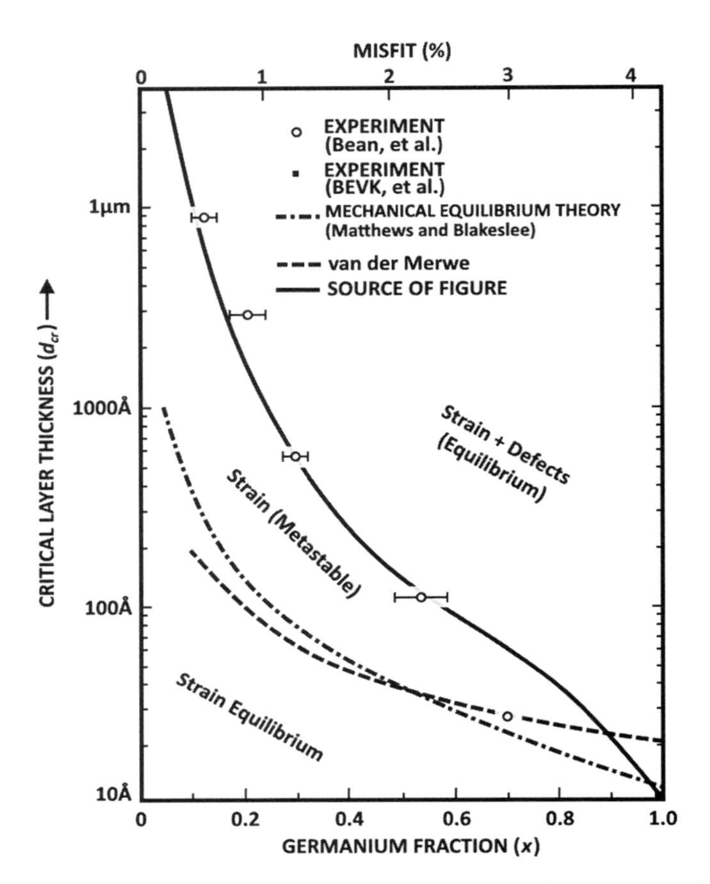

Figure 2.17 The stability–strain diagram for a Ge_xSi_{1-x} layer on a Si substrate. Reprinted from R. People and J. Bean, *Appl. Phys. Lett.* **49**, 229 (1986), with the permission of AIP Publishing.

by temperature effects or other external perturbations. Central to the stability of pseudomorphic structures is the question of whether or not the strain energy leads to damage of the materials when the structures are subjected to various forms of external stress and processing. The experience accumulated in this field shows that in the case of small strain energy the heterostructures are stable. For example, in the case of the GaP/GaAsP layered system, the strain energy is about 10^{-3} eV per atom. Since this quantity is rather small in comparison with the energy required to remove the atom from its lattice site, this system can be stable for sufficiently thin layers.

The strain states discussed above are shown in Fig. 2.17 as a function of x for Ge_x Si_{1-x} layers grown on Si substrates. This "phase diagram" – a plot of the critical thickness of the layer versus the Ge fraction – consists of three regions: strained layers with defects at large thicknesses, nonequilibrium (metastable) strained layers without defects at intermediate thicknesses, and equilibrium and stable layers without defects at small thicknesses. According to these results, a stable Ge layer on Si (the largest misfit) cannot be grown with a thickness greater than 10 Å or so.

2.7.2 Si/Ge Heterostructures

Let us consider the Si/Ge system in more detail. This system is very interesting and important because it has opened new horizons for silicon nanotechnology and different Si-based applications. The data of Table 2.8 shows that heterostructures based on Si and Ge materials should be designed so that they are *always* pseudomorphic.

First of all, the stability and quality of these Si/Ge pseudomorphic heterostructures depend strongly on the thicknesses of the strained layers as discussed previously. In fabricating Si/Ge structures, one grows different numbers of Si and Ge atomic monolayers. Thus, layer thicknesses can be characterized by the numbers of these monolayers. Let n and m be the numbers of Si and Ge monolayers, respectively. This system is known as the Si_n/Ge_m *superlattice*. The second important factor that determines the quality of these structures is the *material of the substrate* on which the superlattices are grown. We have discussed the case of Ge_xSi_{1-x} layers grown upon a Si substrate; see Fig. 2.17. For the fabrication of Si/Ge superlattices, the substrates of choice are frequently either Ge_xSi_{1-x} alloys or GaAs.

Let us consider Ge_xSi_{1-x} as a substrate. The elastic energy of a strained system depends on the alloy composition of the substrate. Figure 2.18 illustrates this dependence for different numbers of monolayers for the symmetric case, $n = m$. In accordance with the previous discussion, the elastic energy increases with increasing thicknesses of the strained layers for a given substrate material. Because of this, one employs superlattices with few monolayers: $2 \leq (n, m) \leq 5$. Figure 2.18 also shows a nontrivial strain energy dependence on the alloy composition of the substrate; the minimal strain energy is expected for x in the range from 0.4 to 0.6.

Another important characteristic of pseudomorphic Si/Ge structures is the distribution of the elastic energy over the monolayers of the superlattice. It has been shown that the most homogeneous distribution over layers occurs for the Si–Ge alloy with $x \approx 0.5$. From this point of view, $Si_{0.5}Ge_{0.5}$ substrates are preferable. However, these results depend strongly on the orientation of the substrate material. Often, the direction of growth on the substrate is chosen to be the [001] direction.

2.7.3 Lattice-Matched III–V Heterostructures

Let us return to Fig. 2.14 and discuss lattice-matched heterostructures in more detail. From this figure, we can determine the lattice constants of different compounds. First of all, one can see that the GaAs/AlAs system is really unique because the lattice constants have almost identical values. In order to achieve lattice matching for other cases, it is possible to combine either a binary compound and a

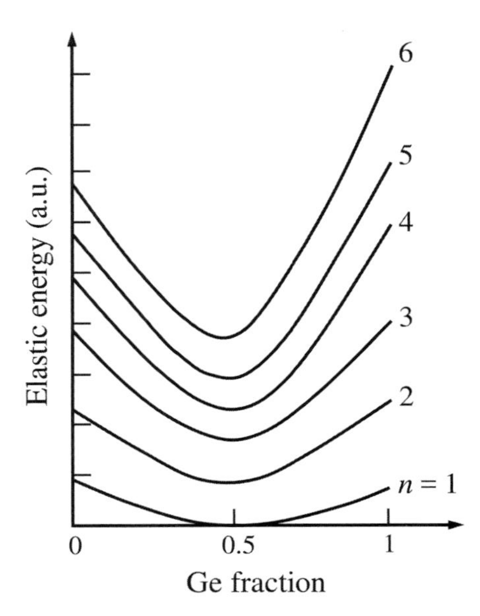

Figure 2.18 The elastic energy of strained Si_n/Ge_n superlattices with different numbers of monolayers n as a function of the Ge fraction in a substrate Ge_xSi_{1-x}. The numbers labeling the curves indicate n.

ternary compound, or two ternary compounds having appropriate ratios of atomic species within each layer. For example, in the case of GaInAs/InP structures, lattice matching is achieved exactly only for $Ga_{0.47}In_{0.53}As$, where the ratio of the Ga to In is 47 to 53 in the GaInAs layer; for other ratios, the GaInAs layer is not lattice-matched with the InP. Moreover, the wide-bandgap $Ga_{0.51}In_{0.49}P$ material is compatible with the narrow-bandgap GaAs material.

In conclusion, the broad range of possibilities for controlling bandgaps and band-offsets both for electrons and for holes, as well as electron and hole effective masses, provides the basis for energy-band engineering. Through such energy-band engineering, it is possible to design and fabricate high-quality heterostructures with designated optical and electrical properties. If one cannot achieve the desired properties using lattice-matched compositions, it is possible to employ strained pseudomorphic structures.

2.8 Wide-Bandgap Materials and Heterostructures

Among the wide-bandgap materials, GaN-based, AlN-based, and InN-based structures – sometimes referred to as III-nitride semiconductors – have found a broad variety of uses in optoelectronic devices. In particular, the bandgaps of these semiconductors open the way to the realization of devices such as light-emitting diodes and lasers with photon energies spanning the range of about 1 eV to several

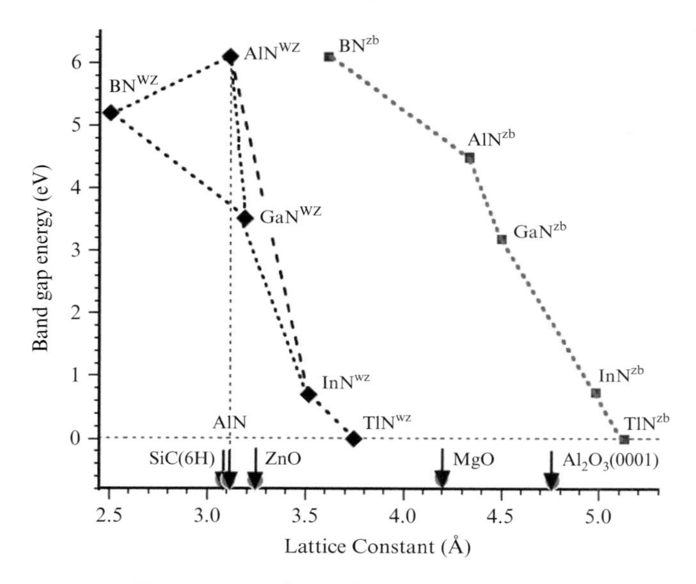

Figure 2.19 Energy gaps of wurtzite (wz) and zincblende (zb) nitride-based semiconductors – AlN, GaN, and InN – as a function of the lattice constants for the materials. For comparison, other nitride-based semiconductors are included, along with other semiconductors with similar lattice constants. SiC(6H) and Al_2O_3 are frequently used as substrates for AlN-, GaN-, and InN-based structures. ZnO has been studied in heterostructures with GaN. Reproduced from Fig. 8.1 of R. Collazo and N. Dietz, "The group III-nitride material class: from preparation to perspectives in photoelectrocatalysis," in *Photoelectrochemical Water Splitting: Issues and Perspectives*, ed. H.-J. Lewerenz and L. M. Peter, RSC Publishing, pp. 193–222 (2013). Reproduced by permission of The Royal Society of Chemistry.

electron-volts, as shown in Fig. 2.19. These so-called III-nitride materials can take on a cubic (zincblende) or hexagonal (wurtzite) structure. As shown in Fig. 2.19, the bandgaps for the zincblende and wurtzite structures are similar. As is also evident from the energy gaps shown in Fig. 2.19, the variations in the bandgaps for $Al_xIn_{1-x}N$, $Al_xGa_{1-x}N$, and $In_xGa_{1-x}N$ deviate from linear relations. This deviation is taken into account by introducing a bowing parameter, b, specific to each of these families.

The cubic (zincblende) crystal structure was discussed previously in this chapter. The wurtzite lattice is depicted in Fig. 2.20.

The wurtzite structure may also be visualized as planes of atoms in hexagonal arrangements with a lattice constant a, which are stacked on each other with a separation c. The direction perpendicular to the hexagonal planes is known as the c-axis. Values for the in-plane (in the hexagonal planes) lattice constant a and the lattice constant c along the c-axis are given in Table 2.9 along with information on the effective masses in AlN, GaN, and InN.

The III-nitride compound semiconductors have been used extensively both in electronic devices such as field-effect transistors fabricated from $Al_xGa_{1-x}N$, GaN,

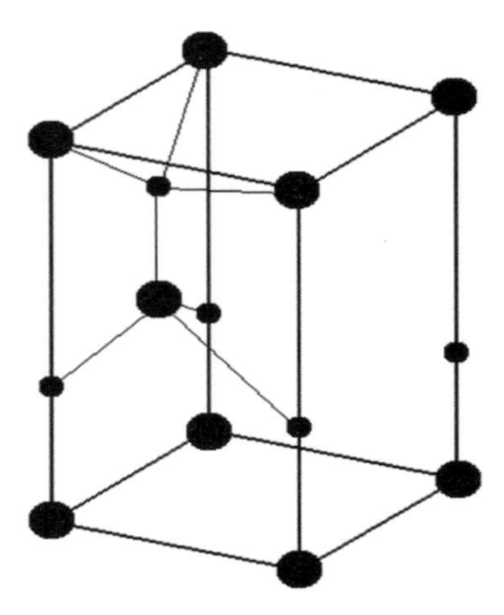

Figure 2.20 The wurtzite lattice structure. The wurtzite structure may also be visualized as planes of atoms in hexagonal arrangements with a lattice constant a, which are stacked on each other with a separation c. The direction perpendicular to the hexagonal planes is known as the c-axis.

Table 2.9 Lattice constants a (in the hexagonal planes) and c (perpendicular to the hexagonal planes) as well as electron effective masses, m_n/m, and heavy-hole and light-hole effective masses, m_{hhz}/m and m_{lhz}/m

	AlN	GaN	InN
a (nm)	0.3110	0.3199	0.3585
c/a	1.606	1.634	1.618
m_n/m	0.25–0.39	0.27(6)	0.11
m_{hhz}/m	2.02–3.13	1.1	0.5–1.63
m_{lhz}/m	0.24–3.53	1.1	0.27

There is considerable uncertainty in the known effective masses for different directions in these wurtzite materials. Additional information is available in *Handbook of Nitride Semiconductors and Devices*, Vol. 1, ed. H. Morkoc (Wiley-VCH Verlag GmbH & Co. KGaA, Weinheim, 2008), and *Semiconductor Group IV Elements and III–V Compounds*, ed. O. Madelung (Springer-Verlag, Berlin, 1991).

and AlN heterostructures and in optoelectronic devices such as light-emitting diodes (LEDs) operating in the ultraviolet and blue regions of the electromagnetic spectrum. Indeed, as shown in Table 2.10, GaN has a bandgap in the ultraviolet region of the electromagnetic spectrum. As we shall see in the following discussion, replacing some of the Ga atoms in GaN with In to form $In_xGa_{1-x}N$ lowers the bandgap so that it corresponds to blue light. The availability of LEDs operating in the blue region of the spectrum has promoted many applications of GaN-based blue LEDs in optical systems where there is a need to cover broad regions of the

Table 2.10 Reported values of the conduction band-edge energy, E_c, and valence band-edge energy, E_v, relative to the vacuum, based on values obtained from density-functional calculations are summarized here for AlN, GaN, and InN. In addition, experimental values for $E_c - E_v$ are given for comparison

	AlN	GaN	InN
E_c relative to vacuum	−0.5 eV	−2.8 eV	−4.9 eV
E_v relative to vacuum	−6.2 eV	−6.0 eV	−5.6 eV
$E_c - E_v$ experimental	6.11−6.2 eV	3.51 eV	0.6 to 0.7 eV

Taken from density-functional calculations and related experimental summaries reported in P. G. Moses, M. Miao, Q. Yan, and C. G. Van de Walle, Hybrid functional investigations of band gaps and band alignments for AlN, GaN, InN, and InGaN, *J. Chem. Phys.* **134**, 084703 (2011).

electromagnetic spectrum. In order to design heterostructure devices employing GaN-, AlN-, and InN-based layers it is essential to know how the conduction band-edge energies, E_c, valence band-edge energies, E_v, and bandgaps of these materials differ from each other. The change in the conduction band-edge energy, E_c, and valence band-edge energy, E_v, at an interface between two materials is known as the offset. These offsets may be determined by referencing all energies to the vacuum level. In order to free an electron at the top of the valence band, it is necessary to add energy until the electron has sufficient energy to escape the attractive interaction with the crystal lattice. If the vacuum energy is set to zero as our reference energy, the magnitude of the energy needed to free an electron from AlN, GaN, or InN is shown in Table 2.10. As discussed previously in this chapter, the energy offsets for the conduction band-edge energy, E_c, and the valence band-edge energy, E_v, may be determined straightforwardly when all energies are referenced (measured) relative to the vacuum level. For example, using the values for E_v given in Table 2.10, it follows that $\Delta E_v(\text{GaN} - \text{AlN}) = 0.2\,\text{eV}$, so that the offset between the valence bands of GaN and AlN at the interface of a GaN/AlN heterostructure (assuming very thin layers so that the structure is pseudomorphic) is 0.2 eV. Unfortunately, this simple picture of band alignments is complicated in many cases by the presence of interfacial states and charge transport across the heterostructure interface, which produce electric fields and bandbending that affect the energies near the interface. However, the technique of referencing the conduction band-edge energies, E_c, and valence band-edge energies, E_v, to the vacuum level provides a useful approach for estimating offsets in many cases. One of the calculated and observed features of the bandgap of ternary and more complex (quaternary and higher) III-nitride-based systems is that the bandgap does not scale linearly with the x value. For example, Fig. 2.21 illustrates how the values of E_c (upper curve) and E_v (lower curve) for $\text{In}_x\text{Ga}_{1-x}\text{N}$ change as functions of x, according to calculations using density functional techniques.

In Table 2.6 we presented the bandgaps of ternary compounds based on As and on Sb. Analogous dependences are observed for the N-based wide-bandgap materials that we present in Table 2.11. The following equation is used in

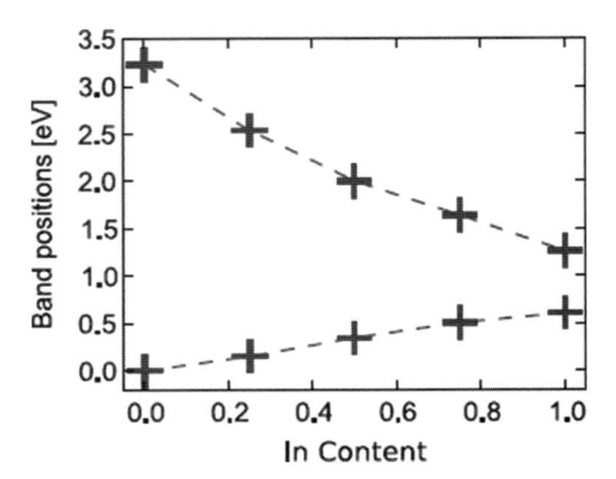

Figure 2.21 The bandgap of $In_xGa_{1-x}N$ as a function of x as calculated using density functional techniques. The upper curve represents $E_c(x)$ and the lower curve represents $E_v(x)$. Reprinted from P. G. Moses and C. G. Van de Walle, *Appl. Phys. Lett.* **96**, 021908 (2010), with the permission of AIP Publishing.

Table 2.11 Bandgaps of nitride-based ternary compounds

Ternary alloy	Bandgap $E_g(AN)$ (eV)	Bandgap $E_g(CN)$ (eV)	Bowing parameter b
$In_xGa_{1-x}N$	0.70 ± 0.05	3.52 ± 0.1	1.6 ± 0.2
$Al_xGa_{1-x}N$	6.10 ± 0.1	3.52 ± 0.1	0.7 ± 0.1
$In_xAl_{1-x}N$	0.70 ± 0.05	6.10 ± 0.1	$3.4x + 1.2$

Reproduced from Fig. 8.1 from R. Collazo and N. Dietz, "The group III-nitride material class: from preparation to perspectives in photoelectrocatalysis," in *Photoelectrochemical Water Splitting: Issues and Perspectives*, ed. H.-J. Lewerenz and L. M. Peter, RSC Publishing, pp. 193–222 (2013). Reproduced by permission of The Royal Society of Chemistry.

Table 2.11 for specifying the dependence of the bandgaps of ternary compounds on composition:

$$E_g(A_xC_{1-x}N) = E_g(AN) \times x + E_g(CN) \times (1-x) - b \times x \times (1-x), \qquad (2.32)$$

where A is replaced by In, Al, and In and C by Ga, Ga, and Al, for $In_xGa_{1-x}N$, $Al_xGa_{1-x}N$, and $In_xAl_{1-x}N$, correspondingly.

The first Brillouin zone for a wurtzite III-nitride is shown in Fig. 2.22. Here $\vec{b}_1$, $\vec{b}_2$, and $\vec{b}_3$ are the reciprocal-lattice vectors in $\vec{k}$-space (wavevector space), and the real space lattice vectors are $\vec{a}_1 = (a/2, -a\sqrt{3}/2, 0)$, $\vec{a}_2 = (a/2, a\sqrt{3}/2, 0)$, and $\vec{a}_3 = (0, 0, c)$. Constructing orthogonal vectors in real space and reciprocal space using the standard formulae, $\vec{b}_1 = 2\pi(\vec{a}_2 \times \vec{a}_3)/\mathcal{V}$, $\vec{b}_2 = 2\pi(\vec{a}_3 \times \vec{a}_1)/\mathcal{V}$, $\vec{b}_3 = 2\pi(\vec{a}_1 \times \vec{a}_2)/\mathcal{V}$, where $\mathcal{V} = |\vec{a}_1 \cdot (\vec{a}_2 \times \vec{a}_3)|$, it follows that the reciprocal-lattice vectors are $\vec{b}_1 = (2\pi/a, -2\pi/a\sqrt{3}, 0)$, $\vec{b}_2 = (2\pi/a, 2\pi/a\sqrt{3}, 0)$, and $\vec{b}_3 = (0, 0, 2\pi/c)$. In terms of these reciprocal-lattice vectors the high-symmetry points have the coordinates given in Table 2.12.

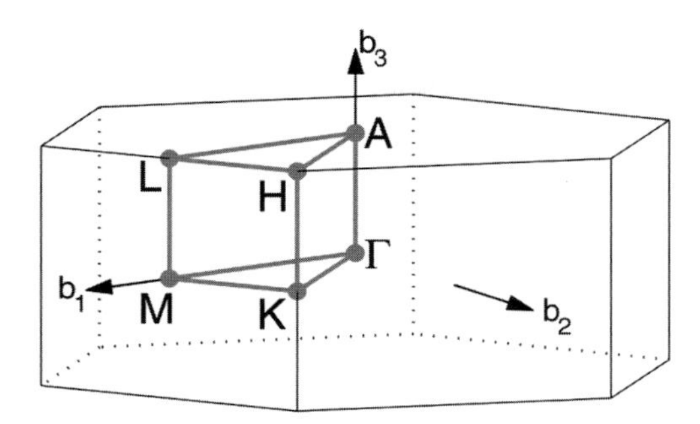

Figure 2.22 The first Brillouin zone for a wurtzite III-nitride. Here $\vec{b}_1$, $\vec{b}_2$, and $\vec{b}_3$ are the reciprocal-lattice vectors in $\vec{k}$-space (wavevector space). Reprinted from Fig. 13 in *Computational Materials Science*, **49**, W. Setyawan and S. Curtarolo, "High-throughput electronic band structure calculations: Challenges and tools," pp. 299–312, Copyright (2010), with permission from Elsevier.

Table 2.12 Symmetry $\vec{k}$-points for the hexagonal lattice

	Multiplier times b_1	Multiplier times b_2	Multiplier times b_3
Γ	0	0	0
A	0	0	1/2
H	1/3	1/3	1/2
K	1/3	1/3	0
L	1/2	0	1/2
M	1/2	0	0

This section on wide-bandgap III-nitrides concludes with a presentation of the energy bands for GaN and AlN. The energy bands for GaN and AlN are presented in Figs. 2.23 and 2.24, respectively. The features of special interest include the direct bandgap in Fig. 2.23(a), where the wurtzite GaN energy band structure has a direct bandgap with a minimum in the conduction band and a maximum in the valence band at the Γ point.

2.9 Quantum Dots

This section deals with the optical properties of quantum dots, which are nanocrystals with confinement in all three dimensions. In practice, stacked arrays of quantum dots with stack thicknesses of tens of nanometers or more may be fabricated and such structures have been used to make optoelectronic devices where optical transitions occur as a result of the electronic structure of these layered arrays. There is still challenging research required in order to make such structures with the large volumes needed to reach the full potential of quantum-dot-based optical devices.

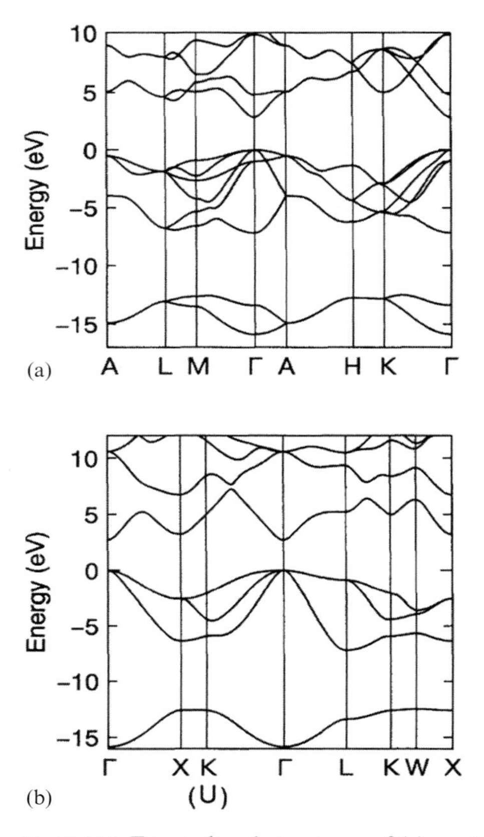

Figure 2.23 Energy band structures of (a) wurtzite GaN and (b) zincblende GaN. The energy scale is chosen so that the top of the valence band at the Γ point has zero energy. Reprinted with permission from Fig. 5 of K. Miwa and A. Fukumoto, *Phys. Rev.* **B 48**, 7897 (1993). Copyright (1993) by the American Physical Society.

Semiconductor quantum dots have a long history going back to Michael Faraday, who performed experiments on chemically synthesized ("bottom-up" fabrication in today's terminology) quantum dots suspended in a solution in which they are synthesized. In the last two decades of the twentieth century, there was a period of rapid progress in the self-assembly of quantum dots on growth surfaces as a result of the advent of layer-by-layer growth techniques such as molecular-beam epitaxy (MBE) as well as metallo-organic chemical vapor deposition (MOCVD). These growth techniques may be used to grow lattice-matched layers on a substrate (such as $Al_xGa_{1-x}As$ on GaAs substrates, or $In_{0.52}Al_{0.48}As/In_{0.53}Ga_{0.47}As$ on the substrate on InP-based substrates). When there is a lattice mismatch, it is possible to grow on the substrate strained layers of the growth material which are stable for some number of monolayers – so-called pseudomorphic layers. However, after the growth thickness exceeds the critical thickness, the lowest energy state corresponds to a state where the pseudomorphic layers evolve to give an array of quantum dots. Such self-assembled quantum dots have been fabricated for a variety of materials;

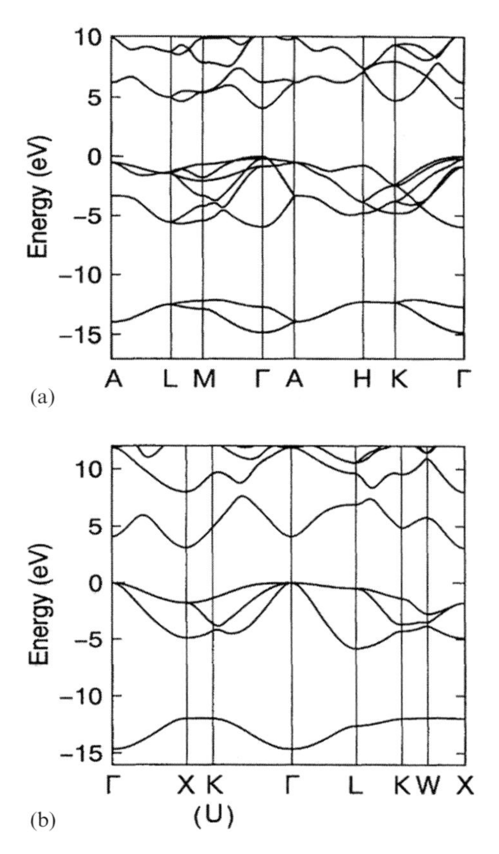

Figure 2.24 Energy band structures of (a) wurtzite AlN and (b) zincblende AlN. The energy scale is chosen so that the top of the valence band at the Γ point has zero energy. Reprinted with permission from Fig. 6 of K. Miwa and A. Fukumoto, *Phys. Rev.* **B 48**, 7897 (1993). Copyright (1993) by the American Physical Society.

they include InAs quantum dots on GaAs, InGaAl quantum dots on GaAs, InSb quantum dots on GaAs, and GaN quantum dots on AlN, to name just a few, and they take on a variety of shapes such as pyramids, truncated pyramids, and truncated hexagonal pyramids. The optical properties of such self-assembled quantum dots have been studied extensively for applications including photodetectors and quantum dot lasers; detailed accounts of these studies may be found in the references at the end of this chapter devoted to semiconductor quantum dots and quantum dot heterostructures.

An example of such a self-assembled system for GaN quantum dots in an AlN matrix is shown in Fig. 2.25.

The image illustrated in Fig. 2.25 was published as part of a study on the influence of AlN overgrowth on the structural properties of GaN quantum wells and quantum dots grown by plasma-assisted molecular-beam epitaxy. The growth of high-quality

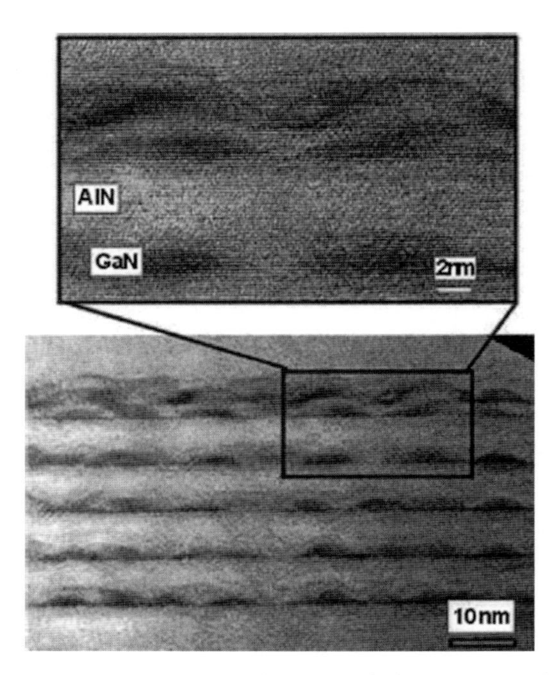

Figure 2.25 High-resolution transmission electron microscope images of a GaN/AlN quantum dot superlattice. In the superlattice the pyramidal structures are the GaN quantum dots. This superlattice was grown at 730°C. If the AlN spacer is thick enough, a smooth AlN top surface is observed. By contrast, if the spacer is very thin (~2 nm), the roughness induced by the GaN quantum dot layer is reproduced at the AlN top surface. Reprinted from Fig. 10 in N. Gogneau, *et al.*, *J. Appl. Phys.* **96**, 1104 (2004) with the permission of AIP Publishing.

quantum dot arrays with a suitable number of layers to fabricate efficient, high-quality quantum dot devices is an ongoing pursuit.

In contrast to self-assembled quantum dots, the colloidal quantum dots introduced at the beginning of this discussion have approximately spherical shapes. They are formed chemically in a solution upon the introduction of suitable chemicals and catalysts in the solution, which may be water, ethane, toluene, etc. For example, CdS quantum dots with diameters of approximately 5 nm may be synthesized in a population of about 10^{16} charged quantum dots per ml by mixing $CdCl_2$ with $Na_2S \cdot 9H_2O$ in a pH 2 solution. The ensuing chemical reaction produces CdS quantum dots in an electrolyte containing Na and Cl ions. The fact that the quantum dots have like charges causes them to repel one another and remain in suspension. The list of types of colloidal quantum dot has been expanding rapidly over the last 25 years as a result of their widespread uses as solution-based nanosensors and nanotags suitable for applications in water-based biological systems.

Synthesis techniques now allow the growth of layered (coated) quantum dots. As reported in Table 2.13, ZnS-capped CdSe nanocrystals were found to be strongly luminescent as a result of the high-quality interface between CdSe and

Table 2.13 Selected semiconductors composing colloidal quantum dots[a]

Compound semiconductor	Bandgap (eV)	Spontaneous polarization (C/m^2) and references
AlN	6.2	−0.081 Andreev and E. P. O'Reilly (2001)
CdS hexagonal	2.4 E_g $(A)^b$ 2.5 E_g $(B)^b$	0.002 Jerphagnon (1970)
CdS cubic	2.55 E_g $(C)^b$ 2.5	
CdSe hexagonal	1.75 E_g $(A)^b$	0.006
CdSe cubic	1.771 E_g $(B)^b$ 2.17 E_g $(C)^b$ 1.9	Schmidt, *et al.* (1997) Hines and Guyot-Sionnest (1996) report high quantum efficiency of strongly luminescing ZnS-capped CdSe nanocrystals
CdTe	1.49	de Paula, *et al.* (1998)
$CuInSe_2$	1.04	Wasim, *et al.* (2000)
$CuIn_3Se_5$	1.21	Wasim, *et al.* (2000)
$CuIn_5Se_8$	1.15	Wasim, *et al.* (2000)
GaN	3.36	−0.029 Zhang, *et al.* (1998); Micic, *et al.* (1999)
$In_{0.48}Ga_{0.52}P$	1.97	Griffin, *et al.* (2000)
PbS	0.41	Nenadovic, *et al.* (1990); Machol, *et al.* (1993)
PbSe	0.27	Wehrenberg, *et al.* (2002)
Si	3.5 for 1 nm diameter direct-bandgap nanoparticle	Smith, *et al.* (2005); behaves as direct-bandgap material for small diameters
TiO_2	3.2	Rajh, *et al.* (2004)
ZnS	3.68	Xu and Schoonen (2000)
ZnO	3.35	−0.07 Jerphagnon (1970)

[a]Based on reported parameters in A. D. Andreev and E. P. O'Reilly, *Phys. Rev.* **B 62**, 15851 (2001); A. M. de Paula, L. C. Barbosa, C. H. B. Cruz, O. L. Alves, J. A. Sanjurjo, and C. L. Cesar, *Superlattices and Microstructures* **23**, 1104 (1998); I. J. Griffin, D. Wolverson, J. J. Davies, M. Emam-Ismail, J. Heffernan, A. H. Kean, S. W. Bland, and G. Duggan, *Semicond. Sci. Technol.* **15**, 1030–1034 (2000); M. Hines, and P. Guyot-Sionnest, *J. Phy. Chem.* **100**, 468 (1996); J. Jerphagnon, *Phys. Rev.* **B 2**, 1091 (1970); J. L. Machol, F. W. Wise, R. C. Patel, and D. B. Tanner, *Phys. Rev.* **B 48**, 2819 (1993); O. I. Micic, S. P. Ahrenkiel, D. Bertram, and A. J. Nozik, *Appl. Phys. Lett.* **75**, 47 (1999); M. T. Nenadovic, M. I. Comor, V. Vasic, and O. I. Micic, *J. Phys. Chem.* **94**, 6390 (1990); T. Rajh, J. Saponjic, J. Liu, N. M. Dimitrijevic, N. F. Scherer, M. Vega-Arroyo, P. Zapol, L. A. Curtiss and M. Thurnauer, *Nano Lett.*, **4**, 1017–1023 (2004); M. E. Schmidt, S. A. Blanton, M. A. Hines, and P. Guyot-Sionnest, *J. Chem. Phys.* **106**, 5254 (1997); A. Smith, Z. H. Yamani, N. Roberts, J. Turner, S. R. Habbal, S. Granick, and M. H. Nayfey, *Phys. Rev.* **B 72**, 205307 (2005); S. M. Wasim, C. Rincon, G. Marin, and J. M. Delgado, *Appl. Phys. Lett.* **77**, 94–96 (2000); B. L. Wehrenberg, C. Wang, and P. Guyot-Sionnest, *J. Phys. Chem.* **B 106**, 10634 (2002); Y. Xu and M. A. A. Schoonen, *American Mineral.* **85**, 543 (2000); C. J. Zhang, H. W. H. Lee, I. M. Kennedy, and S. H. Risbud, *Appl. Phys. Lett.* **72**, 3035 (1998).

[b]The difference between bandgaps (A) and (B) is due to internal crystal fields; the difference between bandgaps (A) and (C) is due to spin–orbit interactions.

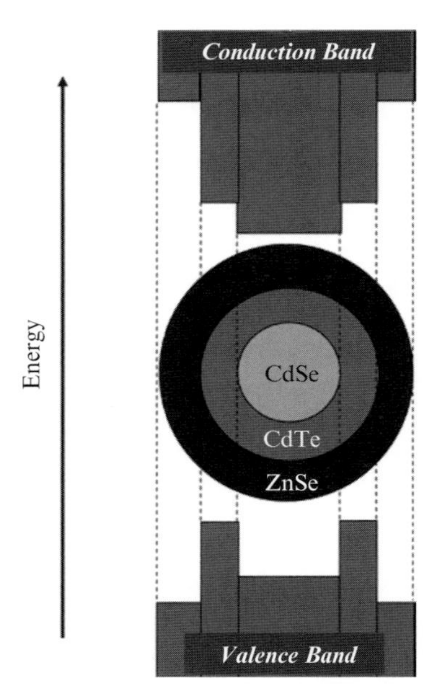

Figure 2.26 One of the possible type-II/type-I nanocrystals obtained by synthesis of layered semiconductor quantum dots. Reprinted (adapted) with permission from Fig. 1 of B. Blackman *et al.*, "Bright and water-soluble near IR-emitting CdSe/CdTe/ZnSe type-II/type-I nanocrystals, tuning the efficiency and stability by growth," *Chem. Mater.* **20**, 4847–4853 (2008). Copyright 2008 American Chemical Society.

ZnS, resulting in a high quantum efficiency of emission relative to absorption of ZnS-capped CdSe quantum dots. These techniques also facilitate the synthesis of CdSe/CdTe/ZnSe type-II/type-I quantum dots, as depicted in Fig. 2.26, which emit in the infrared. In the study reported in Fig. 2.26 it was found that the photoluminescent (PL) efficiency and stability of CdSe/CdTe type-II quantum dots were improved by the growth of ZnSe onto the CdTe surface. This effect is due to the fact that the type-I heterojunction between the CdTe and ZnSe layers causes the confinement of photogenerated electrons and holes within the CdSe and CdTe layers, respectively. Uniform growth of the ZnSe shell onto the CdSe/CdTe core/shell dots was found to be necessary to achieve optimal efficiency and stability. Clearly, the same band engineering techniques that are effective for convention nanostructures are also effective when applied to colloidal semiconductor quantum dots.

2.10 Two-Dimensional Monolayer Crystals (Graphene, Silicene, Germanene, Tinene, Black Phosphorus, Monochalcogenides, Dichalcogenides, Trichalcogenides, and Boron Nitride)

Of special interest in optical applications is the emergence of two-dimensional (2D) crystals with nonzero bandgaps, which can be used to form heterostructures based on van der Waals interactions. Clearly, having nonzero bandgaps opens the

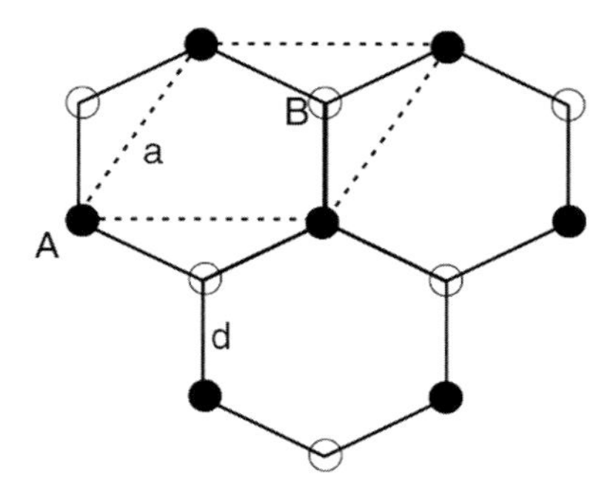

Figure 2.27 The crystal structure of silicene. The lattice is hexagonal, a unit cell is illustrated by the dashed area, and the basis consists of two Si atoms labeled A and B. Reprinted with permission from Lok C. Lew and Yan Voon, "Physical properties of silicene," Chapter 1 of M. J. S. Spencer and T. Morishita (eds.), *Silicene*, Springer Series in Materials Science 235, Springer International Publishing, Switzerland, 2016.

Table 2.14 Structural and electronic parameters for various 2D structures; d is the bond length, a is the hexagonal lattice parameter, Δz is the buckling height, v_F is the Fermi velocity, m^* is the effective mass, and E_g is the bandgap

	C (graphene)	Si (silicene)	Ge (germanene)	Sn (tinene)
a (nm)	2.468	3.868	4.060	4.673
d (nm)	1.425	2.233	2.344	2.698
Δz (nm)	0.00	0.45	0.69	0.85
v_F (10^6 m s^{-1})	1.01	0.65	0.62	0.55
m^*/m	0.000	0.001	0.007	0.029
E_g (meV)	0.0	1.9	33	101

Adapted from Table 1 of L. Matthes *et al.*, "Massive Dirac quasiparticles in the optical absorbance of graphene, silicone, germanene, and tinene," *J. Phys.: Condens. Matter* **25**, 395305 (2013).

way to exploiting these materials for optoelectronic applications. Among these 2D materials are silicene, germanene, tinene, black phosphorus, monochalcogenides, dichalcogenides, trichalcogenides, and boron nitride. These emerging 2D mateials are introduced in this section.

There are several classes of 2D structures that have nonzero bandgaps. The first such class of 2D structures includes silicene, which is Si-based, germanene, which is germanium-based, and tinene, which is tin-based. These 2D structures have hexagonal crystal structures (similar to that of graphene) as illustrated by Fig. 2.27. Table 2.14 summarizes the physical parameters for selected 2D structures.

Beginning in 2014, black phosphorus has been the subject of intensive study, partly in view of its bandgap being of larger magnitude than those of many 2D layers including silicene, germanene, and tinene.

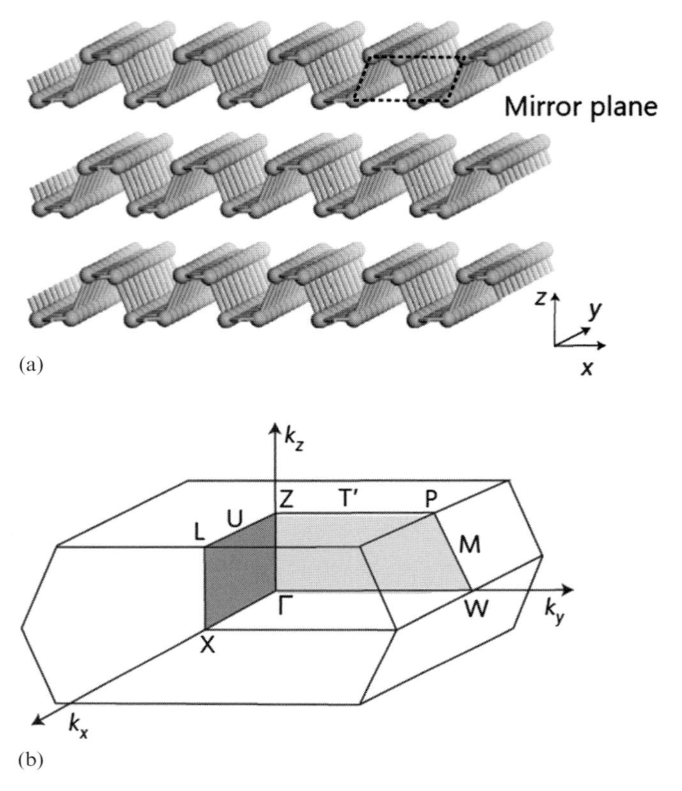

Figure 2.28 (a) The crystal structure of black phosphorus and (b) the three-dimensional Brillouin zone of black phosphorus. The lattice exhibits in-plane atomic buckling along the x-axis, resulting in a two-fold anisotropy along the x- and y- directions; this buckling has a direct influence on its band structure and optical selection rules. The ΓXLUZΓ parallelogram plane in (a). highlights a mirror-reflection symmetry in black phosphorus. Reprinted by permission from Figure 1 (a and b) of Springer, *Nature Nanotechnology*, "Polarization-sensitive broadband photodetector using a black phosphorus vertical p–n junction," Hongtao Yuan, *et al.*, ©2015.

As depicted in Fig. 2.28, the lattice of black phosphorus has an in-plane atomic buckling along the x-axis, resulting in a two-fold anisotropy along the x- and y- directions; as one would expect, it is known that this buckling has a direct influence on the band structure and optical properties of black phosphorus. The ΓXLUZΓ parallelogram plane corresponds to a mirror-reflection symmetry in black phosphorous. One measure of the bandgap of black phosphorus is the number of energy states per unit energy at a given energy – the so-called density of states. The density of states will be discussed in greater detail in Chapter 3. At this point in our discussion, it is emphasized that a region of no states between a lower valence band with some number of states per unit energy, and an upper conduction band with some number of states per unit energy, is indicative of a bandgap, as shown

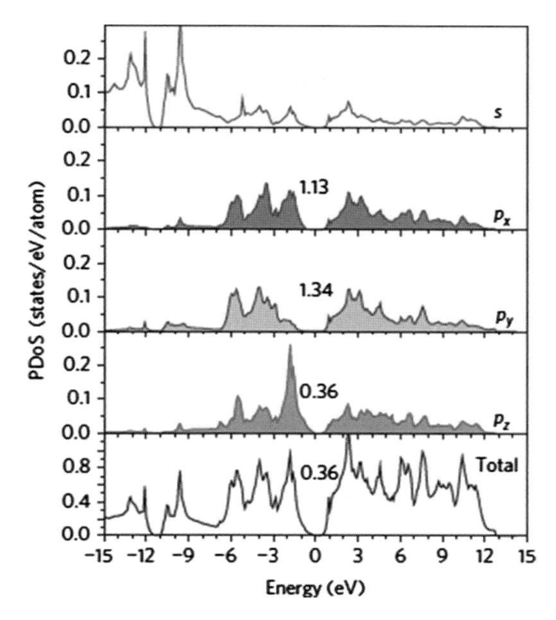

Figure 2.29 Band calculations indicate a direct bandgap of 0.36 eV for bulk black phosphorus. The partial densities of states (PDoS) for s, p_x, p_y, and p_z orbitals are illustrated for black phosphorus. The gap size in electron-volts is indicated for the second through fifth panels by the number at the center of these panels. The total bandgap of 0.36 eV (indicated in the fifth panel) indicates that the gap size for black phosphorus is dominated by that of the p_z orbital along the z crystal axis. Reprinted by permission from Figure 1(e) of Springer, *Nature Nanotechnology*, "Polarization-sensitive broadband photodetector using a black phosphorus vertical p–n junction," Hongtao Yuan, *et al.*, ©2015.

separately – as partial densities of states – in Fig. 2.29 for the s, p_x, p_y, and p_z states in units of number of states per electron-volt per atom.

According to the calculations presented in Fig. 2.29, the bandgap of black phosphorus is expected to be roughly 0.36 eV. These calculations give some insight into the bandgap of black phosphorus. In related experimental studies, a broadband photodetector made of a layered black phosphorus structure exhibits a very broad range of wavelengths, from about 400 nm to 3750 nm, for which the black phosphorus functions as a photodetector. Figure 2.30 depicts the photoresponsivity over the range 400 nm to 1700 nm. The relationship between the wavelength of a photon and the energy of the photon can be easily determined from $E_{\text{photon}} = hf$ and $\lambda f = c$, namely $E_{\text{photon}} = hc/\lambda$. In units of electron-volts and micrometers, $E_{\text{photon}}\ (\text{eV}) = 1.24\,\text{eV}\,\mu\text{m}/\lambda(\mu\text{m})$, so that a photon with a wavelength of 400 nm (0.4 μm) corresponds to an energy of 3.1 eV and a photon with a wavelength of 1700 nm (1.7 μm) corresponds to an energy of 0.73 eV.

A class of 2D structures with bandgap energies close to or larger than those of black phosphorus is based on chalcogenide structures; members of this family include Ga-, Ge-, InSe-, and Sn-monochalcogenides, Mo-, Sn-, and

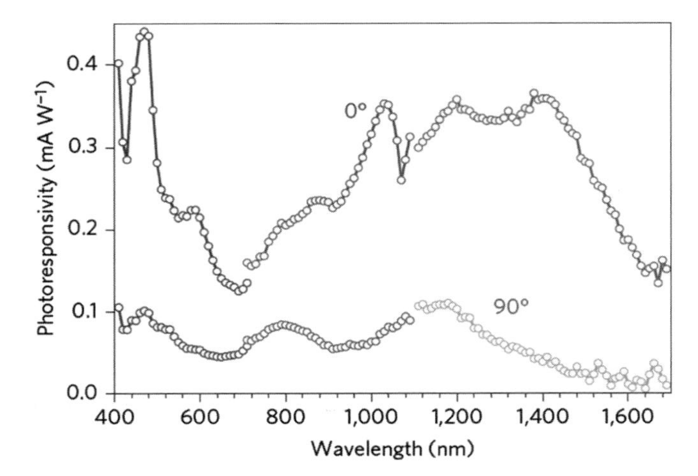

Figure 2.30 Photoresponsivity of a broadband photodetector made of layered black phosphorus structure, where the polarization angle of 0° corresponds to the x crystal axis and 90° corresponds to the y crystal axis. Reprinted by permission from Figure 3 of Springer, *Nature Nanotechnology*, "Polarization-sensitive broadband photodetector using a black phosphorus vertical p–n junction," Hongtao Yuan, *et al.*, ©2015.

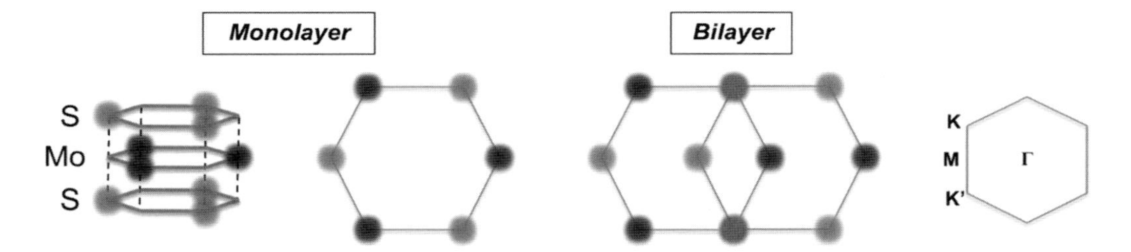

Figure 2.31 Positions of atoms for monolayer and bilayer MoS_2 as well as the locations of selected symmetry points for MoS_2 (at right). Similar structures also apply to the cases where Mo is replaced by W and S is replaced by Se. Adapted by permission from Figure 1 of Springer, *Nature Nanotechnology*, "Control of valley polarization in monolayer MoS_2 by optical helicity," Kin Fai Mak, *et al.*, ©2012.

W-dichalcogenides, and Ti-, Zn-, and Hf-trichalcogenides. Among these, semiconductor transition-metal dichalcogenides, MX_2, where M represents Mo or W and X represents S or Se, have bandgaps spanning the visible and near-infrared spectral regions. Monolayers of these 2D structures have direct bandgaps, which makes them useful in optical and optoelectronic applications. The positions of atoms for monolayer and bilayer MoS_2 as well as the locations of selected symmetry points for MoS_2 are depicted in Fig. 2.31. See also Table 2.15.

The optical properties of many of these 2D structures exhibit resonances caused by bound states of electrons and holes, known as excitons. In the simplest case,

Table 2.15 Effective masses and optical gaps of selected transition-metal dichalcogenides – MoS_2, $MoSe_2$, WS_2, and WSe_2

	MoS_2	$MoSe_2$	WS_2	WSe_2
Effective mass m^*/m	~0.5	~0.6	~0.4	~0.4
Optical gap magnitude (eV)	~2	~1.7	~2.1	~1.75

Adapted by permission from Table 1 of Springer, *Nature Photonics*, "Photonics and optoelectronics of 2D semiconductor transition metal dichalcogenides," Kin Fai Mak, *et al.*, ©2016.

a single electron and a hole bind together with a negative (attractive) binding energy that manifests the exciton as an energy state appearing below the conduction band energy with an energy equal to the conduction band energy minus the exciton binding energy. Multiple exciton energy levels may appear below the conduction band energy, corresponding to the multiple bound-state energies of the exciton. Excitonic features are frequently present in optical spectra and excitons can play an important role in the interaction of light with materials containing excitons. Excitonic effects are known to be exceptionally strong in selected 2D materials, allowing the fabrication of devices that operate at room temperature. In many materials, excitonic effects are not evident at room temperature and they appear only at lower temperatures, in contrast to what occurs in many 2D materials.

As illustrated in Fig. 2.32, strong excitonic effects are present in 2D transition-metal dichalcogenides. For example, the absorption spectrum of a monolayer of MoS_2, presented on the right of Fig. 2.32(a) as a solid line, exhibits pronounced exciton resonances of transitions between two valence band states (split by spin interactions) and the conduction bands. In Fig. 2.32(a) the dashed line shows the absorbance (arbitrary units) in the absence of excitonic effects. Furthermore, MoS_2 is known to have a rich variety of excitons, depicted in Fig. 2.32(b) as an exciton and higher-order excitonic complexes: a two-particle charge-neutral exciton, three-particle charged excitons (trions), and a four-particle bi-exciton.

In closing this section on 2D structures of interest in photonic applications, brief mention is made of the relatively wide indirect-bandgap (~6 eV) material hexagonal boron nitride (hBN). In part, the interest in hBN for optical and optoelectronic applications stems from the coupling of photons with the optical phonons in this polar dielectric. As discussed by G. Cassabois, P. Valvin, and B. Gil ("Hexagonal boron nitride is an indirect bandgap semiconductor," *Nature Photonics* **10**, 262–267 (2016)), high-purity crystals have recently shown the potential of hBN for intense deep ultraviolet emission around 215 nm. G. Cassabois, P. Valvin, and B. Gil have recently addressed the nature of the bandgap of hBN by presenting evidence for an indirect bandgap at 5.955 eV from optical spectroscopy, and they have shown the existence of phonon-assisted optical transitions and have reported an exciton binding energy of about 130 meV.

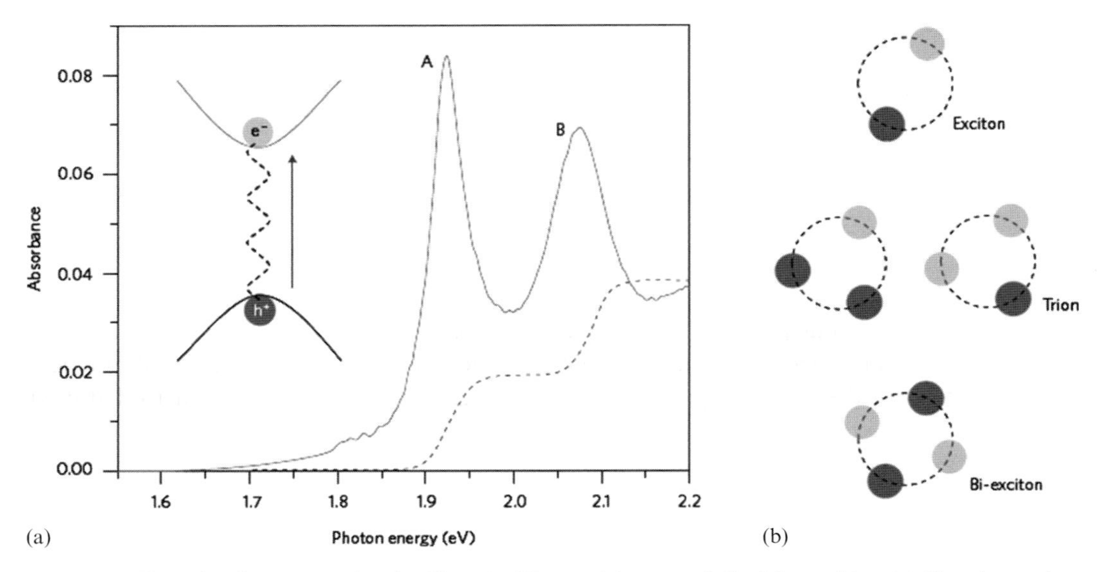

Figure 2.32 The role of strong excitonic effects in 2D transition-metal dicalchogenides. (a) The absorption spectrum of monolayer MoS_2 at 10 K (solid line). A and B correspond to exciton resonances of transitions between two valence band states (split by spin interactions) and the conduction band. The dashed line shows the absorbance (arbitrary units) in the absence of excitonic effects. The inset on the left depicts an optically generated electron–hole pair, forming a bound exciton. (b) The rich exciton population in MoS_2 is depicted here as an exciton and higher-order excitonic complexes: a two-particle charge-neutral exciton, three-particle charged excitons (trions), and a four-particle bi-exciton. Adapted by permission from Figure 2 of Springer, *Nature Photonics*, "Photonics and optoelectronics of 2D semiconductor transition metal dichalcogenides," Kin Fai Mak, *et al.*, ©2016. The solid curve was adapted by the authors of the above-cited paper by permission from Figure 1 of Springer, *Nature Materials*, "Tightly bound trions in monolayer MoS_2," Kin Fai Mak, *et al.*, ©2012.

In summary, a major new direction in the field of optical heterostructures is based on the emergence of 2D crystals with nonzero bandgaps that can be used to form heterostructures based on van der Waals interactions. These nonzero bandgaps have opened the way to exploiting these materials for optical and optoelectronic applications. Among these 2D materials with nonzero bandgaps are silicene, germanene, tinene, black phosphorus, monochalcogenides, dichalcogenides, trichalcogenides, and boron nitride. The study of these materials is still in an early stage, but the optical properties of these structures signify widespread optical and optoelectronic applications.

2.11 Closing Remarks to Chapter 2

We began this chapter with the definition of crystalline materials. We introduced two very important components of crystals, i.e., the electron subsystem and the crystalline lattice. The optical and optoelectronic applications of a material depend on its electronic properties. As a result of the periodicity of crystals, one can interpret the electron energy in terms of energy bands. Depending on the relative positions of the energy bands and their filling by electrons, a material can be described as a dielectric, semiconductor, or metal. We found that, despite the interaction of the electrons with the great number of atoms and ions composing a crystal, one can attribute a quasi-momentum (a wavevector) to an electron and, frequently, in the actual range of the wavevectors, the electron energy can be approximated by a simple dependence on the wavevector. Several new notions that are necessary to describe electrons in these cases were introduced: (1) the positions of the energy minima, (2) the bandgap, and (3) the effective masses.

We analyzed semiconductor alloys and learned that for many cases of practical interest the alloy energy spectra can be described in a manner similar to that used to describe pure crystals. These alloys may be "engineered" to produce considerable variations in the electron parameters.

Next, we explained the formation of discontinuities both in the valence band and in the conduction bands when two materials are brought together to form a heterojunction. This effect provides the practical basis for modifying electron energy spectra, and it is particularly useful for a spatial modulation of the potential profiles experienced by electrons and holes as well as for the creation of various artificial heterostructure-based nanostructures with energy barriers; these nanostructures include quantum wells, quantum wires, quantum boxes, and superlattices.

By analyzing the state-of-the-art techniques in the fabrication of heterostructures, it was established that high-quality heterostructures can be produced using materials with similar crystal properties, for example, materials from the same group. The ratio of the lattice constants of the two materials is a critical parameter for such heterostructures. If these lattice constants almost coincide, as in the case of lattice-matched materials, one can produce a heterostructure without strain due to the absence of lattice mismatch or misfit imperfections. An example of such a lattice-matched system is the AlGaAs/GaAs heterostructure. For lattice-mismatched materials we found that only thin layers can accommodate the lattice mismatch and retain near-perfect crystalline structure. The resultant structures are pseudomorphic strained layers. The strain in pseudomorphic heterostructures leads to a set of new phenomena. In particular, it affects the energy spectra in the strained layers. An example is the Si/Ge heterostructure. We introduced and described so-called wide-bandgap materials, which find many optoelectronic applications where photons have energies greater than those of visible photons. We briefly discussed quantum dot materials. Finally, we introduced and described two-dimensional (2D) material structures that portend many applications in optics and optoelectronics.

Recently, emphasis has been placed on reconfigurable and/or adaptive materials, and nanostructure-based materials fit very well into that category. We indicated the basics of the formation of nanomaterials achieved by combining materials with different bandgaps. Very often this approach is named band structure engineering. In Chapter 3 we will demonstrate that potential profile engineering can be achieved by using selective doping of nanostructures. The built-in electric field allows one to control the distribution of carriers in the material. An applied external voltage can substantially change that distribution, allowing substantial control of the properties of nanostructures and nanostructure-based devices.

2.12 Control Questions

1. Using the lattice constant of silicon, $a = 5.43 \times 10^{-8}$ cm, and the fact that the number of Si atoms per unit volume, a^3, is eight, calculate the number of the atoms per cm^3 and the density of the crystalline silicon (silicon's atomic weight is 28.1 g/mol).
2. Estimate the volume of the first Brillouin zone in $\vec{k}$-space for a simple cubic lattice with the lattice constant $a = 5 \times 10^{-8}$ cm. Assume that the average energy of the electrons is $3k_B T/2$, where k_B and T are Boltzmann's constant ($k_B = 1.38 \times 10^{-23}$ J/K) and the ambient temperature, respectively. Estimate the volume occupied by electrons in $\vec{k}$-space at $T = 300$ K and the effective mass $m^*/m = 0.1$, with m the free-electron mass. Compare these two volumes and discuss whether Eq. (2.17) is valid for electrons with the parameters specified previously.
3. Three electrons with the same energy are placed in three different energy valleys of silicon. The valleys are located at the Δ-points of the [100]-, [010]-, and [001]-axes. Assuming that all three electrons move along the same direction, say [100], and using Eqs. (2.17) and (2.19), find the ratio of the velocities for these electrons.
4. Consider a valence band consisting of light- and heavy-hole branches. Assume that a heavy hole with energy E is transferred to a light-hole state with the same energy E. Find the ratios of the quasi-momenta and the velocities of the hole in the initial and final states.
5. For $Al_xGa_{1-x}As$ alloy, find the composition having an energy bandgap equal to 2 eV, using Table 2.6. For this alloy, determine the effective masses in the Γ and X valleys using Table 2.7.
6. Assume that for some applications it is necessary to use a film of InGaAs of high quality, which can be grown on an InP substrate. By using Eq. (2.20) and data presented in Table 2.8, find the lattice-matched composition of this alloy, its lattice constant, the energy bandgap, and the wavelength of the light corresponding to this bandgap.
7. Assume that the conduction band-offset for an $Al_xGa_{1-x}As$/GaAs heterojunction is 60% of the difference of the bandgaps of these materials. Find the composition of the AlGaAs layer necessary for the resulting heterojunction to have an energy barrier for the electrons equal to 0.3 eV. Calculate the energy barrier for the holes.
8. From the values of the lattice constants given in Table 2.8, explain why it is feasible to grow stable $Al_xGa_{1-x}As$/GaAs and $In_xAl_{1-x}As/In_yGa_{1-y}As$ heterostructures. Also explain why it is difficult to grow stable GaP/SiC and InP/SiC heterostructures.

9. Consider a spherical quantum dot which has a conduction band energy much less than those of the surrounding materials. Treat the electron in this quantum dot as an electron in an infinitely deep potential well, to show that $E_n = n^2\pi^2\hbar^2/2\,m^* a^2$, where a is the diameter of the quantum dot.
10. Using the reciprocal-lattice vectors of the high-symmetry points given in Table 2.12, show that $\vec{b}_1, \vec{b}_2$, and $\vec{b}_3$ – the reciprocal-lattice vectors in $\vec{k}$-space (wavevector space) – for the hexagonal Brillouin zone are $\vec{b}_1 = (2\pi/a, -2\pi/a\sqrt{3}, 0)$, $\vec{b}_2 = (2\pi/a, 2\pi/a\sqrt{3}, 0)$, and $\vec{b}_3 = (0, 0, 2\pi/c)$ when the real space lattice vectors are taken as $\vec{a}_1 = (a/2, -a\sqrt{3}/2, 0)$, $\vec{a}_2 = (a/2, a\sqrt{3}/2, 0)$, and $\vec{a}_3 = (0, 0, c)$. Hint: construct orthogonal vectors in real space and reciprocal space using the standard formulae $\vec{b}_1 = 2\pi(\vec{a}_2 \times \vec{a}_3)/\mathcal{V}, \vec{b}_2 = 2\pi(\vec{a}_3 \times \vec{a}_1)/\mathcal{V}$, and $\vec{b}_3 = 2\pi(\vec{a}_1 \times \vec{a}_2)/\mathcal{V}$, where $\mathcal{V} = |\vec{a}_1 \cdot (\vec{a}_2 \times \vec{a}_3)|$.
11. In device applications it is essential to know the position of the Fermi energy level, E_F, relative to E_c and E_v. Using the Fermi–Dirac distribution of Appendix A, show that $E_F = (E_c + E_v)/2$ when the probability of finding an electron at E_c is equal to the probability of finding a hole at E_v.

3 Electrons in Quantum Structures

3.1 Introduction

In conventional semiconductor devices with one or two types of charge carriers, different operational functions may be achieved by the creation of a junction within the same material, for example, p–n, n^+–n junctions, etc. These junctions are commonly referred to as *homojunctions*, and devices fabricated from such structures are called *homostructure devices*. Typically, the characteristic scales of homojunction devices exceed considerably the de Broglie wavelength of the charge carriers, so in these devices the carriers are described in terms of *semiclassical physics*.

If more than one material is used in a device, this device is called a *heterostructure device*. A heterojunction is formed when two different materials are joined together at a junction. It is now possible to make such heterojunctions extremely abrupt; in fact, with modern techniques for materials growth, heterojunction interfaces may be fabricated with a thickness approaching only one atomic monolayer. The use of heterojunction barriers in such devices facilitates decreasing their characteristic scales to a scale of the order of the de Broglie wavelength of the carriers. In these devices, the carriers are described in terms of *quantum physics* and the heterostructures and devices are called *quantum structures* and *devices*.

As discussed in Chapter 2, there is a rich selection of diverse materials (semiconductors, metals, and insulators) that leads to high-quality interfaces and permits the fabrication of different heterostructures which manifest desired electron characteristics as functions of the applied potential. Below we study the simplest models of quantum heterostructures.

After the discovery of the first two-dimensional crystal, graphene (around 2004), other two-dimensional crystals have been synthesized and fabricated in recent years (bilayered graphene, transition-metal dichalcogenides, black phosphorus, boron nitride, silicene, etc.). The properties of these two-dimensional materials are very different from those of usual bulk crystals. These properties facilitate the use of two-dimensional crystals in numerous nanoelectronic and optoelectronic devices.

In this chapter we present the basic electronic properties of quantum heterostructures and two-dimensional crystals.

3.2 Quantum Wells

We begin with an analysis of the simplest quantum heterostructure – a quantum well. A quantum well can be fabricated by growing a single layer of one material between two other layers of different material(s), which are characterized by wider bandgaps than the central layer, as shown in Fig. 3.1. The band discontinuity provides the confinement of carriers inside the well.

3.2.1 Wave Functions and Energy Subbands

We will use the following idealized form of the heterostructure potential – for the case where the materials surrounding the central layer are the same:

$$V(z) = \begin{cases} 0, & \text{for } |z| \leq L/2, \\ V_{\mathrm{b}}, & \text{for } |z| \geq L/2, \end{cases} \tag{3.1}$$

where V_{b} and L are the depth and the thickness of the well, respectively. This situation is illustrated in Fig. 3.2.

The potential $V(z)$ is a function of the coordinate z only. In this case the electron motion in the two other directions, x and y, is free and can be described by a plane wave. Thus, we can write for the wave function of the electron

$$\psi(x, y, z) = e^{ik_x x + ik_y y} \chi(z). \tag{3.2}$$

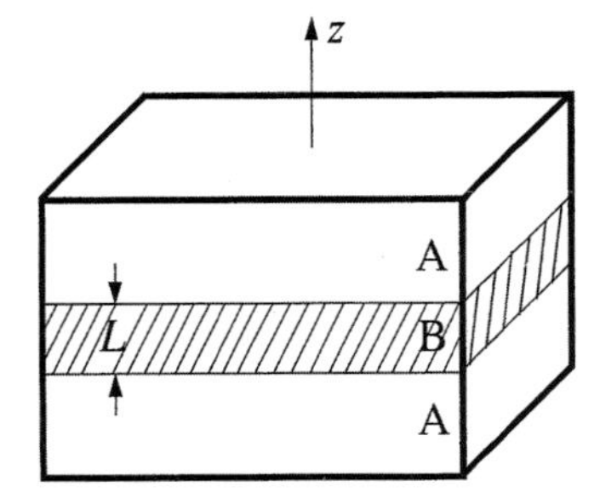

Figure 3.1 A three-layer structure providing electron confinement in the layer B. The layers A are made of a wider-bandgap material; the layer B is made of a narrower-bandgap material.

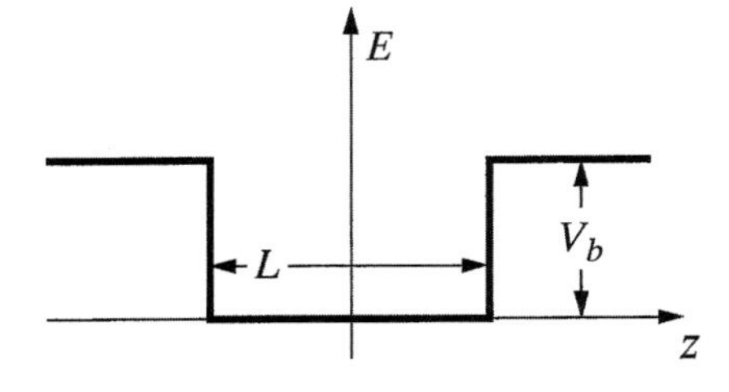

Figure 3.2 The potential energy of an idealized quantum well.

After substitution of this wave function into the Schrödinger equation (see Appendix A, Eq. (A.27)) we obtain the following equation for $\chi(z)$:

$$\left[-\frac{\hbar^2}{2m^*}\frac{\partial^2}{\partial z^2}+V(z)\right]\chi=\left(E-\frac{\hbar^2\vec{k}_{||}^2}{2m^*}\right)\chi,\quad \text{where } \vec{k}_{||}=(k_x,k_y). \tag{3.3}$$

The quantity $\hbar^2\vec{k}_{||}^2/2m^*$ is the kinetic energy of motion of the electron in the x, y plane. Let us define ϵ by

$$\epsilon=E-\frac{\hbar^2k_{||}^2}{2m^*}. \tag{3.4}$$

Since E is the total energy, ϵ is the energy of the motion transverse to the x, y plane. Thus, we have separated the in-plane and transverse variables, and we have reduced the three-dimensional problem to a one-dimensional Schrödinger problem:

$$\left[-\frac{\hbar^2}{2m^*}\frac{\partial^2}{\partial z^2}+V(z)\right]\chi=\epsilon\chi. \tag{3.5}$$

According to the analysis given in Appendix A, we can expect two types of solutions of Eq. (3.5). The first is characterized by a discrete energy spectrum, for $\epsilon < V_b$, of the bound electron states. The second corresponds to a continuous spectrum, for $\epsilon > V_b$, of unbound states. In spite of the discontinuity in the potential at $z=\pm L/2$, χ and $d\chi/dz$ are continuous everywhere, and χ is finite for unbound states or vanishes for bound states at $z\to\pm\infty$.

For $\epsilon < V_b$, outside the well the solution has the form

$$\chi(z)=\begin{cases}Ae^{-\kappa_b(z-L/2)}, & \text{for } z\geq L/2,\\ Be^{\kappa_b(z+L/2)}, & \text{for } z\leq -L/2,\end{cases} \tag{3.6}$$

where $\kappa_b=\sqrt{-2m^*(\epsilon-V_b)/\hbar^2}$. The solution inside the well is a simple combination of plane waves,

$$\chi(z)=C\cos(k_w z)+D\sin(k_w z),\quad \text{for } |z|\leq L/2, \tag{3.7}$$

where $k_w=\sqrt{2m^*\epsilon/\hbar^2}$; A, B, C, and D are arbitrary constants. As a result of the symmetry of the problem, we can choose either even or odd combinations in Eqs. (3.6) and (3.7). Consequently, continuity of the wave function implies that $A=B$ for the even solution and $A=-B$ for the odd solution. Now, we have two constants for odd solutions and two for even solutions in our problem. For example, for even solutions we find

$$\chi(z)=\begin{cases}C\cos(k_w z), & \text{for } |z|\leq L/2,\\ Ae^{\mp\kappa_b(z\mp L/2)}, & \text{for } |z|\geq L/2,\end{cases} \tag{3.8}$$

where the signs "−" and "+" correspond to positive and negative values of z, respectively.

The next step in finding the solution is to match the functions and their derivatives at the points $z=\pm L/2$. For example, for even solutions we obtain from Eq. (3.8) the following system of algebraic equations:

$$C\cos(k_{\mathrm{w}}L/2) = A \quad \text{and} \quad Ck_{\mathrm{w}}\sin(k_{\mathrm{w}}L/2) = A\kappa_{\mathrm{b}}. \tag{3.9}$$

This system of linear homogeneous equations has solutions if the corresponding determinant is zero:

$$\tan(k_{\mathrm{w}}L/2) = \kappa_{\mathrm{b}}/k_{\mathrm{w}}. \tag{3.10}$$

An analogous relationship can be found for odd solutions:

$$\cot(k_{\mathrm{w}}L/2) = -\kappa_{\mathrm{b}}/k_{\mathrm{w}}. \tag{3.11}$$

The algebraic equations (3.10) and (3.11) can be solved numerically, but it is more instructive to analyze them graphically. We can transform Eqs. (3.10) and (3.11) into the following results:

$$\cos(k_{\mathrm{w}}L/2) = \pm k_{\mathrm{w}}/\kappa_0, \quad \text{for } \tan(k_{\mathrm{w}}L/2) > 0, \tag{3.12}$$

$$\sin(k_{\mathrm{w}}L/2) = \pm k_{\mathrm{w}}/\kappa_0, \quad \text{for } \cot(k_{\mathrm{w}}L/2) < 0, \tag{3.13}$$

where $\kappa_0 = \sqrt{2m^* V_{\mathrm{b}}/\hbar^2}$. The signs "+" and "−" in Eq. (3.12) are to be chosen when $\cos(k_{\mathrm{w}}L/2)$ are positive or negative, respectively. The same is valid for "+", "−", and $\sin(k_{\mathrm{w}}L/2)$ in Eq. (3.13) for the odd solutions.

The left-hand, $\mathcal{L}$, and right-hand, $\mathcal{R}$, sides of Eqs. (3.12) and (3.13) can be displayed on the same plot as the functions of k_{w}; see Fig. 3.3. The portions of the curve in the intervals $[2\pi(l-1)/L, 2\pi(l-1/2)/L]$ correspond to even solutions and those in the intervals $[2\pi(l-1/2)/L, 2\pi l/L]$ to odd solutions; here $l = 1, 2, 3, \ldots$ The right-hand sides of Eqs. (3.12) and (3.13) are linear functions with a slope equal to κ_0^{-1}. The left-hand side is a *cosine* or *sine* function. The intersections of the left-hand sides with the linear function give us values $k_{\mathrm{w},n}$, for which our problem has solutions satisfying the necessary conditions. All solutions are represented in the first quadrant if both positive and negative values of $\mathcal{L}$ and $\mathcal{R}$ are plotted in the first quadrant.

In order to analyze the results we note that the problem is characterized by two independent parameters: the depth of the well, V_{b}, and the width, L. We can fix one

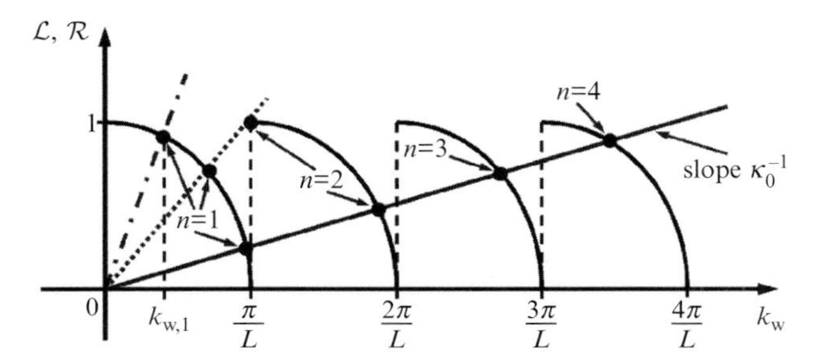

Figure 3.3 Graphical solution for Eqs. (3.12) and (3.13). The solutions correspond to the intersections of the line and the curves. The dot-dashed line corresponds to small κ_0, resulting in only one solution of Eq. (3.12). The dotted line corresponds to a critical value of κ_0 when the second level appears in the well as a result of the solution of Eq. (3.13). The solid line corresponds to an intermediate value of κ_0, leading to four solutions.

of these parameters and vary the other. Let us vary the depth of the well $V_{\rm b}$, i.e., the parameter κ_0. In this case the left-hand side of Eq. (3.12) and the corresponding curves $\mathcal{L}$ in Fig. 3.3 do not change, but the slope of the line, $k_{\rm w}/\kappa_0$, is controlled by κ_0. We can see that, at small κ_0, when the slope is large, as shown by the dot-dashed line in Fig. 3.3, there is only one solution $k_{{\rm w},1}$, corresponding to small k_0. This first solution exists at any κ_0 and gives the first energy level, $\epsilon_1 = \hbar^2 k_{{\rm w},1}^2/2m^*$. As κ_0 increases, a new energy level occurs at $k_{\rm w} = \kappa_0 = \pi/L$, as shown by the dotted line in Fig. 3.3, with energy slightly below $V_{\rm b}$ ($\epsilon_2 \approx V_{\rm b}$). When κ_0 increases further, the first and second levels become deeper and a third level occurs in the well, and so on. Indeed, new levels occur when the parameter $\sqrt{(2m^* V_{\rm b} L^2)/(\pi^2\hbar^2)}$ becomes an integer, so that the number of levels with energy $\epsilon < V_{\rm b}$ is

$$1 + {\rm Int}\left[\sqrt{\frac{2m^* V_{\rm b} L^2}{\pi^2 \hbar^2}}\right], \tag{3.14}$$

where Int[x] indicates the integer part of x. Figure 3.4 depicts the number of bound states as a function of the well thickness for two specific values of $V_{\rm b}$ and for effective masses close to those of electrons and holes in AlGaAs/GaAs structures.

Explicit expressions for the energies of bound states can be found easily for two extreme cases. The first corresponds to a very shallow well, when only one level exists: $\epsilon_1 \approx m^* L^2 V_{\rm b}^2/2\hbar^2$. The second case is related to an infinitely deep well, when $\kappa_0 \to \infty$; in this case, the slope of the linear function in Fig. 3.3 tends to zero and the solutions are $k_{\rm w} = \pi n/L$, or

$$\epsilon_n = \frac{\hbar^2\pi^2 n^2}{2m^* L^2}, \quad n = 1, 2, 3, \ldots \tag{3.15}$$

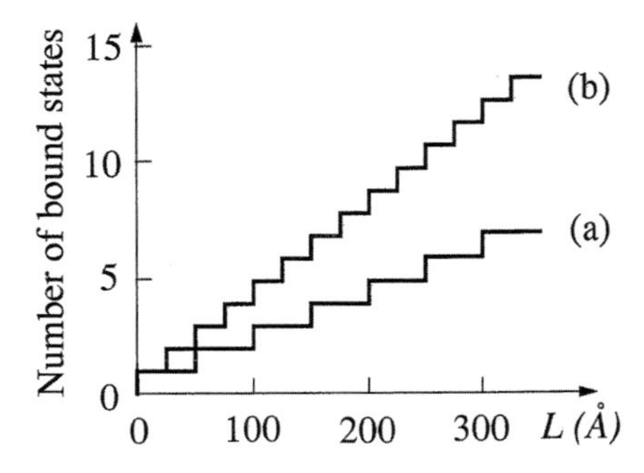

Figure 3.4 The number of bound states of a square well plotted as a function of the well thickness: (a) $V_{\rm b} = 224$ meV, $m^* = 0.067m$; (*b*) $V_{\rm b} = 150$ meV, $m^* = 0.4m$. Republished with permission of John Wiley and Sons Inc., from Fig. 3 on page 6 of *Wave Mechanics Applied to Semiconductor Heterostructures*, G. Bastard, 1988; permission conveyed through Copyright Clearance Center, Inc.

We can see that for the latter case the separations between levels increase when n increases. Let us apply the last expression to a square quantum well with an effective mass $m^* = 0.067m$ (as for GaAs) and $L = 125$ Å. We obtain $\epsilon_1 \approx 35\,\text{meV}$, $\epsilon_2 \approx 140$ meV, etc.

One can see that the energy spacings between levels for the quantum wells with thicknesses of the order of 100 Å are tens to hundreds of meV. These values should be compared with the Fermi energy of a degenerate two-dimensional electron gas as given by Eq. (A.35) at a low temperature, or with the thermal (average) energy of the electrons at high temperature, $k_B T$. For instance, at room temperature the thermal energy is about 26 meV. This comparison shows that only the lowest energy levels can be occupied by thermal electrons under typical device operational conditions.

Another frequently encountered well-like potential is that of an idealized single heterojunction. As a rule, the most important portion of this potential is near a heterojunction and, as illustrated in Fig. 3.5, it has a triangle-like shape formed by both a discontinuity of the electron (hole) band and an electrostatic field of electrons and/or remote ionized impurities. If the potential is characterized by high barriers, the lower energy levels may be studied with some accuracy by applying the following approximation:

$$V(z) = \begin{cases} eFz, & \text{for } z > 0, \\ \infty, & \text{for } z \leq 0. \end{cases} \tag{3.16}$$

Here F is the electrostatic field. For this triangular model of the quantum well, the Schrödinger equation for the transverse component of the electronic wave function has the following form inside the well:

$$\left[-\frac{\hbar^2}{2m^*} \frac{\partial^2}{\partial z^2} + eFz \right] \chi = \epsilon \chi. \tag{3.17}$$

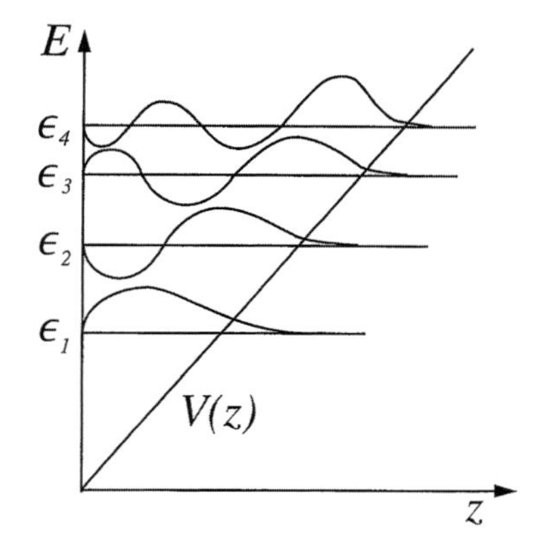

Figure 3.5 Triangular shape of the well potential, subband energies, and wave functions.

Outside the well, at $z \leq 0$ we set $\chi(z) = 0$ because the high barrier prevents significant penetration of electrons into the barrier region. Equation (3.17) is known as the Airy equation and its solutions are given by a special form of the Airy function $\mathcal{A}i$:

$$\chi(z) = \text{constant} \times \mathcal{A}i\left[\left(\frac{2m^*}{\hbar^2 e^2 F^2}\right)^{1/3} (eFz - \epsilon)\right]. \tag{3.18}$$

The special function $\mathcal{A}i[p]$ can be presented in an analytical form for two extreme cases: one is for $p \gg 1$,

$$\mathcal{A}i(p) = \frac{1}{2p^{1/4}} \exp\left(-\frac{2}{3}p^{3/2}\right), \tag{3.19}$$

and the second is for $p < 0,\ |p| \gg 1$,

$$\mathcal{A}i(p) = \frac{1}{|p|^{1/4}} \sin\left(\frac{2}{3}p^{3/2} + \frac{\pi}{4}\right). \tag{3.20}$$

Accordingly, at $z \to \infty$, the solutions (3.18) vanish for any ϵ, but the requirement for the wave function to vanish at $z \leq 0$ gives us the following condition: the argument of the function of Eq. (3.18) at $z = 0$ must coincide with one of the zeros of the Airy function, p_n. As a result we obtain the following quantization of the energy ϵ:

$$\epsilon_n = -\left(\frac{e^2\hbar^2 F^2}{2m^*}\right)^{1/3} p_n. \tag{3.21}$$

The first root of the Airy function is $p_1 \approx -2.35$. For $n \gg 1$ it is possible to use the approximation of Eq. (3.20) to find the zeros:

$$p_n = -\left[\frac{3\pi}{2}(n + 3/4)\right]^{2/3}. \tag{3.22}$$

Hence, at large n the energy, ϵ, is proportional to $n^{2/3}$. This result implies that, in contrast with the previous case, the distance between levels decreases when the energy increases. Using Eqs. (3.21) and (3.22), one can evaluate the positions of the lowest energy levels for the AlGaAs/GaAs junction:

$$\epsilon_1 = 3.94 \times 10^{-5} F^{2/3}\ (\text{eV}); \quad \epsilon_2 = 6.96 \times 10^{-5} F^{2/3}\ (\text{eV}). \tag{3.23}$$

Here, the energies are in units of electron-volts and the electric field is in units of V/cm. It is very simple to find the relation between the electrostatic field F and the electron concentration inside the triangular well. Actually, in the limit of negligibly small concentrations of ionized impurities in a narrow-bandgap semiconductor, i.e., in the region $z > 0$, Gauss's law gives $F = 4\pi e N_s/\kappa$, where N_s is the electron area (surface) concentration and κ is the dielectric constant of the semiconductor. If we take the surface electron concentration, N_s, to be 10^{12} cm^{-2} and take $\kappa = 13$, as for the AlGaAs system, we obtain an electric field $F = 1.39 \times 10^5$ V/cm and the energies are $\epsilon_1 = 106$ meV and $\epsilon_2 = 190$ meV. The separation between the first two levels is almost 100 meV. So, even at room temperature only the lowest level will be populated.

Thus, in contrast to the case of motion of three-dimensional electrons in bulk semiconductors, the confinement of electrons in one dimension results in the creation of energy subbands, ϵ_n, which contribute to the energy spectrum:

$$E_{n,\vec{k}_{||}} = \epsilon_n + \frac{\hbar^2}{2m^*}\left(k_x^2 + k_y^2\right). \tag{3.24}$$

Here ϵ_n is the quantized energy associated with the transverse (perpendicular to the heterostructure) confinement. Thus, two quantum numbers, one discrete, n, and the other continuous, $\vec{k}_{||}$, are now associated with each electron subband. At fixed n, the continuum range of $\vec{k}_{||}$ spans the energy band, which is usually referred to as a *two-dimensional subband*; see Fig. 3.6. One can interpret the subbands of Eq. (3.24) as being a set of minima in the electron energies. The existence of such a set of minima drastically changes some characteristics of the electron system. For instance, an impurity creates a series of energy levels under the electron band in the three-dimensional case. In a quantum well, each subband generates a series of impurity levels.

If the temperature and the electron concentration are such that only the lowest energy level is filled, free motion of the electron is possible only in the x, y plane, i.e., in two directions. This system is frequently referred to as a *two-dimensional electron gas*. The behavior of a two-dimensional electron gas differs strongly from that of a bulk crystal. We shall now give an analysis of the density of states for electrons in a quantum well.

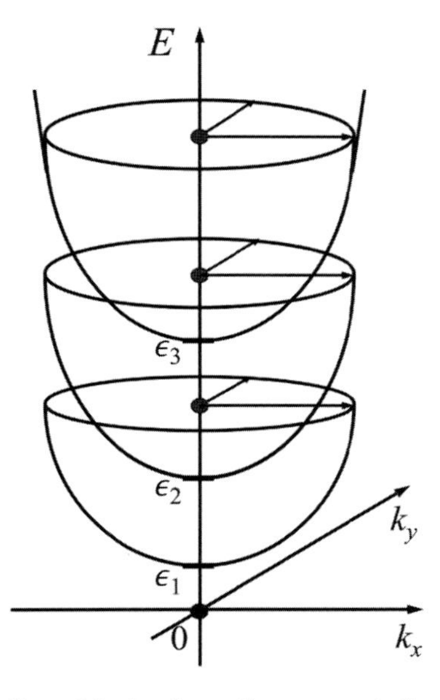

Figure 3.6 A schematic representation of the energy spectrum of the two-dimensional electrons in the first three subbands.

3.2.2 The Density of States of a Two-Dimensional Electron Gas

From the results presented in the last section we see that the electron energy spectrum in a quantum well is complex and consists of a series of subbands. The distances between subbands are determined by the profile of the potential, while inside each subband the spectrum is continuous and these continuous spectra overlap. For these complex spectra it is convenient to introduce a special function known as the density of states, $\varrho(E)$, which gives the number of quantum states $d\mathcal{N}(E)$ in a small interval dE around energy E:

$$d\mathcal{N} = \varrho(E)dE. \tag{3.25}$$

If the set of quantum numbers corresponding to a certain quantum state is designated as ν, the general expression for the density of states is defined by

$$\varrho(E) = \sum_{\nu} \delta(E - E_{\nu}), \tag{3.26}$$

where E_{ν} is the energy associated with the quantum state ν. In our case the set of quantum numbers includes a spin quantum number, s, a quantum number, n, characterizing the transverse quantization of the electron states, and a continuous two-dimensional vector, $\vec{k}_{||}$. Hence, $\nu \equiv \{s, n, \vec{k}_{||}\}$. There is a two-fold spin degeneracy of each state ($s = \pm 1/2$), so that

$$\varrho(E) = 2 \sum_{n,k_x,k_y} \delta\left(E - \epsilon_n - \frac{\hbar^2(k_x^2 + k_y^2)}{2m^*}\right). \tag{3.27}$$

In order to calculate the sum over k_x and k_y, we can define the area of the surface of the quantum well as $S = L_x \times L_y$, where L_x and L_y are the sizes of the quantum well in the x- and y- directions, respectively. If cyclic boundary conditions are assumed in the x- and y- directions, the possible values of k_x and k_y are

$$k_x = 2\pi l_x / L_x, \quad k_y = 2\pi l_y / L_y, \quad l_x, l_y = 0, 1, 2, \ldots \tag{3.28}$$

It is well known that these results are independent of the assumption of cyclic boundary conditions.

Thus, $\Delta k_x = 2\pi / L_x$, $\Delta k_y = 2\pi / L_y$, and we can write

$$\sum_{k_x,k_y}(\ldots) = \frac{L_x L_y}{(2\pi)^2} \int\int dk_x\, dk_y(\ldots). \tag{3.29}$$

By using Eq. (3.29), we transform Eq. (3.27) into

$$\varrho = \frac{m^* L_x L_y}{\pi\hbar^2} \sum_n \int_0^{\infty} d\epsilon_{||}\, \delta(E - \epsilon_n - \epsilon_{||}) = \frac{Sm^*}{\pi\hbar^2} \sum_n \Theta(E - \epsilon_n), \tag{3.30}$$

where $\epsilon_{||} = \hbar^2 k_{||}^2 / 2m^*$, $S = L_x L_y$, and $\Theta(x)$ is the Heaviside step function: $\Theta(x) = 1$ for $x > 0$ and $\Theta(x) = 0$ for $x < 0$. Very often the density of states per unit area, ρ / S, is used to eliminate the size of the sample. Each term in the sum of Eq. (3.30)

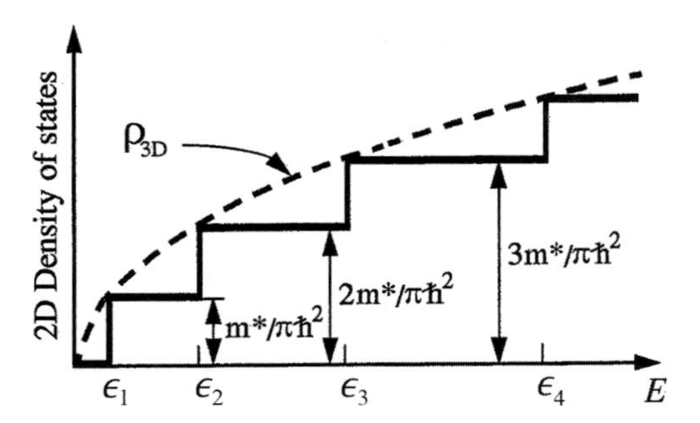

Figure 3.7 The density of states for two-dimensional electrons in an infinitely deep potential well.

corresponds to the contribution from one subband. The contributions of all subbands are equal and independent of energy. As a result, the density of states of two-dimensional electrons exhibits a staircase-shaped energy dependence, each step being associated with one of the energy states, ϵ_n. Figure 3.7 depicts the two-dimensional density of states.

It is instructive to compare these results with the density of states of electrons in bulk crystals:

$$\varrho^{3D} = \left(\frac{m^*}{\hbar^2}\right)^{3/2} \frac{\mathcal{V}}{\pi^2}\sqrt{2E}, \tag{3.31}$$

where $\mathcal{V}$ is the crystal volume: $\mathcal{V} = SL_z$.

It can be seen from Fig. 3.7 that the differences between the two- and three-dimensional cases are most pronounced in the energy regions of the lowest subbands. For large n the staircase function lies very close to the bulk curve $\varrho^{3D}(E)$ and coincides with it asymptotically.

The dramatic changes in the electron density of states caused by the dimensional confinement of crystals to produce low-dimensional systems are manifest in a variety of major modifications in conductivity, optical properties, etc. Indeed, as we shall see, these modifications in the density of states also lead to new physical phenomena.

It is useful to present the Fermi energy and the Fermi wavevector of *two-dimensional electrons* in the low-temperature limit defined by Eqs. (A.34), (A.35), (A.37), and (A.38):

$$E_F = \frac{\pi\hbar^2}{m^*} n_{2D}, \quad k_{F,2D} = (2\pi n_{2D})^{1/2}, \quad T \to 0, \tag{3.32}$$

where n_{2D} is the electron concentration per unit area (compare this with the similar results for three-dimensional electrons given by Eqs. (A.33) and (A.38)).

3.3 Electrons in Single- and Few-Monolayer Crystals

In this section we consider a few examples of single- and few-monolayer crystals, focusing on their electron properties. These properties are dramatically different from those of two-dimensional electrons in quantum wells fabricated from the usual three-dimensional materials, for example with the use of heterojunctions.

3.3.1 Basic Electronic Properties of Graphene

As discussed in Chapter 2, graphene is a single two-dimensional sheet of carbon atoms in a honeycomb lattice. It is a "perfect" two-dimensional electronic material, since carrier motion occurs necessarily in the thinnest two-dimensional layer.

The structural flexibility of graphene is reflected in its electronic properties. The sp^2 hybridization between one s orbital and two p orbitals leads to a trigonal planar structure with the formation of a strong σ bond between carbon atoms that are separated by $a = 1.42$ Å. Note that the σ bond is responsible for the robustness of the lattice structure in all carbon allotropes. These bonds have a filled shell and form a deep valence band. The remaining p orbital (oriented perpendicular to the crystal plane) is responsible for the formation of the electronic band structure in the vicinity of the Fermi level.

The carbon atoms are arranged in a hexagonal structure, as shown in Fig. 3.8. This is a triangular lattice with two atoms per unit cell. The lattice vectors can be written as

$$\vec{a}_1 = \frac{a}{2}\left(3, \sqrt{3}\right), \quad \vec{a}_2 = \frac{a}{2}\left(3, -\sqrt{3}\right).$$

In the corresponding Brillouin zone the reciprocal-lattice vectors are

$$\vec{b}_1 = \frac{2\pi}{3a}\left(1, \sqrt{3}\right), \quad \vec{b}_2 = \frac{2\pi}{3a}\left(1, -\sqrt{3}\right).$$

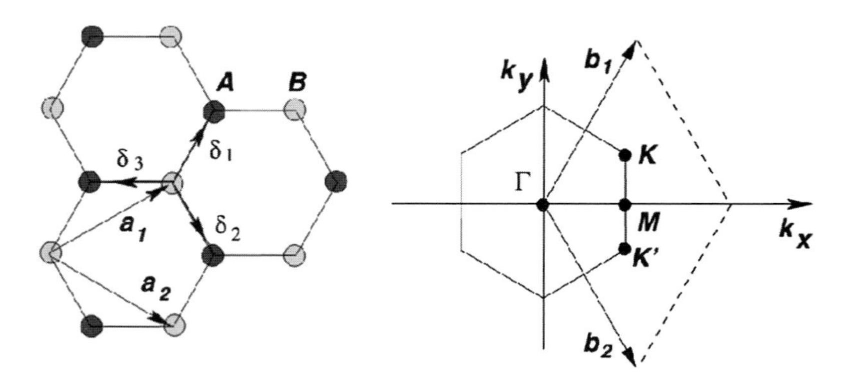

Figure 3.8 The honeycomb lattice and its Brillouin zone. Left: the lattice structure of graphene, made out of two interpenetrating triangular lattices; $\vec{\delta}_1$, $\vec{\delta}_2$, $\vec{\delta}_3$ indicate the nearest neighbors. Right: the corresponding Brillouin zone. The Dirac cones are located at the $\vec{K}$ and $\vec{K}'$ points. Reprinted figure with permission from figure 2 in A. H. Castro Neto, *et al.*, *Rev. Mod. Phys.* **81**, 109 (2009). Copyright 2009 by the American Physical Society.

The two corners of the Brillouin zone

$$\vec{K}=\frac{2\pi}{3a}\left(1,\frac{1}{\sqrt{3}}\right),\quad \vec{K}'=\frac{2\pi}{3a}\left(1,-\frac{1}{\sqrt{3}}\right).$$

are inequivalent and are the so-called Dirac points. Below we will see that they are of importance for the electronic properties of graphene.

Though two-dimensional wavevectors, a Brillouin zone, and an energy band structure for electrons can be introduced very similarly to what can be done in the case of three-dimensional crystals, the energy dispersion $E(k)$ for graphene is quite unusual and can be obtained only by applying microscopic calculations.

One such approach is the "tight-binding" method for the calculation of electronic band structure. In general, this quantum-mechanical method describes the properties of tightly bound electrons in solids. The electrons in this model should be tightly bound to the atoms to which they belong and they should have limited interaction with states and potentials on the surrounding atoms of the solid. As a result, the wave function of an electron will be rather similar to the atomic orbital of the free atom to which it belongs. The energy of the electron will also be rather close to the excitation/ionization energy of the electron in the free atom or ion, while the interaction with potentials and states on neighboring atoms is limited. This interaction is characterized by the interatomic matrix element between the atomic orbitals on nearby atoms. It is also called the nearest-neighbor hopping parameter.

By applying the tight-binding method for the monoatomic layer of carbon atoms, one can obtain the following energy dispersions for the conduction and valence bands:

$$E_{\mathrm{c,v}}(\vec{k})=\pm\gamma_0\sqrt{1+4\cos\left(\frac{k_y a}{2}\right)+4\cos\left(\frac{k_x\sqrt{3}a}{2}\right)\cos\left(\frac{k_y a}{2}\right)},\tag{3.33}$$

where $\vec{k}=\{k_x,k_y\}$ and $\gamma_0\approx 3.2\,\mathrm{eV}$ is the nearest-neighbor hopping parameter. The signs "+" and "−" apply for E_c and E_v, respectively.

In Fig. 3.9 the dependences $E_{\mathrm{c,v}}(k_x,k_y)$ are presented. Being of complex shapes, these dependences have distinct peculiarities for low-energy excitations. First, $E_\mathrm{c}(k_x,k_y)$ and $E_\mathrm{v}(k_x,k_y)$ touch each other at the points K and K′ introduced above. This means that there is no bandgap, i.e., graphene is a *gapless material*. Second, near these points $E_\mathrm{c}(k_x,k_y)$ and $E_\mathrm{v}(k_x,k_y)$ do not have parabolic energy dispersion, but instead have *linear* energy dispersion. In the right-hand part of Fig. 3.9, the linear dispersion is illustrated by the so-called *Dirac cones* for the conduction and valence bands. The linear dispersion means that the charge carriers in this material behave as relativistic particles with zero rest mass. They are often referred to as *massless Dirac fermions*. If we introduce the momentum $\vec{p}$ measured relative to the *Dirac points*, the linear dispersion discussed above takes the form

$$E_{\mathrm{c,v}}(\vec{p})=\pm v_\mathrm{F}|\vec{p}|+\mathcal{O}[p^2/K^2].\tag{3.34}$$

Here, the parameter $v_\mathrm{F}=\sqrt{3}a\gamma_0/2\hbar$ is the so-called Fermi velocity. It is proportional to the lattice constant, a, and to the nearest-neighbor hopping parameter, γ.

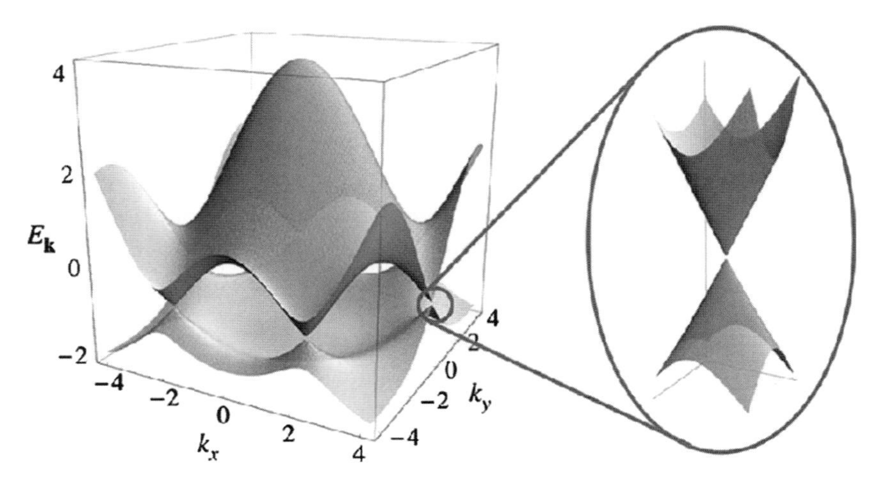

Figure 3.9 Electronic dispersion in the honeycomb lattice. The energy spectrum of Eq. (3.33) (left) and an expanded view of the energy bands close to one of the Dirac points (right), where the bands touch. Reprinted figure with permission from figure 3 in A. H. Castro Neto, *et al.*, *Rev. Mod. Phys.* **81**, 109 (2009). Copyright 2009 by the American Physical Society.

Remember that, for a particle with given energy $E(\vec{p})$, the velocity is defined as $\vec{v} = dE(\vec{p})/d\vec{p}$. Thus, a particle whose energy is presented by Eq. (3.34) moves with the velocity

$$\vec{v} = v_{\mathrm{F}} \frac{\vec{p}}{p}.$$

That is, the absolute value of the velocity is independent of the momentum (and the energy). The Fermi velocity is estimated as 10^8 cm/s (roughly 1/300 of the velocity of light). This is the most striking difference from the usual case of electrons with a finite mass m^*, when the velocity is $v = p/m^* = \sqrt{2E/m^*}$, and the velocity changes considerably with the momentum p or the energy E.

Note that the linear Dirac-like energy dispersion with constant Fermi velocity is obviously an approximate result valid for long-wavelength excitations in the conduction and valence bands of graphene, i.e., at $|\vec{k} - \vec{K}|$, $|\vec{k} - \vec{K}'| \ll |\vec{K}|$, $|\vec{K}'|$. One can give the following estimate: if the deviation of the wavevector from the point K or K$'$ is less than $0.25\,\mathrm{nm}^{-1}$, then the linear Dirac dispersion is valid, for which the energy $|E_{\mathrm{c,v}}| < 0.4\,\mathrm{eV}$. Because major physical effects in solids, as well as their practical applications, are associated with low-energy excitations, we may conclude that the linear Dirac dispersion is adequate for understanding the basic electronic properties of graphene.

Let us calculate the density of states of the electrons both in the conduction band and in the valence band. We can use the definition of the density of states presented in Eq. (3.27), where the index ν enumerates the quantum states. For the case under consideration, the total set of quantum numbers includes the spin, s, the type of Dirac point, $v = \mathrm{K}, \mathrm{K}'$ (this is often called the "valley" number), and the two-dimensional wavevector measured relative to the Dirac points, $\vec{q}$. Because carbon is a light atom, the spin–orbit interaction is weak in graphene and we can accept two-fold degeneracy of the conduction and valence bands, $g_s = 2$. Two inequivalent Dirac points K, K′ give rise to the "valley" degeneracy $g_v = 2$. Now we can perform similar calculations to those which led from Eq. (3.27) to Eq. (3.30). The result is the density of states of the carriers in graphene:

$$\varrho(E) = \sum_{\nu} \delta[E - E_{\nu}] = g_s g_v \frac{S}{(2\pi)^2} \int\int dq_x\, dq_y\, \delta[E - v_F \hbar q]$$
$$= g_s g_v \frac{S}{2\pi} \int q\, dq\, \delta[E - v_F \hbar q] = \frac{2S}{\pi \hbar^2 v_F^2} |E| \,. \tag{3.35}$$

Here S is the area of the graphene sample. The result is valid both for the conduction band, $E > 0$, and for the valence band, $E < 0$. It is worth noting that the density of states for graphene is radically different from the density of states of two-dimensional electrons in quantum wells (compare this with Eq. (3.30)). In particular, $\varrho \to 0$ if $E \to 0$. Numerical calculations of $\varrho(E)$ are presented in Fig. 3.10 for small energies and for the full electron bandwidth.

As for any Fermi particle, the equilibrium distribution function of the low-energy excitations in graphene is given by Eq. (A.28) with the energy of Eq. (3.34). Thus, the probability of finding of an electron with given momentum, p, spin, s, and valley number, v, is

$$\mathcal{F}_{p,s,v} = \frac{1}{1 + \exp[(v_F p - E_F)/k_B T]} \,. \tag{3.36}$$

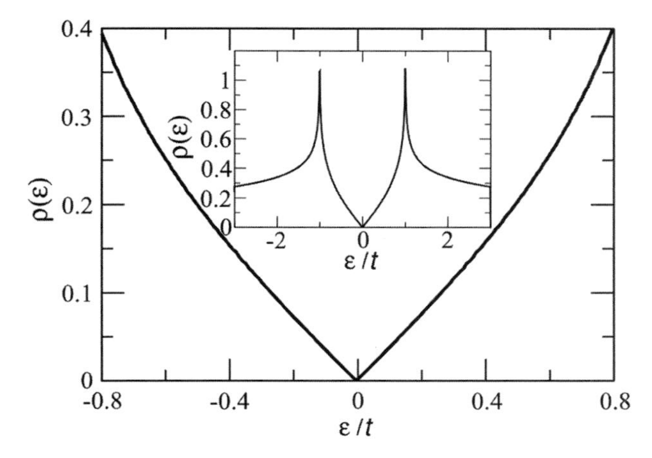

Figure 3.10 The density of states of graphene close to the Dirac point. The inset shows the density of states over the full electron bandwidth, $t = 2.5$ eV. Reprinted figure with permission from figure 2 in S. Das Sarma, *et al.*, *Rev. Mod. Phys.* **83**, 407 (2011). Copyright 2011 by the American Physical Society.

Using the density of states of Eq. (3.35), we can find the relationship between the electron area concentration, n, and the Fermi level, E_F (>0):

$$n = \frac{2}{\pi \hbar^2 v_F^2} \int_0^\infty \frac{E\,dE}{1 + \exp[(E - E_F)/k_B T]}. \tag{3.37}$$

In doped graphene at low temperatures, the electrons can be degenerate. For such a case, from Eq. (3.37) we obtain the Fermi momentum, p_F, the Fermi wavevector, q_F, and the Fermi energy, E_F:

$$p_F = \hbar q_F = \hbar \sqrt{\frac{4\pi n}{g_s g_v}} = \hbar \sqrt{\pi n}, \quad E_F = \hbar v_F \sqrt{\frac{4\pi n}{g_s g_v}} = \hbar v_F \sqrt{\pi n}. \tag{3.38}$$

Setting, for example, $n = 10^{12}\,\text{cm}^{-2}$, we obtain $E_F = 0.12\,\text{eV}$, which sufficiently exceeds the thermal energy $k_B T$ even at room temperature ($\approx 0.026\,\text{eV}$). That is, even at modest concentrations, the electrons in graphene are degenerate up to room temperature.

In pristine (undoped) graphene the Fermi level precisely coincides with the Dirac points, i.e., $E_F = 0$. The concentrations of electrons and holes are equal and dependent on the temperature T as follows:

$$n = p = \frac{\pi (k_B T)^2}{6\hbar^2 v_F^2}. \tag{3.39}$$

For example, at room temperature, $n = p = 7.75 \times 10^{10}\,\text{cm}^{-2}$.

In conclusion, we emphasize the significant differences between the electronic properties of graphene and those of two-dimensional semiconductor heterostructures.

The carrier confinement in graphene is ideally two-dimensional, because the graphene layer is one atomic monolayer thick. For two-dimensional semiconductor structures, the electron dynamics are two-dimensional until the electrons remain in a fixed subband. Actually, two-dimensional semiconductors are quasi-two-dimensional systems, which have a certain width in the third direction. When higher subbands become populated at large carrier density, such semiconductor systems are no longer two-dimensional.

Two-dimensional semiconductor systems typically have considerable bandgaps, so that to realize two-dimensional electrons and two-dimensional holes it is necessary to use different electron-doped or hole-doped structures. In contrast, graphene is a gapless semiconductor, and a change of charge of the current carriers occurs at the Dirac point; the transition from electrons to holes (or vice versa) can be realized in a single graphene structure.

Two-dimensional semiconductors become insulating when the Fermi level enters the bandgap. As a result of the gapless character of graphene, the Fermi level is practically always either in the conduction band or in the valence band; i.e., graphene is always conductive.

In monolayer graphene the dispersion is linear, while two-dimensional semiconductors have a quadratic energy dispersion. This leads to substantial differences between the transport properties of these two systems. In graphene, extremely high mobility is achieved due to the fact that the electron rest mass is zero.

3.3.2 Two-Dimensional Transition-Metal Dichalcogenide Crystals

As introduced in Chapter 2, layered transition-metal dichalcogenides of MX_2 type, where M = Mo, W, Nb, Re, Ti, Ta, etc., and X = S, Se, Te, etc., have been studied extensively both experimentally and theoretically. The transition-metal dichalcogenides are characterized by layered crystal structures with two-dimensional layers composed of strong X—M—X intralayer covalent bonding, while the bonding between layers is of the van der Waals type, i.e., the interlayer bonding is very weak. This strong bonding anisotropy leads to unusual two-dimensional electronic structures. From a qualitative point of view, all of the members of the whole class of MX_2 crystals have very similar properties. The most studied structure is molybdenum disulfide, MoS_2. Thus, our discussion will focus on molybdenum disulfide.

Single-layer MoS_2 is a typical two-dimensional semiconductor of the layered transition-metal dichalcogenide family. Single layers can be extracted from bulk crystals of MoS_2 using, for example, the micromechanical cleavage technique commonly associated with the production of graphene. Other methods of achieving a single layer include lithium-based intercalation and liquid-phase exfoliation. After thin layers of MoS_2 have been formed, they can be transferred to different substrates.

Bulk crystals of MoS_2 are *indirect-bandgap* semiconductors with a bandgap of about 1.2 eV. Owing to quantum confinement of the electrons, decreasing the number of layers in mesoscopic MoS_2 structures gives rise to a transformation into a *direct-gap* semiconductor with a bandgap of 1.8 eV. Being an ultrathin semiconductor with a finite bandgap, single-layer MoS_2 becomes a material complementary to graphene, which, as we discussed above, is, in principle, a gapless material.

In contrast with graphene, in a monolayer of MoS_2 the atoms are not situated in a plane: two layers of sulfur atoms in a two-dimensional hexagonal lattice are stacked over each other. Each Mo atom sits in the center of a trigonal prismatic cage formed by six sulfur atoms, as shown in Fig. 3.11(a). The top view of the MoS_2 monolayer, shown in Fig. 3.11(b), indicates that the natural and stable structure of this monolayer is a honeycomb lattice similar to graphene. The shaded region in Fig. 3.11(b) corresponds to one primitive cell. The unit cell parameter is $a = 3.12$ Å, and the vertical separation between sulphur layers is 3.11 Å. Despite the similarity of the top view to graphene, from Fig. 3.11(a) it is obvious that there is no inversion symmetry in monolayer MoS_2. This breaking of inversion symmetry changes dramatically the electronic properties of MoS_2 in comparison with graphene. Another important difference between MoS_2 and graphene is a strong spin–orbit coupling that originates from the properties of the heavy transition-metal atoms. Thus, in contrast to graphene, for all MX_2 two-dimensional crystals the spin physics can play an important role, whereas in graphene all electron states can be considered as two-fold spin degenerate.

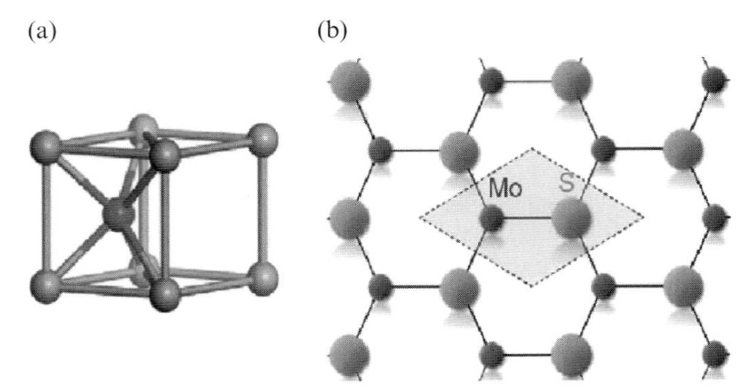

Figure 3.11 The crystal structure of monolayer MoS_2. (a) The coordination environment of Mo (darker sphere) in the structure. Sulfur is indicated by the lighter spheres. (b) A top view of the monolayer MoS_2 lattice, emphasizing the connection to a honeycomb lattice similar to graphene. Reprinted by permission from figure 1 in Springer, *Nature Commun.*, "Valley-selective circular dichroism of monolayer molybdenum disulphide," Ting Cao, *et al.*, Copyright 2012.

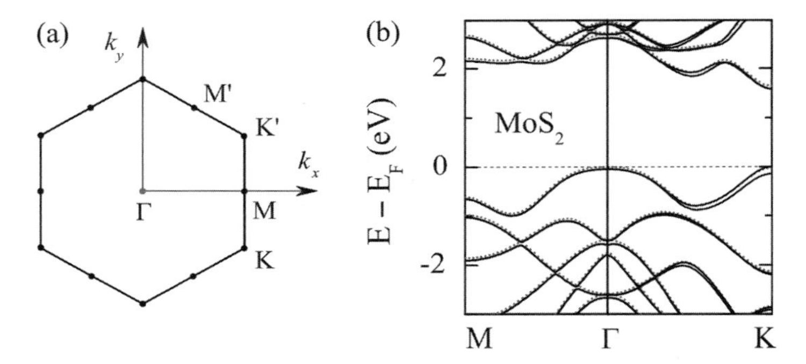

Figure 3.12 (a) The first Brillouin zone of the MoS_2 monolayer. (b) The electronic band structure of the MoS_2 monolayer with (solid lines) and without (dotted lines) inclusion of the spin–orbit interaction. Reprinted figure with permission from figures 1c and 2a in Z. Y. Zhu, *et al.*, *Phys. Rev.* **B 84**, 153402 (2011). Copyright 2011 by the American Physical Society.

The first Brillouin zone of the MoS_2 monolayer is very similar to that of graphene, as shown in Fig. 3.12(a), where the high-symmetry points Γ, K, K′, M, M′ are indicated. In Fig. 3.12(b), the detailed electronic band structure is presented for wavevectors changing along the directions $\Gamma \rightarrow$ M and $\Gamma \rightarrow$ K. When MoS_2 is thinned to the monolayer limit, the material acquires direct bandgaps located exactly at the corners of the Brillouin zone, at the K and K′ points, as can be seen in Fig. 3.12(b). Note that, if the vector $\vec{K}$ corresponds to a K point in the Brillouin zone, the vector $-\vec{K}$ corresponds to a K′ point. At these points the energies E_c and E_v have a minimum and a maximum, respectively. The effective masses for

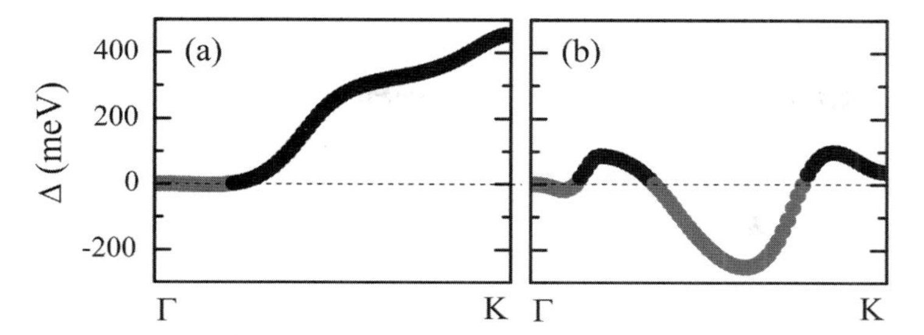

Figure 3.13 The spin splitting as a function of the wavevector changing along the line $\Gamma \to K$ for the WSe_2 monolayer. (a) The uppermost valence band. (b) The lowermost conduction band. The sign of the parameter Δ corresponds to the spin orientation. Reprinted figure with permission from figure 4 in Z. Y. Zhu, *et al.*, *Phys. Rev.* **B 84**, 153402 (2011). Copyright 2011 by the American Physical Society.

these points are finite and estimated as $0.4m$ and $0.6m$ for the c and v zones, respectively, with m the mass of a free electron. Thus, the energy dispersions of the MoS_2 monolayer are as for typical semiconductor materials; the only difference is that the wavevector is two-dimensional.

For MoS_2, the spin–orbit splitting is noticeable in Fig. 3.12(b), especially for the valence band. Other group VI transition-metal dichalcogenides forming monolayers, $MoSe_2$, WS_2, and WSe_2 have electronic properties very similar to those of MoS_2. However, the spin–orbit interaction and, as a consequence, the valley–spin coupling are considerably greater for heavy transition-metal atoms. For example, for a monolayer of WSe_2, the spin splitting of the valence band at the K, K′ points is equal to 0.456 eV. The dependences of the spin splitting, Δ, for both energy bands in WSe_2, are depicted in Fig. 3.13 as functions of the wavevector. The spin splitting is large in the vicinity of the K points (up to hundreds of meV), which is of primary importance because the carriers (electrons and holes) are accumulated in valleys near these points. The spin splitting for the conduction band is less than that for the valence band and is a nonmonotonic function of the wavector. Interestingly, the sign of the spin projection in the direction of the wavevector also changes along the Brillouin zone, as shown in Fig. 3.13. These observations provide evidence of the strong interaction between the spin and the translational motion of the crystalline electron.

This strong interaction of the spin and the translational motion, in combination with the previously mentioned inversion-symmetry breaking, gives rise to a new effect – strong valley–spin coupling in the valence band. Indeed, consider some $\vec{K}$ point for which there is spin splitting of the valence band, Δ. For split valence states (the upper and the lower) the spin states are different, so the spin projections on the wavevector are of different signs. That is, for the point $-\vec{K}$, i.e., for the K′ point, the splitting Δ is the same, but the directions of the spin for the upper and lower split states have opposite signs in comparison with those for the K point. This statement is illustrated in Fig. 3.14, where the conduction band, the split valence bands, and the spin polarization are shown schematically. Thus, the points K and K′ and the corresponding valleys in the valence band are sufficiently inequivalent.

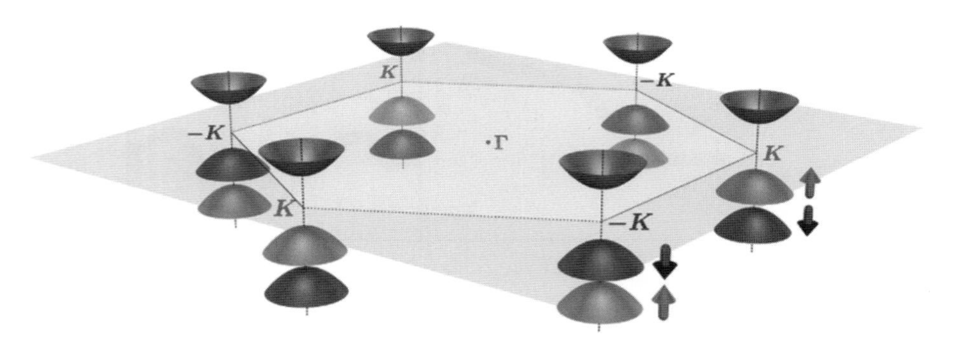

Figure 3.14 A schematic drawing of the band structure and spin at the band edges located at the K, K′ points. Reprinted figure with permission from figure 1c in D. Xiao, *et al.*, *Phys. Rev. Lett.* **108**, 196802 (2012). Copyright 2012 by the American Physical Society.

The valley–spin coupling leads to some important conclusions. First, both spin and valley relaxations are suppressed because the flipping of each index, either the spin index or the valley index, alone is forbidden. The splittings for MoS_2, $MoSe_2$, and WS_2 are respectively 1.48 eV, 0.183 eV, and 0.426 eV. Second, the v−c interband transitions are governed by *chiral selection rules*, i.e., the optical selection rules become valley- and spin-dependent. In particular, valley and spin indices can be selectively excited by optical fields of different circular polarizations and frequencies.

In summary, the transition-metal dichalcogenide two-dimensional crystals are direct-bandgap materials, with bandgaps above 1 eV. The structures of the monolayers are not flat, as for graphene, and provide the phenomenon of inversion-symmetry breaking. The minima of the conduction band and the maxima of the valence band occur at the high-symmetry points K, K′, as illustrated in Fig. 3.14. At these points electrons in the conduction and valence bands are characterized by their effective masses. They can be called *massive Dirac fermions*. The effective masses are relatively large, which leads to low electron and hole mobilities. As a result of the strong spin–orbit interaction in the transition-metal atoms, the spin-related effects in the monolayers are large and the spin-splitting of the valence band may be as much as 0.15–0.4 eV at the K, K′ points. Because of the lack of inversion symmetry, the electron spin states are sufficiently different at the K and K′ points. This gives rise to a new effect, valley–spin coupling, which generates unusual transport and optical electron properties.

3.3.3 Black Phosphorus of Thickness a Few Monoatomic Layers

In bulk black phosphorus, the atoms are strongly bonded in-plane, forming layers, while interlayer interaction through van der Waals forces is weak in a manner similar to that in graphite, as indicated in Subsection 3.3.1. However, in graphite every carbon atom is coupled to three neighboring atoms through sp^2 hybridization, whereas

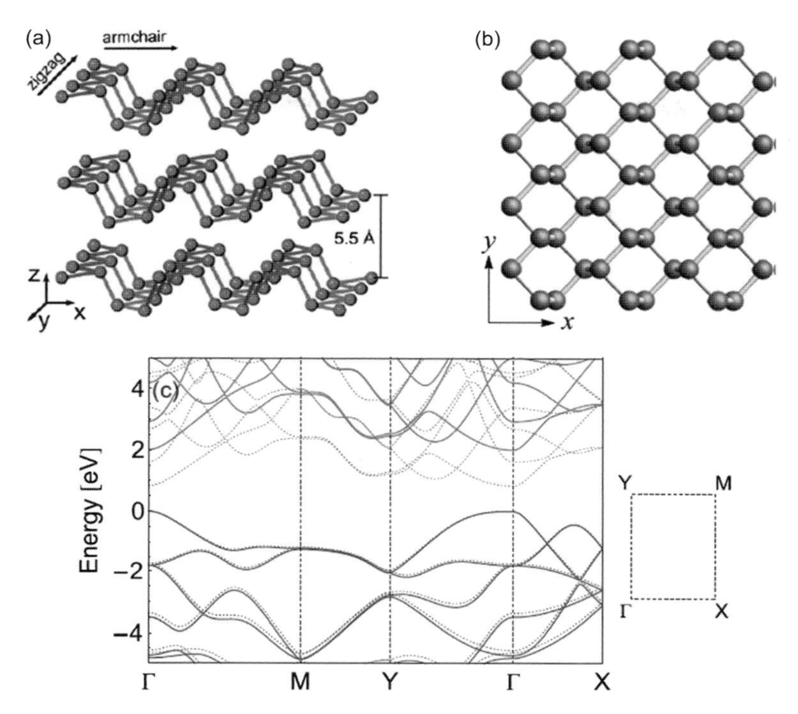

Figure 3.15 Two different perspectives on the black phosphorus lattice. (a) Lattice representation for few-layer phosphorus where the x-axis is perpendicular to the ridge direction and the y-direction is parallel to the ridge direction. (b) A top view of the monolayer phosphorus lattice. (c) The calculated band structures of monolayer phosphorus. The solid and dotted lines show two different methods of calculation. The energy scale is set to zero at the top of the valence band. Reprinted figure with permission from figure 1 in V. Tran, *et al.*, *Phys. Rev.* **89**, 235319 (2014). Copyright 2014 by the American Physical Society.

in black phosphorus the phosphorus atoms are of the $3s^2 3p^3$ valence-shell configuration, i.e., they have five valence electrons for bonding. Every phosphorus atom bonds to three neighboring phosphorus atoms through sp^3 hybridization, which leads to the arrangement of the phosphorus atoms in a *puckered honeycomb* lattice, as shown in the schematic diagrams in Figs. 3.15(a) and (b). One can see that such an atomic arrangement stipulates in-plane anisotropy with two inequivalent directions, as indicated in Fig. 3.15(a): parallel to the atomic ridges (zigzag direction) and perpendicular to the ridges (armchair direction). The bulk black phosphorus has an electronic band structure characteristic of a direct-bandgap semiconductor with a bandgap of about 0.36 eV.

The weak coupling between atomic layers in phosphorus allows one to apply mechanical exfoliation methods to isolate atomically thin black phosphorus from bulk layered crystals and obtain two-dimensional black phosphorus crystals of thickness a single, or a few, atomic layers. It was found that, independently of the number of layers, the material remains of direct-bandgap type, with the minimal

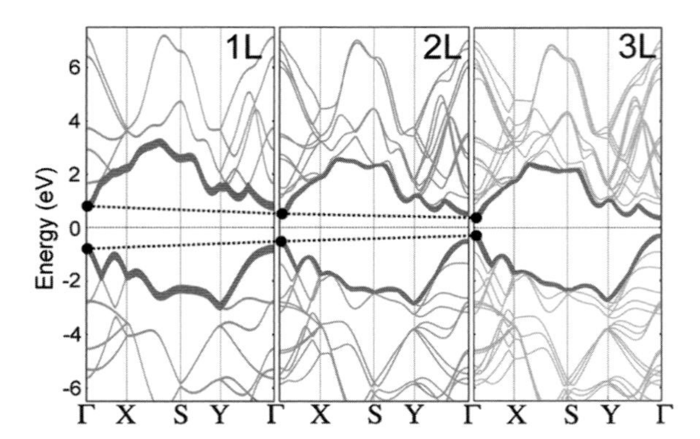

Figure 3.16 The calculated band structure of monolayer, bilayer, and trilayer black phosphorus. Adapted with permission from figure 1b in A. Castellanos-Gomez, *J. Phys. Chem. Lett.* **6**, 4280 (2015). Copyright 2015 American Chemical Society.

energy gap at the center of the Brillouin zone (at the Γ point, in contrast with semiconducting transition-metal dichalcogenides, for which the minimal bandgap occurs at the K points; these materials are of direct-bandgap type only for single-layer crystals). The electron band structure of single-layer black phosphorus is shown in Fig. 3.15(c).

It is important that the bandgap strongly depends on the number of layers in the case of black phosphorus: the bandgap increases when the number of layers decreases, as illustrated in Fig. 3.16. It is accepted that this thickness-dependent bandgap can be explained by invoking the quantum confinement of the electrons in the out-of-plane direction. For atomically thin black phosphorus the bandgap is about 2 eV. The strong thickness dependence of the phosphorus bandgap promises an exceptional tunability: in the same material, by varying the sample thickness one can change the energy gap and optical properties in a wide spectral range that cannot be covered in any other material. Additionally, the anisotropic properties of black phosphorus can be exploited in polarization-sensitive electronic and optoelectronic devices.

3.4 Quantum Wires

In Section 3.2 we showed that the transition from a three-dimensional electron gas to a two-dimensional electron gas is due to quantization of the electron motion in one direction. As a result, the electron is characterized by two degrees of freedom. The two-dimensional electron gas can be created by imposing a one-dimensional confining potential $V(z)$, which can be formed by one plane heterojunction or by layered heterostructures.

In order to make the transition from a two-dimensional electron gas to a one-dimensional electron gas, the electrons should be confined in two directions and only

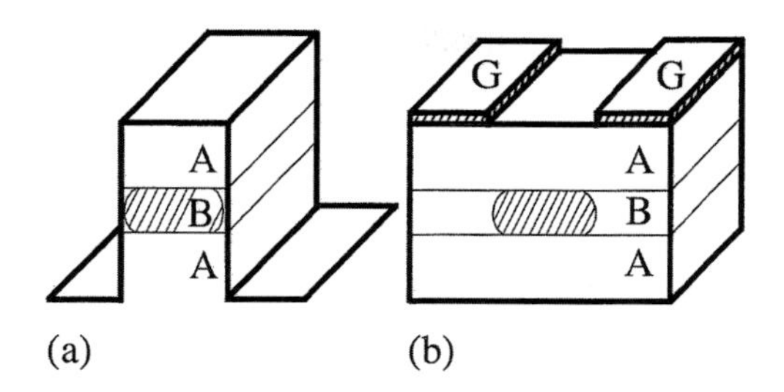

Figure 3.17 Two of the simplest examples of structures providing electron confinement in two dimensions: (a) using an etching technique and (b) using the split-gate technique.

one degree of freedom should remain. That is, one should design a two-dimensional confining potential $V(y,z)$. The x-direction remains the only one available for free-electron propagation.

There are several modern technological methods for the formation of such structures. The simplest of them are illustrated by Fig. 3.17. The case in Fig. 3.17(a) (compare this with Fig. 3.1) corresponds to etching of the top layer A, the confining material B, and part of bottom layer A. In the case of Fig. 3.17(b) two gates are deposited on the top layer A. A negative potential applied to these gates controls the lateral electron confinement.

3.4.1 Wave Functions and Energy Subbands

Let us consider general features of electrons in heterostructures with a confining potential $V(y,z)$. According to the method of the separation of variables, solutions of the Schrödinger equation (A.27) can be sought in the form

$$\psi(x,y,z) = e^{ik_x x}\chi(y,z), \tag{3.40}$$

so that the equation for the transverse electron wave function, $\chi(y,z)$, is

$$\left[-\frac{\hbar^2}{2m^*}\left(\frac{\partial^2}{\partial y^2}+\frac{\partial^2}{\partial z^2}\right)+V(y,z)\right]\chi(y,z) = \epsilon\chi(y,z), \tag{3.41}$$

where $\epsilon = E - \hbar^2 k_x^2/(2m^*)$ is the electron energy in the two transverse directions. If one can find a solution, $\chi_i(y,z)$, of Eq. (3.41) corresponding to the discrete energy ϵ_i, one can obtain the total energy of the electrons in a form analogous to Eq. (3.24):

$$E = \epsilon_i + \frac{\hbar^2 k_x^2}{2m^*}, \tag{3.42}$$

where k_x is now a *one-dimensional vector*. Comparison of Eqs. (3.42) and (3.24) reveals important properties of *one-dimensional subbands*. The wave function $\chi_i(y,z)$ corresponding to the discrete energy level, ϵ_i, is localized in some area of the y,z plane. This means that the electrons of this quantum state i are confined in the y- and z- directions around the minimum V_0 of the potential $V(y,z)$ and they are able to propagate along the x-axis only. Such artificial systems where electrons are guided along only one direction are called quantum wires.

The simplest case for which the two-dimensional Schrödinger problem can be solved is given by an infinitely deep rectangular potential,

$$V(x,z) = \begin{cases} 0, & \text{for } 0 \le y \le L_y,\ 0 \le z \le L_z, \\ \infty, & \text{for } y \le 0,\ z \le 0,\ y \ge L_y,\ z \ge L_z. \end{cases} \tag{3.43}$$

Here L_y and L_z are the transverse dimensions of the wires. For this case – of an infinitely deep potential – the electron wave function $\chi(y,z)$ can be represented as a product of functions depending separately on y and z:

$$\chi(y,z) = \chi(y)_{n_1}\chi(z)_{n_2}. \tag{3.44}$$

For each of the directions, the solutions of the one-dimensional Schrödinger problem have the same form, namely

$$\chi_{n_1}(y) = \sqrt{\frac{2}{L_y}}\sin\left(\frac{\pi y n_1}{L_y}\right),\ \chi_{n_2}(z) = \sqrt{\frac{2}{L_z}}\sin\left(\frac{\pi z n_2}{L_z}\right),\quad n_1,\ n_2 = 1,2,3,\ldots, \tag{3.45}$$

and the quantized energy, ϵ_i, of the transverse motion of the electrons is

$$\epsilon_{n_1,n_2} = \frac{\hbar^2\pi^2}{2m^*}\left(\frac{n_1^2}{L_y^2} + \frac{n_2^2}{L_z^2}\right). \tag{3.46}$$

These results are applicable for quantum wires, which may be fabricated from two-dimensional electron systems by patterning one-dimensional features in the plane of the electron gas through the use of nanoscale lithographic techniques, with subsequent processing by etching, etc.; see Fig. 3.17(a).

Several other methods provide supplementary confinement of two-dimensional electrons in the second direction; as examples, such one-dimensional confinement of a two-dimensional electron gas may be accomplished by the application of an external voltage, a strain-induced potential, etc. The case illustrated by Fig. 3.17(b) is known as the *split-gate technique*, and it is used very frequently. For this case the confining external potential can be approximated by a parabolic dependence, so that

$$V(y,z) = V_1(y) + V_2(z),\ V_1(y) = \frac{1}{2}\left(\frac{\partial^2 V_1}{\partial y^2}\right)_{y=0} y^2. \tag{3.47}$$

If the potential $V_2(z)$ provides high barriers at $|z| = L_z/2$, the solution of the problem is given by the following expressions:

$$\chi_{n_1,n_2}(y,z) = \text{constant} \times e^{-\alpha^2 y^2}\mathcal{H}_{n_1}(\alpha y)\sin\left(\frac{\pi z n_2}{L_z}\right), \tag{3.48}$$

$$\epsilon_{n_1,n_2} = \hbar\omega(n_1 + 1/2) + \frac{\hbar^2\pi^2 n_2^2}{2m^* L_z^2}. \tag{3.49}$$

Here we have introduced the notations

$$\omega = \sqrt{\frac{2}{m^*}\left(\frac{\partial^2 V_1}{\partial y^2}\right)_{y=0}}, \quad \alpha = \frac{m^*\omega}{\hbar}, \quad \mathcal{H}_n(y) = (-1)^n e^{y^2}\frac{d^n}{dy^n}e^{-y^2},$$

where $\mathcal{H}_n$ are the Hermite polynomials.

The examples presented for the potentials defined by Eqs. (3.43) and (3.47) demonstrate that the energy levels which arise in quantum wires are strongly dependent on the form of the confining potentials. However, the following important conclusions are quite general: the additional confinement of electrons leads to an increase of the lowest energy level; and there is a two-fold set of levels for quantum wires. Two quantum numbers, n_1 and n_2, now characterize any energy level corresponding to the transverse directions. The total energy of electrons, defined by Eq. (3.42), also includes the kinetic energy of one-dimensional electron propagation characterized by the one-dimensional wavevector k_x.

3.4.2 The Density of States for a One-Dimensional Electron Gas

Let us calculate the density of states of a one-dimensional electron gas. In the definition of Eq. (3.26), ν enumerates all of the possible quantum states of the system. In the case of quantum wires, $\nu = \{s, n_1, n_2, k_x\}$. Rewriting Eq. (3.26) in the form

$$\varrho(E) = \sum_{n_1, n_2} \varrho_{n_1, n_2}(E), \tag{3.50}$$

one can calculate the contribution to the density of states from a single subband:

$$\varrho_{n_1 n_2}(E) = 2\sum_{k_x} \delta\left(E - \epsilon_{n_1, n_2} - \frac{\hbar^2 k_x^2}{2m^*}\right), \tag{3.51}$$

where the factor of 2 is due to the spin, s. The sum has to be calculated in the same way as for Eq. (3.27):

$$\begin{aligned} \varrho_{n_1 n_2}(E) &= \frac{2L_x}{\pi}\int_0^\infty dk_x\, \delta\left(E - \epsilon_{n_1, n_2} - \frac{\hbar^2 k_x^2}{2m^*}\right) \\ &= \frac{L_x}{\pi}\sqrt{\frac{2m^*}{\hbar^2}}\frac{1}{\sqrt{E - \epsilon_{n_1, n_2}}}\Theta(E - \epsilon_{n_1, n_2}). \end{aligned} \tag{3.52}$$

Here L_x is the length of the wire, and the factor 2 comes from the fact that the summation in Eq. (3.51) is from $-\infty$ to $+\infty$ and we replace it by integration from 0 to ∞. A plot of $\varrho(E)$ for one-dimensional electrons is shown schematically in Fig. 3.18.

Let us compare the density of states of the one-dimensional electron gas of Eq. (3.52) with the density of states of the two-dimensional gas of Eq. (3.30). The characteristics of these two densities of states are very different. Instead of the

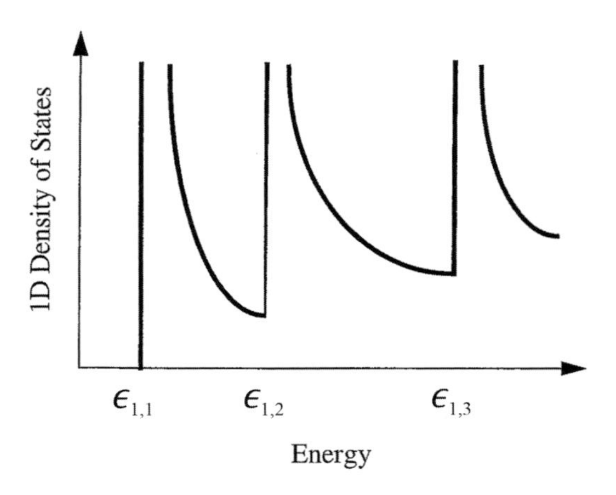

Figure 3.18 The density of states for one-dimensional electrons.

step-function behavior of the two-dimensional case, the one-dimensional density of states, $\varrho(E)$, diverges at the bottom of each subband and then decreases as the kinetic energy increases. This behavior is very remarkable because it leads to a whole set of electrical and optical effects peculiar to quantum wires.

3.5 Quantum Dots

So far we have considered semiconductor heterostructures where an electron is confined in one or two directions. This leads to quantization of the electron spectrum, resulting in two- or one-dimensional energy subbands. This drastically changes the density of states. But still there is at least one direction for free propagation of the electron, along the structure parallel to the barriers associated with the confining potentials; these structures can be used for electronic devices based on this transport.

The advances in semiconductor technology allow one to go further and fabricate heterostructures where all existing degrees of freedom of electron propagation are quantized. These so-called quantum dot, or quantum box, systems are like artificial atoms, and they demonstrate extremely interesting behavior.

Before we begin our study of such *zero-dimensional systems*, let us emphasize briefly that the trend toward further lowering of the dimensionality of the electron gas portends useful applications both in electronics and in optoelectronics. Let us imagine that one may quantize the electron states in all three possible directions and obtain a new physical object – a *pseudoatom*, or *macroatom*. Many questions concerning the usefulness of these objects for applications would naturally arise. And these doubts are understandable, because most electronic systems employ electric voltages and electric currents. A fundamental question arises: what is the current through a macroatom? Valid answers are as follows. First, there exists the possibility of passing an electric current through an artificial atom due to tunneling of

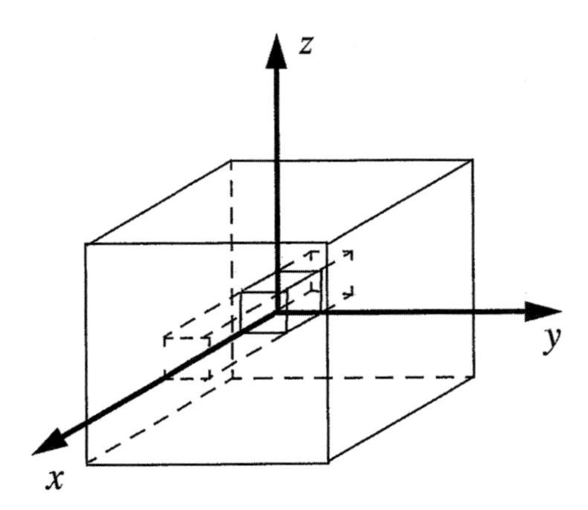

Figure 3.19 A quantum box embedded in a matrix.

electrons through quantum levels of the atom. Second, electrical methods of control of functions are not the only methods possible. The control can also be realized by light, sound waves, etc. Hence, there are several possible ways to achieve the necessary control of useful functions of zero-dimensional devices.

3.5.1 Wave Functions and Energy Levels

When we are considering the energy spectrum of a zero-dimensional system, we have to study the Schrödinger equation (A.27) with a confining potential which is a function of all three coordinates and confines the electron in all three directions. The simplest potential, $V(x, y, z)$, of this type is

$$V(x,y,z) = \begin{cases} 0, & \text{inside the box,} \\ +\infty, & \text{outside the box,} \end{cases} \tag{3.53}$$

where the box is restricted by the following conditions: $0 \leq x \leq L_x$, $0 \leq y \leq L_y$, $0 \leq z \leq L_z$; see Fig. 3.19. For this case, one can write down the solutions of the Schrödinger equation (A.27) immediately:

$$\psi_{n_1,n_2,n_3}(x,y,z) = \sqrt{\frac{8}{L_x L_y L_z}} \sin\left(\frac{\pi x n_1}{L_x}\right) \sin\left(\frac{\pi y n_2}{L_y}\right) \sin\left(\frac{\pi z n_3}{L_z}\right), \tag{3.54}$$

$$E_{n_1,n_2,n_3} = \frac{\hbar^2 \pi^2}{2m^*}\left(\frac{n_1^2}{L_x^2} + \frac{n_2^2}{L_y^2} + \frac{n_3^2}{L_z^2}\right), \tag{3.55}$$

where $n_1, n_2, n_3 = 1, 2, 3, \ldots$ Of fundamental importance is the fact that E_{n_1,n_2,n_3} is the total electron energy, in contrast with previous cases, where the solution for the bound state in a quantum well and quantum wire gave us only the energy spectrum associated with transverse confinement. Another unique feature is the presence of three discrete quantum numbers resulting straightforwardly from the existence of three directions of quantization. Thus, we obtain three-fold discrete energy levels and wave functions localized in all three dimensions of the quantum box.

Generally, all of the energies are different, i.e., the levels are not degenerate. However, if two or all three dimensions of the box are equal, or their ratios are integers, some levels with different quantum numbers coincide. Such a situation results in degeneracy: two-fold degeneracy if two dimensions are equal and six-fold degeneracy for a cube. This discrete spectrum in a quantum box and the lack of free-electron propagation are the main features distinguishing quantum boxes from quantum wells and quantum wires. As is well known, these features are typical for atomic systems.

The similarity with atoms is easily seen from another example of a confining potential: the case of a spherical dot. The wave functions in a spherical confining region such as that for a spherical quantum dot are well known in quantum mechanics to yield Bessel-function solutions. These Bessel functions are used in many cases to model quantum dots with more complex confinement geometries. As an example, Bessel-function solutions are presented in reference books on quantum dot heterostructures which deal primarily with non-spherical quantum dots that self-assemble on a growth surface, such as those discussed in Chapter 2. At the very least, these solutions of the Schrödinger equation for a particle trapped in a spherical confining potential provide a starting point for gaining insights into the nature of optical transitions in quantum dots and thus have to be considered. Herein, the wave function calculation is presented for a spherical quantum dot.

In the simplest case, the potential of a spherical quantum dot can be assumed to be

$$V(r)=\begin{cases}0, & \text{for } r\leq R,\\ V_{\mathrm{b}}, & \text{for } r\geq R.\end{cases} \tag{3.56}$$

Here r is the magnitude of the radius vector, or coordinate, and R is the radius of the spherical dot.

From quantum mechanics it is known that under spherical symmetry the solutions of the Schrödinger equation can be expressed by separating the angular and radial dependences in the form

$$\psi(r,\theta,\phi)=\mathcal{R}(r)\,Y_{l,m}(\theta,\phi). \tag{3.57}$$

Here r, θ, and ϕ are the spherical coordinates, $Y_{l,m}(\theta,\phi)$ are the well-known spherical functions, and l and m are quantum numbers corresponding to the angular momentum and its projection on the z-axis, respectively. For the radial function $\mathcal{R}(r)$, the Schrödinger equation is

$$\left[-\frac{\hbar^2}{2m^*}\frac{\partial^2\chi(r)}{\partial r^2}+V_{\mathrm{eff}}(r)\right]\chi(r)=E\chi(r), \tag{3.58}$$

where

$$\chi(r)=r\mathcal{R}(r) \quad \text{and} \quad V_{\mathrm{eff}}(r)=V(r)+\frac{\hbar^2 l(l+1)}{r^2}. \tag{3.59}$$

Thus, as a result of the spherical symmetry the problem reduces formally to a one-dimensional equation. The effective potential $V_{\text{eff}}(r)$ depends on the quantum number l, but does not depend on m. Accordingly, the energy levels must be degenerate with respect to the quantum number m, and the number of degenerate states is equal to $2l+1$. The energy is a function of two quantum numbers: the principal quantum number, n, which comes from Eq. (3.58), and the angular-momentum quantum number, l. One can find a general analysis of the problem with the potential $V_{\text{eff}}(r)$ in most textbooks on quantum mechanics. Below we present only results for the simplest cases; however, additional results are presented in Chapters 5 and 7, where spherical quantum dots and their applications are discussed.

If $l=0$, one can obtain easily the solutions of Eq. (3.58) with the potential of Eq. (3.56):

$$\psi(r)=\begin{cases} A\sin(k_{\text{w}}r/r),\ k_{\text{w}}=\sqrt{2m^*E}/\hbar, & \text{if } r<R, \\ Be^{-\kappa_{\text{b}}r}/r,\ \kappa_{\text{b}}=\sqrt{2m^*(V_{\text{b}}-E)}/\hbar, & \text{if } r>R. \end{cases} \tag{3.60}$$

The matching of the functions (3.60) and their derivatives at $r=R$ gives us equations very similar to those studied for the one-dimensional problem. As a result, one can find an equation for the energy similar to Eq. (3.13):

$$\sin(k_{\text{w}}R)=\pm\sqrt{\frac{\hbar^2}{2m^*V_{\text{b}}}}k_{\text{w}}. \tag{3.61}$$

Only the roots of this equation satisfying the condition $\cot(k_{\text{w}}R)<0$ can be chosen. We can repeat the analysis given in the previous subsections for a rectangular one-dimensional potential well of finite depth. The solutions of Eq. (3.61) are identical to the even states of the previous case represented by Eq. (3.13). Thus, we may apply the same qualitative analysis using Fig. 3.3. Only the portions of the curve in the intervals $[2\pi(i-1/2)/L, 2\pi i/L]$, $i=1,2,3,\ldots$, in Fig. 3.3 should be taken into consideration. The analysis shows that a level exists inside the spherical well if

$$V_{\text{b}}\geq\frac{\pi^2\hbar^2}{8m^*R^2}. \tag{3.62}$$

Thus, a potential well must be large enough or deep enough to confine the electron. If the potential well is very deep, one can obtain solutions of Eq. (3.61):

$$k_{\text{w}}R=\pi n;\quad E_{n,l=0}=-V_{\text{b}}+\frac{\hbar^2\pi^2n^2}{2m^*R^2}, \tag{3.63}$$

where we have shifted the origin of the energy coordinate so that $E_{n,l=0}<0$ corresponds to bound states. The series of levels with $l=0$ is very similar to that for the quantum well problem.

If the angular momentum $l>0$, the problem can be analyzed for a large depth of the well, $|V_{\text{b}}|\rightarrow\infty$. In this case the function $\chi(r)$ defined by Eq. (3.59) is nonzero only in the interval $0\leq r\leq R$, where it obeys the following equation:

$$\frac{d^2\chi}{dr^2}+\left[k_{\text{w}}^2-\frac{l(l+1)}{r^2}\right]\chi(r)=0. \tag{3.64}$$

Introducing a dimensionless variable $z = k_{\mathrm{w}}r$ and the new function $\varphi = \chi(z)/\sqrt{z}$, we obtain the so-called Bessel equation,

$$\frac{d^2\varphi}{dz^2} + \frac{1}{z}\frac{d\varphi}{dz} + \left[1 - \frac{(l+1/2)^2}{z^2}\right]\varphi = 0.$$

This differential equation is known to have two linearly independent solutions, which are designated the *Bessel*, $J_{l+1/2}(z)$, and *Neumann*, $N_{l+1/2}(z)$, functions, respectively. Since the Neumann functions diverge as $z \to 0$, they are inadmissible as solutions in quantum mechanics, which means that the solutions to Eq. (3.64) are the Bessel functions, $J_{l+1/2}$. The first three such functions are given by

$$J_{1/2}(z) = \sqrt{\frac{2}{\pi z}}\sin z,$$

$$J_{3/2}(z) = \sqrt{\frac{2}{\pi z}}\left[\frac{\sin z}{z} - \cos z\right],$$

$$J_{5/2}(z) = \sqrt{\frac{2}{\pi z}}\left[\left(\frac{3}{z^2} - 1\right)\sin z - 3\frac{\cos z}{z}\right].$$

In general, the radial wave functions have the form

$$\mathcal{R}\{r\} = \sqrt{\frac{2\pi k_{\mathrm{w}}}{r}}J_{l+1/2}(k_{\mathrm{w}}r). \tag{3.65}$$

According to Eq. (3.57), the total wave functions include the spherical functions $Y_{l,m}(\theta, \phi)$. We present a few of the lowest-order spherical functions:

$$Y_{00} = \left(\frac{1}{4\pi}\right)^{1/2}, \quad Y_{11} = \frac{1}{2}\left(\frac{3}{2\pi}\right)^{1/2}\sin\theta\, e^{i\phi}, \quad Y_{1,-1} = \frac{1}{2}\left(\frac{3}{2\pi}\right)^{1/2}\sin\theta\, e^{-i\phi}.$$

The roots of the equation $J_{l+1/2}(k_{\mathrm{w}}R) = 0$ give the energy levels. Now they are dependent on the angular momentum l. In the theory of atomic spectra one refers to the spherical states with $l = 0$ as s states and the states with $l = 1, 2, 3, \ldots$ as p, d, f, ... states, respectively. An analysis of the roots of the Bessel functions gives us the following series of energy levels in spherical quantum dots:

$$1\mathrm{s}(2);\ 1\mathrm{p}(6);\ 1\mathrm{d}(10);\ 2\mathrm{s}(2);\ 1\mathrm{f}(14);\ 2\mathrm{p}(6);\ldots \tag{3.66}$$

The number in each set of parentheses shows the degeneracy of the states (electron spin is taken into account). One can compare this series with that of the electron in the Coulomb potential $V = -e^2/r$:

$$1\mathrm{s}(2);\ 2\mathrm{s}(2);\ 2\mathrm{p}(6);\ 3\mathrm{s}(2);\ 3\mathrm{p}(6);\ 3\mathrm{d}(10). \tag{3.67}$$

Thus, there is a similarity between the spectra of atoms and spherical quantum dots. Certainly, the high degree of degeneracy will normally be broken in any real situation, and many of the degenerate energy levels will split. The actual picture is thus quite complicated, but the same is true for the atomic spectra of all but the simplest atoms.

3.5.2 The Density of States for Zero-Dimensional Electrons

According to the definition of Eq. (3.26), in the case of quantum boxes or quantum dots the spectra are discrete; accordingly, the density of states is simply a set of δ-shaped peaks:

$$\varrho(E) = \sum_{\nu} \delta(E - E_{\nu}), \tag{3.68}$$

where $\nu = (n_1, n_2, n_3)$. For an idealized system, the peaks are very narrow and infinitely high, as illustrated in Fig. 3.20. In fact, interactions between electrons and impurities as well as collisions with phonons – quanta of the lattice vibrational modes – bring about a broadening of the discrete levels and, as a result, the peaks for physically realizable systems have finite amplitudes and widths. Nevertheless, the major trend of sharpening of the spectral density dependences as a result of lowering the system dimensionality is a dominant effect for near-perfect structures at low temperatures.

From the previous examples of a potential energy $V(x, y, z)$, which creates the quantum box, it is evident that the structure of the energy spectra of the boxes (dots) is complex, especially for large quantum numbers. For each particular case the spectrum can be calculated numerically. However, it is usually straightforward to make reasonable estimates for the total number of energy levels inside a quantum box or dot if this number is large.

Several times in the previous discussions we have exploited the number of states of one-dimensional free motion in a spatial region of length ΔL, in an interval of wavevectors Δk,

$$\Delta\varrho^{1D} = 2\frac{\Delta L\, \Delta k}{2\pi}. \tag{3.69}$$

One can generalize this expression to the three-dimensional case:

$$\Delta\varrho^{3D} = 2\frac{\Delta x\, \Delta k_x\, \Delta y\, \Delta k_y\, \Delta z\, \Delta k_z}{(2\pi)^3}. \tag{3.70}$$

Let us consider an arbitrary potential $V(x, y, z) < 0$. If one fixes some point of space $\vec{r}$ and calculates the number of states corresponding to the small volume

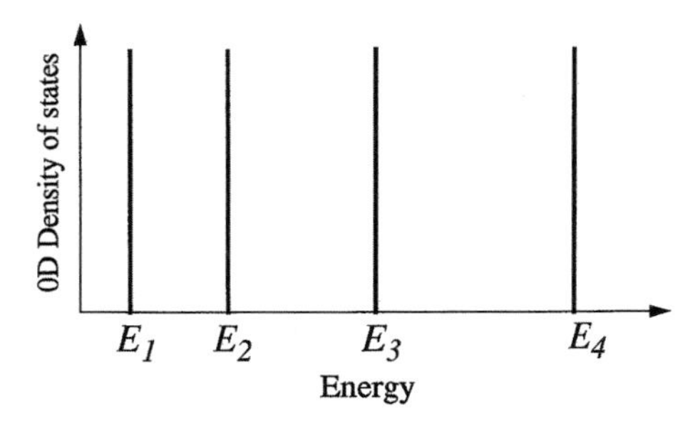

Figure 3.20 The density of states for zero-dimensional electrons.

$\Delta x\,\Delta y\,\Delta z$ around $\vec{r}$, one has to apply Eq. (3.70) and take into account all possible values of the electron wavevector within the region of confinement:

$$0 \leq k(\vec{r}) \leq k_{\max}(\vec{r}). \tag{3.71}$$

Here, $k_{\max}(\vec{r})$ can be defined from the semiclassical approach, for which the total energy is

$$E = \frac{\hbar^2 k^2}{2m^*} + V(\vec{r}). \tag{3.72}$$

For bound states $E \leq 0$, so $k_{\max}(\vec{r}) = \sqrt{2m^* |V(\vec{r})|/\hbar^2}$. Integration over all possible k which satisfy Eq. (3.71) gives the number of states in the volume $\Delta x\,\Delta y\,\Delta z$:

$$\Delta \varrho^{3D} = \frac{k_{\max}^3\, \Delta x\,\Delta y\,\Delta z}{3\pi^2}. \tag{3.73}$$

Now, by summing over all classically allowed electron coordinates, the total number of energy states inside a quantum box is found to be

$$N_t = \frac{2\sqrt{2} m^{*3/2}}{3\pi^2 \hbar^3} \int dx\, dy\, dz |V(\vec{r})|^{3/2}. \tag{3.74}$$

For example, for a box with potential given by Eq. (3.53) and with a finite potential depth, V_b, one obtains

$$N_t = \frac{2\sqrt{2} m^{*3/2}}{3\pi^2 \hbar^3} |V_b|^{3/2} L_x L_y L_z. \tag{3.75}$$

As a numerical example, taking $L_x = L_y = L_z = 100$ Å, $V_b = 0.2$ eV, and $m^* = 0.067m$, it follows that the total number of energy levels inside this box is $N_t = 75$. The actual number of electrons inside the box is less than N_t; the amount of this reduction is determined by the level of impurity doping. It is possible to control the number of localized carriers by applying an external voltage. Indeed, it is possible to change this number from a few electrons to tens of electrons.

3.6 Coupling between Quantum Wells

When we studied a single quantum well, we found that the electron wave function is nonzero in the barrier region. Accordingly, the electron has some finite probability of penetrating into the barrier layer of a heterostructure. This effect may be illustrated by calculating the total probability of finding the electron in the barrier layer:

$$\mathcal{P}_b = \int_{|z|>L/2} dz |\chi(z)|^2, \tag{3.76}$$

where the integration is performed over the barrier layer. The results are shown in Fig. 3.21. This "tunneling" behavior is one of the important manifestations of quantum mechanics and it leads to a range of physical phenomena which are exploited

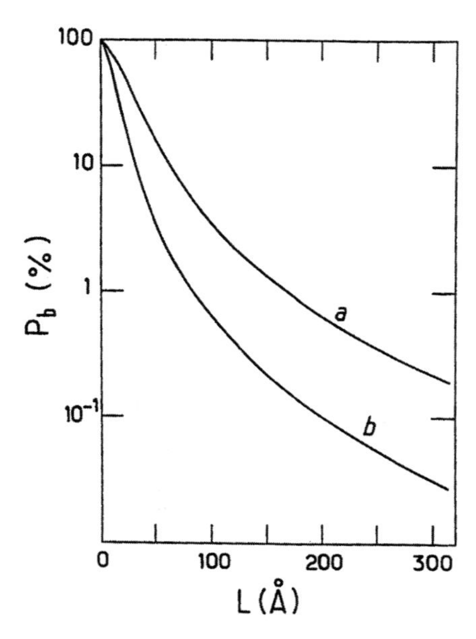

Figure 3.21 The probability of finding the electrons in the barrier layer as a function of the well thickness L. The electron state is assumed to be the ground state of the well. For curve a, $V_b = 224$ meV, $m^* = 0.067m$. For curve b, $V_b = 150$ meV, $m^* = 0.4m$. Republished with permission of John Wiley and Sons Inc., from Fig. 2 on page 5 of *Wave Mechanics Applied to Semiconductor Heterostructures*, G. Bastard, 1988; permission conveyed through Copyright Clearance Center, Inc.

in various electronic and optoelectronic devices. Next we will study the role of tunneling as a mean of providing the coupling between quantum wells.

3.6.1 Double-Quantum-Well Structures

Let us consider the *double-quantum-well* structure, for the case where the two wells, for simplicity, have identical rectangular shapes and are characterized by the same depth, V_b, and width, L. Suppose the distance between the wells is d, as in Fig. 3.22, and the wells are situated symmetrically with respect to the plane $z = 0$. Let us study the z-directed, one-dimensional propagation of the electron across the barrier. For such a *double-well system* the Schrödinger equation (A.27) of the transverse motion has the simple form

$$\left[-\frac{\hbar^2}{2m^*}\frac{\partial^2}{\partial z^2} + V(z + d/2) + V(z - d/2)\right]\chi(z) = \epsilon\chi(z). \tag{3.77}$$

The term $V(z + d/2)$ represents the potential of the left well centered at $z = -d/2$, while the term $V(z - d/2)$ corresponds to the right well. Here $V = V_b > 0$ in the barrier and V is zero inside the wells.

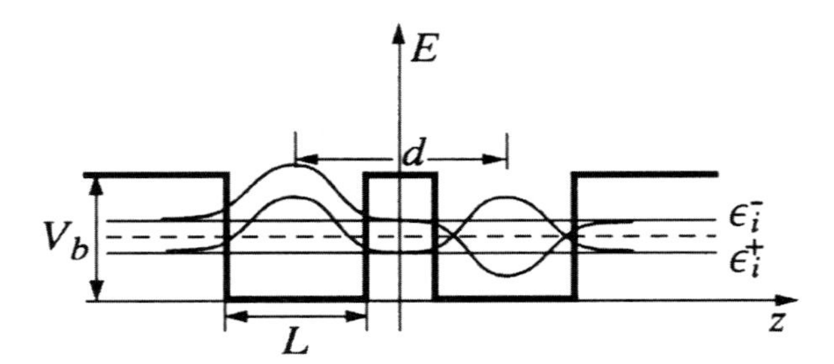

Figure 3.22 Energy levels and wave functions for a double-well heterostructure. The dashed line corresponds to the position of the energy level, ϵ_i, in the separate quantum wells. For coupled wells, the splitting of this level is marked by $\epsilon_i^\pm$. The wave functions of these levels are presented schematically.

One can represent the wave functions of the double-well system through a superposition of the solutions of the single-well problem, $\chi_\nu(z)$:

$$\chi(z) = \sum_\nu [A_\nu \chi_\nu(z - d/2) + B_\nu \chi_\nu(z + d/2)], \tag{3.78}$$

where ν runs over all possible bound and continuum states of the single-well problem. However, if the separation between wells is large enough that the expected tunneling from one well to another has a small probability, we can restrict ourselves to only one level, say, i, and take from Eq. (3.78) the combination

$$\chi(z) = A_i \chi_i(z - d/2) + B_i \chi_i(z + d/2). \tag{3.79}$$

This is the wave function for the electron states with energy near one of the bound states ϵ_i of the separated wells. Our goal now is to find the coefficients A_i and B_i. Because the distance d is supposed to be large, to a first approximation one can conclude that the electron state ϵ_i is two-fold degenerate. The electron in state i may occupy either the left well or the right well, with equal probability. In order to correct the energies and find the coefficients A_i and B_i, we use the following procedure of perturbation theory for degenerate states. We substitute the wave function of Eq. (3.79) into Eq. (3.77). Then we multiply this equation first by $\chi_i^*(z - d/2)$ to obtain one equation and then by $\chi_i^*(z + d/2)$ to obtain another equation. Next, we integrate the left-hand and right-hand sides of both of these equations and, as a result, we find two homogeneous linear equations for A_i and B_i:

$$(\epsilon_i + s_i - \epsilon)A_i + ((\epsilon_i - \epsilon)r_i + t_i)B_i = 0,$$

$$((\epsilon_i - \epsilon)r_i + t_i)A_i + (\epsilon_i + s_i - \epsilon)B_i = 0,$$

where r_i, s_i, and t_i are constants which will be defined precisely in the following discussion. This system has nonzero solutions only for the following two energies $\epsilon = \epsilon_i^\pm$:

$$\epsilon_i^\pm = \epsilon_i + \frac{s_i}{1 \pm r_i} \pm \frac{t_i}{1 \pm r_i}. \tag{3.80}$$

The quantities r_i, s_i, and t_i represent three integrals which characterize electron tunneling in a double-well structure:

$$\begin{aligned} r_i &\equiv \int_{-\infty}^{\infty} dz\, \chi_i^*(z-d/2)\chi_i(z+d/2), \\ s_i &\equiv \int_{-\infty}^{\infty} dz\, \chi_i^*(z-d/2)V(z+d/2)\chi_i(z-d/2), \\ t_i &\equiv \int_{-\infty}^{\infty} dz\, \chi_i^*(z-d/2)V(z-d/2)\chi_i(z+d/2). \end{aligned} \tag{3.81}$$

The first integral is called an "overlap integral," the second a "shift integral," and the last a "transfer integral." Owing to the weak penetration of the electron wave function into the barrier region, the overlap integral is small in comparison with 1, so we can neglect the corresponding terms in Eq. (3.80) and simplify the equations for the energies to yield the form

$$\epsilon_i^{\pm} = \epsilon_i + s_i \pm t_i. \tag{3.82}$$

The diagram in Fig. 3.23 depicts the shift of the energy levels and clarifies the meaning of the shift and transfer integrals. Since s_i and t_i are always less than zero, the first is responsible for shifting the levels down and leads to an energy decrease from the initial energy position of the single well. Likewise, the second determines the splitting of the initial two-fold degenerate level. For the lowest level, which is designated as ϵ_i^+ in Eq. (3.80), the combination in Eq. (3.79) is symmetric, i.e., $A_i = B_i$. The upper level, ϵ_i^-, is antisymmetric, with $A_i = -B_i$. From Eqs. (3.81) one can see that both shifting and splitting are caused by the tunneling phenomenon. The wave functions of these two-well states are sketched in Fig. 3.22. The same considerations are valid for any electron states, i, localized in coupled quantum wells.

Though the approximation of Eq. (3.79) is easy to interpret, its applicability is limited to the case of "thick" and "high" barriers. Meanwhile, our simple model of two identical rectangular wells can be solved exactly without any approximations. The method of solution is similar to that for the rectangular well. Here we present only the expression which defines the energies of the system:

$$2\cos(k_{\rm w}L) + \left(\frac{\kappa_{\rm b}}{k_{\rm w}} - \frac{k_{\rm w}}{\kappa_{\rm b}}\right)\sin(k_{\rm w}L) \pm \left(\frac{\kappa_{\rm b}}{k_{\rm w}} + \frac{k_{\rm w}}{\kappa_{\rm b}}\right)e^{-\kappa_{\rm b}d}\sin(k_{\rm w}L) = 0, \tag{3.83}$$

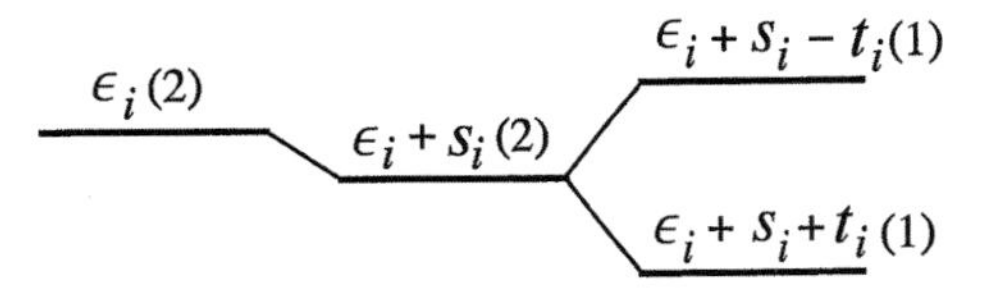

Figure 3.23 A diagram of an energy-level modification in a double-well heterostructure when the shift and transfer integrals are taken into account. The degeneracy of the levels is indicated in parentheses.

where $k_{\rm w}$ and $\kappa_{\rm b}$ are the same as in Eqs. (3.6) and (3.8). The minus and plus signs in the equation correspond, respectively, to the symmetric and antisymmetric solutions which were discussed previously. For thick barriers between wells, in which case the last term of Eq. (3.83) can be dropped, the equation becomes equivalent to Eqs. (3.12) and (3.14), that is, to the equations of the single-well problem. If the distance between the wells d is finite, the last term of the equation shows the splitting of each single-well level and reconstruction of the spectrum. When d is zero, a new set of levels coincides with that of the single-well problem for the case in which the well thickness is $2L$.

Note that Eq. (3.83) gives the energies of the transverse electron motion in a double-well structure. The total energies include the in-plane kinetic energy:

$$E^{(\pm)}_{i,\vec{k}_{||}} = \epsilon^{(\pm)}_i + \frac{\hbar^2}{2m^*}\left(k_x^2 + k_y^2\right).$$

The corresponding wave functions are

$$\Psi^{(\pm)}_{i,\vec{k}_{||}} = \frac{1}{\sqrt{2(1 \pm r_i)}} e^{i(k_x x + k_y y)} \left(\chi_i(z - d/2) \pm \chi_i(z + d/2)\right).$$

The major result of this section is the existence of a special type of interaction between quantum wells; that is, there is a coupling associated with tunneling between adjacent quantum wells. This interaction is caused by the quantum nature of electron propagation and shows that localized electrons can be found at the same time inside both wells. It also shows that the electrons are propagating in the plane x, y.

Let us discuss briefly this in-plane motion and a crossover to the classical case. If at some initial instant of time we localize the electron inside one of the wells by constructing some wave packet, this wave packet will not be an eigenfunction of the stationary equation (3.77). As a result, the wave packet initially localized in the left well will tunnel into the right well; next, the packet will tunnel back into the left well. This cyclic process will be repeated over and over again until some inelastic scattering process causes "collapse" of the wave packet. The characteristic time of the tunneling process, $\tau_{\rm t}$, can be calculated using the transfer integral, t_i, from Eq. (3.81). It is useful here to recall that in quantum physics there exists the uncertainty relation for time and energy: the accuracy of measurement of the energy, ΔE, during a finite time, Δt, must satisfy the following inequality: $\Delta E \times \Delta t \geq \hbar/2$. According to this uncertainty relation we can estimate $\tau_{\rm t}$ as $\tau_{\rm t} \approx \hbar/\Delta\epsilon \approx \hbar/t_i$. On the other hand, as just discussed, these quantum effects associated with free-particle propagation disappear if there is some scattering process. Let us characterize the scattering process by a relaxation time, $\tau_{\rm r}$. Thus, we can say that, if the tunneling time $\tau_{\rm t}$ is considerably less than the relaxation time, $\tau_{\rm r}$, then the electron belongs to both wells. In the opposite case, $\tau_{\rm t} \gg \tau_{\rm r}$, quantum effects vanish and the electron propagates through one of the wells with infrequent transitions to the other well.

The results obtained for the double-quantum-well problem will be exploited further to study the case of multiple-quantum-wells. But, before we close this section, we point out that double-quantum-well structures have attracted much attention because of interesting phenomena in electron parallel transport and various electronic and optoelectronic device applications, such as the velocity-modulated transistor, the generation of high-frequency microwave radiation, the unipolar cascade laser and others, which will be discussed in due course.

3.6.2 Two-Monolayer Crystals

Advanced techniques, which facilitate the production of monolayer two-dimensional crystals as well as their transfer onto different substrates, can be applied to fabricate crystals composed of two, three, etc., monolayers. Such artificial multi-monolayer systems can possess electronic properties significantly different from those of a single monolayer. Below we shall focus on a two-graphene-layer structure that is known as *bigraphene*.

Bilayer graphene is intermediate between graphene monolayers and bulk graphite. Two graphene layers can be deposited with different relative arrangements. We consider here the bilayer structure with so-called *AB stacking*, which is characteristic for 3D graphite (see Fig. 3.24). The distance between the nearest atoms in graphene is $a = 1.4\,\text{Å}$ and the interlayer distance is $d = 0.33\,\text{nm}$. The difference between these distances results in significant weakening of interlayer covalent coupling in comparison with intralayer covalent coupling. The tight-binding method can be used to calculate the energy-band structure of this carbon-atom bilayer. Besides the intralayer hopping parameter $\gamma_0 \approx 3.2\,\text{eV}$, which was defined for the monolayer graphene in Section 3.3.1, a set of similar parameters $\gamma_1, \ldots, \gamma_4$ can be introduced to describe the interlayer interaction. All these parameters are much less than γ_0. Estimates of these parameters have shown that the most important parameter is $\gamma_1 \approx 0.4\,\text{eV}$. On performing the tight-binding calculations with two parameters γ_0, γ_1 near the K, K$'$ points one obtains the following energy dispersions for the low-energy electronic structure of bilayer graphene:

$$E_{\text{c,v}}(p) = \pm\sqrt{\gamma_1^2/2 + U^2/4 + p^2 v_{\text{F}}^2 \pm \sqrt{\gamma_1^4/4 + p^2 v_{\text{F}}^2\left(\gamma_1^2 + U^2\right)}}, \qquad (3.84)$$

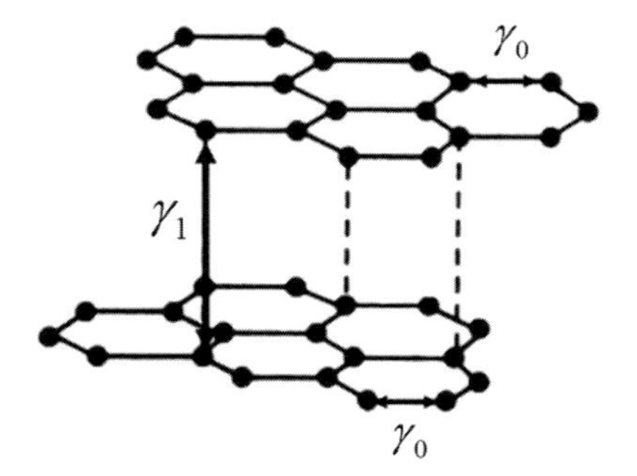

Figure 3.24 The lattice structure of bigraphene. γ_0 and γ_1 are the corresponding hopping parameters for the tight-binding model.

with v_F the Fermi-velocity parameter of the graphene defined via γ_0 (see Eq. (3.34)). Here, the voltage bias across the layers, U, is introduced to account for control of the energy dispersion by external fields. As follows from Eq. (3.84), in the bigraphene structure there are four bands – two conduction bands, $E_c^{1,2}(p)$, and two valence bands, $E_v^{1,2}(p)$.

First consider the case when the external voltage is absent, $U=0$. The energy dispersion curves of pristine bigraphene are presented in Fig. 3.25(a). Note that the lower conduction band, $E_c^1(p)$, and the upper valence band, E_v^1, coincide at $p \to 0$, i.e., bigraphene is a *gapless material*:

$$E_c^1 \approx \frac{v_F p^2}{\gamma_1}, \quad E_v^1 \approx -\frac{v_F p^2}{\gamma_1}. \tag{3.85}$$

For the two other bands, we obtain

$$E_c^2 \approx \gamma_1 + \frac{v_F p^2}{\gamma_1}, \quad E_v^2 \approx -\gamma_1 - \frac{v_F p^2}{\gamma_1}. \tag{3.86}$$

All four bands are of parabolic character, with effective masses $|m^*| = \gamma_1/2v_F^2$. Since v_F and γ_1 were estimated, we find that $|m^*| \approx (0.03\text{–}0.05)m$, i.e., the effective masses in bigraphene are small. Equations (3.85) and (3.86) are valid for small momenta: $p \ll g/v_F$. In the opposite limit, $p \gg \sqrt{g}/v_F$, it is found that $E_{c,v}^{1,2} \propto v_F p$, exactly as in the monolayer case. From the general result of Eq. (3.84) and Eqs. (3.85) and (3.86), it can be seen that the interlayer coupling γ_1 plays the major role in the formation of the bigraphene electronic spectra.

The method which we have applied to obtain hybridization of the electronic states of two graphene layers is very similar to that used in the previous subsection for double-well structures. The interlayer hopping parameter γ_1 plays just the same role as the transfer integral t_i defined in Eq. (3.81). Because in each of the initial carbon monolayers two electronic states, E_c, E_v, were taken into account, four states were obtained – two conduction bands and two valence bands, with energies given by Eqs. (3.85) and (3.86).

As has been emphasized, bigraphene should be considered as a single two-dimensional system quite distinct from double-layer graphene, which can be a composite system consisting of two single layers of graphene, separated by some distance. An example is graphene–boron nitride–graphene with one or several monolayers of dielectric boron nitride. The energy dispersion in such double-layer graphene is still massless Dirac-like, because the interlayer coupling is weak: the interlayer distance is a few nanometers and the interlayer binding is of the van der Waals type instead of covalent binding. In contrast, bigraphene has a quadratic band dispersion with a fixed thickness (interlayer separation) of 0.3 nm similar to that of graphite.

Now consider the case when the bigraphene crystal is placed in an external electric field perpendicular to the crystal. As a result, some potential difference arises between the two graphene monolayers. Define the potential difference as U. Equation (3.84) is relevant to just this situation. According to this equation, a dramatic reconstruction of the electron energy spectra of the bigraphene occurs, as illustrated

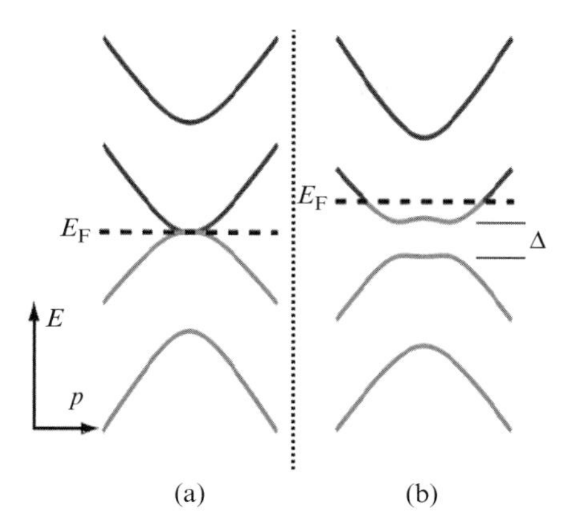

Figure 3.25 Conduction and valence band energy dispersions for pristine bigraphene near the K, K′ points. (a) Pristine bigraphene. (b) Bigraphene in an electric field. The Fermi level is indicated.

in Fig. 3.25(b). First, the electric field induces a finite bandgap between the lower conduction band and the upper valence band, Δ. It can be easily found that for a low voltage, $U \ll \gamma_1$. Indeed, under this criterion Eq. (3.84) is simplified to the form

$$E_{c,v} = \pm \left[U - \frac{2Uv_F^2}{\gamma_1^2} p^2 + \frac{v_F^4 p^4}{2U\gamma_1^2} \right]. \tag{3.87}$$

From this expression it follows that the minimal separation between the two bands is

$$\Delta = 2U - \frac{4U^3}{\gamma_1^2}; \tag{3.88}$$

this occurs at $p = \sqrt{2}U/v_F$. A second feature of these spectra is that the lower branch of the conduction band and the upper branch of the valence band have nonmonotonous dependences and there are portions of these dependences with so-called *negative* effective masses. The negative mass means that a particle moves against an applied force. Generally, the existence of charge carriers with negative effective mass can bring about an instability in electric fields and can be used for the generation of high frequencies.

It is easy to estimate the amplitude of the electric field F necessary to generate a given energy bandgap in the bigraphene, as follows. If the distance between the graphene layers is d, the potential difference is $U = dF$; U can be found from Eq. (3.84), if the energy bandgap Δ is specified. For example, on specifying $\Delta = 0.2$ eV we find $U = 0.25$ V. Using $d = 0.33$ nm, we obtain the necessary field $F = 7.6 \times 10^6$ V/cm. Currently, for bigraphene a bandgap of about 0.3 eV has been experimentally realized.

In conclusion, bigraphene possesses the rich physical properties of low-energy excitations. The energy scale of sufficiently modified excitations is determined by the interlayer hopping parameter γ_1, i.e., it is about 0.3 eV. Being a stable two-dimensional crystal, bigraphene is considered to be very important for nanoelectronic and far-infrared optoelectronic applications because of the possibility of controlling its properties, in particular, the ability to open a bandgap by applying an electric field.

3.7 Superlattices

Modern semiconductor technology facilitates the fabrication not only of single- or double-quantum-well structures, but also of perfectly regular periodic systems of layers. Each layer can act either as a potential well or as a barrier for electrons. If these layers alternate and the number of them is large enough, the system can be considered an artificial one-dimensional lattice with a period exceeding the period of the bulk crystal lattice. Such heterostructures are called *superlattices*. Superlattices have many interesting features and are used in different types of electronic and optoelectronic devices. In this section, we will study the basic properties of superlattices.

An example of a superlattice is shown in Fig. 3.26, where the wells and the barriers have simple rectangular forms. The theory of real superlattices is not simple, because real wells and barriers have complex profiles. In addition, the effective masses can be different in the wells and barriers. Moreover, as the thicknesses of wells and barriers are reduced to a few atomic monolayers, the commonly used effective-mass approximation ceases to be applicable.

To understand the basic behavior of superlattice structures, we will use the experience accumulated in dealing with the problem of periodic structures in crystals and then illustrate the major properties of the structure presented in Fig. 3.26.

This structure can be described by the Kronig–Penney model, which is a principal model for one-dimensional crystals in solid-state physics. The only essential difference is that we need to modify the Kronig–Penney model by using the effective mass instead of the free-electron mass. According to the Kronig–Penney model, any energy level of a single well, ϵ_i, splits into a series of N levels, where N is the number of the wells in the superlattice, as depicted in Fig. 3.27(a). The physical reason for this splitting may be viewed as being electron tunneling among the wells, as was described in detail in the previous section. For an N-well system with high and thick barriers between the wells, to a first approximation the energy levels coincide with

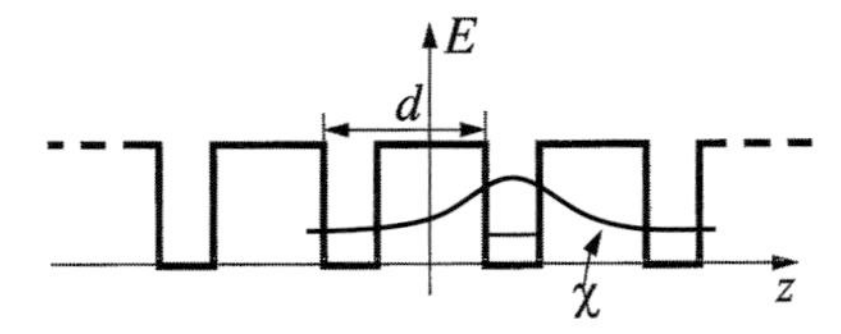

Figure 3.26 A schematic representation of the potential energy of a superlattice with period d. An energy level and a wave function are shown for a single well.

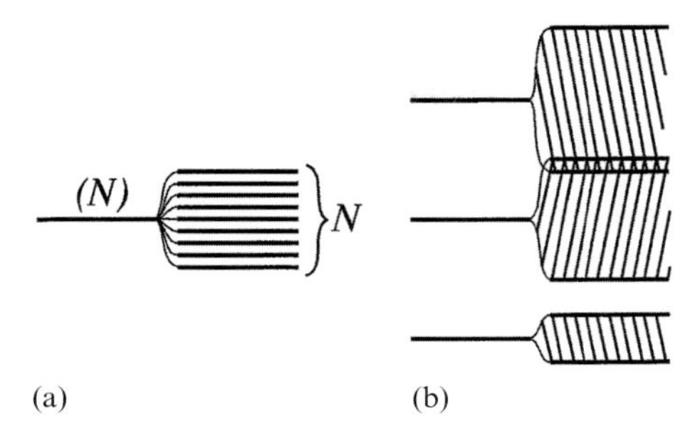

Figure 3.27 Formation of superlattice minibands: (a) splitting of a single level of separated quantum wells into an energy band due to tunneling coupling; and (b) minibands originating from different energy levels of separated wells.

their single-well positions, but they are N-fold degenerate. Tunneling between the wells breaks the degeneracy and results in a splitting into N levels with the wave functions common for all the wells. This implies that in any of these states one can find electrons inside any well. It is worth noting that the width of the energy region occupied by these levels is independent of the number of wells N, if N is large. This energy region is the so-called *energy band* and it depends only on the tunneling characteristics of the system. Generally, energy bands appear for any system which exhibits this tunneling-induced coupling and it does not require periodicity. Indeed, disordered condensed materials are also characterized by energy bands. However, periodicity brings about additional nontrivial behavior of the system. Indeed, periodicity leads to the condition that the physical characteristics of the system do not change when the electron is shifted by exactly one period or an integral number of periods. Applying this requirement to the wave function, we get a direct analog of the Bloch theorem (see Eqs. (2.4) and (2.11)):

$$\chi(z+nd) = C_n \chi(z) = e^{iqnd} \chi(z), \tag{3.89}$$

where d is the period of the superlattice. The latter equation introduces a new parameter in the wave function, namely, the one-dimensional wavevector q. Wave functions with different values of q obey different transformations in accordance with Eq. (3.89). Therefore wave functions corresponding to different states of the same energy band are characterized by different values of q. Thus, q is a new quantum number of the system. In addition, we can introduce a hypothetical length of the superlattice, Nd, and exploit the cyclic boundary conditions

$$\chi(z+Nd) = \chi(z). \tag{3.90}$$

This yields the relationship $e^{iqNd} = 1$ and $q = (2\pi/Nd)p,\ p = 1, 2, \ldots, N$. The latter relationship implies that the total number of quantum states arising as a result of the coupling of the N identical, uncoupled single-well levels is the same as the number of wells in the superlattice. The spacing between neighboring values of q is equal to

$2\pi/Nd$. It is customary to introduce the following interval for q:

$$\left\{-\frac{\pi}{d}, \frac{\pi}{d}\right\}, \tag{3.91}$$

which gives the *first Brillouin zone* for the electrons in a superlattice.

As just discussed, the energy band originates from the N isolated levels of the N single quantum wells as the well separation is decreased and the tunneling coupling among the wells is thereby increased. Now it is clear that each energy level of an isolated single quantum well evolves into an energy band containing N levels when N such isolated quantum wells are coupled through tunneling effects to form a superlattice; see Fig. 3.27(a). Owing to the periodicity each band is characterized by a one-dimensional wavevector q, so that for the ith energy level we obtain the band $\epsilon_i(q)$. Since the tunneling probability increases with increasing level index, i, the width of the bands also increases with i. The narrowest band is the lowest one. Frequently, the higher bands in superlattices overlap. This situation is sketched in Fig. 3.27(b).

Important features of superlattices are the bandgaps and the new man-made dispersion curves $\epsilon_i(q)$ which characterize the energy states of the superlattice. The bandgaps, of course, define the energy intervals where propagating states, the Bloch waves, do not exist. In these intervals electrons cannot propagate in the superlattice.

3.7.1 Wave Functions and Energy Dispersion in Superlattices

The dispersion relation, $\epsilon_i(q)$, for the model presented in Fig. 3.26 can be calculated in the so-called tight-binding approximation when the Bloch wave functions are constructed from the wave functions of the single-well problem with the potential of Eq. (3.1). Thus, the wave function for the superlattice band i which originates from the energy level i can be written as

$$\psi_{i,q}(z) = \frac{1}{\sqrt{N}} \sum_{n=1} e^{iqnd} \chi_i(z - nd). \tag{3.92}$$

Substituting the wave function of Eq. (3.92) into the Schrödinger equation with the superlattice potential $V(z) = \sum_n V(z - nd)$ and taking into account the interactions of just the nearest-neighbor wells, it follows that the energy of the ith subband is

$$\epsilon_i(q) = \epsilon_i + s_i + 2t_i \cos(qd), \tag{3.93}$$

where s_i and t_i are given by Eqs. (3.81). The dispersion relation $\epsilon_i(q)$ is shown schematically in Fig. 3.28(a) (note that t_i is negative). One can compare this result with those for the double-well problem studied previously. The dispersion curve, $\epsilon_i(q)$, reveals the downshift in energies, which is twice that for the double-well problem: in the present case, the two neighboring wells for a given well are taken into account. Instead of the previous splitting into two levels with energies separated by $2|t_i|$, there is an entire continuum energy band of width $\Delta\epsilon_i = 4|t_i|$. One can see that the width does not depend on the number of wells in the superlattice. The last term of Eq. (3.93) gives us the dispersion relation for the ith energy band. These

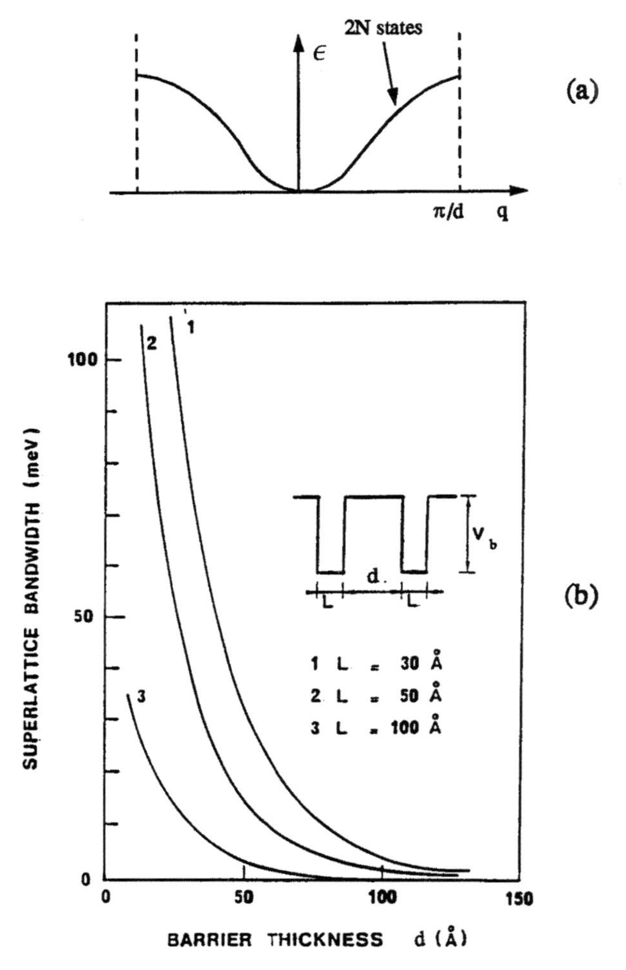

Figure 3.28 (a) The dispersion relation of electrons in a superlattice, Eq. (3.93). (b) The bandwidth of the lowest miniband for the GaAs/$Al_xGa_{1-x}As$ superlattice as a function of the barrier thickness ($x = 0.2$, $V_b = 212$ meV). Part (b) after Fig. 5 in G. Bastard, *Acta Electron.* **25**, 147 (1983).

superlattice bands are frequently referred to as *minibands*. The bandwidths of minibands are finite and determined by the transfer integral, and they depend strongly on the superlattice parameters as illustrated in Fig. 3.28(b).

So far, we have studied only the transverse propagation of an electron, i.e., propagation along the axis of the superlattice. The total electron energy including the energy associated with the in-plane propagation is

$$E_i(\vec{k}, q) = \epsilon_i + s_i + \frac{\hbar^2 k_{||}^2}{2m^*} - 2|t_i| \cos(qd), \tag{3.94}$$

where $\vec{k}_{||}$ is the two-dimensional wavevector in the x, y plane. This formula demonstrates how dramatically the electron spectrum is modified and controlled by an artificial lattice.

Near the bottom and the top of the minibands it is possible to simplify further the energy spectrum of Eq. (3.94). Using the series expansion of $\cos(qd)$ leads to the following approximation around the point $q=0$:

$$E_i(\vec{k},\ q) = (\epsilon_i + s_i - 2|t_i|) + \frac{\hbar^2 q^2}{2M_i} + \frac{\hbar^2 k_{||}^2}{2m^*}, \tag{3.95}$$

where $M_i = \hbar^2/(2d^2|t_i|)$. The effective mass, M_i, in the z-direction is determined by the transfer integral, in accordance with the tunneling character of the transverse propagation. Equation (3.95) reveals that the electron propagation is extremely anisotropic. The anisotropy is controlled by the thickness of the layers and the height of the barriers. Another interesting aspect of the dispersion curve of Eq. (3.94) is the existence of a portion of the dispersion curve with *negative effective mass* where the second derivative of ϵ_i with respect to q changes its sign at $q=\pi/2d$. Near the top of the miniband, the effective mass associated with transverse propagation is equal to $-M_i$. The existence of this negative effective mass means that the electron is propagating in a direction opposite to the applied force.

Let us estimate the bandwidth and effective mass for the miniband. Assuming $L=5$ nm and $d=10$ nm we find $\Delta\epsilon_i = 4t \approx 15$ meV. Hence, $M_i = 4.4m$, where m is the free-electron mass, i.e., the mass of an electron in a superlattice is roughly one order of magnitude larger than typical effective masses of bulk crystals, while the widths of the minibands are relatively small. This is why low temperatures are preferable for applications of miniband effects.

The finite bandwidth allows the possibility of observing and applying novel effects that take place in any periodic system, when an electric field is applied. If there is no electron scattering, the electron gains energy in an electric field until it reaches the top of the band, where q reaches the boundary of the Brillouin zone. Then, the electron "reflects" back and continues its propagation in the opposite direction, decelerating until it reaches the bottom of the band. After that, it accelerates again. This process repeats again and again. Thus, an electron in a miniband of finite bandwidth oscillates in real space and in momentum space. These oscillations are known as *Bloch oscillations*. Electron scattering destroys these oscillations. The oscillations of charged electrons lead to the emission of electromagnetic radiation. In the quantum picture discussed previously, the energy miniband is composed of a set of discrete energy levels. In the presence of an electric field, band bending lifts the energy degeneracies that produce the superlattice minibands, and the energy spectrum of a superlattice with large N becomes an energy ladder of N discrete levels. Such an energy spectrum is known as a *Stark ladder*. Transitions between the levels lead to emission or absorption of electromagnetic radiation. These processes have been known for many decades, but in bulk crystals the bandwidths are so large that electrons cannot reach the top of the band in the ballistic regime, i.e., without scattering. Because the bandwidth can be controlled and ballistic propagation along the superlattice axis can occur, superlattices provide a unique possibility for observing these effects.

3.7.2 The Density of States

Let us calculate the density of states for superlattices, $\varrho(E)$, where E is the total electron energy; we shall apply Eq. (3.26). The set of quantum numbers for an electron state includes the spin, s, the quantum number of the superlattice band, i, the wavevector along the axis of the superlattice, q, and the two-dimensional in-plane wavevector, $\vec{k}_{||}$. Thus, from Eqs. (3.26) and (3.94) we obtain

$$\varrho(E) = \sum_{s,i,q,\vec{k}_{||}} \delta\left[E - \epsilon_i - s_i + 2|t_i|\cos(qd) - \frac{\hbar^2 k_{||}^2}{2m^*}\right]. \tag{3.96}$$

The sums over s and $\vec{k}_{||}$ can be calculated in the manner discussed in previous sections; the result is (compare with Eq. (3.30))

$$\varrho(E) = \frac{L_x L_y m^*}{\pi\hbar^2} \sum_{i,q} \Theta\left[E - \epsilon_i - s_i + 2|t_i|\cos(qd)\right]. \tag{3.97}$$

Hence, when the total energy is less than that at the bottom of the ith miniband, $E < \epsilon_i + s_i - 2|t_i|$, the contribution to the density of states from the ith miniband is zero. For $E > \epsilon_i + s_i + 2|t_i|$ the contribution to $\varrho(E)$ from the ith miniband is equal to the total number of states of the miniband, $\varrho_0(E)$, i.e., the number of wells in the superlattice, N, multiplied by the density of states in one well, as depicted in Fig. 3.29. For intermediate situations, one has to integrate the Θ-function over q in the region $-q_E < q < q_E$, where

$$q_E = \frac{1}{d}\arccos\left[\frac{\epsilon_i + s_i - E}{2|t_i|}\right].$$

As a result, one obtains the contribution of the ith miniband to the density of states in the form

$$\Delta\varrho_i = \frac{L_x L_y m^*}{\pi\hbar^2} N \begin{cases} 0, & E < \epsilon_i + s_i - 2|t_i|, \\ (1/\pi)\arccos[(\epsilon_i + s_i - E)/2|t_i|], & \epsilon_i + s_i - 2|t_i| < E < \epsilon_i + s_i + 2|t_i|, \\ 1, & E > \epsilon_i + s_i + 2|t_i|. \end{cases} \tag{3.98}$$

Figure 3.29 illustrates the energy dependence of $\varrho(E)$. One can see that the density of states in the superlattice exhibits a more complex form than the staircase of a single quantum well. The propagation of the electron across the barriers makes each step of the staircase smoother in comparison to that of N noncoupled single quantum wells.

We have considered a very long superlattice; indeed, the previous discussion has been based on the assumption that $N \to \infty$. For most multiple-quantum-well systems about 10 quantum wells are necessary for the system to behave as a superlattice.

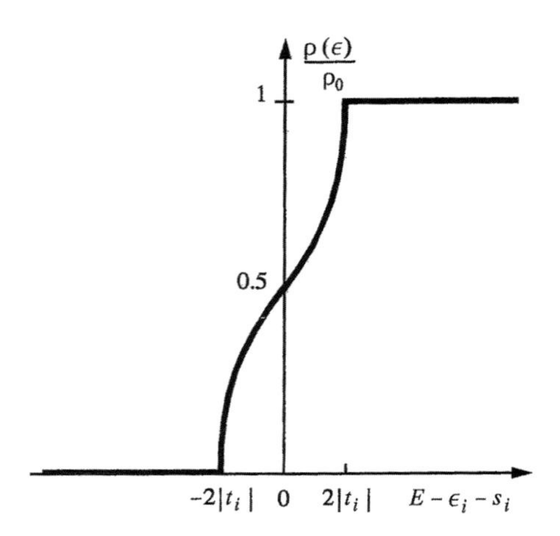

Figure 3.29 The density of states ϱ of a superlattice as a function of the electron energy; $\varrho_0 = m^* NL_xL_y/\pi\hbar^2$. Republished with permission of John Wiley and Sons Inc., from Fig. 12 on page 26 of *Wave Mechanics Applied to Semiconductor Heterostructures*, G. Bastard, 1988; permission conveyed through Copyright Clearance Center, Inc.

In particular, the widths of the minibands saturate for superlattices having $N > 10$. It must be emphasized again that, if the quantum wells and the barriers contain only a few atomic layers, the effective-mass approximation is not applicable for the description of the transverse electron motion in superlattices.

3.8 Excitons in Quantum Structures

Thus far in this chapter, we have studied the one-electron properties of quantum heterostructures and have neglected the interaction between carriers. However, this interaction is important and leads to fundamentally new properties of the materials and heterostructures. In this section we begin to study many-particle effects as embodied in the simplest two-particle case of excitons. These excitons are of great importance for most of the electrical and optical properties of semiconductors and devices. Indeed, our brief discussion of excitons in Chapter 2 reveals that excitonic effects can be pronounced in two-dimensional material structures.

3.8.1 Excitons

As we know already, the usual energy-band structure of bulk materials consists of conduction and valence bands separated by a bandgap, E_g. In dealing with the optical spectra of semiconductors (see Fig. 3.30), this energy can be associated with the fundamental edge of the spectrum because phototransitions between the bands are possible only when the photon energy $\hbar\omega > \hbar\omega_g = E_g$; here ω_g is the threshold frequency corresponding to the *fundamental absorption edge*. The band-to-band

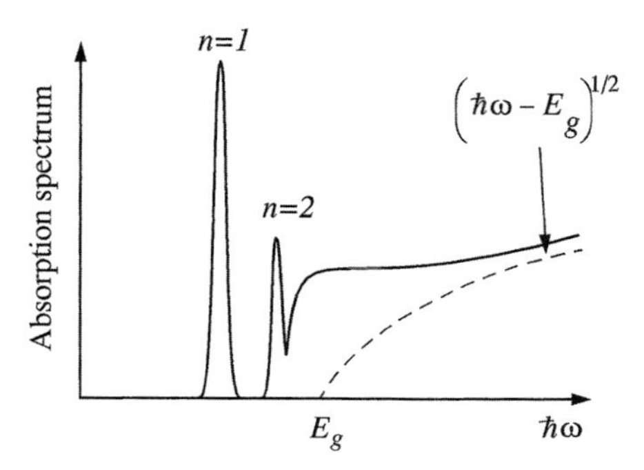

Figure 3.30 Optical spectra of a semiconductor near the fundamental energy-band edge at E_g.

absorption spectrum is shown by the dashed line in Fig. 3.30 for the case when the Coulomb interaction of an electron and hole is disregarded. In fact, one usually observes a series of more or less narrow lines at frequencies $\omega < \omega_g$, which are especially pronounced at low temperatures. They are represented by solid lines in Fig. 3.30. These lines are attributed to *excitons*. An exciton is a quasi-particle, which is formed by the Coulomb interaction between an electron and a hole. This coupled pair can propagate through the crystal as a single particle. An exciton also has internal degrees of freedom, associated with the relative distance between the electron and hole composing the exciton. These internal degrees of freedom are quantized, i.e., there are different quantum states of the exciton, namely the lowest energy states, which correspond to a fixed distance between the electron and hole, and a number of excited states. This quantization produces the previously mentioned series of peaks in the optical spectrum below the fundamental energy band edge. The exciton energies depend on the energy spectra of the electrons and holes, particularly on their effective masses, the dielectric constants of the materials, etc. This results in different exciton energies for different crystals. This picture is idealized and it is applicable to such crystals as Si and Ge as well as to III–V and II–VI compounds, etc. Furthermore, Fig. 3.30 clearly shows changes in the absorption spectrum above the fundamental edge at $\hbar\omega > E_g$. These changes are due to the Coulomb correlation of free (uncoupled) electrons and holes generated under photon absorption.

An exciton is analogous to the bound state of an electron at a charged impurity, because the interaction has the same Coulomb character. But, in contrast to the impurity case, all states of the electron–hole pair are excited states. Indeed, the ground state of a crystal corresponds to the valence band being filled entirely by electrons and there are, accordingly, no electrons in the conduction band and no holes in the valence band, as shown in Fig. 3.31. Any state of the crystal with a hole in the valence band and an electron in the conduction band is an excited state, i.e., the exciton corresponds to an excitation of the crystal. Therefore, as an excited state of the crystal, the exciton has a finite lifetime.

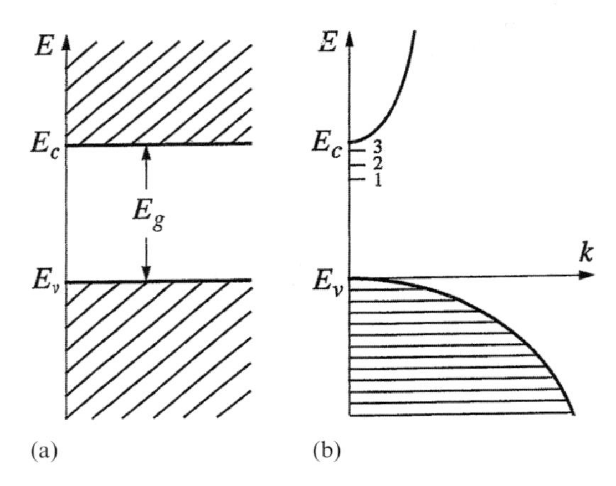

Figure 3.31 (a) Conduction and valence energy bands of a crystal. (b) Dispersion dependences for electrons and holes, and exciton levels, $1, 2, 3$. In the ground state of a crystal all energy states of the valence band are filled, while all states of the conduction band are empty.

We start from the simplest bulk material band structure as depicted in Fig. 3.31, where both the valence band and the conduction band are characterized by simple parabolic dispersion relations:

$$E_{\mathrm{c}}(k) = E_{\mathrm{g}} + \frac{\hbar^2 k_{\mathrm{e}}^2}{2m_{\mathrm{e}}} \quad \text{and} \quad E_{\mathrm{v}}(k) = -\frac{\hbar^2 k_{\mathrm{h}}^2}{2m_{\mathrm{h}}}, \tag{3.99}$$

where m_{e} and m_{h} are the effective masses at the center of the conduction and valence bands, respectively.

Let us take one electron from the completely filled valence band and put it in the conduction band. This leads to an excited state of the crystal. The process corresponds schematically to the following transition: $(v, k_{\mathrm{v}}) \to (c, k_{\mathrm{c}})$, where k_{v} and k_{c} are the wavevectors of the electron in the valence band and in the conduction band, respectively. An empty place in the valence band may be considered to be a hole with positive mass m_{h}, positive charge, and wavevector $k_{\mathrm{h}} = -k_{\mathrm{v}}$. The energy of the excited state is thus

$$E_{\mathrm{g}} + E_{\mathrm{c}}(k_{\mathrm{c}}) + |E_{\mathrm{v}}(k_{\mathrm{h}})| \equiv E_{\mathrm{g}} + \frac{\hbar^2 k_{\mathrm{e}}^2}{2m_{\mathrm{e}}} + \frac{\hbar^2 k_{\mathrm{h}}^2}{2m_{\mathrm{h}}}. \tag{3.100}$$

Equation (3.100) does not include the following Coulomb interaction energy between the electron and the hole:

$$V_{\mathrm{Cl}} = -\frac{e^2}{\kappa\, |\vec{r}_{\mathrm{e}} - \vec{r}_{\mathrm{h}}|},$$

where κ is the dielectric permittivity of the crystal, $\vec{r}_{\mathrm{e}}$ is the coordinate of the electron, and $\vec{r}_{\mathrm{h}}$ is the coordinate of the hole.

The pair of particles, electron and hole, can be described by *the two-particle Schrödinger equation*:

$$(H_{\mathrm{e}} + H_{\mathrm{h}} + H_{\mathrm{Cl}})\Psi(\vec{r}_{\mathrm{e}}, \vec{r}_{\mathrm{h}}) \equiv \left(\frac{\hat{\vec{p}}_{\mathrm{e}}^{\,2}}{2m_{\mathrm{e}}} + \frac{\hat{\vec{p}}_{\mathrm{h}}^{\,2}}{2m_{\mathrm{h}}} + V_{\mathrm{Cl}} \right) \Psi(\vec{r}_{\mathrm{e}}, \vec{r}_{\mathrm{h}})$$
$$= (E - E_{\mathrm{g}})\Psi(\vec{r}_{\mathrm{e}}, \vec{r}_{\mathrm{h}}), \tag{3.101}$$

where $H_{\mathrm{e,h}} = \hat{\vec{p}}_{\mathrm{e,h}}^{\,2}/(2m_{\mathrm{e,h}})$ are the Hamiltonians of the free electron and hole, respectively. The momentum operators are $\hat{\vec{p}}_{\mathrm{e,h}} = -i\hbar\, \partial/\partial \vec{r}_{\mathrm{e,h}}$. It is easy to see that the total Hamiltonian of the free electron and hole has the energy given by Eq. (3.100) as an eigenvalue.

Let us introduce the relative coordinate $\vec{r}$,

$$\vec{r} = \vec{r}_{\mathrm{e}} - \vec{r}_{\mathrm{h}}, \tag{3.102}$$

and the coordinate $\vec{R}$ of the center of mass of the electron–hole pair:

$$\vec{R} = \frac{m_{\mathrm{e}}\vec{r}_{\mathrm{e}} + m_{\mathrm{h}}\vec{r}_{\mathrm{h}}}{m_{\mathrm{e}} + m_{\mathrm{h}}}. \tag{3.103}$$

Then we can rewrite the Schrödinger equation (3.101) in the form

$$\left(\frac{\hat{\vec{P}}^{\,2}}{2(m_{\mathrm{e}} + m_{\mathrm{h}})} + \frac{\hat{\vec{p}}^{\,2}}{2m_{\mathrm{r}}} - \frac{e^2}{\kappa r} \right) \Psi(\vec{r}, \vec{R}) = (E - E_{\mathrm{g}})\Psi(\vec{r}, \vec{R}), \tag{3.104}$$

where we have taken

$$m_{\mathrm{r}} = \frac{m_{\mathrm{e}} m_{\mathrm{h}}}{m_{\mathrm{e}} + m_{\mathrm{h}}}, \quad \hat{\vec{P}} = -i\hbar \frac{\partial}{\partial \vec{R}}, \quad \text{and} \quad \hat{\vec{p}} = -i\hbar \frac{\partial}{\partial \vec{r}}. \tag{3.105}$$

Here, m_{r} is the reduced mass of the electron–hole pair, $\hat{\vec{P}}$ is the momentum operator of the pair as a whole, and $\hat{\vec{p}}$ is the internal momentum operator of the electron–hole pair.

The Schrödinger operator is the sum of terms which are functions of either $\vec{r}$ or $\vec{R}$ only. This is why the internal and center-of-mass variables are separable and the wave function is a product:

$$\Psi = \frac{1}{\sqrt{V}} \exp(i\vec{K}\vec{R})\psi(\vec{r}). \tag{3.106}$$

The total exciton energy, E, is a sum:

$$E = E_{\mathrm{g}} + \frac{\hbar^2 K^2}{2(m_{\mathrm{e}} + m_{\mathrm{h}})^2} + E_{\mathrm{ex}}, \tag{3.107}$$

where the second term is the kinetic energy of the exciton. The important feature of the system under consideration is that the exciton with effective mass $(m_{\mathrm{e}} + m_{\mathrm{h}}) = M^{\mathrm{ex}}$ *can propagate in the crystal as a plane wave* with a wavevector $\vec{K}$.

The exciton internal interaction energy, E_{ex}, and the wave function ψ are related by the equation

$$\left[-\frac{\hbar^2 \nabla_r^2}{2m_r} - \frac{e^2}{\kappa r}\right]\psi(r) = E_{ex}\psi(r), \tag{3.108}$$

where ∇_r is an operator in the relative coordinate $\vec{r}$. Equation (3.108) is similar to the well-known equation for the hydrogen atom. The only difference is that the reduced mass of the electron–nucleus system is replaced by the reduced mass of the electron–hole pair, m_r. Therefore, the solution to the hydrogen-atom problem may be adopted and it follows immediately that the exciton wave function of the ground state has the form

$$\psi(r) = \frac{1}{\sqrt{\pi (a_0^{ex})^3}} \exp(-r/a_0^{ex}),$$

where the exciton radius and ground-state energy are

$$a_0^{ex} = \frac{\hbar^2 \kappa}{m_r e^2} \quad \text{and} \quad E_{ex} \equiv -Ry^{ex}, \text{ where } Ry^{ex} = \frac{m_r e^4}{2\kappa^2 \hbar^2}. \tag{3.109}$$

To further exploit the analogy with the hydrogen atom, the Rydberg constant $Ry = me^4/2\hbar^2$, which determines the ionization energy of the hydrogen atom, is replaced by an effective Rydberg constant Ry^{ex} appropriate to the solid-state case of an impurity or exciton. From Eq. (3.109) it follows that the exciton radius a_0^{ex} is proportional to the dielectric constant κ and inversely proportional to the reduced effective mass. These results predict a large value of a_0^{ex}, which is generally greater than 100 Å for III–V compounds.

To give an idea of the values of the exciton energies, in Fig. 3.32 the energies of the lowest (ground) exciton states, $E_0 = |E_{ex}|$, are presented versus the bandgap energy for different bulk semiconductors. The electron effective mass increases with increasing bandgap; this explains the increase in the exciton ground-state energy with increasing bandgap. As emphasized in Chapter 2, excitonic effects are pronounced for many two-dimensional materials.

3.8.2 Excitons in Quantum Wells

The results of the previous subsection corresponded to the case of a bulk crystal. Let us apply the same approach to the study of excitons in double heterostructures. To be specific, we consider a structure which consists of a layer of material B embedded between two semi-infinite materials A. For excitons in bulk crystals it was found that the properties of the exciton depend strongly on the characteristics of the valence and conduction bands. In the case of a heterostructure, the excitonic properties depend on the parameters in both materials, A and B. There exist at least two types of energy-band diagram in heterostructures.

In type-I heterostructures the conduction band discontinuity $V_{b,e}$, or the *band offset*, and that of the valence band, $V_{b,h}$, are such that electrons and holes are confined

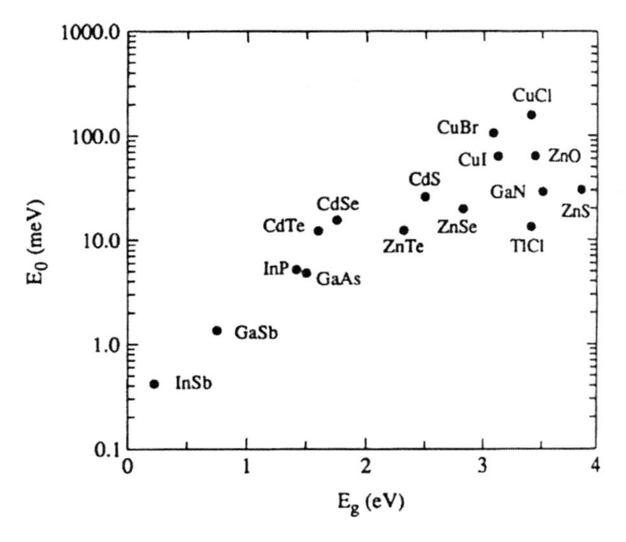

Figure 3.32 Experimental values for the exciton coupling energy E_0 versus the bandgap E_g. Reprinted by permission from figure 9.3 of Springer, *Semiconductor Optics*, 4th edition, by C. F. Klingshirn, Copyright 2012.

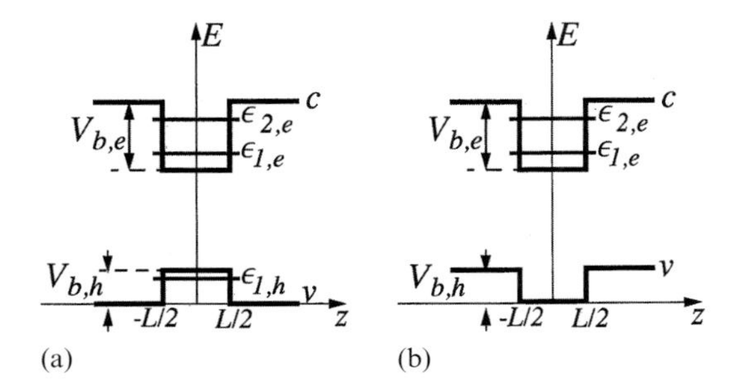

Figure 3.33 Two types of band diagrams of double heterostructures: (a) type I; and (b) type II.

in the same layer, as shown in Fig. 3.33(a); here the example is for AlGaAs/GaAs quantum structures.

In type-II heterostructures each layer of the structure confines only one type of carrier as illustrated in Fig. 3.33(b). An example is given by the InAs/GaSb system: the electrons are confined in the InAs layer, while the holes are confined in the GaSb layers.

An exciton in a type-I heterostructure has many features in common with the ordinary bulk-like exciton. However, type-II heterostructures present a qualitatively different physical picture in which the electrons and holes of the electron–hole pairs are spatially separated, the so-called *interface exciton*.

Type-I Structures

First, we shall consider type-I structures. The Schrödinger equation should include the kinetic energies of both the electron and the hole, with the potentials $V_e(r_e)$ and $V_h(r_h)$ describing the wells for the two particles,

$$V_e(z) = \begin{cases} 0, & \text{for } |z| \le L/2, \\ V_{b,e}, & \text{for } |z| \ge L/2, \end{cases} \tag{3.110}$$

$$V_h(z) = \begin{cases} 0, & \text{for } |z| \le L/2, \\ V_{b,h}, & \text{for } |z| \ge L/2, \end{cases} \tag{3.111}$$

and the energy of interaction between the electron and the hole. Figure 3.33(a) illustrates the band structure for the system under consideration. Thus the Hamiltonian is

$$\mathcal{H} = E_g + \frac{\vec{p}_{||e}^{\,2}}{2m_e} + \frac{\vec{p}_{||h}^{\,2}}{2m_h} + V_e(\vec{r}_e) + V_h(\vec{r}_h) - \frac{e^2}{\kappa\,|\vec{r}_h - \vec{r}_e|}. \tag{3.112}$$

If we fix some electron and hole states inside the well, say $\chi_{n,e}(z_e)$ and $\chi_{l,h}(z_h)$, and neglect the electron–hole interaction, we can express the wave function of the pair in the form

$$\Psi(\vec{r}_e, \vec{r}_h) = \frac{1}{S}\exp(i(\vec{k}_{||e}\vec{\rho}_e + \vec{k}_{||h}\vec{\rho}_h))\chi_{n,e}(z_e)\chi_{l,h}(z_h),$$

where $\vec{k}_{||e}$ and $\vec{k}_{||h}$ are the two-dimensional wavevectors of each of the particles, and $\vec{\rho}_e$ and $\vec{\rho}_h$ are two-dimensional coordinates of the electron and the hole, respectively; S is the area of the quantum well layer B. The energy corresponding to this wave function is

$$E = E_g + \epsilon_{n,e} + \epsilon_{l,h} + \frac{\hbar^2 k_{||e}^2}{2m_e} + \frac{\hbar^2 k_{||h}^2}{2m_h}.$$

It is clear that the last two equations describe decoupled and uncorrelated motion of the electron and the hole.

Since the quantum well potentials break the translational symmetry in the z-direction, instead of the transformations of Eqs. (3.102) and (3.103) one can introduce new coordinates in the x, y plane:

$$\vec{R}_{||} = \frac{m_e\vec{\rho}_{e||} + m_h\vec{\rho}_{h||}}{m_e + m_h} \quad \text{and} \quad \vec{\rho} = \vec{\rho}_{e||} - \vec{\rho}_{h||}. \tag{3.113}$$

Both vectors characterize in-plane propagation. The z coordinates are still not transformed. As a result the Hamiltonian of Eq. (3.112) takes the form

$$\mathcal{H} = \frac{\hat{\vec{P}}_{||}^2}{2M^{ex}} + \frac{\hat{\vec{p}}_{||}^{\,2}}{2m_r} - \frac{e^2}{\kappa\sqrt{\rho^2 + (z_e - z_h)^2}} + \frac{\hat{p}_{z,e}^2}{2m_e} + \frac{\hat{p}_{z,h}^2}{2m_h} + V_e(z_e) + V_h(z_h), \tag{3.114}$$

where $\hat{\vec{P}}_{||}$, $\hat{\vec{p}}_{||}$, $\hat{p}_{z,e}$, and $\hat{p}_{z,h}$ are momentum operators defined as in Eq. (3.105). One can again apply the previously used approach of Eq. (3.106) by introducing partial factorization of the wave function $\psi(\vec{r}_e, \vec{r}_h)$:

$$\Psi(\vec{r}_e, \vec{r}_h) = \frac{1}{\sqrt{S}} \exp(i\vec{K}_{\parallel}\vec{R}_{\parallel})\psi(z_e, z_h, \vec{\rho}), \tag{3.115}$$

where $\vec{K}_{\parallel}$ characterizes the in-plane center-of-mass wavevector of the pair. The function $\psi(z_e, z_h, \vec{\rho})$ can be simplified if we consider some particular subbands of the electron and the hole and neglect the contributions of all other subbands. For an electron from the nth subband and a hole of the lth subband, we can approximate the wave function by

$$\psi(z_e, z_h, \vec{\rho}) = \chi_{n,e}(z_e)\chi_{l,h}(z_h)\phi(\vec{\rho}), \tag{3.116}$$

where $\phi(\vec{\rho})$ describes the relative motion of the electron and the hole.

The physical meaning of this approximation is that only the in-plane propagation of the particles is correlated as a result of the Coulomb interaction, while the transverse propagation is independent of the electron–hole coupling. Such an approximation is valid if the coupling energies of the particles are small in comparison with the energy of separation between subbands for both types of particles. In other words, the width of the well is supposed to be smaller than the effective Bohr radius. Note that each combination of subbands (n, e) and (l, h) gives the set of coupled "ground" and excited states of the pair. Frequently the states of most physical interest are those formed from the lowest subbands of the electron and the hole.

Substituting the function ψ into the Schrödinger equation and integrating over coordinates z_e and z_h, the effective potential energy is found to have the form

$$V_{\mathrm{eff}}^{(\mathrm{ex})}(\rho) = -\frac{e^2}{\kappa}\int\int dz_e\, dz_h \frac{|\chi_{1,e}(z_e)|^2|\chi_{1,h}(z_h)|^2}{\sqrt{\rho^2 + (z_e - z_h)^2}}.$$

The Schrödinger equation with such a potential cannot be solved exactly. Instead, we need to use an approximate method to find solutions. The simplest such approach is the variational method. The idea of this method is based on the fact that, for an arbitrary Schrödinger equation,

$$H\psi(\vec{r}) = E\psi_i(\vec{r}),$$

the integral (the functional)

$$\mathcal{E}\{\psi\} = \int d\vec{r}\, \psi^*(\vec{r})H\psi(\vec{r})$$

reaches the minimal value if $\psi(\vec{r})$ is an exact normalized solution of the initial Schrödinger equation for some quantum state. Indeed, this minimal value of the functional exactly equals the energy of this quantum state.

To realize the variational method one may select a particular shape of a *trial function* $\psi(\vec{r}, \alpha, \beta, \ldots)$, with any number of parameters $\alpha, \beta, \ldots$ Usually the shape of the wave function is selected according to physical considerations (the expected shape of ψ, symmetry, etc.). Then one calculates the functional and obtains it as a function of the parameters: $\mathcal{E}\{\psi\} = \mathcal{E}(\alpha, \beta, \ldots)$. Since the true function should extremize the value of $\mathcal{E}$, the parameters $\alpha, \beta, \ldots$ can be determined from the extremum conditions

$$\frac{\partial \mathcal{E}}{\partial \alpha} = 0, \quad \frac{\partial \mathcal{E}}{\partial \beta} = 0, \ldots$$

In fact, only the wave function of the lowest energy (ground state) results in the lowest minimum of $\mathcal{E}$. Such a procedure gives the best approximation to the wave function of the selected shape. In general, the variational principle gives a very powerful method for the calculation of eigenfunctions and eigenvalues.

Returning to the problem of excitons in two-dimensional quantum structures, we write down the functional which should be minimized:

$$\mathcal{E}_{\text{ex}}\{\psi\} = \int\int dx\, dy \left(\frac{\hbar^2}{2m_{\text{r}}} (\nabla_2 \psi)^2 + V_{\text{eff}}^{(\text{ex})}(\rho)\psi^2 \right),$$

where, as before, ∇_2 is a two-dimensional operator. We choose a simple trial function of the form

$$\phi(\rho) = \frac{1}{\lambda}\sqrt{\frac{2}{\pi}} e^{-\rho/\lambda}, \tag{3.117}$$

which allows one to find the energy of the exciton in the quantum well and its radius, λ. In the approximation of Eqs. (3.116) and (3.117) the internal degrees of freedom of the electron and hole are separated from their transverse propagation; moreover, the transverse propagation of the electron and that of the hole are completely separate from each other and uncorrelated. A more sophisticated approximation could include some correlation of the transverse degrees of freedom of the electron and hole, like

$$\phi(\rho, z) = C \exp\left[-\frac{\sqrt{\rho^2 + (z_{\text{e}} - z_{\text{h}})^2}}{\lambda} \right]. \tag{3.118}$$

In fact, for both trial functions the results can be obtained only by numerical calculations. In Fig. 3.34 the exciton binding energies, E_{ex}, are plotted for the two previously considered approximate cases as functions of the quantum well layer thickness, L. The binding energy increases in thin layers and, in the extreme case $L \to 0$, it reaches a value four times that in the bulk materials. The second approximation of Eq. (3.118) predicts a greater binding, or equivalently coupling energy, for large thickness. Because both trial functions contain only one variational parameter λ, this result clearly indicates that there is a correlation between the transverse and the in-plane propagation of the pair. Note that only the second trial function yields the correct limit in the case $L \gg a_0^{\text{ex}}$, where a_0^{ex} is defined by Eq. (3.109). In this limit the exciton takes on bulk-like behavior.

Thus for a type-I semiconductor heterostructure there are exciton states originating from a combination of the electron and the hole subbands. If the energy separations between subbands are substantially larger than the energy of the exciton, only the in-plane propagation of the electron and that of the hole exhibit strong correlation. This physical situation is typical for III–V compounds. In this case the transverse propagation characteristics are similar to those of a decoupled electron

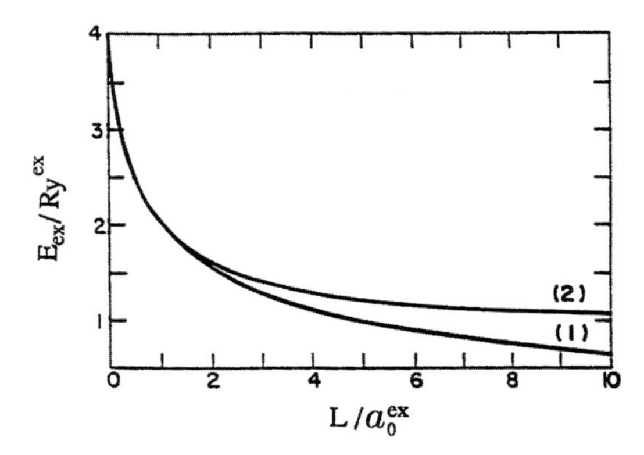

Figure 3.34 The exciton coupling energy for the type-I heterostructure calculated for two approximations of the trial function: curve (1) corresponds to Eq. (3.117), and curve (2) corresponds to Eq. (3.118). Reprinted figure with permission from figure 2 in G. Bastard, *et al.*, *Phys. Rev.* **B 26**, 1974 (1982). Copyright (1982) by the American Physical Society.

and hole, but there is some correlation between the transverse and in-plane components of the propagation. An important feature of the heterostructure under consideration is that, for the lowest subbands, the exciton (coupling) energies exceed those of excitons in bulk materials. Owing to these larger exciton energies the excitons are present up to room temperature, in contrast to the case of bulk materials, where they are washed out at room temperature. However, in some bulk materials, such as ZnO, the bulk exciton binding energy is about 60 meV and these bulk excitons are present at room temperature.

Type-II Structures

The diagram for electron and hole bands for this case is shown in Fig. 3.33(b). One can see that the embedded layer A confines only electrons and serves as a barrier for holes. In the previous case the coupling of the electron and the hole affects, mainly, only the in-plane propagation of the pair. However, in type-II heterostructures, the Coulomb interaction has to dramatically modify the transverse propagation of the hole in order to couple the pair. This is the main difference in excitonic behavior between type-I and type-II heterostructures.

In order to take into account this fact, let us represent the wave function ψ of Eq. (3.115) in the form

$$\psi(\rho, z_e, z_h) = A\chi_{n,e}(z_e) \sum_{\vec{k}_h} C_{\vec{k}_h} \chi_{\vec{k}_h}(z_h) e^{-\rho/\lambda}, \tag{3.119}$$

where A is the normalization constant. The second factor describes the transverse motion of the electron. The fourth factor is the hole wave function, which should be localized due to attraction by the localized electron, though it is constructed from unbound valence band wave functions of holes. The coefficients $C_{\vec{k}_h}$ are to be determined from the Schrödinger equation. The last factor describes the in-plane relative

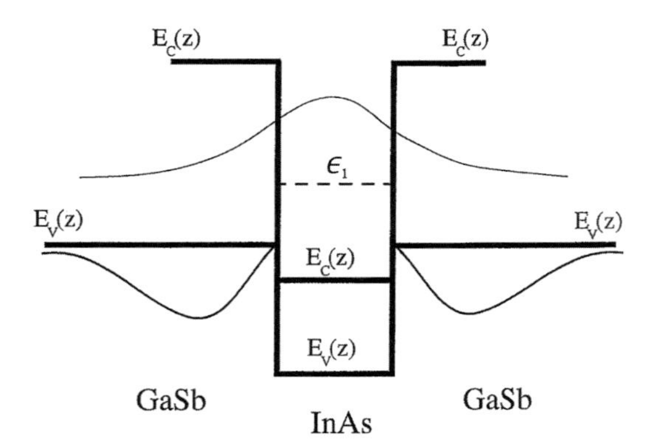

Figure 3.35 The band diagram of the GaSb/InAs/GaSb double heterostructure and the wave functions of the electron and hole composing the exciton in this heterostructure of type II. After Fig. 10 on p. 137 of *Wave Mechanics Applied to Semiconductor Heterostructures*, G. Bastard, Halsted Press, 1988.

motion of the electron and hole. Let us point out that this superposition of the *non-localized wave functions* $\chi_{\vec{k}_h}(z_h)$ should result in *confined states* for holes. This is completely analogous to composing a localized wave packet from plane waves. The results of calculations for the GaSb/InSb/GaSb double heterostructure are shown in Fig. 3.35. One can see that the hole (lower wave function) is indeed confined near the electron-well layer, though the electron and hole wave functions are separated in space. Obviously, this separation leads to a decrease in the exciton coupling energy, E^0_{ex}. The bulk exciton in the InSb crystal is characterized by $Ry^{ex} = 1.3$ meV and $a_0^{ex} = 370$ Å. The coupling energy of the interface exciton is about 0.6–0.8 that of the bulk value, which is much smaller than the coupling in the type-I structure. Thus, we can conclude that the excitonic effects in type-II quantum wells are suppressed substantially relative to those in type-I structures.

Before we close this section devoted to exciton effects, we should make a remark concerning the exciton picture in the case of a complex valence band, as is typical for III–V compounds. A real valence band contains two types of energy dispersion, $E_v(k_v)$: one for heavy holes and the other for light holes. Both of these types of particle can be excited in such crystals. The electrons from the conduction band can couple with both types of hole, causing the formation of *two types of excitons*. In such quantum structures, heavy- and light-hole states are mixed and as a consequence the exciton physics becomes even more complicated. All of these specific features are critical in establishing the optical spectra of quantum structures.

3.9 Nanostructure-Based Materials Are Reconfigurable Nanomaterials

All of the nanostructures considered in this chapter are based on band-structure engineering whereby electrons or holes or both are confined in quantum wells, or in quantum wires or quantum dots that create potential wells for them. In the conclusion of Chapter 2 we mentioned that the properties of quantum structures can be substantially modified by external electric fields. Selective doping of nanostructured materials increases their sensitivity to the field and facilitates the design of adaptive (or reconfigurable) nanomaterials with properties controlled by an applied external voltage.

There are many realizations of such materials and we will give here just a few examples of reconfigurable nanomaterials. Adaptive devices based on those materials will be discussed in Chapter 8.

3.9.1 Control Carrier Concentration in Quantum Structures

Perhaps the simplest example of a reconfigurable nanostructure is the gated hetostructure of n-AlGaAs/p-GaAs, shown in Fig. 3.36. A triangular potential forms the quantum well that was discussed in Section 3.2.1 (see Fig. 3.5). If the AlGaAs barrier is thin (of thickness less than about 60 nm depending on doping), the first quantized level in the well is above the Fermi level in the metal and the well is depleted, see Fig. 3.37(a). For a wider barrier, see Fig. 3.37(b), the concentration in the well is determined by the doping of the AlGaAs and by the thickness of the spacer, d_{sp}, that separates the doped region of the AlGaAs from the heterostructure interface. As depicted in Fig. 3.38, the application of a negative voltage to the metallic gate of the structure shown in Fig. 3.37(b) reduces the concentration of the electrons in the quantum well and can deplete it completely for properly chosen parameters. A positive voltage applied to the gate increases the concentration in the quantum well (see the curve for $d_{sp} = 0$ in Fig. 3.38), and for the structure that is shown in Fig. 3.37(a) it would switch the quantum well from a depletion mode to a population mode.

Gate control is used widely in transistors that are based on a two-dimensional electron gas (2DEG) since depletion of the quantum well leads to transistor turn-off. This is why the structures shown in Figs. 3.37(a) and (b) leads to normally-off and normally-on transistors, respectively. It is important to realize that a change of the electron concentration in the quantum well results not only in the modulation of

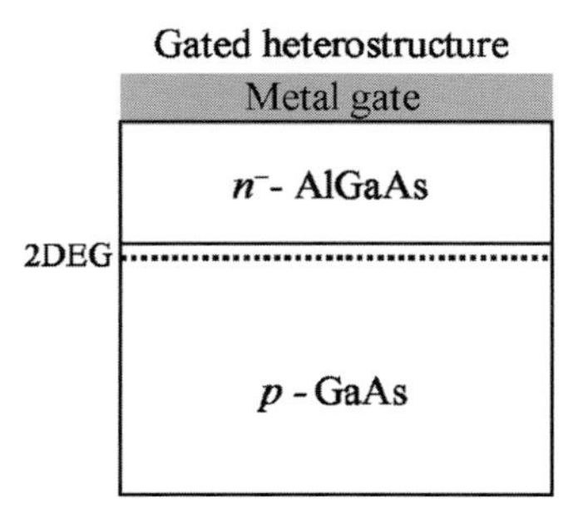

Figure 3.36 An example of a gated heterostructure with two-dimensional electron gas (2DEG).

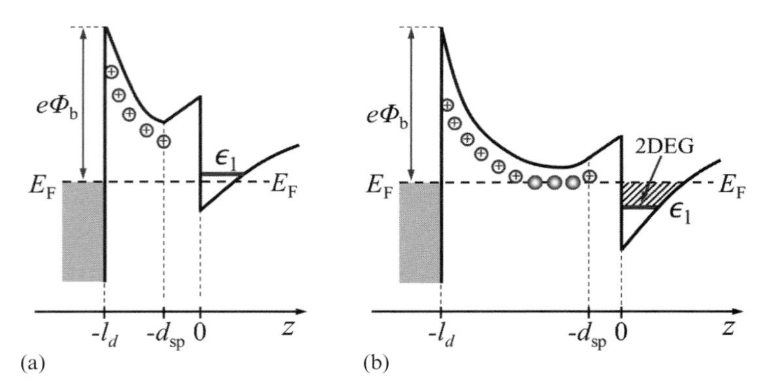

Figure 3.37 Conduction band diagram for metal/n-AlGaAs/p-GaAs heterostructures. The built-in Schottky voltage controls the depletion region under the metallic gate. It results in (a) a depletion of the quantum well for a narrow barrier and (b) filling of the quantum well, with doping controlling the concentration in the well, for a wide barrier (60 nm or wider).

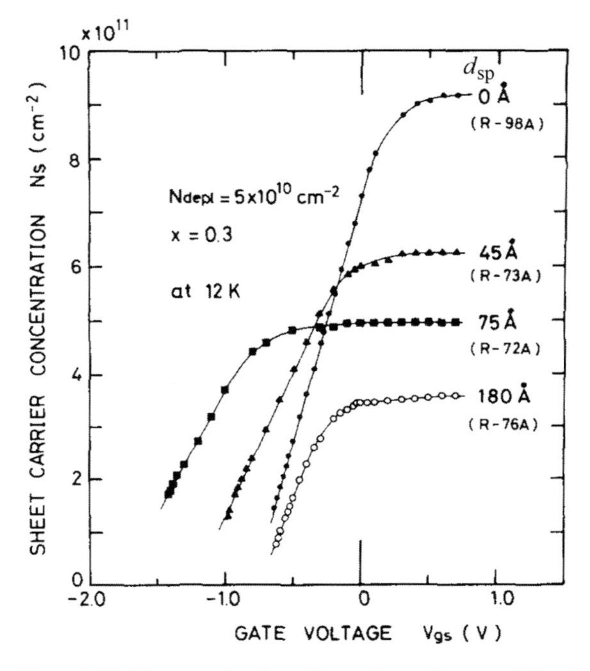

Figure 3.38 Measured gate-voltage dependences of the channel density of two-dimensional electrons in $Al_{0.3}Ga_{0.7}As$/GaAs structures at $T = 12$ K and various spacer thicknesses d_{sp}. All samples are doped with $N_D = 4.6 \times 10^{17}$ cm^{-3}, except for sample R-76A, which has $N_D = 9.2 \times 10^{17}$ cm^{-3}. Reprinted from figure 3 in K. Hirakawa *et al.*, *Appl. Phys. Lett.*, **45**, 253 (1984), with the permission of AIP Publishing.

the conductivity of the two-dimensional channel but also in the modification of its optical properties. Increasing the electron concentration leads to an increase in the interband absorption frequency, as electrons from the valence band cannot go into states that are already occupied by the electrons in the conduction band. At the same

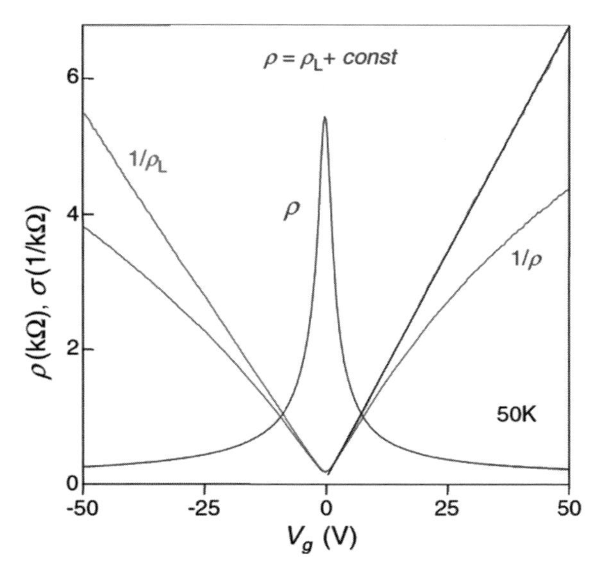

Figure 3.39 The measured resistivity and conductivity of a single graphene layer as functions of the gate voltage. SiO_2 of thickness 300 nm was used as a dielectric between the gate and the graphene layer placed on the oxidized Si wafer. Reprinted figure with permission from figure 1 in Morozov, *et al.*, *Phys. Rev. Lett.* **100**, 016602 (2008). Copyright 2008 by the American Physical Society.

time it leads to an increase of the intraband absorption intensity and a reduction of the intraband absorption frequency, due to increased carrier concentration, as will be discussed in Chapters 5 and 8.

Graphene as a zero-bandgap material has a substantial advantage over the conventional semiconductor quantum wells since a change of polarity of the applied voltage will switch the concentration in the graphene between electrons and holes. Figure 3.39 shows the practically linear dependence of the conductivity of a single graphene layer, $\sigma_L = 1/\rho_L = 1/(\rho - 100\,\Omega)$, on the gate voltage, which is explained by the linear increases of electron and hole concentrations in graphene for positive and negative voltages, respectively. To evaluate the resistivity of the single graphene layer, ρ_L, a constant resistivity of $\rho_S = 100\,\Omega$ was subtracted from the resistivity, ρ, of the sample. The concentration of electrons and holes can be changed from zero to $\sim 2 \times 10^{12}\,\text{cm}^{-2}$. This facilitates building a p–n junction using p and n gates separated by an intrinsic region of length $2l$, as shown in Fig. 3.40(a). The p–n junction can be induced in a single graphene layer (not shown) or in a multiple graphene layer structure, as shown in Fig. 3.40(a). The advantage of the electrically induced p–n junction is based on the possibility of controlling the concentration of electrons and holes as well as the length of the intrinsic region. The in-plane p–n junction can be used for electromagnetic energy generation and detection, and, in contrast to the case of a semiconductor p–n junction, where the energy of emitted/absorbed

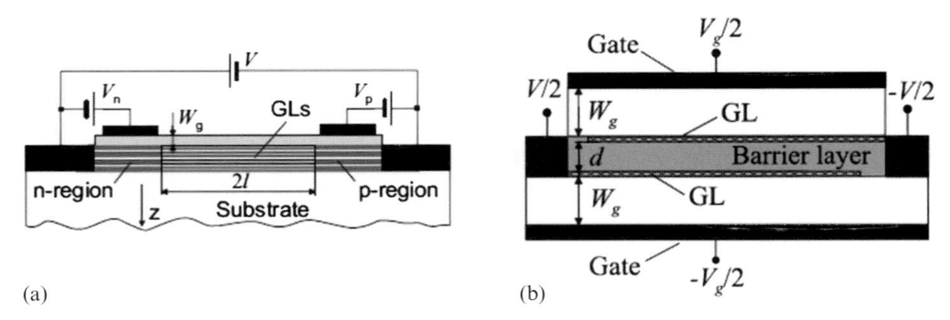

Figure 3.40 (a) A lateral p–n junction in graphene layers (or a single layer, not shown) induced by gates separated by distance W_g and (b) a vertical p–n junction between graphene layers contacted independently. Reprinted figure (a) with permission from figure 1a in M. Ryzhii, *et al.*, *Phys. Rev.* **B 82**, 075419 (2010) Copyright 2014 by the American Physical Society. Figure (b) reprinted from figure 1b in M. Ryzhii *et al.*, *J. Appl. Phys.* **115**, 024506 (2014), with the permission of AIP Publishing.

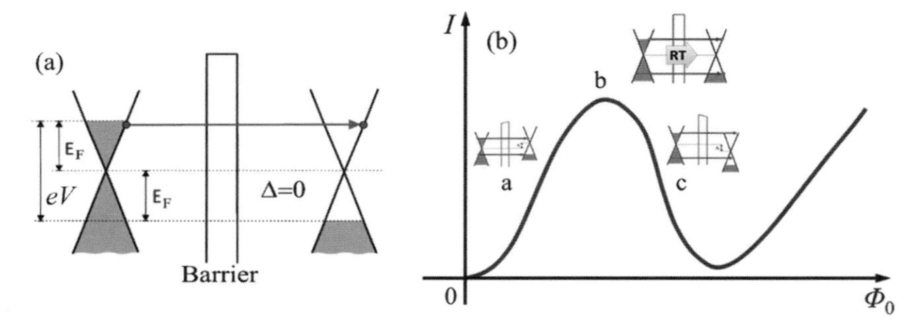

Figure 3.41 (a) Resonant tunneling in a vertical p–n graphene heterojunction and (b) the current–voltage characteristic of a vertical p–n junction between contacted independently graphene layers contacted independently as shown in Fig 3.40(b). Figure (a) reprinted from figure 3a in M. Ryzhii *et al.*, *J. Appl. Phys.* **115**, 024506 (2014), with the permission of AIP Publishing.

photons is determined by the bandgap and ranges from the infrared to the ultraviolet spectrum, zero-bandgap graphene can be used in the THz frequency range. Even more beneficial is a vertical p–n junction of two graphene layers separated by a barrier layer like boron nitride or silicon carbide. As depicted in Fig. 3.40(b), the top and bottom graphene layers are gated by potentials of equal and opposite sign. As a result, the upper layer is populated by electrons and the bottom layer is populated by holes, and their concentration is controlled by the voltage. The graphene layers are contacted independently and the voltage between the left and right contacts changes the relative position of the Dirac point in the layers. If the Dirac points in both graphene layers coincide, as shown in Fig. 3.41(a), resonant tunneling occurs from the n-layer to the p-layer. The magnitude of the current at resonance is determined by the position of the Fermi levels, which are controlled by the gates. If the Dirac points are out of resonance, resonant tunneling is impossible and the current decreases with increasing shift Δ between the Dirac points as shown on the inserts in Fig. 3.41(b). As a result, the current–voltage characteristic has an N-type shape as shown in Fig. 3.41(b). As indicated on the inserts "a" and "b," tunneling with conservation of energy would not conserve momentum (the momentum for the corresponding state on the left is different from that on the right), and additional scattering on impurities and/or defects would need be involved to take care of the conservation of momentum. Such a process has a smaller probability than the resonant tunneling shown on the insert "b." We will discuss photon-assisted tunneling in Chapter 8, where it will be shown that the involvement of a photon in the tunneling process can satisfy energy and momentum conservation and can be used in detectors on portion "a" of the current–voltage characteristic presented in Fig. 3.41, or in emitters/lasers on portion "b" of the same current–voltage characteristic.

3.9.2 Asymmetrically Doped Double-Quantum-Well Structures

We have discussed the control of concentration by a gate for the case when the material of the barrier is doped and electrons move from the impurities to the lower states in the quantum well. In many applications, the quantum wells are doped and the doping controls the carrier concentration. The electrons can fill in either the lowest subband (see Figs. 3.6 and 3.33) or two or more subbands, changing the energy that is required to ionize electrons from the quantum well into the continuum above the barrier (Fig. 3.33). In the case of a double quantum well (see Fig. 3.22), asymmetric doping of the structure in addition to controlling the total concentration of electrons facilitates the reconfiguration of the wave functions and the ionization energy by the external voltage. As an example, Fig. 3.42 demonstrates such a possibility for a structure with two 6.5 nm-thick GaAs layers separated by the 3.1 nm-thick barrier of an $Al_{0.2}Ga_{0.8}As$ layer; the barriers on the left and the right of the double quantum wells have the same height, as they are from the same $Al_{0.2}Ga_{0.8}As$ material. As was discussed in Section 3.6.1, both the lowest quantum level and the second level are split into two levels. The thicknesses indicated above were chosen to ensure that the second level of isolated quantum wells is in the vicinity of the barrier height, as shown in Fig. 3.42(b). At equilibrium, due to doping by Si donors in the right well only, 95% of the electrons are in the right well and 5% of the electrons move to the left well. As a result of the transfer of electrons from the right well with donors into the undoped left well there is a built-in electric field that leads to a difference of about 20 meV between the left and right barrier heights of the double-quantum-well structures. A negative voltage (Fig. 3.42(a)) of about a 30 mV drop across the double quantum well increases the electric field, and the difference in the barrier heights reaches about 50 meV. Now practically all of the electrons are in the right well (99%). A positive applied voltage (Fig. 3.42(c)) of the same magnitude reverses the relative heights of barriers and the concentration of electrons in the left well increases noticeably (by up to 24%). The distance between the split levels is smaller than for the negative applied voltage. Moreover, there is a substantial difference in the wave functions for the three cases shown in Fig. 3.42. Modification of the positions of the levels, of the wave functions, and of the relative barrier heights changes the absorption spectrum of the double quantum well.

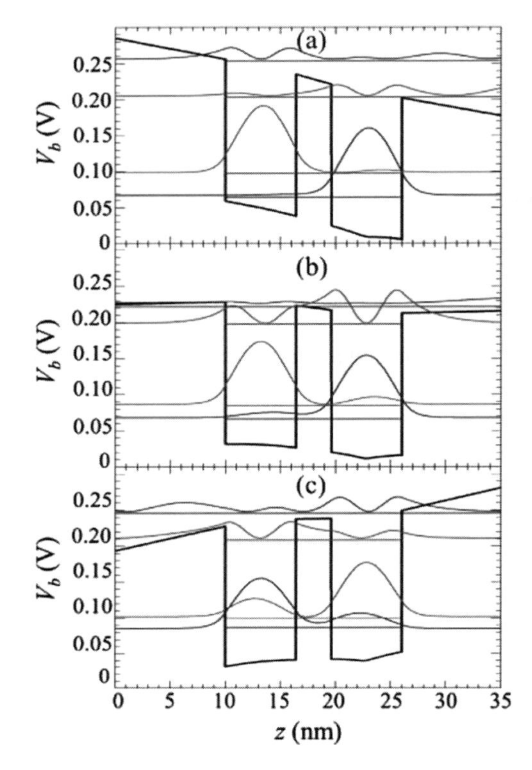

Figure 3.42 Self-consistently calculated potentials and wave functions of the ground and excited states in a double-quantum-well structure at $T = 70$ K under (a) negative bias, (b) zero bias, and (c) positive bias. After Fig. 3 in J. K. Choi, *et al.*, *Jpn. J. Appl. Phys.* **51**, 074004 (2012). Copyright 2012 The Japan Society of Applied Physics.

We will discuss in Chapter 8 how this reconfigurable structure can be used in adaptable infrared detectors with voltage control of the registered frequency.

3.9.3 Reconfigurable Nanomaterials Based on Quantum Dots

We have discussed how an applied voltage can reconfigure nanostructure-based materials by changing the positions of the quantum levels, the electron wave functions, and the electron concentration in the quantum structures. The lifetime of electrons excited from quantum structures is also an important parameter for optoelectronic applications. Here we briefly discuss the possibility of changing the lifetime in quantum-dot-based materials by several orders of magnitude by means of an applied voltage. Figure 3.43 shows schematically the potential barriers created around the negatively charged dots. The charge of the dots is created by the electrons that have left donors that are placed outside of the dots. The potential barrier height, V_{m}, shown in Fig. 3.43, is determined by the number of electrons in the dots, which is controlled by the doping of the barrier material. The electrons excited from

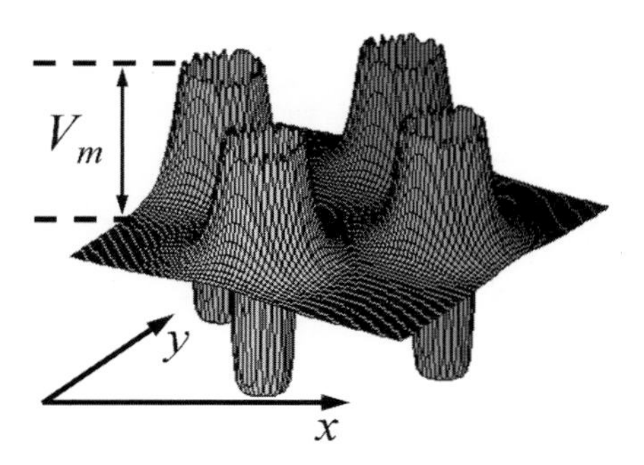

Figure 3.43 A schematic representation of the potential barriers around negatively charged dots.

the dots by external radiation lose their energy and go to the lowest energy between the dots; i.e., they will be in the valleys that are schematically depicted in Fig. 3.43. To be captured back into the dot, an electron should overcome the barrier V_{m}. Since the electron energy is above that of the barrier, the electron can be captured into the dot by losing energy due to the emission of a phonon. The time of capture (the lifetime of the electron outside of the dots) is proportional to the number of dots per unit volume (the concentration of dots), N_{d}, and the volume of a dot, $4\pi a^3/3$, and can be approximated by the following equation:

$$\frac{1}{\tau_{\mathrm{capt}}} = \pi N_{\mathrm{d}}\, a^3 \,\frac{1}{\tau_\varepsilon}\, \exp\left(-\frac{eV_{\mathrm{m}}}{k_{\mathrm{B}}T}\right). \tag{3.120}$$

Here τ_ε is the energy relaxation time. The exponential factor becomes important for $eV_{\mathrm{m}} \geq k_{\mathrm{B}}T$. If $eV_{\mathrm{m}} \gg k_{\mathrm{B}}T$, the lifetime becomes exponentially large compared with the case of uncharged dots with $V_{\mathrm{m}} = 0$.

The dependence of the capture time on the external electric field is shown in Fig. 3.44(a) and the dependence on the barrier height in Fig. 3.44(b) for spherical quantum dots of radius $a = 12$ nm with a distance between the centers of the quantum dots of $b = 72$ nm, temperature 300 K, and concentration of quantum dots $N_{\mathrm{d}} = 1/b^3$. These results were obtained by the Monte Carlo simulation of electrons captured in InGaAs quantum dots that are embedded in the GaAs matrix. As the barrier height increases, the capture time increases by more than two orders of magnitude and can be as long as 50 ns. As the electric field increases above 1 kV/cm, the capture time decreases. This is caused by the increase in energy of the electrons in the high-field region, which is known as the hot-electron phenomenon. As the energy of the electrons becomes comparable to the barrier height they can overcome the barrier and can be captured into the dot.

Control of the lifetime of the electrons is important in many optoelectronic applications and especially in the adaptable infrared photodetectors considered in Chapter 8. Indeed, in a high electric field, the lifetime of photoexcited electrons is short, the detector is fast, and they can be used for fast scanning of an area of

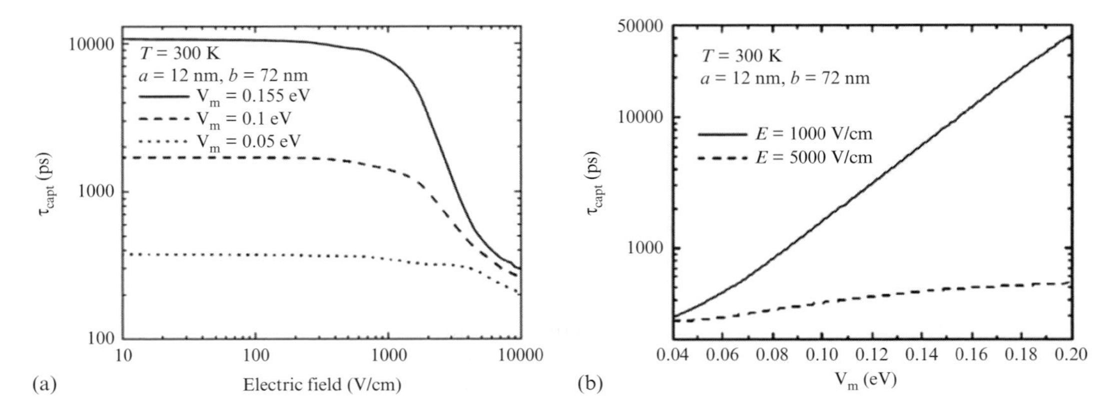

Figure 3.44 The electron capture time (a) as a function of the external electric field for various potential barrier heights and (b) as a function of the barrier height for two values of the electric field. From Figs. 3 and 4 in *Intern. J. High Speed Electron. Syst.*, L. H. Chien, *et al.*, **18**, 1013 (2008); permission conveyed through Copyright Clearance Center, Inc.

interest. If an object is detected in the area of interest, then switching to a lower electric field allows one to get more detailed information about the object. The larger lifetime in a low electric field increases the responsivity of the photodetector.

3.10 Closing Remarks to Chapter 3

In this chapter we have studied the electronic properties of different quantum structures: quantum wells, two-dimensional crystals, double quantum wells, two-monolayer crystals, quantum wires, quantum dots, and superlattices. To display the principal features of electrons in such structures we simplified the problem by restricting ourselves to modeling heterostructures by step-like potentials and assuming a single effective mass for electrons throughout the artificial structures. For these idealized systems, we found drastic changes in the electron spectra. With appropriate dimensional confinement, one, two, or even three electron degrees of freedom become quantized. Instead of the continuous energy bands inherent for bulk materials, we found a series of two-, and one-dimensional energy subbands for quantum wells and quantum wires, respectively, and a series of discrete levels for quantum dots. For superlattices we found a splitting of the bulk energy band into a series of one-dimensional minibands. At temperatures lower than or comparable to the subband energy separation, these minibands lead to pronounced effects. The *physics of low-dimensional electrons* has essentially taken on the status of an entirely new field of physics due to the advent of nanodimensional heterostructures. In superlattices the electron motion becomes highly anisotropic and as a result the effective mass along the superlattice axis is enhanced substantially.

We studied the electronic properties of a monatomic carbon layer, graphene. In graphene the carrier confinement is ideally two-dimensional. Graphene is a gapless semiconductor, so the change of charge of the current carriers occurs at the Dirac point, and the transition from electrons to holes (or vice versa) can be realized in a single graphene structure. The dispersion of monolayer graphene is linear and the electrons can be treated as *massless Dirac fermions*. This leads, in particular, to extremely high electron mobility in graphene.

In contrast to graphene, the two-dimensional transition-metal dichalcogenide crystals are direct-bandgap materials, with the bandgap above 1 eV. The structures of the monolayers are not flat, as for graphene, and exhibit the phenomenon of inversion-symmetry breaking. The electrons in the conduction and valence bands are characterized by the effective masses. They can be described as *massive Dirac fermions*. The effective masses are relatively large, which leads to low electron and hole mobilities. Because of the strong spin–orbit interaction in the transition-metal atoms, the spin-related effects are large in such monolayers. The spin-splitting of the valence band may be as much as 0.15–0.4 eV, and electron spin states are significantly different for different valleys. This gives rise to a new effect: valley–spin coupling, which generates unusual transport and optical electron properties.

Black phosphorus as a single layer and in thicknesses of a few atomic layers complements the list of two-dimensional direct-bandgap crystals. Owing to the strong thickness dependence of the properties of black phosphorus, in the same material one can change the energy gap and optical properties over a wide spectral range by varying the sample thickness. The strongly anisotropic properties of black phosphorus can be exploited in polarization-sensitive electronic and optoelectronic devices.

We show that two-monolayer graphene – bigraphene – has rich physical properties associated with low-energy excitations. Being a stable two-dimensional crystal, bigraphene is considered to be very important for nanoelectronic and far-infrared optoelectronic applications because of the possibility of controlling its properties, in particular, the ability to open a bandgap by applying an electric field.

Also, we studied the energy spectra of excitons in idealized quantum heterostructures. We found that lowering the dimensionality of the electrons causes an increase in the binding energies of excitons. Furthermore, each of the low-dimensional subbands generates its own series of excitons.

By simplifying the description of quantum heterostructures, we revealed the principal features of dimensionally confined heterostructures. In fact, in real heterostructures and devices there exist other, specific, properties that are important for device applications as well as for understanding the physics of such nanostructures. In Chapters 4 and 5, we will study these properties as well as particular parameters for different materials and heterostructures.

In this chapter, we used several different models for the potential energy of carriers in order to describe the principal quantum effects occurring in quantum wells, quantum wires, quantum dots, and superlattices. Various books on quantum mechanics, that are listed at the end of our book, supplement our treatment of one-, two-, and three-dimensional models which have exact solutions and allow extensive analysis of electron properties.

3.11 Control Questions

1. Impose the boundary conditions for continuity of the wave function and continuity of the derivative of the wave function to derive Eqs. (3.12) and (3.13).
2. In the limit of thick barriers show that the equation which defines the energy levels of a double-well system defined by Eq. (3.83) reduces to the equation which defines the energy levels of a single quantum well.
3. Consider a 100 Å wide GaAs quantum well with very thick $Al_{0.3}Ga_{0.7}As$ barriers. For this heterostructure, the conduction band edge of the GaAs well is about 225 meV lower in energy than that of the barrier regions. On the other hand, Eq. (3.15) predicts an infinite number of bound states with $\epsilon_1 = 55$ meV, $\epsilon_2 = 220$ meV, and $\epsilon_3 = 495$ meV, etc. Thus, ϵ_3 for the infinitely deep well exceeds the conduction band offset of the $Al_{0.3}Ga_{0.7}As/GaAs/Al_{0.3}Ga_{0.7}As$ system. Discuss, on the basis of these results, the limits of applicability of Eq. (3.15) for realistic heterostructure devices.
4. For a quantum wire with finite barriers, use the continuity of the electron wave function and the continuity of the derivative of the electron wave function to determine whether a separable treatment as in Eq. (3.44) is applicable to the case of a quantum wire with finite barriers.
5. Consider a spherical quantum dot of GaAs with a surrounding $Al_{0.3}Ga_{0.7}As$ barrier. What is the minimum radius for such a quantum dot if there exists a bound state inside this spherical well?
6. For the zero-dimensional system defined by the potential of Eq. (3.53), it is possible to use the technique of separation of variables to straightforwardly obtain the wave functions of Eq. (3.54) and the eigenenergies of Eq. (3.55). For a more realistic case where the potential discontinuity is finite, the wave functions are not equal to zero at the boundaries of the quantum box. Consider the formulation of the boundary conditions at the corners of the quantum box. Is it still possible to use the technique of separation of variables in the case when the potential discontinuity is finite at the quantum box boundaries?

4 Light–Semiconductor Materials Interaction

4.1 Introduction

Interactions of electromagnetic radiation with semiconductors include processes of emission, absorption, and scattering of light as well as various nonlinear optical phenomena. These processes have always been major subjects of study in solid-state physics. The optics of semiconductors is an important topic in the modern physics of solid-state devices.

In this chapter we begin to study the properties of light and light–semiconductor interactions. We start with a brief review of the basic concepts of electromagnetic fields. We define the classical characteristics of electromagnetic fields such as the energy, intensity, and density of states, and we introduce the concept of the quanta of these fields – photons. Analyzing electromagnetic fields in free space and in optical resonators, we show that resonators drastically change the structure and the behavior of electromagnetic fields. These resonators facilitate control of the spatial distribution of the fields, i.e., they enable one to carry out electromagnetic field engineering. Next we study the interaction of light with matter and define three main optical processes: spontaneous emission, stimulated emission, and stimulated absorption. Calculations of probability amplitudes, total rates of phototransitions and the absorption (amplification) coefficient conclude the principal part of our treatment.

After this preparatory discussion we review the optical properties of bulk semiconductors and define the major optical characteristics of semiconductors with emphasis on the specifics of direct- and indirect-bandgap semiconductors.

4.2 Electromagnetic Waves and Photons

Before presenting a description of the optical properties of semiconductors, we shall recall the basic equations of the electromagnetic fields in condensed matter, dielectrics, semiconductors, and metals, which are the *macroscopic* Maxwell equations. The macroscopic Maxwell equations are for electric and magnetic fields averaged over distances much larger than the interatomic distances for crystals. Details of the derivation of the macroscopic equations from microscopic equations can be found in the textbooks on electromagnetics cited in the Further Reading list for Chapter 4 at the end of the book.

The Maxwell equations of macroscopic electromagnetism can be presented in the following form:

$$\vec{\nabla}\vec{\mathcal{D}} = \rho, \qquad \vec{\nabla}\vec{\mathcal{B}} = 0, \tag{4.1}$$

$$\vec{\nabla} \times \vec{\mathcal{E}} = -\frac{\partial \mathcal{B}}{\partial t}, \quad \vec{\nabla} \times \vec{\mathcal{H}} = \vec{\mathcal{J}} + \frac{\partial \vec{\mathcal{D}}}{\partial t}. \tag{4.2}$$

In these equations $\vec{\mathcal{E}}$ and $\vec{\mathcal{D}}$ are the electric field and the dielectric displacement, $\vec{\mathcal{H}}$ and $\vec{\mathcal{B}}$ are the magnetic field strength and the magnetic field (magnetic induction), and ρ and $\vec{\mathcal{J}}$ are the charge density and the current density, respectively. Then, $\vec{\nabla} = \{\partial/\partial x, \partial/\partial y, \partial/\partial z\}$, and $\vec{a}\vec{s}$ and $\vec{a} \times \vec{s}$ denote the scalar and vector products, respectively.

The macroscopic fields $\vec{\mathcal{E}}$ and $\vec{\mathcal{D}}$, as well as $\vec{\mathcal{H}}$ and $\vec{\mathcal{B}}$, are linked via the polarization $\vec{\Pi}$ and magnetization $\vec{\mathcal{M}}$ by the "material" relationships

$$\begin{aligned} \vec{\mathcal{D}} &= \epsilon_0 \vec{\mathcal{E}} + \vec{\Pi} = \epsilon_0 \kappa \vec{\mathcal{E}}, \\ \vec{\mathcal{H}} &= \frac{1}{\mu_0}\vec{\mathcal{B}} + \vec{\mathcal{M}}, \end{aligned} \tag{4.3}$$

where $\vec{\Pi}$ and $\vec{\mathcal{M}}$ are the polarization and magnetization of the material, respectively, while ϵ_0 and μ_0 are the electric permittivity and magnetic permeability of a vacuum ($\epsilon_0 = 8.854 \times 10^{-12}$ F/m, $\mu_0 = 1.257 \times 10^{-6}$ H/m). Below, we will consider only nonmagnetic media, neglecting the magnetization $\vec{\mathcal{M}}$; i.e., the relative magnetic permeability is assumed to be 1. Instead, electric polarization effects contributing to the dielectric displacement $\vec{D}$ of Eq. (4.3) will be the focus of our attention. In the equation for $\vec{D}$, κ is the dielectric permittivity of the medium (we limit ourselves to analysis of isotropic media, thus κ can be treated as a scalar). The last necessary material relationship links the current density j and the electric field $\mathcal{E}$:

$$\vec{\mathcal{J}} = \sigma \vec{\mathcal{E}}, \tag{4.4}$$

with σ the conductivity of the medium.

For many cases considered later, the algebraic relationships (4.3) and (4.4) with constants κ and σ are adequate for description of the phenomena. However, for some cases we should use more general linear relationships between $\vec{\mathcal{D}}$ and $\vec{\mathcal{E}}$, as well as between $\vec{\mathcal{J}}$ and $\vec{\mathcal{E}}$:

$$\vec{\mathcal{D}}(t) = \epsilon_0 \int_{-\infty}^{t} dt'\, \kappa(t-t')\vec{\mathcal{E}}(t') = \epsilon_0 \int_{0}^{\infty} d\tau\, \kappa(\tau)\vec{\mathcal{E}}(t-\tau), \tag{4.5}$$

$$\vec{\mathcal{J}}(t) = \int_{-\infty}^{t} dt'\, \sigma(t-t')\vec{\mathcal{E}}(t') = \int_{0}^{\infty} dt'\, \sigma(t-t')\vec{\mathcal{E}}(t-\tau). \tag{4.6}$$

Here, instead of constant parameters κ and σ, we have introduced time-dependent *response* functions $\kappa(\tau)$ and $\sigma(\tau)$. The form of Eqs. (4.5) and (4.6) takes account of the effect of the electric field $\vec{\mathcal{E}}$ on the polarization and the current at all previous instants, $t' \le t$. By presenting the fields $\vec{\mathcal{E}}(t)$, $\vec{\mathcal{D}}(t)$ as a superposition of components of the angular frequency, ω,

$$\vec{\mathcal{E}}(t) = \int d\omega\, \vec{\mathcal{E}}_\omega e^{-i\omega t}, \quad \vec{\mathcal{D}}(t) = \int d\omega\, \vec{\mathcal{D}}_\omega e^{-i\omega t},$$

we find the following important relations in the frequency domain:

$$\vec{\mathcal{D}}_\omega = \epsilon_0 \kappa(\omega) \vec{\mathcal{E}}_\omega. \tag{4.7}$$

A similar relationship is valid for the Fourier components of the current, $\mathcal{J}(t)$:

$$\vec{\mathcal{J}}_\omega = \sigma(\omega) \vec{\mathcal{E}}_\omega. \tag{4.8}$$

In Eqs (4.7) and (4.8), $\kappa(\omega)$ and $\sigma(\omega)$ are the Fourier transformations of the response functions $\kappa(\tau)$ and $\sigma(\tau)$. Note that in the general case $\kappa(\omega)$ and $\sigma(\omega)$ are complex-valued functions.

If, for a given material, the functions $\kappa(\omega)$ and $\sigma(\omega)$ are known, then by using the Maxwell equations (4.1) and (4.2), we can derive all the parameters of the electromagnetic waves, i.e., the joint electric and magnetic fields oscillating both in space and in time, and the interactions of these fields with the material.

4.2.1 Electromagnetic Fields, Modes, and Photons in Free Space

In the simplest, homogeneous, case, solutions of the Maxwell equations (4.1) and (4.2) can be found in the form of plane waves. For example, the electric field of the electromagnetic wave of frequency ω can be written as

$$\vec{\mathcal{E}}(\vec{r}, t) = \vec{\xi} F_0 \cos(\vec{q}\vec{r} - \omega t), \tag{4.9}$$

where F_0 is the amplitude of the electric field, $\vec{\xi}$ is the polarization vector of the wave, and $\vec{q}$ is the wavevector associated with the wavelength,

$$\lambda = 2\pi/q.$$

Alternatively, it is possible to use a complex form of the plane wave:

$$\vec{\mathcal{E}}(\vec{r}, t) = \vec{\xi} F_0 e^{-i(\vec{q}\vec{r} - \omega t)}, \tag{4.10}$$

but only the real or the imaginary part of Eq. (4.10) has physical meaning. This complex form is convenient if nonlinear effects are not important. For a uniform nonmagnetic dielectric, the magnetic field $\vec{\mathcal{H}}$ may be expressed in terms of $\vec{\mathcal{E}}$ as

$$\vec{\mathcal{H}}(z, t) = \frac{1}{\omega \mu_0} \vec{q} \times \vec{\mathcal{E}}. \tag{4.11}$$

As we mentioned earlier, the relative permeability of analyzed crystals is equal to unity, because they are *nonmagnetic* materials. For free space, the vector $\vec{\xi}$ is always perpendicular to $\vec{q}$; if $\vec{q}$ is fixed, the electric field has, in general, two projections in the plane perpendicular to the vector $\vec{q}$, which correspond to the *two possible polarizations of the electromagnetic wave.* The energy of the wave can be characterized by the density of the electromagnetic energy,

$$W = \epsilon_0 \kappa \overline{\mathcal{E}^2(t)} = \frac{1}{2}\epsilon_0 \kappa F_0^2, \tag{4.12}$$

where $\overline{\mathcal{E}^2(t)}$ represents the time average of $\mathcal{E}^2(t)$. Introducing the speed of light in vacuum as

$$c = \frac{1}{\sqrt{\epsilon_0 \mu_0}},$$

we can define *the intensity* of the wave as the energy flux through unit area perpendicular to the wavevector $\vec{q}$:

$$\mathcal{I} = \frac{c}{\sqrt{\kappa}} W = \frac{c}{2}\epsilon_0 \sqrt{\kappa} F_0^2. \tag{4.13}$$

In Eqs. (4.12) and (4.13), the dielectric permittivity of the material, κ, should be calculated at the frequency ω. The wavevector, $\vec{q}$, and the frequency, ω, and are related through the dispersion relationship

$$\omega = \omega_q \equiv \frac{c}{\sqrt{\kappa}} q. \tag{4.14}$$

It is assumed in this section that the material medium is uniform and isotropic, and that the dissipation of the energy of the field is negligible in most cases of interest. Equations (4.9)–(4.13) are associated with the *classical description* of electromagnetic fields.

In free space (in vacuum), $\kappa = 1$. Equation (4.1) for a plane electromagnetic wave and the relation between the electric and magnetic field vectors of Eq. (4.3) are just as valid for free space as they are for a dielectric medium. The frequency of an electromagnetic wave, ω, remains the same in free space and in a dielectric, while the wavelength and the wavevector of the wave are different according to Eq. (4.14). For the wavelength and the wavevector of a wave propagating in a dielectric, we will use the designations λ and q, respectively, in the following discussion. We designate the wavelength of a wave in free space, which is often used in optics, as

$$\lambda_0 = \frac{2\pi c}{\omega}.$$

According to quantum physics, electromagnetic radiation consists of an infinite number of *modes*, each of which is characterized by a wavevector $\vec{q}$ and a specific polarization $\vec{\xi}$. Each mode $\{\vec{q}, \vec{\xi}\}$ may be described in terms of a harmonic oscillator of frequency ω_q. Correspondingly, the energy separation between levels of this quantum-mechanical oscillator is

$$\hbar\omega_q = \frac{\hbar c}{\sqrt{\kappa}} q. \tag{4.15}$$

The oscillator can be in the non-excited state, which manifests the so-called ground-state or *zero-point* vibrations of the electromagnetic field. The existence of this zero-point energy is a purely quantum-mechanical phenomenon. The oscillator can be

Table 4.1 Comparison between classical and quantum quantities

Classical quantity	Corresponding quantum quantity
Density of optical energy W	Photon number $\bar{N} = W\mathcal{V}/\hbar\omega$
Optical intensity $\mathcal{I}(\vec{r})$	Photon flux density $\mathcal{I}(\vec{r})/\hbar\omega$
Total optical power $\mathcal{P}$	Photon flux $\mathcal{P}/\hbar\omega = Nc$

excited to some energy level. Let the integer $N_{\vec{q},\vec{\xi}}$ be a quantum number of this level; then the energy of the electromagnetic field associated with the oscillator in mode $\{\vec{q}, \vec{\xi}\}$ is

$$W_{q,\vec{\xi}}\mathcal{V} = \left(N_{q,\vec{\xi}} + \frac{1}{2}\right)\hbar\omega_q, \tag{4.16}$$

where $W_{q,\vec{\xi}}$ is the energy density of the mode and $\mathcal{V}$ is the volume of the system (say, a crystal). One refers to the number of excited levels, $N_{\vec{q},\vec{\xi}}$, as the number of quanta, or the *number of photons*, in the mode under consideration. Note that each of these modes is characterized by space- and time-dependent functions $\vec{\mathcal{E}}(\vec{r}, t)$ and $\vec{\mathcal{H}}(\vec{r}, t)$. Any given electromagnetic field can be described by a set of such photon numbers.

Because the quantum picture has to coincide with the classical picture for a large number of photons, i.e., when $N_{\vec{q},\vec{\xi}} \gg 1$, one must match Eqs. (4.12) and (4.16) in this limit. From this comparison, it is possible to find the relations of the classical amplitude of the electric field F_0 and of the intensity of the wave to the number of photons in the corresponding mode:

$$F_0 = \sqrt{2\hbar\omega_q N_{q,\vec{\xi}}/\epsilon_0\kappa\mathcal{V}}, \tag{4.17}$$

$$\mathcal{I}_{\vec{q},\vec{\xi}} = \frac{c}{\sqrt{\kappa}}\hbar\omega_q N_{q,\vec{\xi}}/\mathcal{V}. \tag{4.18}$$

Moreover, the relation between electromagnetic waves and photons is an example of the wave–particle duality which is characteristic for photons. A comparison of the different characteristics of electromagnetic fields in the classical and quantum pictures is given in Table 4.1.

It is very important that the different modes of an electromagnetic field do not interact with each other as reflected by the linear character of the field in free space; indeed, the equations of electromagnetism are linear equations. Generally, an interaction between these modes is possible only in special media. Such media are called *nonlinear optical media*.

Let us calculate the number of modes inside the frequency interval $d\omega$. We consider a uniform dielectric medium with dimensions much larger than the wavelength of the light, λ. It is known that the number of wavevectors inside the elementary interval $\vec{q}, \vec{q} + d\vec{q}$ is

$$\mathcal{V}\frac{dq_x\, dq_y\, dq_z}{(2\pi)^3}.$$

For an isotropic medium, we can transform this expression to a spherical coordinate system:

$$\mathcal{V}\frac{4\pi q^2}{(2\pi)^3}\,dq = \mathcal{V}\frac{\kappa^{3/2}\omega^2}{2\pi^2 c^3}\,d\omega,$$

where the left part of the equation corresponds to the number of modes for which modulus of the wavevector is in the interval $(q,\ q+dq)$. Taking into account the two independent polarizations of the waves, we find that the density of electromagnetic modes expressed in terms of the number of all modes per unit interval of frequency is

$$\nu(\omega) = \mathcal{V}\frac{\kappa^{3/2}\omega^2}{\pi^2 c^3}. \tag{4.19}$$

In terms of the wavelength, the mode number per unit wavelength interval is

$$\nu(\lambda) = \mathcal{V}\frac{8\pi}{\lambda^4}. \tag{4.20}$$

Equations (4.19) and (4.20) show that the density of the modes increases rapidly when ω increases or, equivalently, as λ decreases. For example, for the same interval $d\lambda$, the number of modes in the middle of the infrared range ($\lambda_0 \approx 5\times 10^{-4}$ cm) differs by four orders of magnitude from that in the visible region ($\lambda_0 \approx 5\times 10^{-5}$ cm). Consider a numerical example for $\hbar\omega \approx 1$ eV, i.e., $\omega \approx 1.5\times 10^{15}\,\text{s}^{-1}$, at $\kappa = 1$ and $\mathcal{V} = 1\,\text{cm}^3$. For this case one obtains $\nu(\omega) = 0.7\times 10^{-3}$ s. Increasing the density of states of the electromagnetic field has very important consequences: a decrease in the radiative lifetimes and an increase in the scattering rates of light in the short-wavelength range.

Exercise 4.1. How many photons per unit volume are there in a GaAs crystal, when a linearly polarized electromagnetic wave with $\lambda_0 = 10^{-6}$ m and $F_0 = 10^2$ V/m propagates in the crystal?

Solution. The wave propagating through the GaAs crystal is a plane wave:

$$\vec{\mathcal{E}}(\vec{r},t) = \vec{\xi}F_0 \exp[-i(\vec{q}\vec{r} - \omega t)], \tag{4.21}$$

where $\omega = cq/\sqrt{\kappa}$, $q = 2\pi/\lambda$, and κ is the dielectric constant of the crystal (for our case $\kappa = 12.9$), so $\omega = 2\pi c/\lambda\sqrt{\kappa}$. The energy of the wave per unit volume (see Eq. (4.12)), i.e., the density of the energy, is

$$W = \epsilon_0\kappa\vec{\mathcal{E}}^2(t) = \frac{1}{2}\epsilon_0\kappa F_0^2. \tag{4.22}$$

since $|\vec{\xi}| = 1$. The wave (4.21) is monochromatic; the density of the photons $\mathcal{N}$ of this fixed mode should be calculated in accordance with Table 4.1 as $W/\hbar\omega$:

$$\mathcal{N} \equiv \frac{N}{\mathcal{V}} = \frac{W}{\hbar\omega} = \frac{\epsilon_0\kappa^{3/2}\lambda}{4\pi c\hbar}F_0^2. \tag{4.23}$$

Numerical evaluation of Eq. (4.23) gives $\mathcal{N} \approx 10^{13}$ photons/m^3.

The formulae in this text are written in the SI system of units as is widely used in electronics and optoelectronics. At the same time, we follow a practice whereby the electric field is given in V/cm, which deviates slightly from the pure SI system of units since the meter is the unit of distance in the SI system instead of the centimeter, which is the unit of distance in the Gaussian system. We, like the authors of the overwhelming majority of texts on this subject, use electron-volts for energy: $1\ \mathrm{eV} = 1.602 \times 10^{-19}\ \mathrm{J} = 1.602 \times 10^{-12}$ erg, where J and erg are the units of energy in the SI and Gaussian systems of units, respectively.

4.2.2 Photons in Nonuniform Dielectric Media

In the previous discussion we introduced the modes of the electromagnetic field and, consequently, described the photons for free space and for uniform dielectric media with dimensions much greater than the electromagnetic wavelengths. In fact, Eqs. (4.19) and (4.20) correspond to a very large box-like resonator. In the general case, the modes of an electromagnetic field and photons may be introduced by the following method. Let a dielectric medium be characterized by a dielectric permittivity which is dependent on the space coordinates, $\kappa(\vec{r})$. Let us assume that the medium is embedded in some completely or partially reflective enclosure which is called a resonator. Maxwell's wave equations determine all of the harmonic modes for the electromagnetic fields, $\mathcal{E}_\nu$ and $\mathcal{H}_\nu$, in the system, where ν is the set of discrete or continuous parameters associated with the solutions. Each solution is characterized by a frequency ω_ν. In the absence of absorption of electromagnetic energy all the frequencies ω_ν are real quantities. One may consider any of these solutions as a mode of the field. Thus, in this way we obtain the total mode structure of the field. Note that, in the general case, the solutions of Maxwell's equations can differ considerably from the plane wave solution of Eq. (4.9). The mode structure, the frequency, the density of states, and other characteristics are strongly dependent on the type of dielectric medium, its geometry, and the properties of the optical resonator.

As an example of a spatially nonuniform system, let us consider a semiconductor heterostructure with variations in the composition and bandgap in one direction, say in the z-direction, as illustrated in Fig. 4.1(a). This variation in the bandgap causes a change in the dielectric constant (see Fig. 4.1(b)). In the narrow-bandgap GaAs, the structure has a larger optical density than it does in the wider-bandgap AlGaAs. For this case, the system of modes is different from that for plane waves. In particular, there are the modes localized within the narrow-bandgap-material region – *the waveguide modes*. Of course, the amplitudes of these modes decay far away from the layer of the narrower-bandgap material, as shown in Fig. 4.1(c). Similar structures and localized modes are used in modern heterostructure lasers.

We already know that, according to the theory of the quantization of the electromagnetic field, each of the previously considered modes represents a particular oscillator, or photon, with energy $\hbar\omega_\nu$ ($\nu \equiv \{\vec{q}, \vec{\xi}\}$). An excitation of this oscillator is characterized by the number of photons, N_ν, and according to Eq. (4.16) the energy of this mode is related to N_ν as follows:

$$W_\nu \mathcal{V} = \hbar\omega_\nu \left(N_\nu + \frac{1}{2} \right).$$

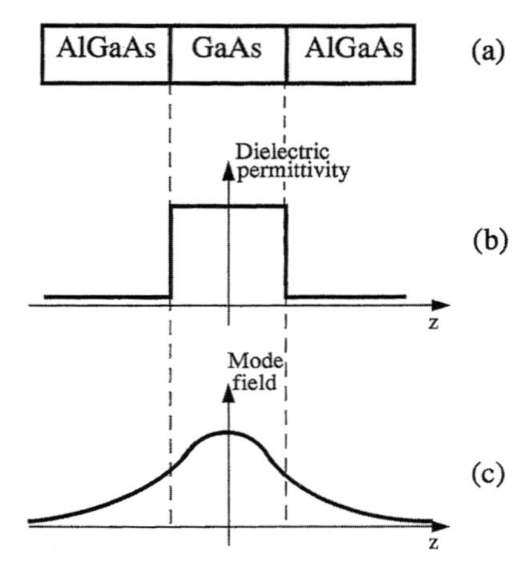

Figure 4.1 A double GaAs/AlGaAs heterostructure as a waveguide. (a) The scheme of the structure. (b) The coordinate dependence of the dielectric permittivity. (c) A sketch of the spatial profile of the fundamental optical mode localized near the GaAs narrow-bandgap layer.

The solutions of the Maxwell equations, $\vec{\mathcal{E}}_\nu(\vec{r})$ and $\vec{\mathcal{H}}_\nu(\vec{r})$, on the basis of which these photons are introduced, are frequently referred to as the *form factors of the photon*. The form factor determines the spatial distribution of the field for a fixed mode, and it represents a very important characteristic of the mode. The probability $P(\vec{r})d\mathcal{V}$ of observing a photon at a point $\vec{r}$ is proportional to the density of electromagnetic energy $W(\vec{r})$, which can be calculated using the form factor. For example, let the form factor of a certain mode be a standing wave $\mathcal{E}(z) = F_0 \sin(2\pi z/\lambda)$. Then at the points $z_n = \lambda n/2$ (the nodal points) the probability of observing the photon is equal to zero; $P = 0$. An atom placed at a point where $\mathcal{E} = 0$ *does not interact with this mode*. At any point where the form factor does not vanish, the atom interacts with the field and can absorb or emit a photon corresponding to the form factor.

These properties of electromagnetic fields are very important for tailoring light–matter interactions because they facilitate the determination of the desired distributions of the fields as well as the conditions resulting in efficient interaction with the electron subsystem. The distributions of the electric and magnetic fields are important for semiconductor lasers, photodetectors, waveguides, and other optoelectronics devices.

4.2.3 Optical Resonators

Let us consider an optical resonator. In the simplest case the resonator consists of plane or curved mirrors, which provide repeated reflections and some kind of "trap" for the light – a cavity – in the region between the mirrors. The optical waves which can be trapped in this cavity compose the resonator modes. A universal

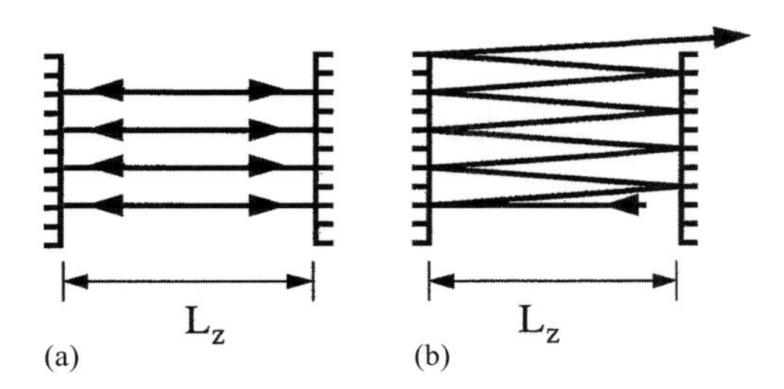

Figure 4.2 Longitudinal and transverse modes in a planar Fabry–Pérot resonator. (a) Longitudinal modes. Strictly perpendicular light rays do not escape from the resonator. (b) Transverse modes. Slightly inclined light rays eventually escape from the resonator and have poor quality factors.

characteristic of the resonator mode is the *quality factor Q*, which can be defined as

$$Q = \omega \times \frac{\text{field energy stored in the cavity}}{\text{power dissipated in the resonator}}. \tag{4.24}$$

The dissipation of electromagnetic energy is caused by many factors: absorption by the mirrors or by matter inside the cavity, light transmission through the mirrors, light scattering, radiation passing out of the resonator as a result of diffraction, etc. For different modes, the quality factors differ and, accordingly, the dissipated power will, in general, be different for each mode.

We would like to draw attention to the very high density of electromagnetic modes in the optical spectrum. This high mode density makes possible the emission of a large number of modes with different frequencies, polarizations, and all possible propagation directions. To avoid this type of extremely incoherent excitation of modes, one can employ so-called *open optical resonators*. The simplest open resonator consists of two plane mirrors parallel to each other, which are finite in their transverse dimensions; an example is given by the so-called Fabry–Pérot etalon, illustrated in Fig. 4.2. In this resonator, most of the modes propagating through the cavity are lost in a few transversals of the cavity since the mirrors are inclined with respect to the mode-propagation directions. This implies that most of the modes – the so-called *transverse modes* (with slightly inclined rays) – have a very low quality factor. Only the waves propagating perpendicular to the mirrors can be reflected back and travel from one mirror to another without escaping from the resonator. These waves correspond to the so-called *longitudinal modes* (with rays strictly perpendicular to the mirrors) of the resonator. Thus, for finite dimensions of mirrors only longitudinal modes can have a high quality factor. Their loss and diffraction are caused by absorption by the mirrors, by transmission through the mirrors, and by wave diffraction on the sides of the mirrors. The latter losses can be made much smaller than those arising from other loss mechanisms. Thus, the open resonator provides a means for strong discrimination of modes. Relatively few of these modes have a high quality factor, Q. According to the definition in Eq. (4.24), it is possible to accumulate light energy in these high-quality modes.

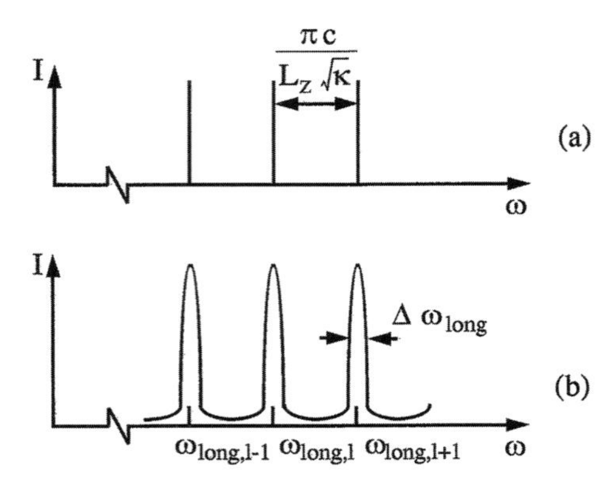

Figure 4.3 A schematic representation of the spectral dependences of intensities of the resonator modes. (a) An ideal lossless resonator. (b) A resonator with finite losses. In (a), waves exist only at precise resonance frequencies. A resonator with losses sustains waves at all frequencies, but attenuation damps the wave magnitudes for frequencies away from resonances.

For ideal mirrors with 100% reflection, the longitudinal modes for which $q_x = q_y = 0$, as in Fig. 4.2(a), are standing waves,

$$\vec{\mathcal{E}} = \vec{\xi}\mathcal{E}_0 \sin(q_z z)\ \cos(\omega t), \tag{4.25}$$

where $\vec{\xi}$ is perpendicular to the resonator axis, z. Since the electric field should vanish at the surfaces of the mirrors, at $z = 0$ and at $z = L_z$ one obtains

$$q_z = \frac{\pi}{L_z} l, \quad l = 1, 2, 3, \ldots \tag{4.26}$$

Thus, instead of a continuous spectrum of electromagnetic waves as in the case of free space, one gets an infinite set of equally separated frequencies, as illustrated in Fig. 4.3(a):

$$\omega_{\text{long},l} = \frac{\pi c}{L_z\sqrt{\kappa}} l. \tag{4.27}$$

The separation between the frequencies of the longitudinal modes depends on the resonator length, L_z.

It is easy to take into account finite reflection and transmission of the mirrors and losses in the cavity. Let us consider an "external" wave traveling through the resonator, say from left to right. Let its amplitude at a particular point of the resonator be F_0. At the same point after a double reflection, we find a wave with the same propagation direction but with amplitude $re^{2i\delta} \times F_0$, where r is the attenuation factor due to the finite mirror transmission, absorption in the cavity, etc., and $\delta \equiv (\omega\sqrt{\kappa}/c)L_z$ is the change of the wave's phase after a double traversal of the resonator. We obtain a similar result after the next double traversal: $re^{2i\delta} \times (re^{2i\delta}) \times F_0$. In fact, the amplitude of the electromagnetic field can be found as a superposition of these waves:

$$F = F_0 + re^{2i\delta}F_0 + r^2e^{4i\delta}F_0 + \cdots = \frac{F_0}{1 - re^{2i\delta}}.$$

Thus we can write the light intensity in the resonator as

$$I = \frac{I_{\max}}{|1 - re^{2i\delta}|^2} = \frac{I_{\max}}{1 + (2F/\pi)^2 \sin^2(\omega\sqrt{\kappa}/c)L_z} \frac{1}{(1-r)^2}, \tag{4.28}$$

where $F \equiv \pi\sqrt{r}/(1-r)$ is a parameter called the *finesse* or *contrast* parameter of the resonator. From Eq. (4.28) we see that the intensity is a periodic function of the frequency, ω, and it has maxima when ω satisfies the resonance conditions of Eq. (4.27). The peaks have a full width at half maximum equal to

$$\Delta\omega_{\text{long},m} = \frac{\pi c}{L_z\sqrt{\kappa}}.$$

Thus, instead of a set of infinitely narrow and high lines as in Fig. 4.3(a), we obtain the mode structure presented in Fig. 4.3(b). In Eq. (4.28) the value $I_{\max}$ is still arbitrary. It is determined by the method of generation of the electromagnetic fields, which can be an external light accumulated in the cavity or light generated by emission inside this cavity. The ratio between the maximum and minimum of the intensity is independent of these methods and is given by

$$\frac{I_{\min}}{I_{\max}} = \frac{1}{1 + (2F/\pi)^2}.$$

Thus, the Fabry–Pérot resonator allows one to discriminate between transverse and longitudinal modes and provides high quality factors for the latter.

The density of states for the longitudinal modes is

$$\nu_{\text{long}}(\omega) = 2\frac{L_z\sqrt{\kappa}}{\pi c}, \tag{4.29}$$

where we have taken into account the two possible polarizations of the wave. This density of states is independent of the light frequency. For $L_z = 1$ cm, we find that the separation between the modes, or *internode spacing*, $\Delta\omega_{\text{long},l}$, is $9.4 \times 10^{10}\ \text{s}^{-1}$ and that $\nu_{\text{long}} = 2 \times 10^{-11}$ s. (Compare the latter value with the previously presented estimate for free space.)

Typical resonator lengths for semiconductor applications are a few hundreds of micrometers. For these resonators, the intermode spacing is still less than the spectral range of emission. To decrease the number of excited modes, it is necessary to increase the intermode spacing to the extreme limit, $L_z = \lambda/2$, i.e., half of the light wavelength in the medium. Such a resonator is called a *microresonator or microcavity*. Modern semiconductor technology makes possible the fabrication of these resonators. We will consider microresonators in the next chapter.

4.2.4 Photon Statistics

For most cases of optoelectronic applications, the electromagnetic fields are far from equilibrium. However, it is instructive to recall from physics that, as a result of interaction with a black body characterized by temperature T, electromagnetic radiation can also come into equilibrium and be characterized by the same temperature. According to the Planck distribution, the number of photons in some chosen mode at equilibrium equals

$$N^{(\mathrm{eq})}_{\vec{q},\vec{\xi}} = \frac{1}{e^{\hbar\omega_q/k_B T} - 1}. \tag{4.30}$$

For the resonator considered above, which is formed by two mirrors, as presented in Fig. 4.3, this formula gives the photon numbers if q and ω_q are determined by Eqs. (4.26) and (4.27). Equations (4.12), (4.15), and (4.30) allow us to evaluate all of the equilibrium properties of the radiation: the spectral density of the energy, the total energy, etc.

4.3 Light Interaction with Matter: Phototransitions

4.3.1 Photon Absorption and Emission

Among the different processes for the interaction of electromagnetic fields and matter, here we review briefly the three major processes: absorption, spontaneous emission, and stimulated emission.

In order to visualize these processes, let us consider a simple two-level system with energies E_1 and E_2 as depicted in Fig. 4.4. The different occupancies of the energy levels of this system correspond to particular states of a system of charged particles, e.g. electrons. The charged particles interact with the electromagnetic field. This interaction is, of course, associated with transitions between quantum states of the system. These transitions are frequently referred to as *phototransitions*. According to quantum theory, the system changes its energy as a result of interaction with electromagnetic waves with frequency

$$\omega = (E_2 - E_1)/\hbar. \tag{4.31}$$

If the lowest energy level, E_1, is occupied, the wave can excite the system into an upper level, E_2, and the electromagnetic energy must decrease. One can describe this process as the *absorption of one photon* because the energy of the electromagnetic field decreases by $E_2 - E_1$. If the upper level, E_2, is occupied, the system can make a transition to level E_1 as a result of interaction with the electromagnetic field. The electromagnetic energy thereby increases by an amount $E_2 - E_1$. This

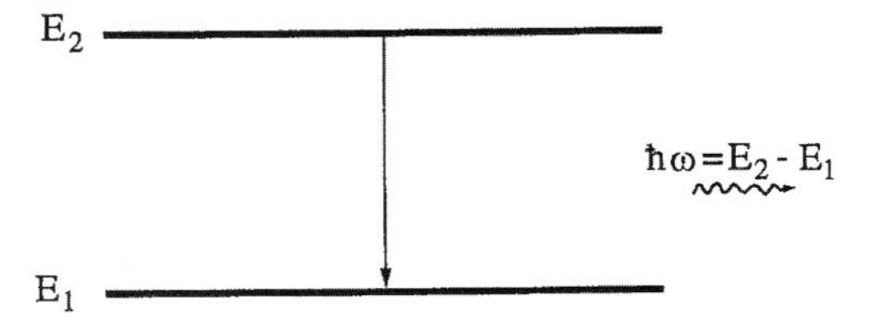

Figure 4.4 A two-level system and the spontaneous emission of a photon.

process represents the *emission of a photon* with energy $\hbar\omega$. When activated by an external electromagnetic wave, the latter process is called *stimulated emission*. It is important for stimulated emission that each emitted photon has energy, direction, polarization, and even phase coinciding precisely with those of the stimulating wave.

Both absorption and stimulated emission can be described as being the result of an interaction with a classical electromagnetic wave. The rates of these processes are proportional to the intensity of the wave. According to Eq. (4.18), this implies that these rates are proportional to the number of photons of given frequency, ω, wavevector, $\vec{q}$, and polarization, $\vec{\xi}$. In this subsection we omit the mode indices in the designation of the frequency. Thus, the rates are proportional to the number of the photons in a given mode, $N_{\vec{q},\vec{\xi}}$:

$$R_{\text{abs}} = B_{12} N_{\vec{q},\vec{\xi}} n_1, \tag{4.32}$$

$$R_{\text{st.em}} = B_{21} N_{\vec{q},\vec{\xi}} n_2, \tag{4.33}$$

where n_1 and n_2 are the numbers of particles in the system occupying levels 1 and 2, respectively. B_{12} and B_{21} are kinetic coefficients describing these processes; the physical significance of these coefficients will be addressed in subsequent discussions. It is easy to see that the two processes of absorption and emission are insufficient to describe the whole picture of interaction between the radiation and matter. For example, let us apply these two processes only for thermal equilibrium conditions, where the ratio of the populations of the two levels is

$$\frac{n_2}{n_1} = e^{-(E_2-E_1)/k_B T} = e^{-\hbar\omega/k_B T}; \tag{4.34}$$

for the sake of simplicity we assume that the degeneracy equals 1 for each level. Using Eqs. (4.30) and (4.32)–(4.34), one can see that $R_{\text{abs}} \neq R_{\text{st.em}}$ at any temperature T. This result is in contradiction to the expected equilibrium between the system and the field. According to the Einstein theory, there is an additional quantum radiative transition in the system with *spontaneous emission of a photon of the same mode*. The rate of this process is

$$R_{\text{sp.em}} = A_{21} n_2, \tag{4.35}$$

where A_{21} is the coefficient or rate of spontaneous emission. The spontaneous process does not depend on the intensity of the electromagnetic wave and takes place even in the absence of this wave. According to quantum electrodynamics, the excited material system spontaneously emits a photon due to interaction with the zero-point vibrations of electromagnetic fields. These zero-point vibrations are related to the fact that the ground-state energy of a harmonic oscillator of frequency ω is $\hbar\omega/2$ and not zero as would be the case for a classical harmonic oscillator.

In contrast to the case of stimulated emission, a photon produced by the spontaneous process has an arbitrary phase. Moreover, this process produces photons with different directions of $\vec{q}$ and polarizations, but fixed energy, i.e., it produces photons of different modes.

Now we can apply the results of Eqs. (4.32), (4.33), and (4.35) to thermal equilibrium. Under equilibrium conditions the total rate of photon emission has to be equal to the rate of photon absorption, thus we will have

$$R_{\text{abs}} = R_{\text{sp.em}} + R_{\text{st.em}}. \tag{4.36}$$

Using the Planck formula of Eq. (4.30) and the ratio of n_2/n_1 of Eq. (4.34), and substituting the expressions for R_{abs}, $R_{\text{sp.em}}$, and $R_{\text{st.em}}$ into Eq. (4.36), one finds the relation

$$B_{21} - A_{21} = (B_{12} - A_{21})e^{\hbar\omega/k_B T}. \tag{4.37}$$

Because this relation has to be satisfied at any arbitrary temperature T, one obtains two equalities:

$$A_{21} = B_{21} = B_{12}. \tag{4.38}$$

Thus, we have established the existence of three basic processes of resonant interaction of radiation and matter: absorption, stimulated emission, and spontaneous emission of photons of an arbitrary mode. Moreover, we have found the relations between the coefficients determining the rates of these processes. It is worth emphasizing that all three processes are related to interactions with photons of the same mode.

As shown previously, an optical resonator effectively enhances the mode structure in the cavity. Combining this observation with the basic optical processes discussed previously, we conclude that these processes are not immutable properties of an atom, molecule, or any other system. In fact, the processes, including spontaneous emission, are consequences of the matter–field coupling, and they would disappear if there were no proper resonant states of the electromagnetic field. For example, in the mirror box considered previously, an atom does not emit or absorb photons if its frequency as given by Eq. (4.31) differs from the resonance-mode frequencies. This result is not very significant for resonators with dimensions much larger than the resonance wavelength, but it becomes very important for microresonators with large intermode frequency differences.

The sum of the stimulated and spontaneous emission rates, as determined by Eqs. (4.33) and (4.35), gives the total emission rate of photons for a fixed mode:

$$R_{\text{em}} = A_{21}(1 + N_{\vec{q},\vec{\xi}})n_2. \tag{4.39}$$

From this equation, one can see that for the case of a fixed mode for which the number of photons, $N_{\vec{q},\vec{\xi}}$, is sufficiently larger than 1, stimulated emission dominates over spontaneous emission. However, there is spontaneous emission of a great number of other modes with the same frequency but with different directions of $\vec{q}$ and different polarizations. This total spontaneous emission can be the dominant radiative process, even if stimulated emission is the most important process for a particular mode.

Now, let us compare the absorption and stimulated emission by calculating the rate of increase of the number of photons in some fixed mode:

$$\left(\frac{dN_{\vec{q},\vec{\xi}}}{dt}\right)_{\mathrm{st}} \equiv R \equiv R_{\mathrm{st.em}} - R_{\mathrm{abs}} = B_{21}N_{\vec{q},\vec{\xi}}(n_2 - n_1). \tag{4.40}$$

This result shows that if

$$n_2 - n_1 > 0, \tag{4.41}$$

stimulated emission dominates over absorption. Evidently, under equilibrium conditions, the opposite is true as $n_1 > n_2$.

The inequality of Eq. (4.41) is the criterion for *population inversion*. If a population inversion is achieved, electromagnetic waves with the resonance frequency can be amplified when passing through the material medium. This process of amplification of the radiation due to population inversion is the key mechanism underlying the operation of a LASER (*Light Amplification by Stimulated Emission of Radiation*).

In the previously obtained equations, the coefficients A_{21}, B_{21}, and B_{12}, which describe the interaction of waves and matter, are linked by the two equalities of Eq. (4.38), but at least one of these coefficients must be calculated independently. At this point, we end our discussion of the simple two-level model and start to consider calculations for more realistic systems.

4.3.2 Calculation of Phototransition Probabilities

We assume that there exist electron states with energies E_i, where i can take on discrete as well as continuous values. For simple, but also the most important, cases, we can use the *dipole approximation* for the energy of the light–matter interaction:

$$V_{\mathrm{int}} = -\vec{D}\vec{\mathcal{E}}, \tag{4.42}$$

where

$$\vec{D} = \sum_n \vec{D}_n \equiv e\sum_n \vec{r}_n Z_n \tag{4.43}$$

is the dipole moment of the system, $Z_n = 1$ for holes and -1 for electrons, and $\vec{\mathcal{E}}$ is the electric field of a wave that has the form of Eq. (4.9). In Eq. (4.43), the sum extends over all charges of the system whose coordinates are $\vec{r}_n$. We consider here only phototransitions involving electrons (holes) and do not analyze the interaction of light with the lattice.

For typical physical conditions, the photon wavevector $\vec{q}$ can be neglected in Eq. (4.9) because the wave amplitude varies over distances which usually are substantially larger than any characteristic scale of the electrons.

In the framework of the semiclassical description of electromagnetic waves, we can calculate the transition probability per unit time between any two states of the system, say between the initial state i and the final state f. For such a process, the interaction potential of Eq. (4.42) acts as a perturbation potential of the system. According to the Fermi golden rule, the probability of transition of one of the electrons from state i to state f is

$$P_{i\to f}=\frac{2\pi}{\hbar}|\langle f|\bar{V}_{f,i}|i\rangle|^2\delta(E_f-E_i-\hbar\omega). \tag{4.44}$$

Here, $\langle f|\bar{V}_{f,i}|i\rangle$ is the matrix element of the perturbation calculated for the wave function of the initial state, $|i\rangle$, and the conjugated wave function of the final state, $\langle f|$ (see Eq. (4.46)). The time-dependent perturbation is assumed to take the form

$$V=\bar{V}^*e^{-i\omega t}+\bar{V}e^{i\omega t}.$$

In the case under consideration we write V_{int} as

$$V_{\text{int}}=-\frac{1}{2}F_0\xi\vec{D}\left(e^{i\omega t}+e^{-i\omega t}\right).$$

Now the transition probability is

$$P_{i\to f}=\frac{\pi}{2\hbar}e^2|\langle f|F_0\xi\vec{D}|i\rangle|^2\delta(E_f-E_i-\hbar\omega), \tag{4.45}$$

where $\langle f|\xi\vec{D}|i\rangle$ is the matrix element calculated for the wave functions of the states i and f. The following shorthand notation is used for the matrix element of an arbitrary function, A, calculated on wave functions Ψ_i, Ψ_f:

$$\int\Psi_f^*A\Psi_i\,d\vec{r}\equiv\langle f|A|i\rangle, \tag{4.46}$$

i.e., $|i\rangle\equiv\Psi_i$ and $\langle f|\equiv\Psi_f^*$, where i, f are sets of quantum numbers.

If $\mathcal{F}(E_\nu)$ denotes the average occupancy of some state ν of the system by electrons (see Chapter 3 for a discussion of the distribution functions of electrons), one can write the *rate of phototransitions* from state i to state f in the form

$$R_{i\to f}=P_{i\to f}\mathcal{F}(E_i)\left[1-\mathcal{F}(E_f)\right], \tag{4.47}$$

where the last multiplier is the number of "empty" places (empty states) which can be populated as a result of the phototransition. If the i state corresponds to the lower energy, then Eq. (4.47) determines the rate of absorption of radiation, i.e., the rate of decrease of the number of photons. Similarly, one can write the rate of the inverse process, $f\to i$:

$$R_{f\to i}=P_{f\to i}\mathcal{F}(E_f)[1-\mathcal{F}(E_i)], \tag{4.48}$$

with

$$P_{f\to i}=\frac{\pi}{2\hbar}e^2|\langle i|F_0\xi\vec{r}|f\rangle|^2\delta\left(E_f-E_i-\hbar\omega\right). \tag{4.49}$$

The δ-functions in Eqs. (4.45) and (4.49) ensure that the initial and final states are separated by the photon energy, $\hbar\omega$. Equation (4.48) determines the rate of increase of radiation intensity, i.e., the rate of increase of the number of photons in a given mode. Finally, by summing Eqs. (4.47) and (4.48) over all possible initial and final states i and f, we can obtain the total rate of increase of the number of photons in the fixed mode,

$$R = \sum_{i,f} \left(R_{f \to i} - R_{i \to f} \right)$$
$$= \frac{\pi e^2}{2\hbar} \sum_{i,f} |\langle i | F_0 \vec{\xi} \vec{r} | f \rangle|^2 \delta \left(E_f - E_i - \hbar\omega \right) \left[\mathcal{F}(E_f) - \mathcal{F}(E_i) \right]. \quad (4.50)$$

Here we have taken into account that the squares of the matrix elements in Eqs. (4.45) and (4.49) are equal. One can see that Eq. (4.50) is an exact analogy of Eq. (4.40), which describes the same quantity for the simple two-level model. Thus, the rate of increase (decrease) of the number of photons, R, is proportional to the difference between the occupancies of the electron states between which the phototransition occurs.

For a spatially homogeneous crystal, Eq. (4.50) allows one to calculate another important characteristic of the system, namely the amplification (absorption) coefficient of the electromagnetic wave.

The rate of change in the energy density of electromagnetic waves can be expressed through the rate of increase of the photon number, R:

$$\frac{dW}{dt} = \frac{\hbar\omega}{\mathcal{V}} R. \quad (4.51)$$

Let us assume that the wave propagates along the z-axis. Then, one can rewrite Eq. (4.51) in terms of the derivative with respect to z:

$$\frac{dW}{dt} = \frac{dW}{dz} \frac{c}{\sqrt{\kappa}}. \quad (4.52)$$

Eliminating F_0^2 from Eqs. (4.12) and (4.50), we can introduce

$$\alpha \equiv -\frac{1}{W} \frac{dW}{dz} \equiv \frac{\sqrt{\kappa}}{c} \frac{\hbar\omega R}{W\mathcal{V}} = \frac{R}{(c/\sqrt{\kappa})N}. \quad (4.53)$$

According to Eq. (4.13), we can rewrite Eq. (4.53) as

$$\frac{dI}{dz} = -\alpha I. \quad (4.54)$$

From (4.54) it is obvious that the coefficient α characterizes an absorption or an amplification of the light. If the coefficient α is positive, it can be called the absorption coefficient. If the coefficient is negative, $-\alpha$ can be called the *amplification coefficient*, or the *gain coefficient*. In an explicit form for the case of plane waves, this coefficient is given by

$$\alpha = -\frac{4\pi^2 e^2 \omega}{c\sqrt{\kappa}\mathcal{V}} \sum_{i,f} |\langle i | \vec{\xi} \vec{r} | f \rangle|^2 \delta \left(E_f - E_i - \hbar\omega \right) \left[\mathcal{F}(E_f) - \mathcal{F}(E_i) \right]. \quad (4.55)$$

Thus, Eq. (4.55) gives the decrement or increment of the electromagnetic wave intensity in space due to phototransitions in the system interacting with the wave:

$$I(z) = I_0 \times \begin{cases} e^{-\alpha z}, & \alpha > 0, \\ e^{|\alpha| z}, & \alpha < 0, \end{cases} \tag{4.56}$$

where I_0 is the wave intensity at $z = 0$. So, we emphasize that absorption of the light occurs if $\alpha > 0$, whereas amplification of the light is possible if $\alpha < 0$. Some texts introduce the designation $\gamma = -\alpha$ as the amplification coefficient at $\alpha < 0$.

It follows from these results that the properties of phototransitions depend on the wave functions and energies of the electron states involved in the process. The electron occupancies of the states are also important. The initial and final states can belong to different electron bands, or to impurities and a band, or to the same band. Depending upon these factors, different mechanisms of absorption and emission of light are possible, and we discuss those below.

4.4 Optical Properties of Bulk Semiconductors

In this section, we shall consider the mechanisms of absorption and emission of photons in semiconductors. Among them, the most important are as follows.

1. *Interband phototransitions (band-to-band transitions)*. Absorption of a photon can result in the creation of an electron–hole pair. The inverse process is radiative electron–hole recombination resulting in the emission of a photon, as illustrated in Fig. 4.5(a).
2. *Impurity-to-band transitions*. In doped semiconductors, an absorbed photon can result in a transition between the bound state of an impurity – a donor or acceptor – and the conduction or valence band, as illustrated in Fig. 4.5(b) for the case of acceptors.
3. *Free-carrier transitions (intraband transitions)*. An absorbed photon can transfer its energy to an electron or a hole, thereby increasing the energy of the electron or

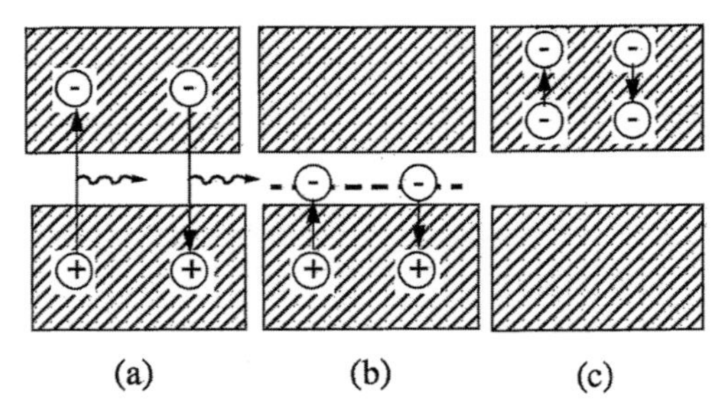

Figure 4.5 Schematics of the absorption and emission of photons in a semiconductor. (a) Band-to-band transitions in GaAs. (b) The absorption of a photon leads to a valence-band-to-acceptor-level transition. (c) A free-carrier transition within the conduction band.

hole within the same band, as illustrated in Fig. 4.5(c). It is important to note that free-carrier transitions require the participation of a "third particle" (another carrier, defect, phonon, etc. to ensure conservation of the momentum as photons have very small wavevectors compared with electron (hole) wavevectors). Inverse processes, such as emission of photons, also occur. Note that the intraband transition probability decreases with increasing photon energy because a larger momentum transfer is required from the "third particle."

4. *Excitonic transitions*. The absorption of a photon can lead to the formation of an electron and a hole in coupled states; these weakly bound electron–hole states are known as excitons. An annihilating exciton can produce a photon.
5. *Phonon-involving phototransitions*. Long-wavelength photons can be absorbed in the excitation of lattice vibrations, i.e., in the process of creating phonons. This mechanism does not involve electrons, but it can result in an overlap of this type of absorption with intraband phototransitions; for example, phonon absorption occurs for the regions from 0.02 to 0.07 eV for GaAs and from 0.1 to 0.2 eV for Si, while free-carrier transitions correspond to energies of less than 0.3 eV for both materials.

The aforementioned phototransitions contribute to the overall absorption in a wide spectral region from far infrared up to ultraviolet spectra.

Figure 4.6 illustrates absorption coefficients for band-to-band transitions for different bulk materials. From this figure, one can see that the absorption increases sharply in the short-wavelength region. Let E_g be the bandgap of the semiconductor. The material is practically transparent for $\hbar\omega < E_g$, where absorption is small.

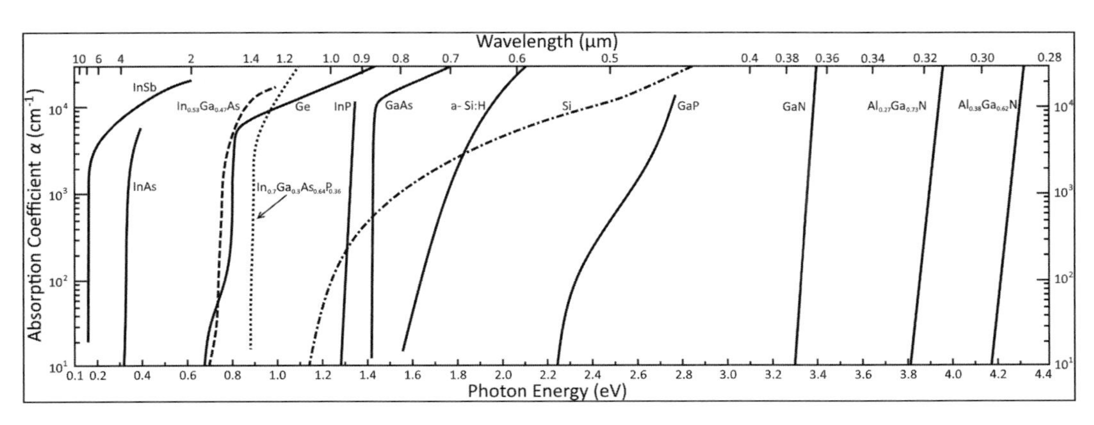

Figure 4.6 Absorption coefficient versus photon energy and wavelength for interband phototransitions in various semiconductors. Data were taken for GaN, $Al_{0.27}Ga_{0.73}N$, and $Al_{0.38}Ga_{0.62}N$ from J. F. Muth *et al.*, *MRS Internet J. Nitride Semicond. Res.* **4**, 51, 502 (1999), for $In_{0.53}Ga_{0.47}As$, $In_{0.7}Ga_{0.3}As_{0.64}P_{0.36}$, G_e, a-Si:H, and Si from G. Barbarino *et al.*, "Silicon photo multipliers detectors operating in Geiger regime: an unlimited device for future applications" in *Photodiodes: World Activities in 2011*, ed. J. W. Park (2011), and for InSb, InAs, InP, GaAs, and GaP from G. E. Stillman, *et al.*, *IEEE Trans. Electron Devices* **31**, 1643–1655. See also related data in Tables 2.11 and 2.12 as well as in Figs. 2.19, 4.17, 4.18, and 7.30.

The situation changes sharply to strong absorption for $\hbar\omega > E_g$ such that $\hbar\omega_g = E_g$ corresponds to the *absorption edge*. The shape of the absorption edge depends significantly on the structure of the electron bands. Direct-bandgap semiconductors such as GaAs have a more abrupt absorption edge and a larger absorption value than do indirect-bandgap materials, of which Si provides an example. In the following sections we will consider in detail the phototransition mechanisms which have just been introduced.

4.4.1 Interband Emission and Absorption in Bulk Semiconductors

As we can see from Fig. 4.6, the contribution of interband transitions to the intensities of emission and absorption processes sharply increases in the spectral region corresponding to the bandgap energy. We can introduce the so-called *bandgap wavelength*, or the cut-off wavelength, $\lambda_g = 2\pi c\hbar/E_g$. If E_g is given in electron-volts, the bandgap wavelength in micrometers is

$$\lambda_g = \frac{1.24}{E_g}. \tag{4.57}$$

The values of E_g and λ_g for various III–V semiconductor materials are apparent from the curves plotted in Fig. 4.6. One can see that interband transitions in III–V compounds cover a wide range from infrared to visible spectra. Optical activity in this spectral region is crucial for all optoelectronic applications of these materials.

A photon absorbed during an interband transition excites an electron from the valence band to the conduction band; i.e., it creates an electron–hole pair as depicted in Fig. 4.7(a). The inverse process – the phototransition of an electron from the conduction band to the valence band – is referred to as the *radiative recombination* (annihilation) of an electron and a hole; this is depicted in Figs. 4.7(b) and (c). According to the general properties of phototransitions highlighted in Section 4.3.1, there exist two such processes: spontaneous and stimulated emission, as

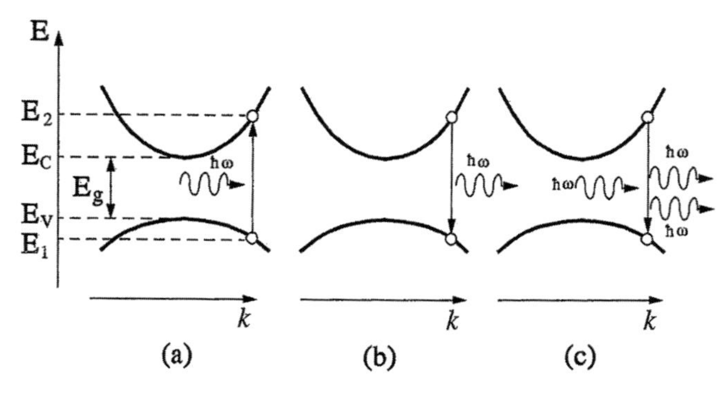

Figure 4.7 (a) The absorption of a photon results in electron–hole generation: when an electron leaves the valence band, the lack of an electron can be considered as a *hole* in the valence band. (b) Radiative recombination: an electron can go into the valence band if the corresponding state is empty, i.e., if there is a hole in this state. (c) Stimulated emission.

illustrated by Figs. 4.7(b) and (c), respectively. In order to analyze these interband transitions, we use the general expression for the rates of the absorption and emission of Eq. (4.50). In the case under consideration, the initial state i and the final state, f, should be selected from the states of the conduction and valence bands. For bulk materials, the set of quantum numbers depends on the type of the band (conduction or valence), the wavevector of the carrier (k_c or k_v, respectively), and the spin σ. Since we are considering the simplest case, we neglect effects associated with a change of the spin states so that spin conservation during phototransitions is assumed. For example, in the case of interband light absorption we should set $i=\{v, k_v, \sigma\}$ and $f=\{c, k_c, \sigma\}$ in Eq. (4.50).

Before we can present further calculations, it is necessary to make a general remark concerning electron wave functions and interband transitions. So far, we have studied processes involving electron states of the same energy band, i.e., *intraband processes*. We have implicitly relied on the effective-mass method (see Subsection 3.1.3), and on the corresponding Schrödinger equation and its solutions. It is now appropriate to recall that the *true wave functions* of electrons in a bulk crystal have the form

$$\Psi_{\vec{k}}(\vec{r}) = \frac{1}{\sqrt{\mathcal{V}}} e^{i\vec{k}\vec{r}} u_{\vec{k}}(\vec{r}), \tag{4.58}$$

where $\mathcal{V}$ is the crystal volume, $\vec{k}$ is the electron wavevector, and $u_{\vec{k}}(\vec{r})$ is the Bloch function with the specific property that it is periodic with the period of the crystal lattice and therefore repeats itself in each unit cell of the crystal. We normalize the Bloch functions by requiring that

$$\frac{1}{\Omega} \int_{\Omega} u^*_{\vec{k}}(\vec{r}) u_{\vec{k}}(\vec{r}) d\vec{r} = 1, \tag{4.59}$$

where Ω is the volume of the unit cell of the crystal. Any electron states not representable by the plane wave functions in Eq. (4.58) can be written as a linear combination of such plane waves:

$$\Psi(\vec{r}) = \int B(\vec{k}) e^{i\vec{k}\vec{r}} u_{\vec{k}}(\vec{r}) d\vec{k}. \tag{4.60}$$

For electron states which are smooth within each lattice period, one can approximate $\Psi(\vec{r})$ as

$$\Psi(\vec{r}) = u_{\vec{k}_0}(\vec{r}) \int B(\vec{k}) e^{i\vec{k}\vec{r}} \, d\vec{k} \equiv u(\vec{r}) F(\vec{r}), \tag{4.61}$$

where $u_{\vec{k}_0}(\vec{r}) \equiv u(\vec{r})$ is the Bloch function of the electron with the minimum (maximum) energy. The last transformation is based on the key assumption that within a given energy band the Bloch function is a weak function of $\vec{k}$ in the vicinity of the band edge. Therefore, the actual wave function consists of two multipliers: the first is the Bloch function at the extremum of the corresponding energy band; and the second, $F(\vec{r})$, is the so-called *envelope function.*

The difference between intraband and interband processes is now clear. The former involves only one Bloch function $u_{\vec{k}_0}(\vec{r})$ and it does not appear explicitly in the final result. For example, let a potential Φ be responsible for the transitions between long-wavelength electron states in the same energy band, say b: thus $\{b,\nu\}\to\{b,\nu'\}$, where ν labels the set of intraband quantum numbers. The matrix element of the corresponding process is determined by the following equation:

$$\begin{aligned}\langle b,\nu|\Phi|b,\nu'\rangle &= \int u^*(\vec{r})F^*_{\nu'}(\vec{r})\Phi(\vec{r})u(\vec{r})F_\nu(\vec{r})d\vec{r}\\ &= \int_\Omega u^*(\vec{r})u(\vec{r})d\vec{r}\sum_{\vec{R}_j} F^*_{\nu'}(\vec{R}_j)\Phi(\vec{R}_j)F_\nu(\vec{R}_j)\\ &= \Omega\sum_{\vec{R}_j} F^*_{\nu'}(\vec{R}_j)\Phi(\vec{R}_j)F_\nu(\vec{R}_j)\\ &= \int F^*_{\nu'}(\vec{r})\Phi(\vec{r})F_\nu(\vec{r})d\vec{r}.\end{aligned}$$

Here we have used the normalization condition of Eq. (4.59), $\vec{R}_i$ is the coordinate vector at the center of the ith cell of the crystal, and the sum has been replaced by an integral. From these results, we see that for intraband processes only the envelope functions are needed. A fundamentally different situation occurs in the case of the interband transitions which are studied here.

We now return to an analysis of Eqs. (4.44)–(4.50). If E_c and E_v are the energies of electrons and holes, the δ-function in Eq. (4.45) dictates the energy conservation law:

$$E_c - E_v = \hbar\omega. \tag{4.62}$$

In Eq. (4.55) the matrix element for the phototransition rates is

$$\langle i|\vec{\xi}\vec{D}|f\rangle \equiv \langle v,k_v,\sigma|\vec{\xi}\vec{D}|c,k_c,\sigma'\rangle,$$

with wave functions

$$\Psi_{v,k_v} = \frac{1}{\sqrt{\mathcal{V}}}u_{v,\vec{k}_{max}}(\vec{r})e^{i\vec{k}_v\vec{r}},$$

$$\Psi_{c,k_c} = \frac{1}{\sqrt{\mathcal{V}}}u_{c,\vec{k}_{min}}(\vec{r})e^{i\vec{k}_c\vec{r}},$$

where $u_{v,\vec{k}_{max}}(\vec{r})$ and $u_{c,\vec{k}_{min}}(\vec{r})$ are the Bloch functions of the electrons in the valence and conduction bands, respectively. Addressing the case of a direct-bandgap semiconductor, we let both the minimum of the conduction band and the maximum of the valence band be at the same point in $\vec{k}$-space, $\vec{k}_{min}=\vec{k}_{max}=\vec{k}_0$. In this case, the above matrix element can be written as:

$$\begin{aligned}\langle v,k_v,\sigma|\vec{\xi}\vec{D}|c,k_c,\sigma'\rangle &\equiv \langle v,k_v|\vec{\xi}\vec{D}|c,k_c\rangle\delta_{\sigma,\sigma'}\\ &= \frac{1}{\mathcal{V}}\delta_{\sigma,\sigma'}\int u^*_{v,\vec{k}_0}e^{-i\vec{k}_v\vec{r}}(\vec{\xi}\vec{D})u_{c,\vec{k}_0}e^{i\vec{k}_c\vec{r}}d\vec{r}\\ &= \frac{1}{\mathcal{V}}\delta_{\sigma,\sigma'}\sum_{\vec{R}_i}e^{-i(\vec{k}_v-\vec{k}_c)\vec{R}_i}\int_\Omega u^*_{v,\vec{k}_0}(\vec{\xi}\vec{D})u_{c,\vec{k}_0}d\vec{r}.\end{aligned} \tag{4.63}$$

The last integral is taken over a unit cell of the crystal and the sum runs over all cells. It is important that the integral does not depend on the cell number and is equal to $\Omega(\vec{\xi}D_{\rm vc})$. Here, $D_{\rm vc}$ is the matrix element involving just the Bloch functions. This quantity provides the selection rules for phototransitions. It depends on the symmetries of the corresponding bands. The matrix element of Eq. (4.63) also depends on the polarization of the light with respect to the crystal axes.

In Eq. (4.63), the sum over all cells of the crystal gives, as usual, the Kronecker δ-function:

$$\frac{\Omega}{\mathcal{V}}\sum_{\vec{R}_{\rm i}} e^{-(\vec{k}_{\rm v}-\vec{k}_{\rm c})\vec{R}} = \delta_{k_{\rm v},k_{\rm c}}.$$

Hence, for the matrix element in Eq. (4.63) we obtain

$$|\langle v, k_{\rm v}|\vec{\xi}\vec{D}|c, k_{\rm c}\rangle|^2 = |(\vec{\xi}D)_{\rm c,v}|^2\delta_{k_{\rm v},k_{\rm c}}. \tag{4.64}$$

In the limit $\mathcal{V} \to \infty$ one can replace $\mathcal{V}\delta_{k_{\rm v},k_{\rm c}}$ by $\delta(k_{\rm c} - k_{\rm v})$. Thus we get an important result – *the selection rule for the initial and final momenta of electrons in the valence and conduction bands* participating in the phototransition:

$$\vec{k}_{\rm v} = \vec{k}_{\rm c}. \tag{4.65}$$

Transitions obeying this rule are represented in the E–$\vec{k}$ diagram in Fig. 4.7 by *vertical lines*. Consequently, changes in the wavevector, $\vec{k}$, are neglected during phototransitions. Vertical transitions are possible only in *direct-bandgap materials*, where the minimum of the conduction band and the maximum of the valence band are situated at the same point of $\vec{k}$-space, as has been assumed in our derivation. In the case of indirect-bandgap materials, phototransitions involving only the electron subsystem of the crystal are forbidden. Such indirect transitions are forbidden without the participation of lattice vibrations, i.e., phonons, or other "third-party" particles like impurities, defects, and so on. An adequate phonon momentum or change of third-particle momentum can make *indirect phototransitions* possible.

The criterion of Eq. (4.65) is approximate because we are assuming that the electric field of the electromagnetic wave is almost constant in space. If we use the exact formula for the electric field of the wave as given by Eq. (4.9), we would obtain the *exact selection rule*:

$$\vec{k}_{\rm e} - \vec{k}_{\rm v} = \vec{q}, \quad |\vec{q}| = \frac{2\pi}{\lambda}.$$

Expressing $k_{\rm c,v}$ through the kinetic energies of the electrons in the conduction and valence bands,

$$k_{\rm c,v} = \sqrt{2m_{\rm c,v}E_{\rm c,v}}/\hbar,$$

one can estimate the ratio of the photon wavevector q to the particle wavevector $k_{\rm c,v}$:

$$\frac{q}{k_{\rm c,v}} = \frac{\hbar\omega}{2E_{\rm c,v}}\frac{v_{\rm c,v}}{c}\sqrt{\kappa}.$$

Here $v_{\rm c,v}$ are the electron and hole velocities, respectively, and c is the velocity of light in vacuum. Because of the strong inequality $v \ll c$, one can conclude that this ratio is always very small. Consequently, the selection rule of Eq. (4.65) is sufficiently accurate. Using $k_v = k_c = k$ and combining Eqs. (4.62) and (4.65), we can rewrite Eq. (4.62) in the form

$$\begin{aligned} E_e - E_h &= E_c^0 + \frac{\hbar^2 k^2}{2m_e} - \left(E_v^0 - \frac{\hbar^2 k^2}{2m_h}\right) \\ &= E_g + \frac{\hbar^2 k^2}{2m_r^*} = \hbar\omega, \end{aligned} \tag{4.66}$$

where E_v^0 and E_c^0 represent the energies at the top of the valence band and bottom of the conduction band, respectively. In addition, m_r is the reduced mass of the electron–hole pair:

$$m_r = \frac{m_e m_h}{m_e + m_h}. \tag{4.67}$$

Note that the reduced mass of the pair appears in the conservation law, even though we are considering phototransitions involving unbound electrons and holes. From Eq. (4.66) we can find the wavevector of the electron created in the conduction band during the phototransition:

$$k = \sqrt{\frac{2m_r}{\hbar^2}(\hbar\omega - E_g)}.$$

Taking into account $\delta(E_c - E_v - \hbar\omega)$ in Eq. (4.55) and δ_{k_c,k_v} in Eq. (4.64), and applying the usual procedure, we can calculate the light absorption (amplification) coefficient in the explicit form

$$\begin{aligned} \alpha &= \frac{4\pi^2\omega}{c\sqrt{\kappa}\mathcal{V}}|(\vec{\xi}\vec{D})_{c,v}|^2 \sum_{\sigma,\sigma',k_v,k_c} \delta_{\sigma,\sigma'}\delta_{k_v,k_c}\delta\left[E_c(k_c) - E_v(k_v) - \hbar\omega\right] \\ &\qquad \times \left[\mathcal{F}(E_v(k_v)) - \mathcal{F}(E_c(k_c))\right] \\ &= \frac{4\pi^2\omega}{c\sqrt{\kappa}\mathcal{V}}|(\vec{\xi}\vec{D})_{c,v}|^2\, 2\sum_{k} \delta\left[E_c(k) - E_v(k) - \hbar\omega\right]\left[\mathcal{F}(E_v(k)) - \mathcal{F}(E_c(k))\right] \\ &= \frac{4\pi^2\omega}{c\sqrt{\kappa}\mathcal{V}}|(\vec{\xi}\vec{D})_{c,v}|^2\, 2\frac{V}{(2\pi)^3}\int d\vec{k}\,\delta\left[E_c(k) - E_v(k) - \hbar\omega\right]\left[\mathcal{F}(E_v(k)) - \mathcal{F}(E_c(k))\right]. \end{aligned} \tag{4.68}$$

Here we have used the substitution

$$\sum_{\vec{k}} \to \frac{\mathcal{V}}{(2\pi)^3}\int d\vec{k}.$$

In Eq. (4.68) the functions $\mathcal{F}(E_c)$ and $\mathcal{F}(E_v)$ are the distribution functions of the electrons in the conduction band and the valence band, respectively.

For further calculations, we use the formulae of Eqs. (4.66) and (4.68) and introduce an *effective density of states* $\rho(\omega)$, which provides the number of states of electrons in direct-bandgap semiconductors which can interact with photons in the

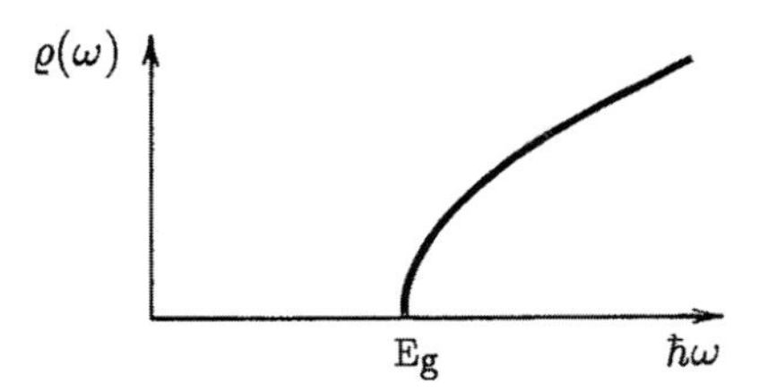

Figure 4.8 The optical density of states in a bulk semiconductor.

frequency range from ω to $\omega + d\omega$:

$$\rho(\omega) = \frac{(2m_r)^{3/2}}{2\pi^2\hbar^2}(\hbar\omega - E_g)^{1/2}, \quad \hbar\omega > E_g. \tag{4.69}$$

Expressing the coefficient α through this equation, we obtain

$$\alpha = \frac{4\pi^2\omega|(\xi\vec{D})_{c,v}|^2}{c\sqrt{\kappa}\hbar}\rho(\omega)\,[\mathcal{F}(E_v(k)) - \mathcal{F}(E_c(k))]. \tag{4.70}$$

We see that the absorption coefficient is proportional to $\rho(\omega)$. Therefore, in view of this proportionality and the functional form of the curve in Fig. 4.8, this quantity is often called the *optical density of states*. Evidently, the optical density of states is an important characteristic because all rates describing basic optical processes depend on $\rho(\omega)$.

We can reveal the frequency dependence of the coefficient α given by Eq. (4.70):

$$\alpha(\omega) = p(\hbar\omega - E_g)^{1/2}[\mathcal{F}(E_v) - \mathcal{F}(E_c)], \tag{4.71}$$

where p is the material constant, which can be expressed via the matrix element of the dipole D and the energies E_c and E_v which satisfy Eq. (4.62). In some cases it is useful to rewrite the expression of Eq. (4.71) in terms of the contributions of both absorption and stimulated emission processes:

$$\alpha(\omega) = \alpha_{abs} - \alpha_{st.em},$$

where

$$\alpha_{abs} = p(\hbar\omega - E_g)^{1/2}\,\mathcal{F}(E_v)(1 - \mathcal{F}(E_c)), \tag{4.72}$$

$$\alpha_{st.em} = p(\hbar\omega - E_g)^{1/2}\,\mathcal{F}(E_c)(1 - \mathcal{F}(E_v)) \tag{4.73}$$

have factors corresponding to those of Eqs. (4.47) and (4.48). Let us mention here that $(1 - F(E_v))$ is the probability that the state at E_v is not occupied by an electron. It is also convenient to express p in terms of the *radiative lifetime* of electrons and holes, τ_R:

$$p = \frac{\sqrt{2}c^2 m_r^{3/2}}{\kappa(\hbar\omega)^2}\frac{1}{\tau_R}, \tag{4.74}$$

with

$$\frac{1}{\tau_R} = \frac{4\omega^3\sqrt{\kappa}}{c^3\hbar}\left|\left(\vec{\xi}D_{c,v}\right)^2\right|.$$

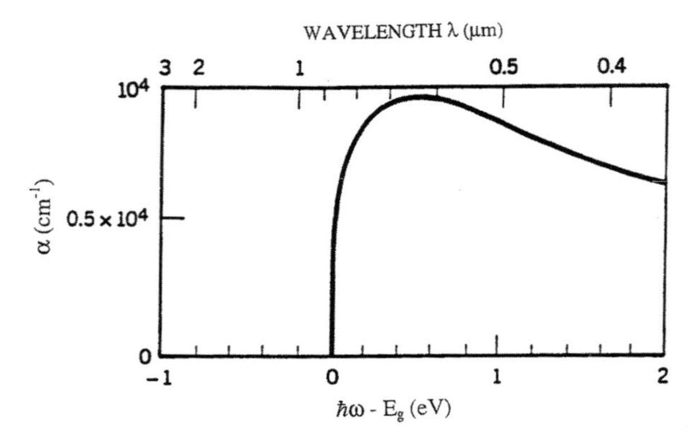

Figure 4.9 Calculated absorption coefficient, α, for GaAs as a function of the photon energy (wavelength) under equilibrium conditions. Only interband phototransitions are taken into account. The exciton absorption is not shown. Republished with permission of John Wiley and Sons Inc., from Fig. 15.2-10 on page 586 of *Fundamentals of Photonics*, B. E. A. Saleh, M. C. Teich, 1991; permission conveyed through Copyright Clearance Center, Inc.

In both formulae, we have set $\omega \approx [E_c - E_v]/\hbar$. From the latter formula it follows that the radiative lifetime decreases with increasing bandgap in a material.

Let us analyze Eq. (4.71) for thermal equilibrium conditions. In this case, the occupancies of energy levels in both bands are described by the same Fermi function,

$$\mathcal{F}(E) = \frac{1}{e^{(E-E_F)/k_B T} + 1},$$

where E_F is the Fermi energy in the material. According to the latter formula,

$$\mathcal{F}(E_v) > \mathcal{F}(E_c)$$

since $E_v < E_c$. Hence, the material absorbs light in accordance with the result $\alpha(\omega) > 0$. If E_F lies within the bandgap and $|E_c^0 - E_F|, |E_F - E_v^0| \gg kT_B$, we can set $\mathcal{F}(E_v) = 1$ and $\mathcal{F}(E_c) = 0$. The absorption coefficient then reduces to

$$\alpha \equiv \alpha_{\text{abs}}(\omega) = \frac{\sqrt{2}c^2 m_r^{3/2}}{\kappa \tau_R} \frac{(\hbar\omega - E_g)^{1/2}}{(\hbar\omega)^2}. \tag{4.75}$$

Figure 4.9 depicts the absorption coefficient, $\alpha_{\text{abs}}(\omega)$, for interband transitions in bulk GaAs at low temperatures; the excitonic absorption is not presented. We see that the interband contribution to the absorption increases sharply above the threshold photon energy ($\hbar\omega = E_g$) and reaches a value of the order of 10^4 cm^{-1}. The spectral region corresponding to Eq. (4.75) is frequently called the *fundamental edge of absorption.* According to Eq. (4.75), the maximum of $\alpha_{\text{abs}}(\omega)$ is reached at $\hbar\omega = 4E_g/3$. The high absorption due to interband phototransitions implies that, in this spectral region, the light is absorbed strongly in the micrometer-thick material layer.

4.4.2 Spectral Density of Spontaneous Emission

From previously discussed general considerations, we know that the Einstein coefficient, A_{21}, describing the spontaneous emission for some fixed mode can be expressed in terms of the coefficient of the stimulated process B_{21} by Eq. (4.33). By applying similar considerations and using Eqs. (4.51)–(4.53), we can obtain the rates of spontaneous emission in the fixed mode from Eqs. (4.71) and (4.72) as follows:

$$R_{\text{sp.em}} = \frac{c}{\sqrt{\kappa}} p(\hbar\omega - E_{\text{g}})^{1/2} \mathcal{F}(E_{\text{c}})\,(1 - \mathcal{F}(E_{\text{v}})).$$

To calculate the spectral density of the rate of spontaneous emission of all possible modes existing in the interval from ω to $\omega + d\omega$, one should multiply $R_{\text{sp.em}}$ by the number of electromagnetic modes per unit volume in the interval $d\omega$, i.e., $(\nu(\omega)/\mathcal{V})d\omega$ (see Eq. (4.19)). Introducing $R_{\text{sp.em}}(\omega) = R_{\text{sp.em}}\,\nu(\omega)/\mathcal{V}$, we obtain

$$R_{\text{sp.em}}(\omega)d\omega = \frac{\kappa\omega^2}{\pi^2 c^2} p(\hbar\omega - E_{\text{g}})^{1/2} \mathcal{F}(E_{\text{c}})(1 - \mathcal{F}(E_{\text{v}}))d\omega. \quad (4.76)$$

Assuming that the Fermi level lies in the bandgap, so that

$$\mathcal{F}(E_{\text{c}})(1 - \mathcal{F}(E_{\text{v}})) = e^{-(E_{\text{c}} - E_{\text{v}})/k_{\text{B}}T} = e^{-\hbar\omega/k_{\text{B}}T},$$

we finally get the basic frequency dependence of the rate of spontaneous interband emission under equilibrium:

$$R_{\text{sp.em}}(\omega) = D_0(\hbar\omega - E_{\text{g}})^{1/2} e^{-\frac{\hbar\omega - E_{\text{g}}}{k_{\text{B}}T}}, \quad (4.77)$$

where D_0 is a constant independent of frequency:

$$D_0 = \frac{\sqrt{2} m_{\text{r}}^{3/2}}{\pi^2 \hbar^2 \tau_{\text{R}}} e^{-E_{\text{g}}/k_{\text{B}}T}.$$

The rate of spontaneous emission $R_{\text{sp.em}}(\omega)$ is shown in Fig. 4.10. The spectrum has a low-frequency edge at $\hbar\omega = E_{\text{g}}/\hbar$ and extends over a width of the order of $2k_{\text{B}}T/\hbar$.

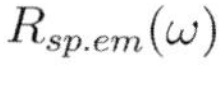

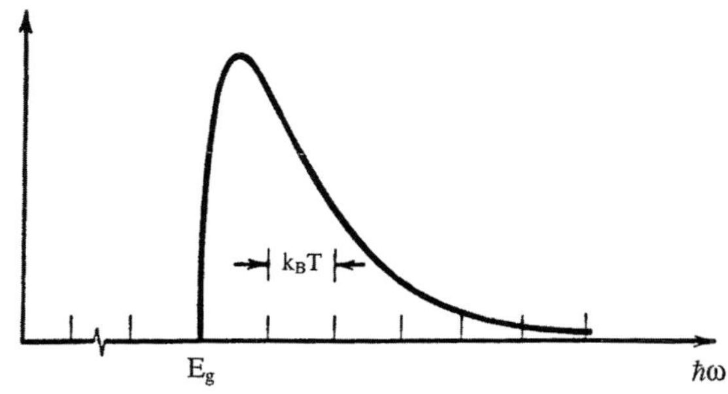

Figure 4.10 The spectral density of the interband spontaneous emission in thermal equilibrium. Republished with permission of John Wiley and Sons Inc., from Fig. 15.2-9 on page 584 of *Fundamentals of Photonics*, B. E. A. Saleh, M. C. Teich, 1991; permission conveyed through Copyright Clearance Center, Inc.

4.4.3 Phototransitions in Semiconductors with Complex Band Structure

We now have equations describing absorption and emission in bulk materials. But we have not taken into account the actual complex structure of the valence band, which is characteristic for III–V compounds, silicon, germanium, etc., i.e., for materials of considerable interest in microelectronics and optoelectronics. Examples of a direct-bandgap semiconductor (GaAs) and an indirect-bandgap semiconductor (Si) are shown in Figs. 2.10(b) and (a), respectively. In Fig. 4.11 we present a sketch of a real band structure of a III–V semiconductor in the vicinity of the bandgap. Such materials have a complex valence band structure. There are three valence bands: the heavy-hole band (hh), the light-hole band (lh), and the split-off the band (sh). This band structure was previously discussed in Chapter 2.

For the bands depicted in Fig. 4.11, the minimum in the conduction band has the same momentum (wavevector) as the maximum in the valence band. In such a case, the semiconductor is referred to as a direct-bandgap semiconductor, as discussed in Chapter 2. In accordance with Eq. (4.56), optical transitions between any of the valence bands and the conduction band can be shown in Fig. 4.11 by practically vertical lines, as depicted in Fig. 4.7. We will discuss these transitions after the next paragraph. It is obvious that in indirect-bandgap semiconductors (see Fig. 2.10(a)), where the lowest conduction band minimum and the valence band maximum under consideration are separated by a wavevector much larger than a photon wavevector, the simple photon absorption processes described above for direct-bandgap semiconductors are not possible (see Fig. 4.6). As is well known, Si has many superior properties and characteristics when used in electronic devices;

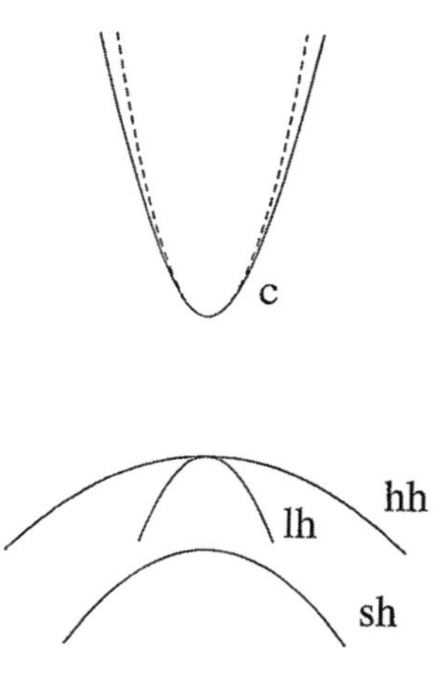

Figure 4.11 A sketch of a real band structure for a III–V semiconductor. A parabolic (full line) and a nonparabolic (dashed line) conduction (c) band are depicted. Three valence bands are shown: heavy-hole (hh), light-hole (lh), and split-off-hole (sh) bands. As discussed previously, the vertical direction represents the energy and the horizontal direction represents the wavevector (linearly related to the momentum) of the carriers.

however, the inability of Si to emit photons, as described above for direct-bandgap semiconductors, has been a great impediment to the use of Si in optoelectronic devices. Most of the research and development underlying optoelectronics has been focused on direct-bandgap materials based on GaAs, InP, and GaN. The optical parameters of these and other semiconductors are given in Table 4.2, which can be found near the end of this chapter. As examples, GaAs-based systems are used in light-emitting diodes near the red part of the spectrum, InP-based systems are used in applications that involve transmission of light in optical fibers that transmit in the near-infrared part of the electromagnetic spectrum, and GaN-based systems are used in light-emitting diodes that operate near the blue and ultraviolet parts of the electromagnetic spectrum as a result of the so-called wide-bandgap structure of GaN.

As we have already indicated earlier, phototransitions in indirect-bandgap materials occur because of the existence of vibrational modes in crystals known as phonons. Indeed, since phonons represent the vibrational modes of the atoms in the crystal, these modes may have wavelengths spanning a range of many lattice constants (so-called zone-center phonons) to roughly a lattice constant (so-called zone-edge phonons). Given this range of phonon wavelengths, it is clear that the phonon wavevectors span the full range of wavevectors across the Brillouin zone. So an electron in the conduction band can make a transition to the valence band if the energy of the emitted photon is sufficient to account for the needed change in the energy of the carrier, and if an emitted or absorbed phonon has the wavevector needed to account for the necessary change in wavevector. Such a process of photon emission with participation of a phonon is known as a second-order process, and it is found that such processes generally have lower probability than the first-order processes involving just a photon. In spite of the lower probability of the second-order optical transitions, indirect-bandgap semiconductors are used in optoelectronic devices. The most important example is the Si-based solar cells.

In most cases, optoelectronic devices employ direct-bandgap semiconductors as their active media. In order to analyze this case, one should slightly generalize Eq. (4.68) by taking into account several valence bands v_j:

$$\alpha(\omega) = \sum_j \alpha_{\mathrm{c},\mathrm{v}_j}(\omega), \tag{4.78}$$

$$\begin{aligned} \alpha_{\mathrm{c},\mathrm{v}_j}(\omega) = 2\frac{4\pi^2\omega}{c\sqrt{\kappa}\mathcal{V}}|(\vec{\xi}\vec{D})_{\mathrm{c},\mathrm{v}_j}|^2 \sum_{\vec{k}_\mathrm{v},\vec{k}_\mathrm{c}} \delta_{\vec{k}_\mathrm{v},\vec{k}_\mathrm{c}}\delta\left(E_\mathrm{c}(k_\mathrm{c}) - E_{\mathrm{v}_j}(k_\mathrm{v}) - \hbar\omega\right) \\ \times \left[\mathcal{F}(E_{\mathrm{v}_j}(k_{\mathrm{v}_j})) - \mathcal{F}(E_\mathrm{c}(k_\mathrm{c}))\right], \end{aligned} \tag{4.79}$$

where $(\vec{\xi}\vec{D})_{\mathrm{c},\mathrm{v}_j}$ is the dipole matrix element calculated with the Bloch functions of the conduction band, c, and the valence band, v_j. The conduction and valence bands are shown in Fig. 4.11. Each band originates from the energy levels of isolated atoms that form the crystal. In this sense, the conduction band originates from s atomic orbitals, while the three valence bands originate from three p atomic orbitals: p_x,

p_y, p_z. Let the corresponding Bloch functions be u_s for the conduction band and u_x, u_y, u_z for the valence bands. This association with atomic orbitals is very useful because it leads to the possibility of understanding the symmetry of the Bloch functions of the particular bands.

Thus, the conduction band Bloch function, u_s, has an even symmetry in all three directions, similar to the spherical symmetry of an s atomic orbital. In the same manner, u_z has odd symmetry along z, but even symmetry in the x- and y- directions. These assignments are analogous to the case of the p_z atomic orbitals. These observations allow us to conclude that the overlap integral vanishes; that is,

$$\langle u_s | u_z \rangle \equiv \int_\Omega u_s^*(\vec{r}) u_z(\vec{r}) d\vec{r} = 0.$$

But the matrix element

$$\langle u_s | D_z | u_z \rangle \equiv \int_\Omega u_s^*(\vec{r}) D_z u_z(\vec{r}) d\vec{r}$$

in the general case is nonzero, and analogously we obtain nonzero matrix elements of D_i for $i = x, y$. All other matrix elements are zero:

$$\langle u_s | D_i | u_j \rangle = 0, \quad i \neq j, \tag{4.80}$$

where i and j are equal to x, y, z. Let us introduce

$$\langle u_s | D_i | u_i \rangle \equiv \mathcal{D}, \quad \langle u_s | D_i | \bar{u}_i \rangle = 0, \tag{4.81}$$

where u_i and $\bar{u}_i$ are the spin-up and spin-down functions. Hence, the spin does not change during dipole phototransitions. Also, we define the constant $\mathcal{D}$ as the basic matrix element.

The electron wave functions in the valence bands are constructed as certain linear combinations of the functions u_j. Let the electron wavevector be $\vec{k}$ and take the z-axis along the vector $\vec{k}$. In this case the linear combinations of the six Bloch functions $u_i, \bar{u}_i$ corresponding to the hh, lh, and sh valence bands can be written, respectively, as

$$\begin{aligned}
u_{hh} &= -\frac{1}{\sqrt{2}}(u_x + iu_y), \quad \bar{u}_{hh} = \frac{1}{\sqrt{2}}(\bar{u}_x - i\bar{u}_y), \\
u_{lh} &= -\frac{1}{\sqrt{6}}(\bar{u}_x + i\bar{u}_y - 2u_z), \quad \bar{u}_{lh} = \frac{1}{\sqrt{6}}(u_x - iu_y + 2\bar{u}_z), \\
u_{sh} &= -\frac{1}{\sqrt{3}}(\bar{u}_x + i\bar{u}_y + u_z), \quad \bar{u}_{sh} = -\frac{1}{\sqrt{3}}(u_x - iu_y - \bar{u}_z).
\end{aligned} \tag{4.82}$$

Here, the prefactors are the normalization constants. These linear combinations constitute a representation known as the *angular-momentum representation*. This representation is very convenient when studying the spin–orbit interaction, which arises from the coupling of the spin and orbital angular momentum. This causes a splitting of the sh band.

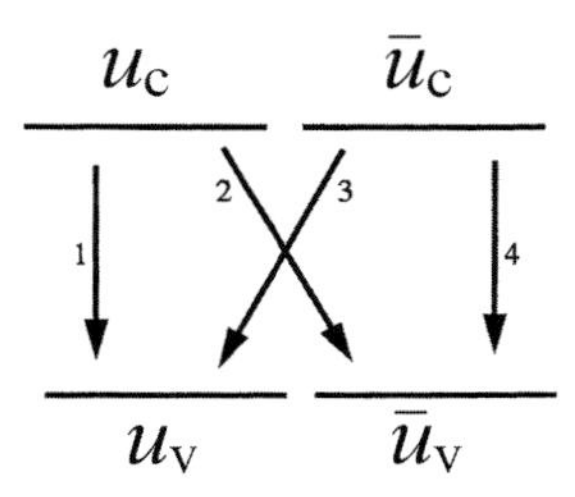

Figure 4.12 The scheme of interband phototransitions: u_c, u_v are the Bloch wave functions of the electrons in the conduction and valence bands, and $\bar{u}_c, \bar{u}_v$ are the complex conjugated functions.

Using Eqs. (4.80)–(4.82) one can obtain more complete and accurate descriptions of the matrix elements for the phototransitions between bands presented in Fig. 4.12. In these results, the constant $\mathcal{D}$ is the only parameter which remains to be either calculated or found from experiment. On the other hand, the angular momentum representation of Eq. (4.82) reveals the strong anisotropy of the electron–light interaction. For further calculations let us return to the general expressions for the absorption coefficient as given by Eqs. (4.78) and (4.79). The summation in Eq. (4.78) runs over all possible initial and final bands:

$$\mathrm{c} \leftrightarrow \mathrm{hh}, \quad \mathrm{c} \leftrightarrow \mathrm{lh}, \quad \mathrm{c} \leftrightarrow \mathrm{sh}.$$

In addition, the sum runs over both transitions $u_c \leftrightarrow u_v$ and $\bar{u}_c \leftrightarrow \bar{u}_v$. Since the wave functions u_v and $\bar{u}_v$ in Eq. (4.82) are states containing both spin polarization orbitals, $u_i, \bar{u}_i$, the transitions $u_c \leftrightarrow \bar{u}_v$ and $\bar{u}_c \leftrightarrow u_v$ become possible. The complete scheme for phototransitions is shown in Fig. 4.12. Taking into account the relations of Eqs. (4.80)–(4.82), we obtain intermediate results for the summation over all possible transitions:

$$\begin{aligned}
|(\vec{\xi}\vec{D})_{\mathrm{c,hh}}|^2 &= \frac{1}{2}|\mathcal{D}|^2\left(|\xi_x + i\xi_y|^2 + 0 + 0 + |\xi_x - i\xi_y|^2\right),\\
|(\vec{\xi}\vec{D})_{\mathrm{c,lh}}|^2 &= \frac{1}{6}|\mathcal{D}|^2\left(|-2\xi_z|^2 + |\xi_x - i\xi_y|^2 + |\xi_x + i\xi_y|^2 + |2\xi_z|^2\right),\\
|(\vec{\xi}\vec{D})_{\mathrm{c,sh}}|^2 &= \frac{1}{3}|\mathcal{D}|^2\left(|-\xi_z|^2 + |-\xi_x + i\xi_y|^2 + |-\xi_x - i\xi_y|^2 + |\xi_z|^2\right).
\end{aligned} \tag{4.83}$$

Here, ξ_i are the projections of the polarization vector of the electromagnetic wave; see Eq. (4.9). Each term within the parentheses corresponds to one of the four transitions shown in Fig. 4.12. Let us assume that the wavevector, $\vec{k}$, is directed along the z-axis. Then, from Eq. (4.83), we obtain

$$\frac{|\vec{\xi}\mathcal{D}|^2}{\mathcal{D}^2} = \begin{cases} 1 - \cos^2\theta & \text{for the hh band,} \\ \frac{1}{3} + \cos^2\theta & \text{for the lh band,} \\ \frac{2}{3} & \text{for the sh band.} \end{cases} \tag{4.84}$$

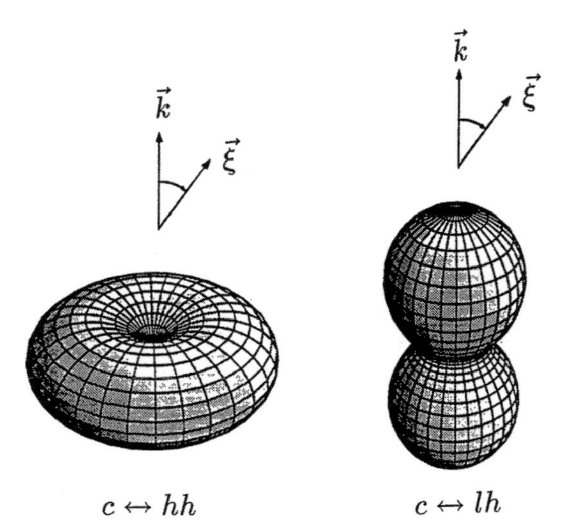

Figure 4.13 The transition strength as a function of the angle between an electron wavevector and the electric field of the light wave. The phototransitions c ↔ hh and c ↔ lh demonstrate different dependences. Reprinted from Fig. 9 on p. 53 of *Quantum Well Lasers*, P. S. Zorry, Jr., *et al.*, "Optical gain in III–V bulk and quantum well semiconductors," pp. 17–96, Copyright (1993), with permission from Elsevier.

We see that transitions to (from) heavy- and light-hole valence bands depend on the angle between the wave polarization and the electron momentum. Figure 4.13 illustrates the corresponding transition strengths. These results indicate that there is a strong polarization dependence of the electron–photon interaction. In contrast, the transition amplitude for the interactions of light with the spin-split valence band electrons is completely isotropic.

In fact, in the three-dimensional case of bulk materials, electromagnetic waves interact with a great number of electrons with all possible $\vec{k}$-vector directions. The average of these results over all possible $\vec{k}$ vectors is independent of the wave polarization for all three types of phototransition:

$$\langle|\vec{\xi}\vec{D}|^2\rangle_{\text{average}} = \frac{2}{3}\mathcal{D}^2.$$

Thus, for bulk materials, one can use this result with Eq. (4.68) with the proper constant $\mathcal{D}$. As we will see in the next chapter, the polarization dependences are strongly pronounced and important for nanostructures.

4.4.4 Excitonic Effects

In our previous discussions we have considered interband phototransitions while ignoring the correlation of electron and hole motion due to the Coulomb interaction. The main result of such a correlation is the formation of a new quasi-particle that is named an *exciton*. As introduced briefly in Chapters 2 and 3, the exciton was described as a bound electron–hole pair. These bound electron–hole pairs have historically been classified as Wannier excitons for the case where the effective separation of the electron and the hole in the pair is greater than a few lattice constants, and as Frenkel excitons when the electron–hole pairs are separated by less than a lattice constant. In many semiconductors, the excitions are of the Wannier type. The existence of excitons results in the appearance of a series of excitonic peaks in the optical spectra below the fundamental edge of interband transitions; that is, the peaks occur for $\hbar\omega < E_g$. These peaks are usually observed at low temperatures, but, for semiconductors such as GaN and ZnO which have large exciton binding energies, exciton peaks (states) are observed at room temperature. In cases where the binding energy is small, washed-out exciton peaks may overlap and create an absorption tail below the fundamental edge that is observable at room temperature.

An important consequence of excitons is the fact that the matrix element of the optical transitions involving excitons is much greater than that of the interband transitions because the electron and hole are close to each other and their wave functions overlap much more than do those of a delocalized electron–hole pair.

The Coulomb correlation of the unbound electron and hole reflects the fact that these particles, if they are generated by photon absorption, move in a mutual Coulomb potential. This Coulomb correlation causes corrections which may be described by a special multiplicative factor (the Sommerfeld factor) related to the probability of phototransition near the optical edge; as a result of this factor, the absorption becomes a sharper function of the photon energy. The spectral region where Coulomb correlations are important can be estimated as

$$0 < E_g - \hbar\omega < E_{ex},$$

where E_{ex} is the binding energy of the exciton. Excitons, i.e., the excited states of the electronic subsystem of the crystals, were analyzed in Chapter 3, Sections 3.8.1 and 3.8.2. The binding energies of the excitons in a variety of group IV, III–V, and II–VI semiconductors are presented in Fig. 3.34 and in Table 4.2 (near the end of this chapter).

Phenomena associated with both excitons and the Coulomb correlations of free particles are important in optical applications of semiconductors. As a result, instead of the simple picture presented in Fig. 4.9, the spectra become complex as shown in Fig. 4.14. As shown in Section 3.8.1, in many cases, the excitonic energy levels may be approximated by the energy levels of a hydrogen-like atom, where the electron and the hole in the exciton have an effective mass given by the reduced mass, m_r, of Eq. (3.105). Let us use this approximation to obtain estimations of the exciton energy levels. For a hydrogen atom, the binding energy is proportional to $-m_0e^4/N^2$, where m_0 is the mass of the free electron, e is the electron charge, and N is the

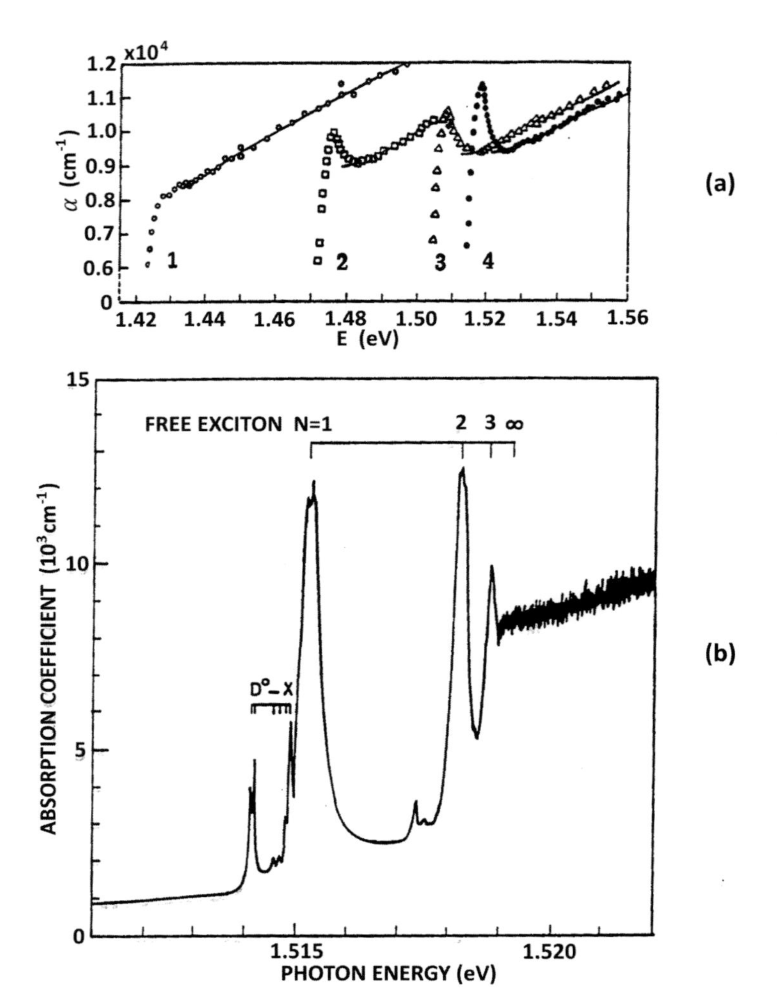

Figure 4.14 Exciton absorption in bulk GaAs. (a) The evolution of the absorption coefficient with temperature: 294 K (curve 1), 186 K (curve 2), 90 K (curve 3), and 21 K (curve 4). (b) The detailed structure of the absorption coefficient at low temperature ($T = 1.2$ K). The peaks marked 1, 2, and 3 correspond to excitation of the three lowest excitonic states. The substructure with below peak 1 (labeled as $D^0 - X$) is the absorption due to excitons localized on impurities. After R. G. Ulbrich and C. Weisbuch, 1976 (unpublished).

principal quantum number ($N = 1, 2, \ldots$). For hydrogen the binding energy equals $-13.6/N^2$ eV. To estimate the binding energy of excitons in a semiconductor material, we can use the same formula, making the substitutions $m_0 \to m_r$ and $e^2 \to e^2/\kappa$ (see Eq. (3.109)), where κ is the dielectric constant of the material. As a result, we estimate that the excitonic energy levels are about a few meV. Measured values of these energies are given in Table 4.2 near the end of this chapter.

4.4.5 Optical Properties of Group-III Nitrides

In view of the practical importance of wide-bandgap group-III nitrides, in this subsection, we briefly consider their optical properties for the example of GaN crystals. The basic material parameters of these crystals were discussed in Section 2.8.

As for other direct-bandgap III–V compounds, many important optical properties of GaN-based materials are determined by carriers with small wavevectors, $\vec{k}$, in the vicinity of the Γ point, which is the center of the Brillouin zone. Conduction-band states for small wavevector, $\vec{k}$, are doubly degenerate with respect to spin and can be characterized by one (for cubic symmetries) or two (for hexagonal symmetries) energy-independent effective masses. The valence band spectrum near the Γ point is more complicated and originates from the six-fold-degenerate state which is designated by Γ_{15}. In zincblende structures, the spin–orbit interaction splits the Γ_{15} state at $\vec{k}=0$, forming the four-fold-degenerate Γ_8 (heavy and light holes) and the doubly degenerate Γ_7 (spin-split hole) levels, as shown schematically in Fig. 4.15. The action of the low-symmetry hexagonal crystal field and the spin–orbit interaction in wurtzite crystals leads to the formation of three distinct levels: Γ_9, upper Γ_7, and lower Γ_7, which we denote as hh (heavy holes), lh (light holes), and sh (spin-split holes), respectively. These valence states generate the A-, B-, and C-type excitons and corresponding lines in photoluminescence experiments on hexagonal crystals.

The energy versus $\vec{k}$ dependences for the valence bands of GaN are presented in Fig. 4.16 (solid lines) for the wavevector $\vec{k}$ being parallel (left) or perpendicular (right) to the crystal c-axis. These dependences differ considerably from the schematic case of Fig. 4.11. One can see that the hole energies depend on the direction of the vector $\vec{k}$, i.e., they are anisotropic and the band splittings reach tens of meV. For comparison, the valence band dispersion relations are also shown for strained GaN (dashed lines), which illustrates that the strain changes the dispersion considerably. Details of the valence bands appear in the absorption and emission spectra of GaN.

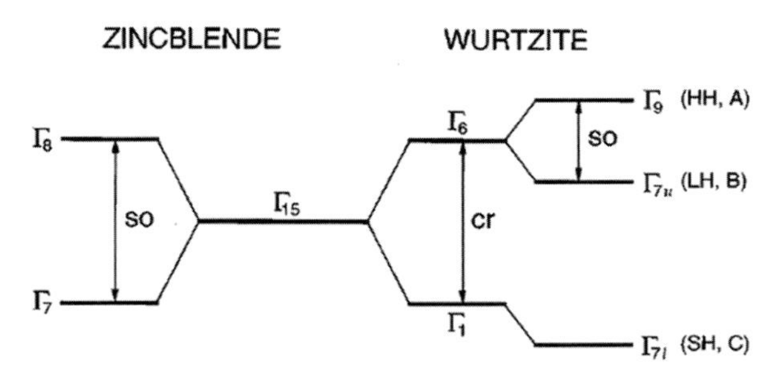

Figure 4.15 The valence band structure in zincblende and wurtzite crystals at the Γ point. Level splitting due to the spin–orbit interaction, SO, is shown for cubic and for hexagonal materials. Reprinted figure with permission from Fig. 1 of Yu. M. Sirenko *et al.*, *Phys. Rev.* **B 55**, 4360 (1997). Copyright (1997) by the American Physical Society.

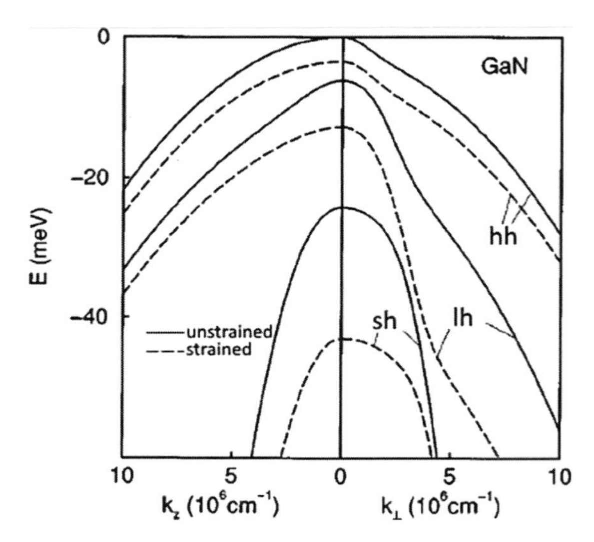

Figure 4.16 The energy dispersion of the hole bands in unstrained (solid) and strained (dashed) bulk GaN for different directions of the wavevector. A biaxial strain of about 0.242% corresponds to the GaN lattice being matched to Al_xGa_{1-x} with $x = 0.1$. Reprinted figure with permission from Fig. 4 of Yu. M. Sirenko *et al.*, *Phys. Rev.* **B 55**, 4360 (1997). Copyright (1997) by the American Physical Society.

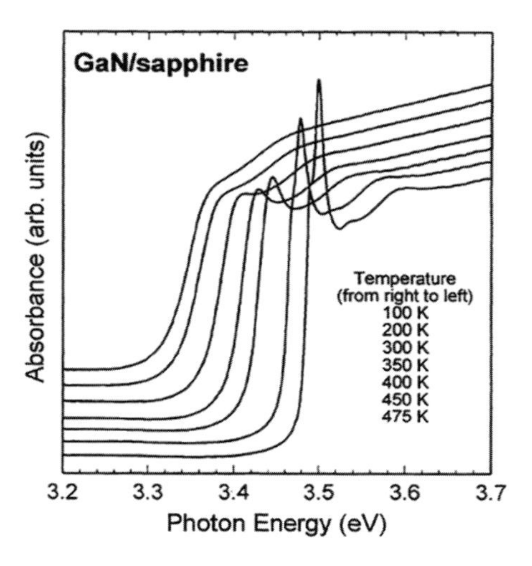

Figure 4.17 Absorption spectra of an epitaxial layer of GaN on sapphire measured at different temperatures. Reprinted from figure 3 of A. J. Fischer *et al.*, *Appl. Phys. Lett.* **71**, 1981 (1997), with the permission of AIP Publishing.

Another peculiarity of the optical spectra of GaN crystals is related to the exciton states, which have energies of about 20 meV (see Table 4.2 near the end of this chapter). These relatively large exciton energies determine the behavior of the light absorption and emission in the vicinity of the band edge of GaN. In Fig. 4.17,

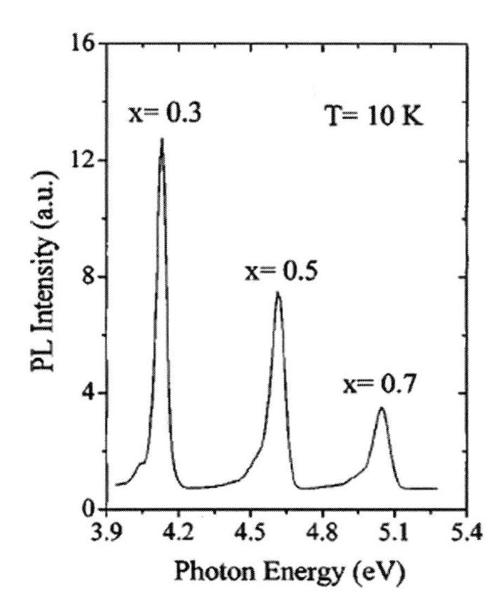

Figure 4.18 Photoluminescence spectra in $Al_xGa_{1-x}N$ alloys with Al concentrations $x = 0.3$, 0.5, and 0.7 measured at 10 K. Reprinted from Fig. 1 in G. Coli *et al.*, *Appl. Phys. Lett.* **80**, 2907 (2002), with permission of AIP Publishing.

absorption spectra of an epitaxial layer of GaN on sapphire measured for a wide temperature interval are presented. Sharp peaks near the absorption edge are clearly seen well above room temperature. For comparison, in GaAs crystals, there is no evidence of excitons at room temperature, as can be seen in Fig. 4.14(a). In GaN at low temperatures, fine spectral features arise from the lowest energy state of the C-exciton related to the split-off valence band (corresponding to the sh band shown in Figs. 4.15 and 4.16).

Emission from the family of group-III nitrides covers a wide spectral region, as can be seen from Table 4.2. Contemporary technologies for the growth of perfect group-III-nitride alloys give the additional possibility of modifying the spectra of emission and absorption by changing the alloy composition. In Fig. 4.18, we present photoluminescence spectra of high-quality $Al_xGa_{1-x}N$ alloys with different Al contents. For a pure GaN crystal the main emission lines correspond to 3.4–3.5 eV. For the presented results these lines are shifted to 4–5 eV i.e., to the deep ultraviolet region. Thus, the use of group-III-nitride alloys, as well as additional strain (as shown in Fig. 4.16), can facilitate the control of optical emission/absorption.

The discussed peculiarities of the spectra of GaN-based materials are important for various optoelectronic applications of GaN, especially for short-wavelength laser diodes.

In addition to the analysis of the optical properties of III–V compounds and group-III nitrides which has already been presented, we collect the values of the bandgap, exciton energies, and refractive indices at room temperature for a

Table 4.2 The bandgap, exciton energies, and refractive indices of selected semiconductors, at room temperature unless noted otherwise

Material	Bandgap (eV), direct bandgap unless noted otherwise	Exciton energy (meV)	Refractive index, n, for photon energies near the bandgap energy
AlN	6.2 (wurtzite)	50	2.5
AlP	2.5 (indirect), 3.6 (direct)	Indirect	3.0
AlAs	2.16 (indirect); $Al_xGa_{1-x}As$ has a direct bandgap for $x \leq 0.4$	Indirect	3.2
AlSb	2.213	Indirect	3.8
CdS	2.42	29	2.6
CdSe	1.84	15	2.5
CdTe	1.6	2.37 at 77 K	~3.2
GaN	3.4 (wurtzite)	21–25	2.8
GaP	2.26	3.5	3.3
GaAs	1.42	4.2	3.6
GaSb	0.726	2	4.0
Ge	0.67 (indirect), 0.9 (direct)	Indirect	4.0
HgTe	0.15 (the energy bandgap of $Hg_{1-x}Cd_xTe$ goes to zero for $x \sim 0.2$)	Bands overlap	Semi-metal
InAs	0.360	1.0	3.8
InN	0.6–0.7 (wurtzite)	4	2.9
In_2O_3	2.6 (indirect), 3.7 (direct)	Indirect	2.05
InP	1.423	4	3.5
InSb	0.172 ($T = 300$ K)	4	4.2
Si	1.12 (indirect), 3.3 (direct)	Indirect	3.6
SnO_2	3.7	130	2.33
ZnO	3.35	60	1.6

The refractive indices for three- and four-element semiconductors (ternary and quaternary semiconductors), are given approximately by a linear interpolation between the values given for the binary (two-element) semiconductors, as illustrated in Section 4.5 for AlGaAs alloys. After B. E. A. Saleh, M. C. Teich, *Fundamentals of Photonics* (Wiley & Sons, New York, 1991) with additional data from the references for this chapter in the Further Reading, as well as from S. Adachi, Optical dispersion relations for GaP, GaAs, GaSb, InP, InAs, InSb, $Al_xGa_{1-x}As$, and $In_{1-x}Ga_xAs_yP_{1-y}$, *J. Appl. Phys.* **66**, 6030 (1989). For the theory of excitons see Donald C. Reynolds and Thomas C. Collins, *Excitons: Their Properties and Uses* (Academic Press, New York, 1981). Selected values of the index of refraction listed here as well as for many other materials can be found at https://refractiveindex.info/, where original sources are referenced.

selection of group IV, III–V, and II–VI semiconductors in Table 4.2. As can be seen from the table, these semiconductors have bandgaps spanning the wide-bandgap ($E_g > 3$ eV with corresponding light wavelengths $\lambda_0 < 400$ nm), mid-bandgap, and narrow-bandgap regions ($E_g < 0.5$ eV with corresponding light wavelengths $\lambda_0 > 2500$ nm).

4.4.6 Refractive Index

Along with the optical processes involving the absorption or emission of real photons, there is another kind of interaction of electromagnetic field with materials that occurs without energy exchange. These processes are usually referred to as *virtual processes*. We can imagine a virtual process such as the absorption of a photon with the creation of an electron–hole pair followed by the instantaneous annihilation of the pair by the emission of an equivalent photon. Thus, real particles are absent, but the process results in a *polarization of the material*. Consequently, virtual processes contribute to the *real part of the dielectric permittivity* κ, or to the *refractive index* of the material,

$$n = \sqrt{\kappa}. \tag{4.85}$$

These virtual processes result in a dispersion dependence of κ or n on the frequency ω. This dispersion relation is important for the design of photonic and optoelectronic devices, particularly those which use optical microcavities, waveguides, and integrated optical elements.

The physics underlying the refractive index is associated with the observation that an electromagnetic wave propagating in a medium with an index of refraction n propagates with a velocity $v = c/n$. Thus, it follows from the Maxwell equations (4.1) and (4.2) that the phase of an electromagnetic wave in vacuum transforms to the phase in a medium with an index of refraction n as follows:

$$\exp\{-i\omega(t - x/c)\} \to \exp\{-i\omega[t - x(n' - in'')/c]\}. \tag{4.86}$$

Here n is presented for the general case, when the refractive index is a complex quantity $n = n' - in''$. Since the power carried by the field, $\mathcal{P}$, is proportional to the field squared, it follows that $\mathcal{P}(x)/\mathcal{P}(0)$ is equal to $\exp(-\alpha x)$, where $\alpha = 2\omega n''/c$ represents the extinction of the power and is accordingly known as the extinction coefficient, or absorption coefficient.

The index of refraction can also be related to reflection. Consider the interface between a vacuum and a material with an index of refraction $n' - in''$. When a photon flux is incident on such an interface, momentum must be conserved. Let R' be the reflected fraction, namley the amplitude reflection coefficient, of the flux; accordingly, the portion $(1 - R')$ is transmitted and conservation of momentum requires that

$$-R' + (1 - R')(n' - in'') = 1.$$

This result may be rewritten as

$$R' = [(n')^2 + (n'')^2 - 2in'' - 1]/[(n' + 1)^2 + (n'')^2].$$

Since the power (or intensity) of the wave (photon flux) goes as the amplitude times its complex conjugate, the sum of the squares of the real and imaginary parts of the momentum conservation equation gives

$$R = [(n' - 1)^2 + (n'')^2]/[(n' + 1)^2 + (n'')^2], \tag{4.87}$$

which can be recognized as the reflection coefficient, R, obeyed by the intensity of the radiation. Thus, the index of refraction relates to the phenomena of reflection and transmission, and the imaginary part of the index of refraction relates to the extinction of the wave, which is related to the absorption in the material.

It is useful to relate the real and imaginary parts of the refractive index, n', n'', to the real and imaginary parts of the dielectric permittivity, κ', κ''. The corresponding relationships can be obtained from the obvious equation

$$(n' - in'')^2 = \kappa' + i\kappa''.$$

The results are

$$\kappa' = (n')^2 - (n'')^2, \quad \kappa'' = 2n'n'', \tag{4.88}$$

$$(n')^2 = \frac{\kappa'}{2} + \frac{1}{2}\sqrt{(\kappa')^2 + (\kappa'')^2}, \quad n'' = \frac{\kappa''}{2n'}. \tag{4.89}$$

The discussion presented in this section is valid for dielectric and semiconductor media when the imaginary parts of the dielectric permittivity, κ', and the refractive index, n'', describe the processes listed at the beginning of Section 4.4 and external currents $\vec{\mathcal{J}}(\omega)$ at optical frequencies are absent or can be neglected. In doped semiconductors and metals these currents contribute to κ'' and n'', which can be taken into account by retaining the term with the current in Eq. (4.2) and using the material relationships (4.4) and (4.6). Such an analysis will be conducted in Chapter 6, Section 6.5.

In order to explain how to estimate the electron contribution to the refractive index, let us recall that the absorption coefficient, $\alpha(\omega)$, and the refractive index, $n(\omega)$, are related by the Kramers–Kronig relation

$$n(\omega) = 1 + \frac{c}{\pi}\int_0^\infty d\omega' \frac{\alpha_{\mathrm{tot}}(\omega')}{(\omega')^2 - \omega^2}. \tag{4.90}$$

In principle, the calculated refractive index depends on the total absorption of the material, $\alpha_{\mathrm{tot}}(\omega)$, over the entire frequency range. The electronic contribution to the refractive index is

$$\Delta n(\omega) \equiv n(\omega, N) - n(\omega, N=0),$$

where N denotes the electron concentration. Now $\Delta n(\omega)$ is given by

$$\Delta n(\omega) = \frac{c}{\pi}\int_0^\infty d\omega' \frac{\alpha(\omega', N)}{(\omega')^2 - \omega^2}, \tag{4.91}$$

where $\alpha(\omega, N)$ is associated with electron absorption and has been estimated in our previous discussion. The results of the refractive-index calculations are shown in Fig. 4.19 for GaAs.

In practice, $\alpha(\omega, N)$ may not be known over the entire range of integration and extrapolation methods are frequently employed to determine $n'(\omega)$. Clearly, the Kramers–Kronig relationship provides a very general method for relating the centrally important index of refraction and the absorption.

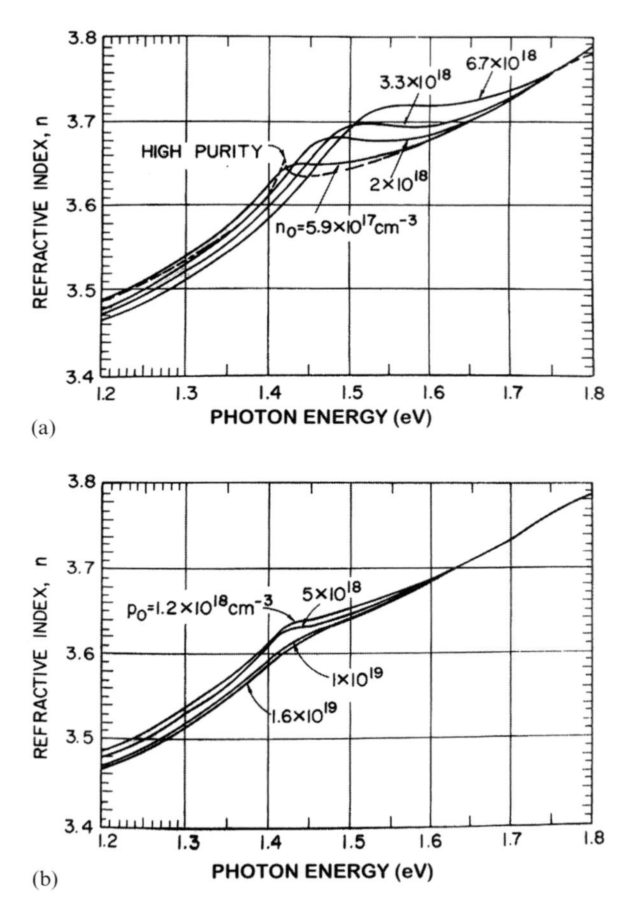

Figure 4.19 The refractive index of high-purity and n-type (a) and p-type (b) GaAs at $T = 300$ K as a function of the photon energy. The energy bandgap is located at the local maximum of the high-purity curve. Reprinted from Fig. 2.5-2 on p. 43 of *Heterostructure Lasers Part A: Fundamental Principles*, H. C. Casey Jr. and M. B. Panish, "Optical fields and wave propagation," pp. 20–109, Copyright (1978), with permission from Elsevier.

Briefly, the dispersive frequency dependence of the refractive indexarises as follows. The major contribution to the refractive index is due to lattice polarization, i.e., the polarization of atoms or ions composing the lattice. The contribution of free electrons is usually small, but this contribution increases in the frequency region corresponding to interband phototransitions. In a high-purity sample, in the vicinity of the point where $\hbar\omega = E_g$, the contribution has an almost resonant character, and can reach an absolute value of about 0.05 (see Fig. 4.19). Figure 4.19 illustrates that in doped samples at high temperatures, the resonance features are smeared out but the dependence on ω due to the electron contribution is still considerable. The refractive indices for different semiconductor materials are collected in Table 4.2.

Another noteworthy feature of the refractive index of the III–V compounds is its dependence on the alloy composition: an increase in the proportion of a narrow-bandgap component in the alloy leads to a larger index of refraction at a fixed frequency. For example, for $Ga_xAl_{1-x}As$ alloys, increasing the proportion of the Ga component decreases the bandgap and increases the refractive index: $n_{Ga_{1-x}Al_xAs} = n_{GaAs} + 0.62x$. This feature is of crucial importance for optoelectronic applications of heterostructures. It makes it possible to create optical waveguides on the basis of heterostructures, and to combine electron confinement and optical confinement within the same narrow-bandgap layers, as discussed in the following chapters.

4.5 Closing Remarks to Chapter 4

In this chapter, we have presented basic concepts for the optics of semiconductors. We started with a consideration of electromagnetic fields and introduced the quanta of these fields – photons. It was emphasized that in optical resonators and nonuniform dielectric media the structure of electromagnetic fields differs considerably from that of free space. As a result, the properties of photons, namely their form factors, polarizations, frequencies, and densities of states, are also different and can be controlled through the use of proper resonators, dielectric waveguides, etc.

We reviewed the interband optics of bulk-like direct-bandgap semiconductors, and calculated the absorption and the spontaneous and stimulated emission. The concept of photon emission and absorption in an indirect-bandgap semiconductor was discussed. Excitonic effects and phototransitions for a real band structure of III–V compounds were analyzed. We briefly considered the refractive index of the materials and the electron contribution to this quantity. A comparison of the bandgaps, exciton binding energies, and refractive indices was made for a variety of group IV, III–V, and III–nitride semiconductors that spans the range of bandgaps from the narrow-bandgap (infrared) portion to the wide-bandgap (ultraviolet) part of the electromagnetic spectrum.

The studied peculiarities of the optical spectra are critically important for various optoelectronic applications including light-emitting devices, laser diodes, and photodetectors of different spectral regions.

4.6 Control Questions

1. Explain where the density of electromagnetic energy at a given amplitude of the electric field is higher: in vacuum or in a medium with dielectric permittivity larger than unity.
2. Explain where the wavelength is larger: in vacuum, or in a medium with dielectric permittivity larger than unity. (Note that the frequency of an electromagnetic wave is conserved at the interface.)

3. Explain where the wavevector is larger: in vacuum, or in a medium with dielectric permittivity larger than unity. (Note that the frequency of an electromagnetic wave is conserved at the interface.)
4. Use Eqs. (4.12) and (4.13) to derive Eq. (4.18).
5. Using Eq. (4.26), obtain an equation for the wavelength that can be maintained in a resonator of length L_z.
6. Using your result from question 5, obtain the difference between the wavelengths corresponding to quantum numbers l and $l + 1$.
7. What is the condition for a population inversion?
8. Obtain an equation for the wavelength of an electromagnetic wave that will be absorbed or emitted by a two-level system. What is the major difference between direct- and indirect-bandgap semiconductors?
9. Does the absorption coefficient depend on the bandgap of a direct-bandgap semiconductor? Explain your answer.
10. Why is the absorption coefficient for indirect-bandgap semiconductors smaller than for direct-bandgap semiconductors?
11. Show that the energies of the bound states of an exciton are

$$-\frac{13.6 m_{\mathrm{r}}}{\kappa^2 N^2 m_0}\ \mathrm{eV},$$

in the case when the exciton may be approximated by a hydrogen-like atom. Here, m_0 is the mass of the free electron, m_{r} is defined in Eq. (4.67), and N is the principal quantum number.

5 Optics of Quantum Structures

5.1 Introduction

The advent of semiconductor quantum structures with their unique properties makes it possible to observe many traditional optical effects in low-dimensional electron systems and to discover their novel features. Furthermore, heterostructures manifest a set of new effects which do not occur in the bulk materials. On the other hand, optical experiments are extremely powerful tools for the characterization of heterostructures, including electron properties, lattice parameters, surface and interface quality, etc. But the most important feature is that the unique optical and electrical properties of heterostructures have opened fundamentally new avenues of development for optoelectronic and novel photonic applications. All of these facts make the optics of heterostructures an important topic in the field of modern solid-state devices.

In this chapter we present the optical properties of quantum structures. We study how electron confinement affects the optical spectra, stimulated emission, and other optical characteristics of quantum structures. Also we review briefly the basic features of the optics of one- and few-monolayer crystals, where the ultimate electron confinement occurs.

5.2 Optical Properties of Quantum Structures

After consideration of the basic interband optical processes in bulk materials, we are ready to study these processes in quantum heterostructures. Specific features of optical processes originate from two basic physical peculiarities. First, the behavior of the electromagnetic waves in heterostructures is different from that in the bulk materials. The spatial nonuniformity affects specific characteristics of the interaction of light with matter, including light propagation, absorption, etc. Second, electrons in quantum structures have energy spectra different from those of electrons in bulk materials. Both factors are analyzed in the following discussions. We introduce several parameters which characterize the interaction of light with matter for different cases. The parameters for light absorption are calculated for type-I heterostructure quantum wells. Various factors affecting the optical properties of low-dimensional electrons, such as the broadening of spectra due to intraband scattering processes, excitonic effects, etc., are analyzed in this section.

5.2.1 Electrodynamics of Heterostructures

Let us start by studying the electrodynamic features of heterostructures in the spectral region which corresponds to band-to-band phototransitions. In bulk materials, one can use the electrodynamics of uniform media and consider the interaction of light with matter as homogeneous in space. In heterostructures, both the refractive index and the energy bandgap, i.e., the fundamental absorption edge, vary spatially. This changes the light propagation and the character of the interaction of light with matter.

For example, a layer with a larger refractive index causes (i) the partial reflection of electromagnetic waves propagating through the layer and (ii) the localization of electromagnetic modes propagating along the layer. As a result, the electromagnetic fields (modes) in quantum structures are substantially different from plane waves. On the other hand, spatial modulation of the bandgap leads to nonuniform absorption and emission of light. In particular, in some spectral regions, narrow-bandgap layers interact with the light, while wide-bandgap layers are transparent to it. Consequently, only certain layer(s) of heterostructures can absorb and emit photons in such a spectral region, i.e., below the fundamental edge of wide-bandgap layers and above that of narrow-bandgap layers.

In order to take into account the foregoing features of electrodynamics we return to the general results of Eqs. (4.45)–(4.50). These equations allow us to consider interaction with arbitrary electromagnetic fields, including ones different from the plane waves of Eq. (4.9). We will exploit the fact that the sizes of quantum structures, L, are always much less than the wavelength λ of the light in the spectral region of interest ($L = 100$–200 Å, $\lambda > 1000$ Å). Equations (4.45) and (4.50), therefore, can be simplified if we assume a particular shape for the electromagnetic fields. Let us consider two cases of light interaction with a quantum well layer: (i) propagation perpendicular to the layers and (ii) propagation along the layers. These two cases are presented in Figs. 5.1(a) and (b). In the first case, the electromagnetic field depends on the z coordinate only, and we can express the electric field as

$$\mathcal{E}(z) = \frac{1}{2}\left(e^{i\omega t} F(z) + e^{-i\omega t} F^*(z)\right).$$

Introduction of a local absorption coefficient is meaningless in this case. The loss or gain of the light energy can be characterized by the *attenuation coefficient*, defined as the change of the light intensity after passing through the layer, $\mathcal{I}_{\rm out}$, relative to the initial intensity, $\mathcal{I}_{\rm in}$:

$$\beta \equiv \frac{\mathcal{I}_{\rm in} - \mathcal{I}_{\rm out}}{\mathcal{I}_{\rm in}}. \tag{5.1}$$

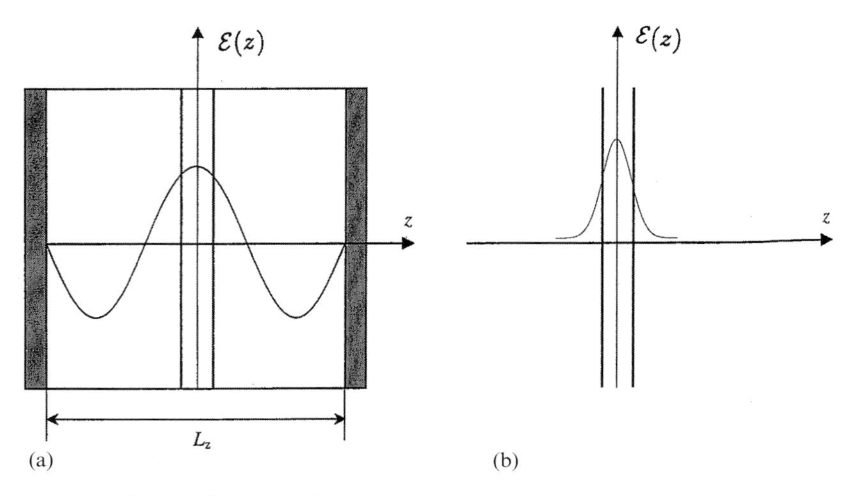

Figure 5.1 Two main types of electromagnetic modes in quantum structures: (a) a standing wave formed by a Fabry–Pérot resonator of size L_z with a quantum structure of size L embedded within it; (b) a wave propagating along the layer and localized predominantly within the quantum structure.

Another useful characteristic of the interaction of light with matter is the decrement (increment) or decay (gain) of the energy of standing waves. Standing waves are formed by the optical resonator in which a heterostructure is embedded. A simple Fabry–Pérot resonator with two plane mirrors is sketched in Fig. 5.1(a). The decrement (increment) of the mode is

$$\gamma \equiv \frac{1}{N}\frac{dN}{dt} = -\frac{1}{N}R, \tag{5.2}$$

where N is the number of photons of the mode under consideration and R is the total rate of photon absorption (emission), given by Eq. (4.50). Since $\lambda \gg L$ as mentioned above, we can rewrite R in the form

$$R = \frac{\pi}{2\hbar}|F(z_{\mathrm{w}})|^2 \sum_{i,f} |\langle i|\vec{\xi}\vec{\mathcal{D}}|f\rangle|^2 \delta\left(E_i - E_f - \hbar\omega\right)\left[\mathcal{F}(E_i) - \mathcal{F}(E_f)\right], \tag{5.3}$$

where z_{w} is the position of the center of the quantum-well layer, and $\vec{\mathcal{D}}$ is the dipole vector. The number of photons N of the fixed mode can be calculated as the total energy of this mode divided by $\hbar\omega$ (see Table 4.1):

$$N = \frac{1}{8\pi\hbar\omega}\int \kappa(z)|F(z)|^2(z)\, d\vec{r} = \frac{S}{8\pi\hbar\omega}\int dz\, \kappa(z)|F(z)|^2, \tag{5.4}$$

where S is the cross-section of the resonator. Thus, the decrement can be written as

$$\gamma = 4\pi^2\omega \frac{|F(z_{\mathrm{w}})|^2}{\int \kappa(z)|F(z)|^2\, dz}\frac{1}{S}\sum_{i,j} |\langle i|\vec{\xi}\vec{\mathcal{D}}|f\rangle|^2 \delta\left(E_i - E_f - \hbar\omega\right)\left[\mathcal{F}(E_i) - \mathcal{F}(E_f)\right], \tag{5.5}$$

where the integral in the denominator extends over the entire resonator of size L_z. For a plane wave of light coming through the quantum-well layer, the attenuation coefficient of Eq. (5.1) can be expressed via γ as

$$\beta = \gamma \frac{L_z}{c} \sqrt{\kappa}, \tag{5.6}$$

where γ can be calculated from Eq. (5.5).

In the simplest case of a narrow quantum well, the changes in the refractive index inside the quantum structure can be neglected and the standing wave is

$$\vec{\mathcal{E}}(z) = \vec{\xi} F_0 \cos(qz)\cos(\omega t), \tag{5.7}$$

where $\vec{\xi}$ is the light polarization vector, which is in the x, y plane, q is a discrete number which depends on the resonator length L_z according to $q = 2\pi n/L_z$, and n is an integer. The form-factor of this mode is $\vec{F}(z) \equiv \vec{\xi} F_0 \cos(qz)$. For the standing wave of Eq. (5.7) the decrement takes the form

$$\gamma = \frac{8\pi^2\omega}{\kappa} \cos^2(qz_{\mathrm{w}}) \frac{1}{SL_z} \sum_{i,j} |\langle i|\vec{\xi}\vec{\mathcal{D}}|f\rangle|^2 \delta\left(E_i - E_f - \hbar\omega\right) \left[\mathcal{F}(E_i) - \mathcal{F}(E_f)\right]. \tag{5.8}$$

As expected, this result depends on the position of the quantum-well layer. Accordingly, if its position in the resonator coincides with a nodal point of the standing wave, the decrement is zero. At an anti-nodal point the decrement reaches its maximum.

Another example is presented in Fig. 5.1(b), where light waves propagate along a heterostructure layer. Basically there are three types of such design. Among them the most interesting design is where a narrow-bandgap layer, possibly a graded layer, localizes the electromagnetic modes. The amplitudes of these modes depend on z and decay far away from the layer. These modes are called *waveguide modes*. One of them is depicted in Fig. 5.1(b). Let the electromagnetic field of this mode be

$$\vec{\mathcal{E}} = \vec{\xi} F(z) \cos(\vec{q}\vec{r} - \omega t), \tag{5.9}$$

where $F(z)$ is the form-factor of the mode. The wavevector $\vec{q}$ is in the x, y plane. For a waveguide mode one can define the electromagnetic energy per unit area of the waveguide plane as

$$w = \frac{1}{4\pi} \int \kappa(z) \vec{\mathcal{E}}^2(z) dz = \frac{1}{8\pi} \int \kappa F(z)^2 \, dz, \tag{5.10}$$

and the number of photons in this mode is

$$N_{\mathrm{w}} = \frac{wS}{\hbar\omega}$$

(compare this with the bulk case presented in Table 4.1). The rate of change of the number of photons in this waveguide mode, R, is given by Eq. (5.3). One can introduce the total intensity of the waveguide mode, $I = wc_{\mathrm{eff}}$, where c_{eff} is the phase velocity of the mode under consideration. Generally, c_{eff} depends on the permittivities of the quantum well layer and its surroundings. Within the simplest approach we can set $c_{\rm eff} \approx c/\sqrt{\kappa}$, neglecting the dependence of the permittivity κ on the z coordinate. Now we can define the absorption (gain) coefficient for the waveguide mode:

$$\frac{1}{I}\frac{dI}{dx} = -\alpha. \tag{5.11}$$

Then α can be expressed as (compare with Eq. (4.55)):

$$\alpha = -\frac{4\pi^2\omega\sqrt{\kappa}}{c}\Gamma\frac{1}{SL}\sum_{i,j}|\langle i|\vec{\xi}\vec{\mathcal{D}}|f\rangle|^2\delta\left(E_f - E_i - \hbar\omega\right)\left[\mathcal{F}(E_f) - \mathcal{F}(E_i)\right], \tag{5.12}$$

where

$$\Gamma \equiv \frac{F^2(z_{\rm w})L}{\int \kappa(z)F^2(z)dz} \tag{5.13}$$

is the so-called *optical confinement factor* and L is the width of the quantum well. The optical confinement factor characterizes the portion of the light energy accumulated within the active layer, where the phototransitions take place. Note that Γ is always less than one. The better the optical confinement, the larger the light absorption or gain.

5.2.2 Light Absorption by Confined Electrons

Now we consider the optical properties of confined electrons. The energy spectrum of a confined system is different from that of a bulk material, because of the additional quantization of both electrons and holes. In general, the equations which we obtained previously should be used in conjunction with the appropriate quantized energy levels as the initial and final states, taking into account also the density of states of these levels. Another important factor is the particular shape of the electron and hole wave functions. These wave functions determine the matrix elements and not only change the magnitudes of the absorption and emission rates, but also lead to new selection rules for the phototransitions.

Let us consider the simplest case of a three-layered semiconductor structure (Fig. 5.2(a)) of type I; type-I structures have band-edge discontinuities such that the same embedded layer provides quantum wells for electrons and holes. Figures 5.2(b) and (c) show the energy levels of both types of carriers. The energy dependence on the in-plane wavevector is shown in Fig. 5.2(c) for each of the low-dimensional subbands. The first obvious conclusion which can be drawn is that there must be a shift of the interband spectra toward a short-wavelength region, since the threshold energy of interband phototransitions can be estimated as

$$\hbar\omega \geq \hbar\omega_{\rm o}{}^{(1)} \equiv E_{\rm g} + E_{{\rm c},1} + E_{{\rm v},1} \equiv E_{\rm g}{}^{{\rm QW},1}. \tag{5.14}$$

Here $E_{{\rm c},i}$ and $E_{{\rm v},j}$ are the positions of the electron and hole two-dimensional subbands, respectively. The indices i and j correspond to the subband numbers. Thus,

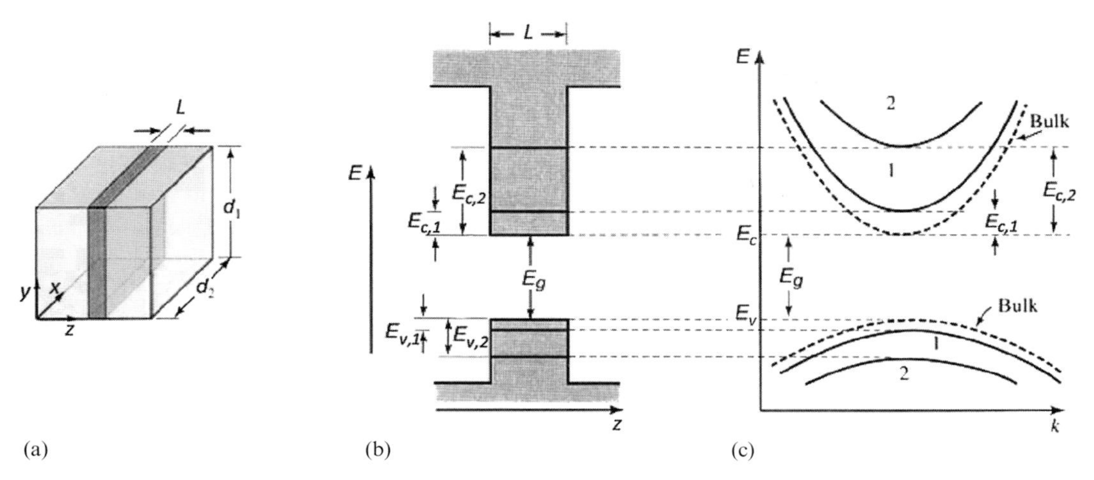

Figure 5.2 (a) The geometry of a three-layer quantum-well structure in which a narrow-bandgap material is embedded within a wider-bandgap material. (b) The band diagram and (c) the dispersion curves of a type-I quantum well. Parabolic dispersion relations are assumed both for the conduction band and for a simple valence band. Two subbands are shown for both bands. Republished with permission of John Wiley and Sons Inc., from Fig. 16.1-24 of *Fundamentals of Photonics*, B. E. A. Saleh and M. C. Teich, Edition 2, Copyright 2007; permission conveyed through Copyright Clearance Center, Inc.

the lowest subbands determine the threshold given by Eq. (5.14). Another conclusion is that the same transition energy $\hbar\omega$ may correspond to different combinations of hole and electron subbands involved in the phototransitions.

In Eqs. (5.3), (5.5), and (5.8) the matrix elements should be estimated from the true wave functions containing envelope functions different from those for bulk electrons (compare these with Eq. (4.58)):

$$\Psi_{\mathrm{v},\vec{k}_\mathrm{v},i} = \frac{1}{\sqrt{S}} u_{\mathrm{v},\vec{k}_0}(r) e^{i\vec{k}_\mathrm{v}\vec{r}} \chi_{\mathrm{v},i}(z),$$

$$\Psi_{\mathrm{c},\vec{k}_\mathrm{c},j} = \frac{1}{\sqrt{S}} u_{\mathrm{c},\vec{k}_0}(r) e^{i\vec{k}_\mathrm{c}\vec{r}} \chi_{\mathrm{c},j}(z),$$

where $u_{\mathrm{v},\vec{k}_0}(r)$ and $u_{\mathrm{c},\vec{k}_0}(r)$ are the Bloch functions, with $\vec{k}_0$ the three-dimensional wavevector corresponding to the minimum (maximum) of the electron (hole) bands involved in phototransition under consideration. In these results and in the following discussion, $\vec{k}_\mathrm{v}$ and $\vec{k}_\mathrm{c}$ are the two-dimensional, in-plane wavevectors, S is the area of the quantum well layer, and $\chi_{\mathrm{v},i}(z)$ and $\chi_{\mathrm{c},j}(z)$ are the envelope wave functions for the quantized transverse motion as discussed in Chapter 3. Thus, the matrix element is (see the definition given in Eq. (4.63))

$$\begin{aligned} \langle v, i, k_\mathrm{v}, \sigma \,|\, \vec{\xi}\vec{\mathcal{D}} \,|\, c, j, k_c, \sigma' \rangle &= \delta_{\sigma,\sigma'} |(\vec{\xi}\vec{\mathcal{D}}_\mathrm{cv})| \frac{\Omega}{S} \sum_{\vec{R}_j} e^{i(\vec{k}_\mathrm{c}-\vec{k}_\mathrm{v})\vec{R}_j} \chi^*_{\mathrm{v},i}(Z_j)\chi_{\mathrm{c},j}(Z_j) \\ &= \delta_{\sigma,\sigma'} \delta_{\vec{k}_\mathrm{v},\vec{k}_\mathrm{c}} |(\vec{\xi}\vec{\mathcal{D}}_\mathrm{cv})| \int dz\, \chi^*_{\mathrm{v},i}(z)\chi_{\mathrm{c},j}(z), \end{aligned} \quad (5.15)$$

where $\vec{\mathcal{D}}_{\rm cv}$ is the interband dipole-matrix element introduced in Chapter 4, σ, σ' are the spin quantum numbers, Ω is the unit-cell volume of the crystal, and $\vec{R}_j = \{X_j, Y_j, Z_j\}$ is the position vector of the *j*th cell. The summation runs over all crystal cells. Equation (5.15) shows that, in addition to the atomic-like dipole-matrix element calculated from the Bloch functions, $\mathcal{D}_{\rm cv}$, the overlap integral of the envelope functions depends on the quantum numbers of the initial and final states of transitions. From Eq. (5.15) the selection rule for *two-dimensional wavevectors* can be derived:

$$\vec{k}_{\rm v} = \vec{k}_{\rm c}. \tag{5.16}$$

This selection rule differs from Eq. (4.65) for bulk crystals, as a result of the lack of translational symmetry in the z-direction, i.e., only in-plane components of the wavevector are conserved during the optical transition. Note that, in comparison with the bulk case, a new factor appears in the matrix element of Eq. (5.15):

$$\int dz\, \chi^*_{{\rm v},i}(z)\chi_{{\rm c},j}(z) \equiv \langle v, j | c, i\rangle, \tag{5.17}$$

which is the overlap integral of the envelope functions from different bands.

Using the matrix element of Eq. (5.15) one can calculate all of the characteristics, β, γ, and α, of Eqs. (5.1), (5.5), and (5.12) for the geometry presented in Fig. 5.1. For example, from Eq. (5.12) we get

$$\alpha = -\frac{4\pi^2\omega\sqrt{\kappa}}{c}\Gamma\frac{1}{SL}\sum_{\sigma,i,\vec{k}_{\rm c};\sigma',j,\vec{k}_{\rm v}} \delta_{\sigma,\sigma'}|(\vec{\xi}\vec{\mathcal{D}}_{\rm cv})|^2\delta_{\vec{k}_{\rm c},\vec{k}_{\rm v}}|\langle v, j|c, i\rangle|^2\delta\left(E_{{\rm c},i} - E_{{\rm v},j} - \hbar\omega\right) \times \left[\mathcal{F}(E_{{\rm c},j}) - \mathcal{F}(E_{{\rm v},i})\right], \tag{5.18}$$

where we assume a nondegenerate valence band (compare this with Eqs. (4.78) and (4.79)). Further transformation of Eq. (5.18) leads to

$$\alpha = -\frac{4\pi^2\omega\sqrt{\kappa}}{c}\Gamma\frac{|(\vec{\xi}\vec{\mathcal{D}}_{\rm cv})|^2}{2\pi^2 L}\sum_{i,j}|\langle v, j|c, i\rangle|^2\int d\vec{k}\,\delta\left(E_{{\rm c},i}(\vec{k}) - E_{{\rm v},j}(\vec{k}) - \hbar\omega\right) \times \left[\mathcal{F}(E_{{\rm v},i}) - \mathcal{F}(E_{{\rm c},j})\right]. \tag{5.19}$$

The energy conservation law following from the δ-function,

$$E_{{\rm c},i}(\vec{k}) - E_{{\rm v},j}(\vec{k}) = \hbar\omega, \tag{5.20}$$

shows that phototransitions can involve different subbands from the valence band and the conduction band.

In the case of parabolic subbands, i.e.,

$$E_{{\rm c},i}(\vec{k}) = E^0_{{\rm c},i} + \frac{\hbar^2 k^2}{2m^*_{\rm c}} \quad \text{and} \quad E_{{\rm v},j}(\vec{k}) = E^0_{{\rm v},j} + \frac{\hbar^2 k^2}{2m^*_{\rm v}}, \tag{5.21}$$

Eq. (5.20) gives the magnitude of the two-dimensional vector $\vec{k}$ corresponding to *vertical transitions* between the v,j and c,i subbands:

$$k = \sqrt{\frac{2m_\mathrm{r}}{\hbar^2}\left(\hbar\omega - E^0_{\mathrm{c},i} - E^0_{\mathrm{v},j}\right)}. \tag{5.22}$$

Finally, we get

$$\alpha = -\frac{4\pi^2\omega\sqrt{\kappa}}{c}\Gamma|(\vec{\xi}\vec{\mathcal{D}})_{\mathrm{cv}}|^2\sum_{i,j}|\langle v,j|c,i\rangle|^2\frac{m_\mathrm{r}^*}{\pi\hbar^2 L}\Theta\left(\hbar\omega - E_{\mathrm{c},i}(\vec{k}) + E_{\mathrm{v},j}(\vec{k})\right)$$
$$\times\left[\mathcal{F}(E_{\mathrm{c},i}) - \mathcal{F}(E_{\mathrm{v},j})\right]. \tag{5.23}$$

Here, m_r is the reduced effective mass of the electron–hole pair (see Eq. (3.105)). The optical density of states is represented by the factors

$$\varrho^{\mathrm{opt}}_{i,j}(\omega) \equiv \frac{m_\mathrm{r}^*}{\pi\hbar L}\Theta\left(\hbar\omega - E_{\mathrm{c},i}(\vec{k}) + E_{\mathrm{v},j}(\vec{k})\right) \tag{5.24}$$

which appear in Eq. (5.23); this optical density of states corresponds to phototransitions between subbands: $\mathrm{v},j \leftrightarrow \mathrm{c},i$. The energies $E_{\mathrm{c},i}(\vec{k})$ and $E_{\mathrm{v},j}(\vec{k})$ must be calculated for the wavevector given by Eq. (5.22). The quantities $\varrho^{\mathrm{opt}}_{i,j}(\omega)$ are step-like functions, which are consistent with the formulae for the density of states of two-dimensional electrons (holes), as discussed in Chapter 3. As the result, the function $\alpha(\omega)$ also has a step-like shape; see Fig. 5.3. When the photon energy exceeds the threshold energy corresponding to transitions between a new pair of electron and hole subbands, a new step in the absorption should be observed. As a result, $\alpha(\omega)$ can be represented as a sum over all pairs of the subbands involved in phototransitions:

$$\alpha(\omega) = \sum_{i,j}\alpha_{\mathrm{sub}}(\omega, i, j), \tag{5.25}$$

$$\alpha_{\mathrm{sub}}(\omega, ij) \equiv -\frac{4\pi^2\omega\sqrt{\kappa}}{c\hbar}\Gamma|(\vec{\xi}\vec{\mathcal{D}})_{\mathrm{cv}}|^2|\langle v,j|c,i\rangle|^2\varrho^{\mathrm{opt}}_{i,j}(\omega)\left[\mathcal{F}(E_{\mathrm{c},i}) - \mathcal{F}(E_{\mathrm{v},j})\right]. \tag{5.26}$$

Schematically $\alpha_{\mathrm{abs}}(\omega)$ is shown in Fig. 5.3 for two quantum well heterostructures with different well widths.

Each contribution $\alpha_{\mathrm{sub}}(\omega, ij)$ is proportional to the overlap integral of the envelope functions $\chi_{\mathrm{c},i}(z)$ and $\chi_{\mathrm{v},j}(z)$. These overlap integrals result in new selection rules. Let us examine the case of infinitely deep quantum structures for electrons and holes. In such quantum structures the eigenfunctions are independent of the effective mass, and form an orthogonal set of functions. The set depends parametrically on the size of the quantum well layer. If we choose the subband numbers $i = j$, the wave functions $\chi_{\mathrm{v},i}(z)$ and $\chi_{\mathrm{c},i}(z)$ are identical, the overlap integral satisfies

$$|\langle v, i|c, i\rangle|^2 = 1,$$

and phototransitions between these subbands are allowed. If $i \neq j$, the envelope functions are orthogonal, $\langle v, i|c, i\rangle = 0$, and the transitions are forbidden. Figure 5.4 shows the allowed and forbidden transitions. For quantum wells with finite barrier

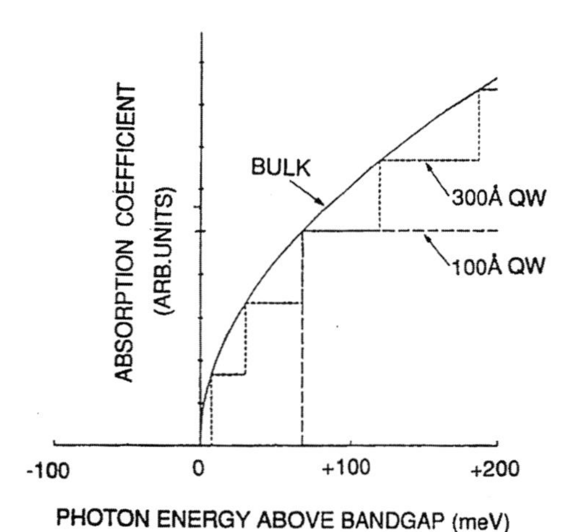

Figure 5.3 A sketch of absorption by multiple subbands of an ideal quantum well. The results for two GaAs quantum wells with well widths of 100 Å and 300 Å are presented. For the narrower well, only one step in the absorption coefficient is shown, whereas five steps are seen in the same frequency range for the wider well. The absorption coefficient for bulk-like material is depicted for comparison. Reprinted figure with permission from Fig. 1 of D. A. B. Miller *et al.*, *Phys. Rev.* **B 33**, 6976, 1986. Copyright (1986) by the American Physical Society.

heights, the wave functions depend on the effective mass, the barrier parameters, etc.; i.e., the two sets of wave functions are different. Thus, when $i \neq j$, orthogonality between the electron and hole functions no longer holds. However, $|\langle v, i|c, i\rangle|^2$ is still of the order of 1, while $|\langle v, i|c, j\rangle|^2 \ll 1$, $i \neq j$. This means that the transition probability is typically very weak or equals zero for $i \neq j$.

Thus we can conclude that quantization of the electron energy leads to major changes in the selection rules and in the intensity of the band-to-band transitions.

5.2.3 Effects of the Complex Valence Band of III–V Compounds

Above we discussed the interband transitions in quantum wells with a simple structure of both the valence and conduction bands; see Eq. (5.21). In fact, for heterostructures based on III–V compounds one should take into account the complex structure of the valence band, which has been analyzed in the previous section.

The energies of the valence subbands are complex functions of the wavevector and depend on the orientation of the quantum well layer with respect to the crystal axes, as discussed in Chapter 4. The left panel of Fig. 5.5 shows these subband energies for a particular AlGaAs/GaAs quantum well for two directions of the in-plane wavevector, [100] and [110]. One can see that the quantization of the holes in a quantum well leads to splitting of the heavy- and light-hole bands (in contrast to the case of bulk GaAs), as well as more complex dispersion dependences giving rise

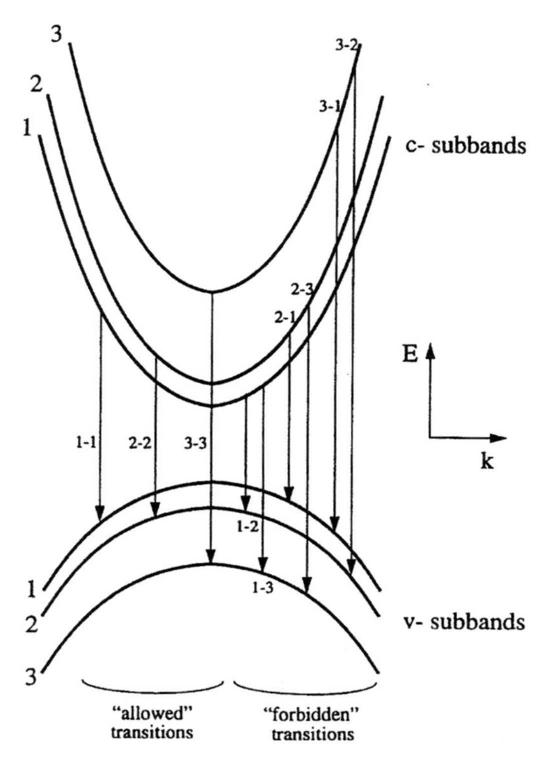

Figure 5.4 Allowed and forbidden interband phototransitions in an ideal quantum well. The vertical and horizontal axes are the energy and the wavevector, respectively. Three subbands for electrons and holes are shown. Vertical lines connecting different subbands indicate all possible transitions. The overlap integrals of subband wave functions of the electrons and holes suppress some of the transitions.

to $\vec{k}$ intervals with negative effective masses ($d^2E(k)/dk^2 < 0$). Given the $E_{c,i}(\vec{k})$ and $E_{v,j}(\vec{k})$ dependences, we can apply Eq. (5.19). For the integral over $\vec{k}$ in Eq. (5.19) one has

$$\frac{1}{2\pi^2 L}\int d\vec{k}\,\delta\left(E_{c,i}(\vec{k}) - E_{v,j}(\vec{k}) - \hbar\omega\right)[\mathcal{F}(E_{v,i}) - \mathcal{F}(E_{c,j})]$$
$$= \frac{1}{\pi L}\frac{k_t}{|d(E_{c,i}(k) - E_{v,j}(k)|/dk|_{k_t}}\Theta\left(\hbar\omega - E_{c,i}(k_t) + E_{v,j}(k_t)\right)$$
$$\times\,[\mathcal{F}(E_{c,i}(k_t)) - \mathcal{F}(E_{v,j}(k_t))], \tag{5.27}$$

where k_t is the absolute value of the wavevector satisfying the energy conservation law of Eq. (5.20). Here the Θ-function comes from the optical density of states corresponding to the phototransitions

$$c, i \leftrightarrow v, j.$$

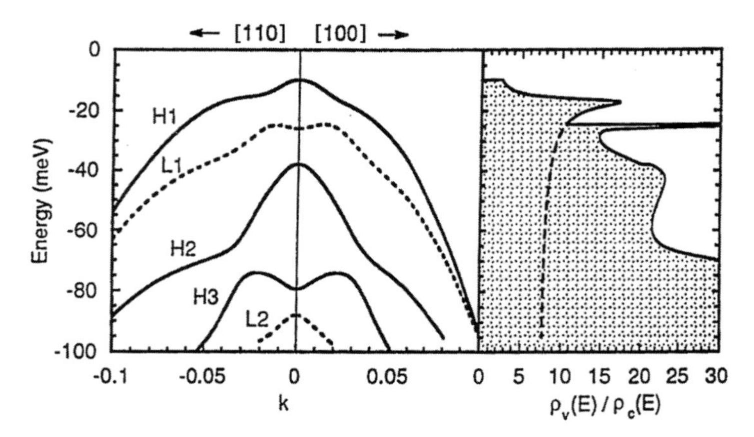

Figure 5.5 Valence subband energies (left) and the density of states (right) in an 80 Å $Al_{0.2}Ga_{0.8}As$/GaAs quantum well for two possible directions, [110] and [100], of the in-plane wavevector, $\vec{k}$. The quantization of the holes in the well of height 95 meV leads to a strong coupling between the heavy and light holes. It results in complex energy subband dependences. The total density of states and the contribution of the heavy holes (dashed line) are on the right, plotted relative to the density of states in the first conduction subband. Reprinted from Figure 17 in *Quantum Well Lasers*, Scott W. Corzine, Ran-Hong Yan, and Larry A. Coldren, Chapter 1 – Optical gain in III–V bulk and quantum well semiconductors, pp. 17–96, Copyright (1993), with permission from Elsevier.

Within the effective-mass approximation of Eq. (5.21) the optical density of states coincides with Eq. (5.24). The nonparabolic dependences $E_{v,j}(\vec{k})$ yield more complex optical densities of states and essentially different transition strengths. Indeed, the density of states of the valence band is presented in the right panel of Fig. 5.5. These data correspond to the energy dependences presented in the left panel of Fig. 5.5. It can be seen that the density of states apparently reflects the $E(\vec{k})$ behavior and, in particular, it causes a nonmonotonic dependence of this density on the hole energy. Peaks in the density of states correspond to the points with zero second derivatives of energy, i.e. $d^2E(k)/dk^2 = 0$. Thus, the real band structure results in more complex dependences of the optical density of states and interband spectra than for the two parabolic bands of Eq. (5.21). These dependences are especially important for laser generation, since they affect the light gain and the spectrum of the laser emission.

5.2.4 Other Factors Affecting the Interband Optical Spectra

There are further physical factors, which have not been taken into account in the previous discussion; however, these factors also affect the interband optical spectra and cause them to differ from the ideal step-like shape shown in Fig. 5.3. One of them is the broadening of the spectra due to *intraband scattering processes*. Indeed,

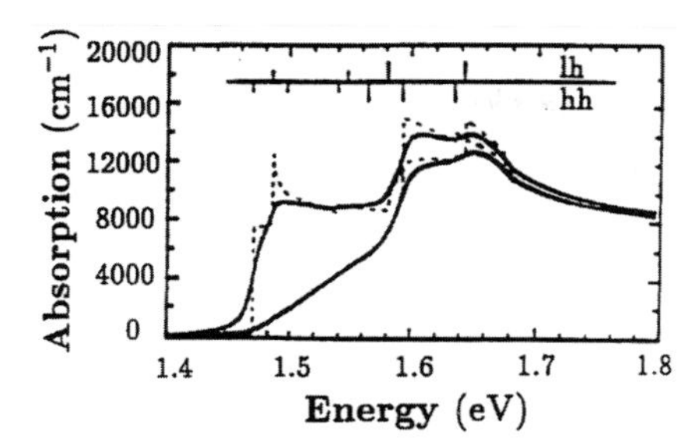

Figure 5.6 Light absorption as a function of the photon energy in an 80 Å $Al_{0.2}Ga_{0.8}As/GaAs$ quantum well. The dashed lines correspond to calculations based on the equations discussed in the text. The solid lines are obtained for a broadening due to scattering processes. At the top of the figure, markers indicate the c ↔ hh and c ↔ lh transitions between the first electron subband and three lowest hole subbands. Reprinted from Figure 23(a) in *Quantum Well Lasers*, Scott W. Corzine, Ran-Hong Yan, and Larry A. Coldren, Chapter 1 – Optical gain in III–V bulk and quantum well semiconductors, pp. 17–96, Copyright (1993), with permission from Elsevier.

intraband scattering processes lead to uncertainties in the electron and hole energies. According to the general uncertainty relations, one can estimate the energy uncertainties as

$$\Delta E_{\mathrm{v,c}} \approx \hbar / \tau_{\mathrm{v,c}},$$

where $\tau_{\mathrm{v,c}}$ is the intraband scattering time in the valence (v) or conduction (c) bands. The energy uncertainties, or, in other words, the broadening of the energy levels, result in broadening of the optical spectra. It is possible to take this effect into consideration by means of a simple procedure: replace the δ-function in Eqs. (5.8) and (5.12) by the broadening function $\Delta(\hbar\omega - E_{\mathrm{v,c}})$. An example of the broadening function is a Lorentzian with a half width at half maximum of Γ:

$$\Delta(\hbar\omega - E_{\mathrm{v,c}}) = \frac{1}{\pi} \frac{\Gamma_{\mathrm{v,c}}}{(\hbar\omega - E_{\mathrm{v,c}})^2 + \Gamma_{\mathrm{v,c}}{}^2}, \tag{5.28}$$

where $\Gamma_{\mathrm{v,c}} \equiv \hbar/\tau_{\mathrm{c}} + \hbar/\tau_{\mathrm{v}}$. (In general, one can consider $\Gamma_{\mathrm{v,c}}$ as being a function of the energies E_{c} and E_{v}, but independent of ω.) The results presented in Fig. 5.6 show that the spectral shape of the intensities of interband transitions becomes smooth and broad in spite of the sharp step-like behavior that is observed for the density of states.

It is known that the scattering rates increase with increasing temperature. Hence, the optical spectra become broader at higher temperatures. Besides the broadening effects, an essential temperature dependence of the spectra is caused by the temperature dependence of the distribution functions, $\mathcal{F}_{\mathrm{c}}$ and $\mathcal{F}_{\mathrm{v}}$.

The optical density of states of the two-dimensional systems of Eq. (5.24) is finite, but small. Thus, to observe the peculiarities of the low-dimensional behavior, one

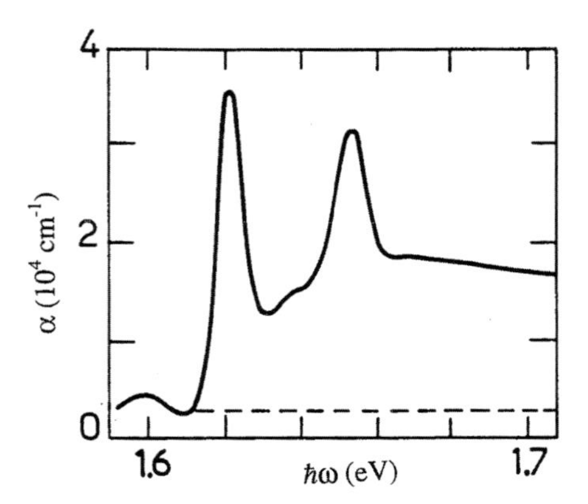

Figure 5.7 The measured absorption spectrum of an AlAs/GaAs multiple-quantum-well structure. Reprinted figure with permission from Fig. 8 of Y. Masumoto, *et al.*, *Phys. Rev.* **B 32**, 4275, 1985. Copyright (1985) by the American Physical Society.

must use high-resolution equipment and employ *multi-quantum-well structures* with roughly 20–100 quantum well layers.

Another factor which affects optical spectra is the presence of *excitonic effects*. Excitonic effects on bulk-material spectra and quantum well spectra have been discussed in Chapters 3 and 4. In quantum wells the exciton energy is greater than that in bulk materials. This leads to an enhancement of the exciton spectral lines, such that they are clearly observed up to room temperature. The absorption spectrum at low temperature is presented in Fig. 5.7 for AlAs/GaAs multiple-quantum-well heterostructures near the fundamental absorption edge. The two pronounced peaks in absorption are due to the excitation of heavy- and light-hole excitons in 76 Å GaAs wells. In Fig. 5.8 the spectrum of another AlGaAs heterostructure is shown for room temperature. The structure consists of 50 periods with 100 Å GaAs wells. The exciton resonances with quantum numbers $n = 1$, 2, and 3 are extremely pronounced. The heavy- and light-hole transitions are resolved for $n = 1$. The large resonance effects related to excitons are used in many applications, as will be discussed in the next chapters.

So far we have considered quantum structures of type I. In type-II semiconductor heterostructures the interband phototransitions are indirect in real space. This leads to a smoothing of the absorption (emission) spectra, similar to that found for indirect transitions in bulk materials. To reliably measure the quantization effects in the absorption spectra of type-II quantum structures, one has to have a multilayer structure with more than 100 layers.

5.2.5 Polarization Effects

In low-dimensional electron systems the effects of light polarization on the optical spectra are much more pronounced than in bulk materials. It has been shown in

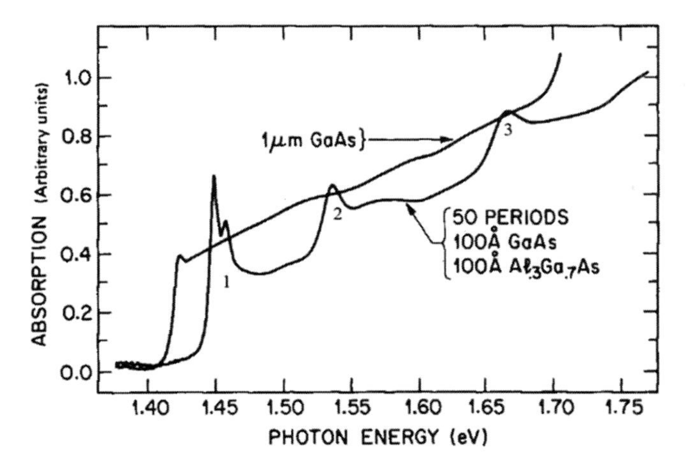

Figure 5.8 The absorption spectrum measured at room temperature for an $Al_{0.3}Ga_{0.7}As/GaAs$ multiple-quantum-well structure. The numbers show the positions of three lowest exciton transitions with $n = 1, 2, 3$. For comparison, the absorption in bulk GaAs is shown. Copyright 1989 from Fig. 6 in S. Schmitt-Rink, D. S. Chemla, and D. A. B. Miller, *Adv. Phys.* **38**, 89 (1989). Reproduced by permission of Taylor and Francis.

the previous chapters that in bulk materials for a fixed wavevector $\vec{k}$ the interaction with photons is strongly polarization-dependent. However, after averaging over all possible directions of the wavevectors in three-dimensional space this dependence disappears in bulk materials, such as Si, Ge and III–V compounds. In fact, it is true for any cubic crystal. A completely different physical situation occurs in quantum heterostructures, where electrons and holes are confined to two, one, or zero directions. Therefore, one can expect strong polarization effects in the optical spectra of these structures.

Let us consider a quantum well in a heterostructure of type I. The confined states can be treated as a sum of two plane waves, resulting in a standing wave with wavevectors parallel and anti-parallel to the z-axis. Thus, we can approximately treat any electron state as a combination of states characterized by the wavevector $\vec{k} = \{\vec{k}_{||}, k_z\}$. Within this approach, the polarization dependence is similar to that of the bulk case studied in Sections 4.4.3. The only difference is that we must average over all possible directions of $\vec{k}$ in the plane of a quantum well. As a result, $\vec{k}_{\text{ave}}$ coincides with $\vec{k}_z$. This means that the polarization dependences of the transition strengths are proportional to those shown in Fig. 4.13. We assume that the phototransitions are near the band edge, where the $\vec{k}$ vector mainly consists of the quantized component k_z. Otherwise, we should take into account additional mixing of the valence bands in the quantum well. If we set $\vec{k} \approx \vec{k}_{\text{ave}} \approx (0, k_z)$, we can use Eq. (4.84). For example, light with polarization perpendicular to the quantum well layer does not cause phototransitions between the conduction band and the heavy-hole subband; i.e., $(\vec{\xi}\vec{\mathcal{D}})_{c-hh} = 0$. In contrast, the interaction of this light

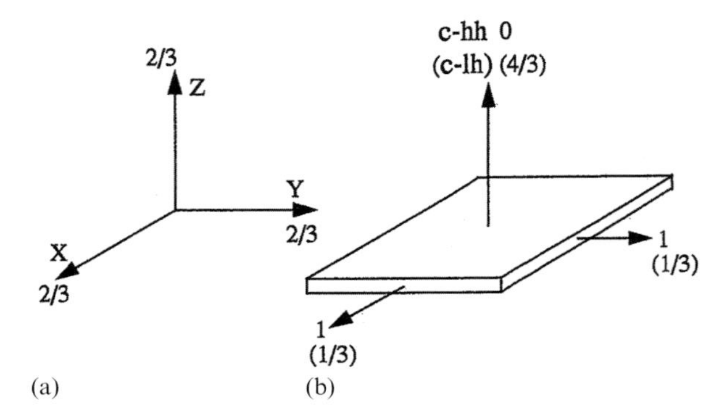

Figure 5.9 Relative transition intensities shown by lengths of the arrows are presented for various valence bands for (a) bulk and (b) quantum well III–V-compound structures. In case (b), the polarization vertical to the quantum well plane corresponds to conduction-band–heavy-hole transitions, while the in-plane polarization corresponds to conduction-band–light-hole transitions.

with the light-hole subband reaches a maximum for perpendicular polarization and one has $(\vec{\xi}\vec{\mathcal{D}})_{c-lh} = \frac{4}{3}\mathcal{D}$. On the other hand, for light with the polarization vector in the quantum well plane the intensity of phototransition to the heavy-hole band increases to the maximum, and $(\vec{\xi}\vec{\mathcal{D}})_{c-hh} = \mathcal{D}$, while the intensity of phototransition to the light-hole band decreases to the minimum with $(\vec{\xi}\vec{\mathcal{D}})_{c-lh} = \frac{1}{3}\mathcal{D}$. These results are illustrated by the diagram in Fig. 5.9, where the bulk case is presented as well for comparison. In these diagrams the relative intensities of phototransitions for different polarizations and types of valence bands are shown.

Consequently, there is a strong polarization dependence of the interaction of light with carriers in confined quantum structures. These effects are important for light-emitting diodes and laser diodes based on quantum structures.

5.3 Intraband Transitions in Quantum Structures

In the previous section we studied the interaction of electrons and holes with light which results in band-to-band phototransitions. In this section we shall analyze *intraband phototransitions*; for this case the free carriers absorb or emit photons while remaining in the same energy band, i.e., either in the conduction band or in the valence band. We are going to study two types of phototransition: those between subbands confined in a quantum well and those from confined subbands to extended electron states. For both cases we calculate the attenuation of light passing across the quantum well layer and the absorption coefficient for the light propagating along this layer. We show that intraband phototransitions are strongly polarization-dependent. We present several experimental and theoretical results for intraband phototransitions in more complex heterostructure systems with two coupled quantum wells or with three coupled quantum wells.

5.3.1 Intraband Absorption and Conservation Laws

We start by recalling that, in an ideal bulk crystal, intraband phototransitions are impossible because the energy and momentum conservation laws,

$$E(\vec{k}) - E(\vec{k}') = \hbar\omega \quad \text{and} \quad \vec{k} - \vec{k}' = \vec{q}\,(\approx 0), \tag{5.29}$$

cannot be satisfied simultaneously; here, $E(\vec{k}) = \hbar^2\vec{k}^2/2m^*$ is the electron kinetic energy, $\vec{k}$ is the three-dimensional momentum, $\omega = cq/\sqrt{\kappa}$ is the light frequency, and $\vec{q}$ is the wavevector of the light. In fact, as shown in the previous chapters, the change in the electron momentum is negligible during the phototransitions. Intraband phototransitions can be induced only by phonons, impurities, and crystal imperfections, which ensure momentum conservation.

In contrast to bulk materials, intersubband phototransitions are allowed in semiconductor heterostructures. Consider a quantum well. Let the x, y axes be in the plane of the quantum well layer. In systems with heterojunctions, a carrier is not characterized by the z component of the momentum. In the proper description, one introduces quantum numbers: a subband number i, the in-plane wavevector $\vec{k} \equiv \{k_x, k_y\}$, and a spin number σ. The subband energy dispersion is given by $E_i(\vec{k})$. Now, instead of Eq. (5.29), for i–j intersubband phototransitions, one can write

$$E_i(\vec{k}) - E_j(\vec{k}') = \hbar\omega, \quad \vec{k} - \vec{k}' \approx 0, \tag{5.30}$$

where

$$E_i(\vec{k}) = E_i^{(0)} + \frac{\hbar^2\vec{k}^2}{2m^*}. \tag{5.31}$$

Consequently, if $\hbar\omega$ coincides with intersubband distances $E_i^{(0)} - E_j^{(0)}$, Eq. (5.30) is satisfied. If one of the electron states involved in a phototransition is in an unconfined, *extended* state, the energy of the transverse motion, E_z, is a true quantum number. Thus, instead of Eq. (5.30), the conservation laws take the form

$$E(E_z, \vec{k}) - E_j(\vec{k}') = \hbar\omega, \quad \vec{k} - \vec{k}' \approx 0, \tag{5.32}$$

where

$$E(E_z, \vec{k}) = V_b + E_z + \frac{\hbar^2 k^2}{2m^*}, \tag{5.33}$$

where V_b is the depth of the quantum well. Equation (5.32) is satisfied at $E_z = E_j^{(0)} + \hbar\omega - V_b$.

We see that both Eq. (5.30) and Eq. (5.32) can be satisfied for certain photon energies. Consequently, intraband phototransitions are allowed and can occur without the presence of phonons, defects, etc. Two types of process corresponding to the cases of Eqs. (5.30) and (5.32) are shown in Figs. 5.10(a) and (b). The type shown in Fig. 5.10(a) corresponds to so-called *intersubband* phototransition. The second type of process may be described as photo-ionization of a quantum well and photo-capture of an electron into the well. Note that, for the intrasubband case, when $i = j$, we return to the conditions given by Eq. (5.29), which forbid intraband phototransitions within the same subband.

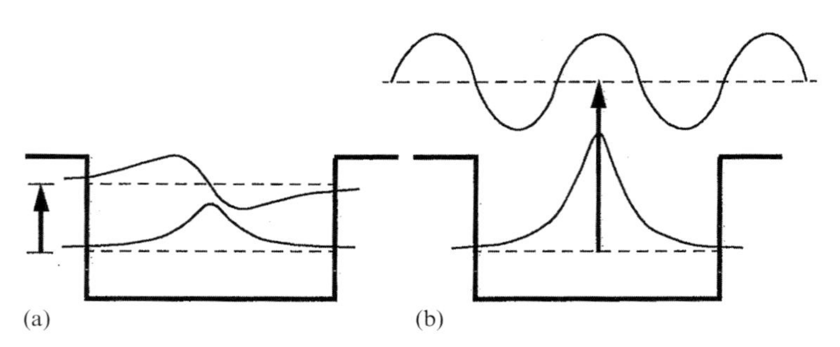

Figure 5.10 The schematics of intraband phototransitions: (a) intersubband transitions and (b) confined-state-to-extended-state transitions.

Taking into account that the energy differences between subbands are tens to hundreds of meV and the depths of quantum wells are of the same order of magnitude, one can see from Eqs. (5.30) and (5.32) that intraband phototransitions correspond to the absorption or emission of *infrared light*.

There is another important feature of intraband phototransitions between extended states in any system with heterojunctions. Since the z component of the electron momentum perpendicular to the plane of a heterojunction is no longer conserved, the energy conservation law can always be satisfied for a certain photon energy. For example, for unconfined electrons the energy is given by Eq. (5.33). It is easy to see that a photon with any energy $\hbar\omega$ can be absorbed or emitted and induce a transition from any state with energy E_z to a state with energy $E_z + \hbar\omega$.

Thus, we may conclude that intraband phototransitions are allowed even in ideal heterostructures. The properties of these phototransitions depend strongly on the heterostructure parameters and may be varied over a wide range (for example, by variation of the quantum well thickness). This makes quantum structures very attractive for applications in different infrared devices such as photodetectors and infrared emitters, particularly infrared lasers, etc.

5.3.2 Intersubband Phototransitions

For intraband (intersubband) processes we can apply general equations for phototransition rates with initial and final quantum number sets denoted by

$$\nu = \{b, i, \vec{k}, \sigma\} \quad \text{and} \quad \nu' = \{b, j, \vec{k}', \sigma'\},$$

where b is the label of the band ($b = \text{v, c}$). Now, the phototransition probability can be written as

$$P_{i,\vec{k},\sigma;j,\vec{k}',\sigma'} = \frac{2\pi}{\hbar} |\langle b, i, \vec{k}, \sigma | \vec{\mathcal{E}}\vec{D} | b, j, \vec{k}', \sigma' \rangle|^2 \delta \left[E_j(\vec{k}') - E_i(\vec{k}) - \hbar\omega \right], \tag{5.34}$$

where $\vec{\mathcal{E}}$ is the electric field of the light wave, and $\vec{D} \equiv e\vec{r}$ is the dipole moment of the electron. To be specific, we suppose that $E_j(\vec{k}') > E_i(\vec{k})$, i.e., we consider the absorption of light. The electric field of a light wave can be written in the form of a plane wave:

$$\vec{\mathcal{E}} = \vec{\xi} F(\vec{r}).$$

For the intraband (intersubband) phototransitions, the Bloch functions are the same for both initial and final electron states. Thus, the matrix elements in Eq. (5.34) are defined only by the *envelope wave functions*. According to Chapter 3, the envelope wave functions of confined electrons have the form

$$\psi_{b,l,\vec{k}}(\vec{r}) = \frac{1}{\sqrt{S}} \chi_l(z) e^{i\vec{k}\vec{r}}, \tag{5.35}$$

where S is the area of the quantum well layer, and $\chi_l(z)$ is the confined wave function for the lth subband, $l = 1, 2, \ldots$ Since the dependence of the electric field on position, $F(\vec{r})$, is very smooth compared with both the quantum well thickness, L, and the electron wavelength, $2\pi/k$, this dependence can be neglected in the matrix element. Then, taking into account the conservation of spin in electric-dipole transitions, we can calculate the matrix element:

$$\begin{aligned}
\langle b, i, \vec{k}, \sigma | \vec{\xi}\vec{r} | b, j, \vec{k}', \sigma' \rangle &= \langle b, i, \vec{k} | \vec{e}\vec{r} | b, j, \vec{k}' \rangle \delta_{\sigma,\sigma'} \\
&= \delta_{\sigma,\sigma'} \frac{1}{S} \int d^3r\, \chi_j^*(z) e^{-i\vec{k}'\vec{r}} (\xi_x x + \xi_y y + \xi_z z) \chi_i(z) e^{i\vec{k}\vec{r}} \\
&= \xi_z \delta_{\sigma,\sigma'} \delta_{k,k'} \int dz\, \chi_j^*(z) z \chi_i(z) \\
&\equiv \xi_z \delta_{\sigma,\sigma'} \delta_{k,k'} \langle \chi_j | z | \chi_i \rangle .
\end{aligned} \tag{5.36}$$

Here we take into account that for $i \neq j$ the integral

$$\int dz\, \chi_j^*(z) \chi_i(z) = 0,$$

because the wave functions of different subbands are orthogonal.

From Eq. (5.36) one can see that, if the polarization vector of the light $\vec{\xi}$ lies in the plane of the quantum well layer (the x, y plane), the matrix element is zero and phototransitions are impossible. If the vector $\vec{\xi}$ has a z component, intersubband processes take place. The probability of phototransitions can be written as

$$P_{i,\vec{k},\sigma;j,\vec{k}',\sigma'} = \frac{2\pi}{\hbar} e^2 |\xi_z|^2 |F(z_{\mathrm{W}})|^2 \delta_{\vec{k},\vec{k}'} \delta_{\sigma,\sigma'} |\langle \chi_j(z) | z | \chi_i(z) \rangle|^2 \delta\left(E_j(\vec{k}') - E_i(\vec{k}) - \hbar\omega \right), \tag{5.37}$$

where $F(z_{\mathrm{W}})$ is the magnitude of the electric field in a quantum well layer situated at $z \approx z_{\mathrm{W}}$. Equation (5.37) gives both of the conservation laws as well as an explicit polarization dependence of the phototransitions. For the case of an infinitely deep well which has wave functions

$$\chi_l(z) = \sqrt{\frac{2}{L}} \sin\left(\frac{\pi l}{L}\right) z, \quad l = 0, 1, 2, \ldots \; (0 < z < L), \tag{5.38}$$

we get the overlap integrals

$$\langle \chi_j(z)|z|\chi_i(z)\rangle = \frac{L}{\pi^2}\left[\frac{\cos(\pi(j-i))-1}{(j-i)^2} - \frac{\cos(\pi(j+i))-1}{(j+i)^2}\right]. \tag{5.39}$$

Thus, we find the expected result: the matrix element is proportional to the thickness of the quantum well L. Equation (5.39) implies some selection rules: for transitions between two subbands with the same parity, the matrix element is zero, i.e, these transitions are forbidden. The transitions are possible only between states with different parities. Another result which is evident from Eq. (5.39) is that the matrix element decreases when the subband numbers i, j increase or their difference, $|i-j|$, increases. The maximum absolute value of the matrix element corresponds to transitions between the lowest subbands:

$$|\langle \chi_2(z)|z|\chi_1(z)\rangle| = \frac{16}{9\pi^2}L. \tag{5.40}$$

For a quantum well with a finite depth, V_b, the corrections to Eq. (5.39) are of the order of $\hbar/L\sqrt{2m^* V_b}$. The inverse of this quantity determines the number of confined states in the well. Consequently, the above analysis is correct if the well contains at least several subbands.

From these results it is straightforward to calculate the absorption rate for the intraband transitions:

$$\mathcal{R}_{abs}(\omega) = \sum_{i,\vec{k},\sigma;j,\vec{k}',\sigma'} P_{i,\vec{k},\sigma;j,\vec{k}',\sigma'}\left[\mathcal{F}_{i,\sigma}(\vec{k}) - \mathcal{F}_{j,\sigma'}(\vec{k}')\right], \tag{5.41}$$

where $\mathcal{F}_{i,\sigma}(\vec{k})$ is the occupation number of the state $\{b, i, \vec{k}, \sigma\}$. For the parabolic subbands of Eq. (5.31) the δ-function in Eq. (5.41) is independent of the wavevector and the sum over $\vec{k}$ can be calculated in an explicit form:

$$\mathcal{R}_{abs}(\omega) = \frac{2\pi}{\hbar}e^2|\xi_z|^2F_0^2(z_w)S\sum_{i;j}|\langle\chi_j|z|\chi_i\rangle|^2\delta\left(E_j^{(0)} - E_i^{(0)} - \hbar\omega\right)[n_i - n_j]. \tag{5.42}$$

Here

$$n_l \equiv \sum_{\vec{k},\sigma}\mathcal{F}_{l,\sigma}(\vec{k})$$

represents the population of the lth subband per unit area, i.e., the concentration of electrons per unit area in that subband. From Eq. (5.42) it can be seen that the absorption spectrum consists of a series of infinitely high and narrow peaks at photon energies corresponding to the intersubband separation energies. This result is a consequence of our assumption that exactly the same parabolic dispersion relations, $E_l(\vec{k})$, apply for all subbands involved in the transition (see Eq. (5.31)). Allowance for energy-broadening mechanisms would change these δ-peaks into lines with finite heights and widths. Equation (5.28) presents a possible broadening function $\Delta(\omega)$, which leads to finite peaks of $\alpha(\omega)$. The parameters $\tau_{c,v}$ in Eq. (5.28) should be replaced by the intraband scattering time τ_{int}. Another cause of the finite peaks is a deviation of $E_i(\vec{k})$ from the parabolic dependence of Eq. (5.31). A certain amount

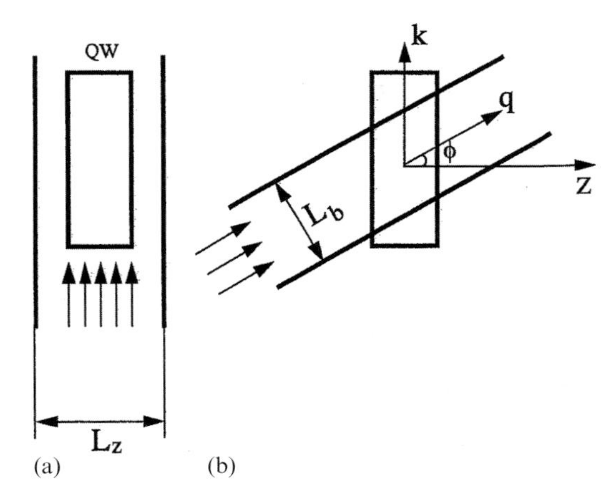

Figure 5.11 Two geometries for intraband absorption of the light beam in a quantum well: (a) the beam with aperture $L_y \times L_z$ propagates along the well layer; and (b) the beam axis is tilted with respect to the well layer, so the area of the aperture is $L_y \times L_b$.

of nonparabolicity always exists in the electron and hole spectra because the kinetic energy with effective mass is only the second term of an infinite series expansion in the electron energy. A real energy dispersion is more complex, as illustrated by Fig. 5.5 for the hole band.

We see that the absorption rate depends directly on the subband population of the quantum wells. Population of the lowest subband can be achieved by means of doping. For one of the most important applications – photodetectors – a special type of selective doping is used: quantum wells are doped, while barriers remain undoped and assure high mobility for the "over-barrier" motion of electrons.

In the previous section we studied some features of the electrodynamics of nonuniform media, particularly of heterostructures. We introduced the main parameters which characterize the interaction of light with matter. All those considerations remain valid for intraband absorption. Thus, we can introduce parameters like β, γ, and α of Eqs. (5.1), (5.2), and (5.11), which can describe light absorption (amplification), keeping in mind infrared-photodetector and infrared-laser applications of intersubband phototransitions. We consider two possible physical situations as illustrated in Fig. 5.11.

The first is light propagation along a quantum well layer, as shown in Fig. 5.11(a), but the light wave is not localized within the layer (compare this with Fig. 5.1). If the transverse dimensions of the light beam are L_y and L_z, we get

$$\frac{1}{\mathcal{I}_x}\frac{d\mathcal{I}_x}{dx} = -\alpha = -\frac{\beta_o}{L_z}, \tag{5.43}$$

where β_o is the attenuation due to a single quantum well.

In Eq. (5.43), $\mathcal{I}_x$ is the total power of the light beam propagating along the x-axis; obviously, the light intensity is $\mathcal{I}_x/L_yL_z$. Next, β_o can be written as

$$\beta_o = \frac{16\pi^2 e^2\omega}{c\sqrt{\kappa}}|\xi_z|^2 \sum_{i;j} |\langle \chi_j|z|\chi_i\rangle|^2 \Delta\left(E_j^{(0)} - E_i^{(0)} - \hbar\omega\right)[n_i - n_j]. \tag{5.44}$$

From Eq. (5.44), one can see that β_o is independent of the aperture of the beam $L_y \times L_z$, while the absorption coefficient α is inversely proportional to L_z.

The second situation corresponds to having the light beam tilted with respect to the layer and having a finite cross-section $L_b \times L_y$, as shown in Fig. 5.11(b). For this case, taking into account a small amount of absorption by one layer, we find for the light attenuation per transit through the quantum well layer

$$\frac{\mathcal{I}_{in} - \mathcal{I}_{out}}{\mathcal{I}_{in}} = \frac{\beta_o}{\cos\phi} \equiv \beta, \tag{5.45}$$

where ϕ is the angle between the light propagation direction and the z-axis. Equation (5.45) is valid if the beam aperture L_b is less than the length of the layer L_x: that is, if $L_b/\cos\phi < L_x$.

According to these results, light with polarization parallel to the heterointerface of the quantum well layer, $\xi_z = 0$, does not interact with electrons in the heterostructure. When light propagates along the layer, as in Fig. 5.11(a), the light polarization can be either in the x, y plane, i.e., $\xi_y = 1, \xi_z = 0$, so that the light does not interact with the heterostructure, or perpendicular to it, i.e., $\xi_y = 0, \xi_z = 1$, so that the light is absorbed. In the case of a tilted light beam, we may neglect the light reflection from the quantum well layer and set $|\xi_z|^2 \approx \sin^2\phi$. Thus, the angular dependence of the attenuation β is

$$\beta \propto \frac{\sin^2\phi}{\cos\phi}. \tag{5.46}$$

The magnitude of the light attenuation by one quantum well is typically small and is usually about 0.01 or less; thus, intersubband phototransitions should be measured in multiple-quantum-well structures with the number of wells ranging from 50 to 100. Thick barriers between quantum wells suppress the tunneling effect as well as the formation of minibands. Figure 5.12 shows an example of intersubband absorption. The GaAs wells are 70 Å wide, and the $Al_{0.27}Ga_{0.63}As$ barriers are 150 Å thick. The doping density of the well equals 2×10^{18} cm^{-3}. The calculated subband energies are $E_1^0 = 47$ meV and $E_2^0 = 171$ meV for a barrier height of 202 meV. The results are given in terms of the absorption coefficient, α, of Eq. (5.43). The attenuation, β_o, due to a single quantum well can be estimated by multiplying α by the thickness of one period of the structure, i.e., 2.2×10^{-6} cm. Thus, the maximum attenuation equals 0.0044. The lineshape of the absorption coefficient has the Lorentzian shape of Eq. (5.28) with a lifetime broadening of about 0.18 ps. The latter is of the order of the intraband relaxation time at room temperature. The result of calculations of the absorption coefficient is presented as the dotted line in Fig. 5.12 for comparison.

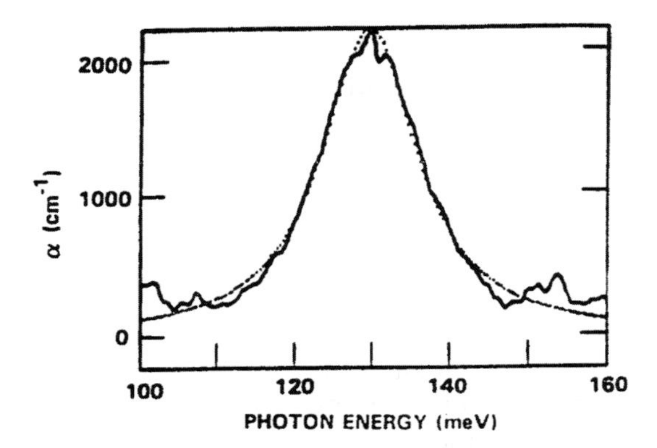

Figure 5.12 The intersubband absorption coefficient of a multiple-quantum-well heterostructure. The solid line represents experimental data, the dotted line exhibits the Lorentzian distribution. Copyright 1993 from Fig. 3.2 of *Semiconductor Interfaces, Microstructures, and Devices. Properties and Applications* by Z. C. Feng. Reproduced by permission of Taylor and Francis Group, LLC, a division of Informa plc.

Above we derived the equations for a single quantum well or a multiple well structure with identical wells and wide barriers between them. Advanced technology facilitates the fabrication of periodic structures with a complex design for each period. For example, a period can consist of two asymmetric coupled quantum wells, or of three coupled wells, etc. The results obtained for a single quantum well can be used for calculation of the light absorption in complex structures. One need only substitute the proper envelope functions $\chi_i(z)$ into the matrix element of Eq. (5.36). Then, on the basis of these matrix elements, one can calculate the parameters of light absorption. Figure 5.13(a) shows results for the two-asymmetric-coupled-well structure and Fig. 5.13(b) shows those for a three-coupled-well structure based on $Al_{0.48}In_{0.52}As/Ga_{0.47}In_{0.53}As$ heterostructures; the conduction band offset is 510 meV. The calculated energy levels and wave functions are also presented. The double-coupled-well structure consists of wells which are 59 Å and 24 Å wide and a coupling barrier with a width of 13 Å. There are three confined electron states, with $E_1 = 102$ meV, $E_2 = 252$ meV, and $E_3 = 373$ meV, in this structure. The experimental results for the absorption coefficient are for a 20-period structure. The areal doping concentration in the well equals 1.2×10^{11} cm^{-2}. Only the lowest level is populated for this doping concentration. The experiment was performed at a temperature of 10 K with multiple passes of the light beam through the heterostructure. The two absorption peaks correspond to intersubband phototransitions: $E_1 \rightarrow E_2$ (the first peak) and $E_1 \rightarrow E_3$ (the second peak).

The triple-coupled-well structure of Fig. 5.13(b) consists of wells with widths of 46 Å, 20 Å, and 19 Å, from left to right respectively, and 10 Å-thick coupling barriers. The wells contain four confined states, with $E_1 = 126$ meV, $E_2 = 242$ meV, $E_3 = 383$ meV, and $E_4 = 494$ meV. The experiments were carried out with a

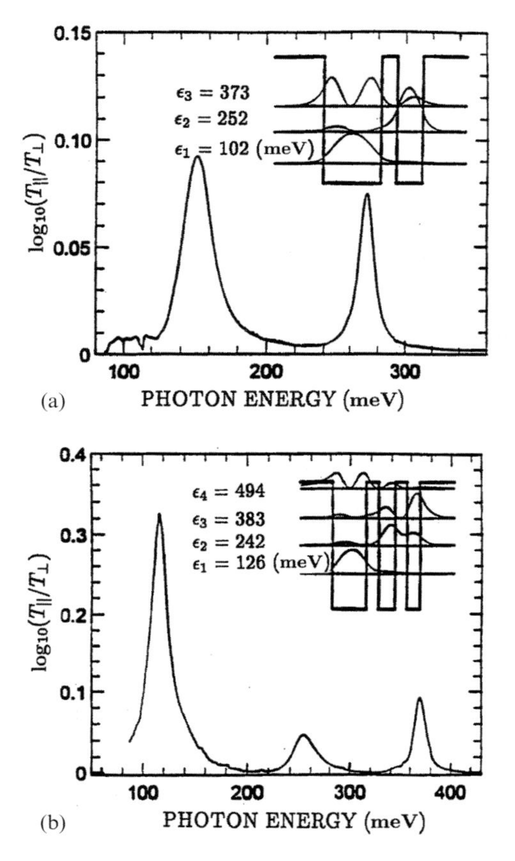

Figure 5.13 The measured intersubband absorption as a function of the photon energy in (a) an asymmetric two-coupled-quantum-well structure and (b) a three-coupled-quantum-well structure. The parameters of the structures are given in the text. The experiments were carried out at $T = 10$ K. The inserts show subband energies and corresponding wave functions. Reprinted figure with permission from Fig. 5 of C. Sirtori *et al.*, *Phys. Rev.* **B 50**, 8663, 1994. Copyright (1994) by the American Physical Society.

40-period structure at an areal doping concentration equal to 3.2×10^{11} cm^{-2}. The three absorption peaks correspond to intersubband transitions from the lowest energy level: $E_1 \to E_2$, $E_1 \to E_3$, and $E_1 \to E_4$.

The results shown in Fig. 5.13 assure one that desirable absorption spectra in the infrared region can be tailored using intersubband phototransitions. Structures with complex periods can be employed in various optoelectronic devices: multi-color photodetectors for light detection in two or more spectral bands, intraband-transition lasers, generators of microwave radiation in the terahertz region, etc.

5.3.3 Phototransitions to Extended States

Let us now consider phototransitions between confined states $\{b, i, \vec{k}, \sigma\}$ and extended states $\{b, E_z, \vec{k}', \sigma'\}$. The transition probability is

$$P_{i,\vec{k},\sigma;E_z\vec{k}',\sigma'} = \frac{2\pi}{\hbar}\delta_{\sigma,\sigma'}|\langle b, E_z, \vec{k}'|\vec{\mathcal{E}}\vec{D}|b, i, \vec{k}\rangle|^2\delta\left(V_{\text{b}} + E_z - E_i^0 - \hbar\omega\right). \tag{5.47}$$

The δ-function in this equation gives the energy conservation law of Eq. (5.32). In the same manner as before, one finds matrix elements of the form

$$\langle b, E_z, \vec{k}'|\vec{\mathcal{E}}\vec{D}|b, i, \vec{k}\rangle = F(z_{\text{w}})\xi_z\delta_{\vec{k},\vec{k}'}\int dz[\chi_{Ez}(z)]^*z\chi_j(z).$$

For confined states, we can use the wave function χ_i given by Eq. (5.38). For the extended states the wave function can be approximated as the plane wave

$$\chi_{E_z}(z) \equiv \chi_K(z) = \frac{1}{\sqrt{L_z}}e^{iKz}, \tag{5.48}$$

with

$$K = \frac{\sqrt{2m^*E_z}}{\hbar} = \frac{2\pi n}{L_z}, \quad n = 0, 1, 2, \ldots,$$

where L_z is the size of the system in the z-direction. L_z does not appear in the final, physically significant results. Here, we have neglected electron reflection from the discontinuities in the potential at $z = 0$ and L. Now, the absorption rate for the case under consideration can be written as

$$\mathcal{R}_{\text{abs}}(\omega) = \frac{2\pi}{\hbar}e^2|\xi_z|^2F^2(z_{\text{w}})\sum_{i;K}|\langle\chi_K|z|\chi_i\rangle|^2\delta\left(V_{\text{b}} + E_z - E_i^{(0)} - \hbar\omega\right) \times\left[\mathcal{F}_i(\vec{k}) - \mathcal{F}_K(\vec{k})\right]. \tag{5.49}$$

The last term in the brackets, $\mathcal{F}_K$, can be omitted if most of the electrons are inside the well. Finally, summation over K gives

$$\mathcal{R}_{\text{abs}}(\omega) = \frac{2m^*e^2}{\hbar^3}|\xi_z|^2F^2(z_{\text{w}})\sum_i\frac{L_z}{K_i(\omega)}|\langle\chi_{K_i}|z|\chi_i\rangle|^2Sn_i, \tag{5.50}$$

where

$$K_i(\omega) \equiv \frac{\sqrt{2m^*E_{z,i}(\omega)}}{\hbar},$$
$$E_{z,i}(\omega) = E_i^{(0)} + \hbar\omega - V_{\text{b}} > 0.$$

In order to avoid cumbersome equations we present the matrix element for the most interesting case, that of phototransitions involving the lowest subband, $j = 1$:

$$|\langle\chi_K|z|\chi_1\rangle| = \pi\sqrt{\frac{2L}{L_z}}L\left|\frac{4KL\cos(KL/2) - (\pi^2 - K^2L^2)\sin(KL/2)}{(\pi^2 - K^2L^2)^2}\right|. \tag{5.51}$$

This matrix element is a nonmonotonic function of E_z, and vanishes in the two limits $KL \to 0$ and $KL \gg 1$. It reaches a maximum at the resonance condition, $KL = \pi$:

$$|\langle \chi_K|z|\chi_1\rangle|_{\max} = \frac{L}{\pi}\sqrt{\frac{L}{2L_z}}. \tag{5.52}$$

It is instructive to compare this result with Eqs. (5.39) and (5.40). There is an important difference between this case and the case studied in the previous section. For confined-state-to-confined-state transitions, the absorption spectrum consists of a series of peaks; for confined-state-to-extended-state transitions the absorption spectrum covers a rather wide band situated at $\hbar\omega > V_b - E_1^{(0)}$, with its maximum at $\hbar\omega \approx V_b$ and a half-width of about $E_1^{(0)}$.

Now, we can use Eqs. (5.50) and (5.51) for calculation of the optical characteristics due to phototransitions to extended states. The attenuation, β_o, is

$$\beta_o = \frac{16\pi^2 e^2 m^* \omega}{\hbar^2 c\sqrt{\kappa}}|\xi_z|^2 L_z \sum_i \frac{1}{K_j(\omega)}|\langle \chi_{K_i}|z|\chi_1\rangle|^2 n_i. \tag{5.53}$$

Since the square of the matrix element of Eq. (5.51) is proportional to $1/L_z$, the attenuation coefficient β_o, as given by Eq. (5.53), is independent of L_z. This result is consistent with the meaning of the quantity β_o: it determines the loss (gain) of the light energy during a single passage through the quantum-well layer and it should not depend on the structure's size in the z-direction.

An example of photo-absorption to extended states is presented in Fig. 5.14. The experimental and calculated (curve (a)) results are shown for an $Al_{0.2}Ga_{0.8}As/GaAs$ multiple-quantum-well structure. The well width is 46 Å, the barrier height is 145 meV. The narrow wells are used to provide the only two-dimensional subband, with energy $E_1 = 68$ meV. One can see from Fig. 5.14 that there is a wide absorption band with a maximum at the resonance condition $KL \approx \pi$, which corresponds to the energy $E_z = 15$ meV above the barriers. The considerable sensitivity of the spectrum to the well thickness is demonstrated by curve (b), which has been calculated with the same parameters except for a thickness of 40 Å. For such a well thickness the only level, E_1, has a larger energy, which leads to a "blue" shift of the absorption spectrum. This result shows that the fabrication of structures with different thicknesses of layers provides the possibility of controlling the absorption spectrum.

Note that intraband phototransitions to extended states correspond to the photo-ionization of a quantum well, i.e., photo-generation of the electrons that would move over the barriers. This type of photo-absorption is similar to the usual photo-ionization of impurities.

Let us stress once again that intraband absorption is strongly polarization-dependent. Table 5.1 briefly summarizes the selection rules for intraband transitions with different polarizations and propagation directions of light. The quantum wells are assumed to be perpendicular to the z-direction.

In conclusion, the absorption spectra corresponding to intraband transitions in quantum heterostructures are very sensitive to the parameters of the structures. This flexibility of tuning into the necessary absorption spectra has many

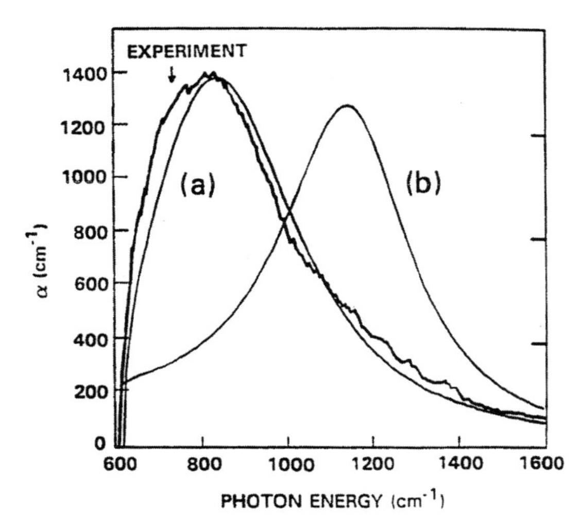

Figure 5.14 Experimental results and calculations for the absorption coefficient of confined-state-to-extended-state phototransitions. The multiple quantum well heterostructure is described in the text. Calculated curves (a) and (b) correspond to well widths of 46 Å and 40 Å, respectively. Copyright 1993 from Fig. 3.4 of *Semiconductor Interfaces, Microstructures, and Devices. Properties and Applications* by Z. C. Feng. Reproduced by permission of Taylor and Francis Group, LLC, a division of Informa plc.

Table 5.1 The selection rules for intersubband phototransitions in quantum wells

Direction of propagation	Polarization		
	ξ_x	ξ_y	ξ_z
z	Forbidden	Forbidden	Impossible
x	Impossible	Forbidden	Allowed
y	Forbidden	Impossible	Allowed

potential applications for infrared photodetecting. The use of intraband transitions in photodetectors is discussed further in Section 8.4.

5.4 Optical Properties of Two-Dimensional (Few-Monolayer) Crystals

The optical properties of two-dimensional crystals provide information on the electronic band structure and other fundamental electronic properties of these materials, and they also pave the way for fundamentally new optoelectronic devices and applications.

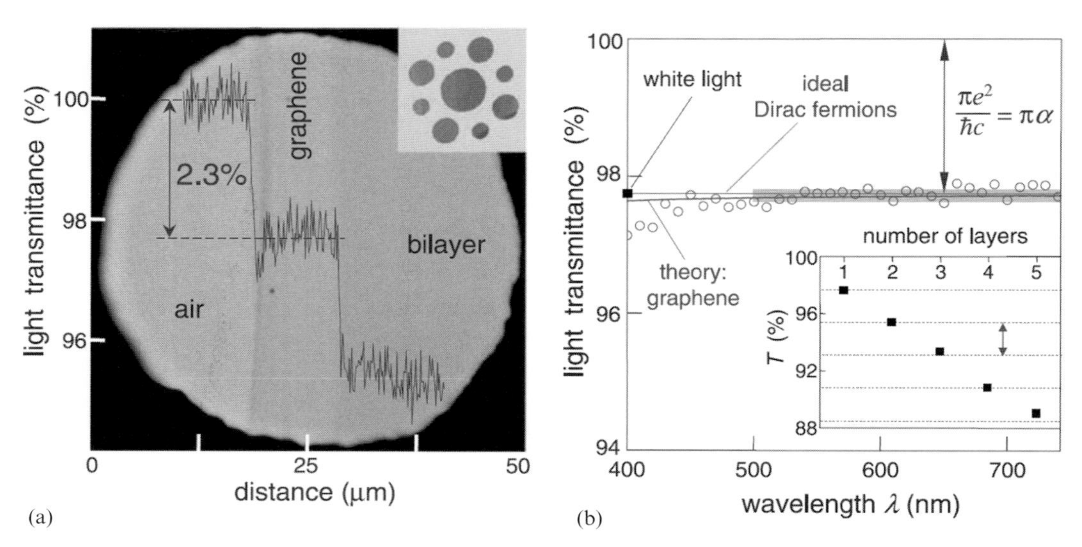

Figure 5.15 (a) A photograph of a 50 mm aperture partially covered by graphene and its bilayer with schematically depicted optical absorption. (b) The transmittance spectrum (open circles) of single-layer graphene in comparison with the theoretical transmission for ideal massless Dirac fermions and real graphene. The inset shows the relative absorption as a function of the number of layers in graphite stacks. From Fig. 1 in R. R. Nair, *et al.*, *Science* **320**, 1308 (2008). Reprinted with permission from AAAS.

5.4.1 Optics of Graphene and Bigraphene

We start with an analysis of the optical properties of graphene. Basic information on this monatomic carbon sheet has been given in Sections 2.9 and 3.3. In Fig. 5.15(a), a photograph of an aperture that is partially open (air) and partially covered by graphene and bilayer graphene is presented. The photograph illustrates the possibility of the direct visualization of graphene and bigraphene. The absorption of graphene and bigraphene is estimated at the levels of 2.3% and 4.6%, respectively. In Fig. 5.15(b), the measured transmittance spectrum of graphene is shown. It is found to be practically independent of the light wavelength. The relative absorption increases linearly with the number of graphene layers, as can be seen in the inset to Fig. 5.15(b). Initially it was not clear why a few monolayers, or even single monatomic sheets, could be seen through an optical microscope. The explanation of this is not trivial and will be presented below.

As was discussed in Section 3.3, the low-energy electronic spectrum of graphene can be understood as follows. There are two inequivalent points in the momentum space, K and K′, where two cones, one for the conduction band and another for the valence band, touch each other. This results in the gapless character of the spectrum and the linear dispersion of both bands in the low-energy region. Applying the classification of the mechanisms of light absorption and emission discussed in Section 4.4, we can state that for graphene there are two main optical processes – intraband (within the same cone) and interband phototransitions within the same

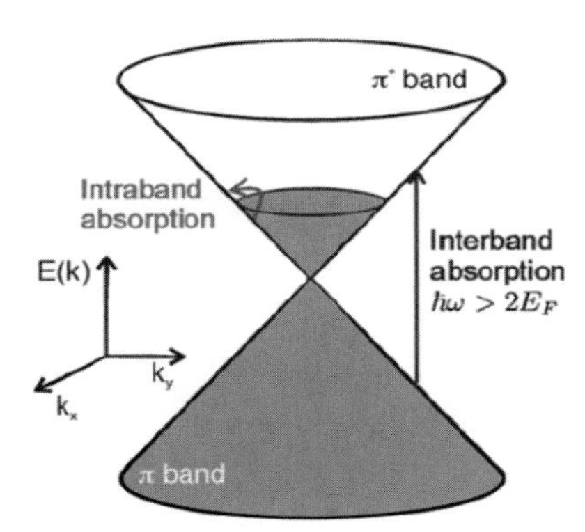

Figure 5.16 The linear dispersion relation of energy, E, vs. wavevector, $\vec{k} = \{k_x, k_y\}$, is illustrated in this diagram along with representative intraband and interband transitions in doped graphene. The nonvanishing Fermi level gives rise to the characteristic onset of the optical interband transition at photon energy $2E_F$. From Fig. 5(b) in M. Orlita and M. Potemski, Dirac electronic states in graphene systems: optical spectroscopy studies, *Semicond. Sci. Technol.* **25**, 063001 (2010). ©IOP Publishing. Reproduced with permission. All rights reserved.

K or K′ point. These processes are schematically depicted in Fig. 5.16. Phototransitions involving electron states from different K points are significantly suppressed, because they require a large change of the electron momentum (wavevector), similarly to the indirect transitions discussed in Chapters 3 and 4 for indirect-bandgap semiconductors.

A qualitative explanation of the interaction of an electromagnetic wave with a one-atom-thick layer is the following. If the wave has an electric-field component parallel to the layer, the electric field affects the electrons, giving rise to interband transitions and/or electron oscillations within a band. For both cases, electron motion induced by this field produces a time-dependent in-plane electric current. For a plane wave of frequency ω, the Fourier component of the induced current density can be represented as $j(\omega) = \sigma(\omega)F_\omega$, where $\sigma(\omega)$ is the conductivity (for high frequencies it is often called the "optical conductivity") and F_ω is the amplitude of the corresponding component of the field. The current density $j(\omega)$, in turn, generates electromagnetic waves. A coherent superposition of them with the incident wave results in the transmitted and reflected waves.

Let us recall that to describe the interaction of light with an active layer of thickness much less than the light wavelength we should introduce the transmission coefficient, $T = \mathcal{I}_{\text{out}}/\mathcal{I}_{\text{in}}$, and the *attenuation coefficient* $\beta = (\mathcal{I}_{\text{in}} - \mathcal{I}_{\text{out}})/\mathcal{I}_{\text{in}}$ (see the discussion in Section 5.2). The reflection coefficient can be expressed via T and β: $R = \mathcal{I}_{\text{r}}/\mathcal{I}_{\text{in}} = 1 - T - \beta$. Here $\mathcal{I}_{\text{in}}$, $\mathcal{I}_{\text{out}}$, and $\mathcal{I}_{\text{r}}$ are the intensities of the incident, transmitted, and reflected light waves, respectively. It can be shown that the transmission coefficient can be expressed via the optical conductivity, $\sigma(\omega)$:

$$T = \frac{1}{|1 + 2\pi\sigma(\omega)/c|^2},$$

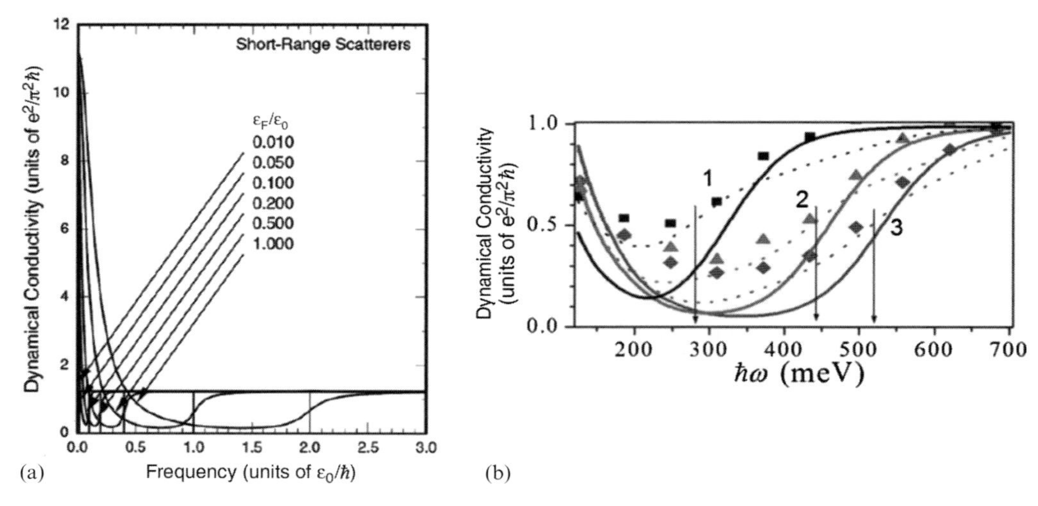

Figure 5.17 (a) Calculations of the real part of the optical conductivity of a homogeneous graphene sample (per valley) as a function of frequency (in arbitrary units, $\varepsilon_0/\hbar$) for different Fermi levels (in arbitrary units of energy, ε_0). (b) Experimental results (sets of squares, triangles and diamonds) for the optical conductivity fit calculations that take into account the inhomogeneity of graphene better (dotted lines) than the curves for homogeneous graphene (solid lines). Figure 5.17(a) after Fig. 4 (a) in T. Ando *et al.*, *J. Phys. Soc. Japan* **71**, 1318 (2002). Licensed under CC BY 4.0. Figure 5.17(b) reprinted with permission from Fig. 6 of F. T. Vasko, *et al.*, *Phys. Rev.* **B 86**, 235424, 2012. Copyright (2012) by the American Physical Society.

where c is the velocity of light. Because $|\sigma(\omega)| \ll c$, we obtain the following simple formula for the transmission:

$$T = 1 - \frac{4\pi\,\mathrm{Re}[\sigma(\omega)]}{c}.$$

Results of calculations of $\mathrm{Re}[\sigma(\omega)]$ for graphene are presented in Fig. 5.17. Both interband and intraband transitions were taken into account. Remarkably, for the Dirac electrons in graphene the theory predicts the universal conductivity $\sigma_0 = e^2/4\hbar$ in the limit of high energies. This explains why the transmission coefficient is independent of frequency in the optical range shown in Fig. 5.15(b). If the Fermi energy approaches the Dirac point ($E(\vec{k}) = 0$), the optical conductivity is equal to σ_0 for any finite frequency. Importantly, the interband transitions are blocked at energies below twice the Fermi energy ($2E_\mathrm{F}$); this effect can easily be understood from the schematic diagram shown in Fig. 5.16.

Figure 5.17(a) shows the results of calculations of the real part of the optical conductivity for a homogeneous graphene sample when only short-range scattering is important. Figure 5.17(b) shows that calculations that take into account inhomogeneity and imperfections fit the experimental results substantially better than do calculations of absorption in homogeneous samples. Indeed, the experimental results show a substantially wider smearing of the absorption edge of the interband absorption due to the inhomogeneity and imperfections of graphene that lead to a long-range scattering of carriers. On changing the Fermi energy a pronounced

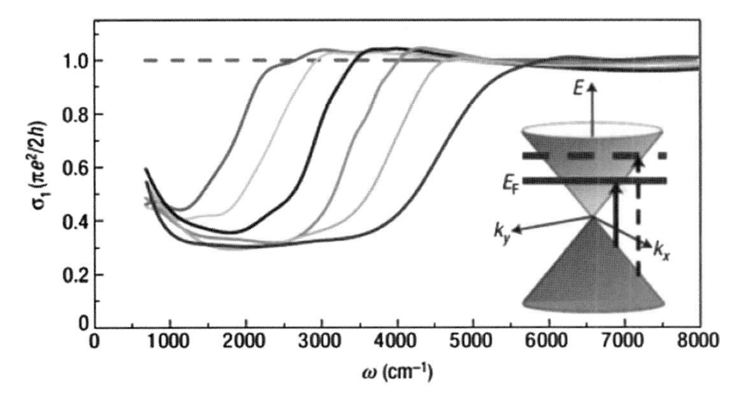

Figure 5.18 The experimentally determined real part of the optical conductivity of graphene on a Si/SiO_2 substrate for different positions of the Fermi level. Reprinted by permission from figure 2(b) in Springer, *Nature Phys.*, "Dirac charge dynamics in graphene by infrared spectroscopy," Z. Q. Li, *et al.*, Copyright 2008.

modification of the optical properties of graphene is observed. The Fermi energies for three cases presented in Fig. 5.17(b) are shown by vertical arrows. The inhomogeneity and imperfections also lead to an increase in intraband absorption, which is not shown in Fig. 5.17(b). The second important effect of inhomogeneity is in closing the gap between interband and intraband absorption, which is extremely pronounced in Fig. 5.17(a).

Experimentally, the optical response of graphene placed on Si/SiO_2 substrates has been investigated in the mid-infrared range both for the reflection mode and for the transmission mode. The optical conductivity, $\sigma(\omega)$, extracted from these experiments (see Fig. 5.18) is found to be in agreement with theoretical predictions, which in addition to Fig. 5.15(b) amounts to important evidence of the Dirac-like electronic spectrum in graphene.

Now we briefly consider the optical properties of graphene bilayers, i.e., bigraphene. The analysis of the electronic structure of bigraphene given in Chapter 3, in Subsection 3.6.2, showed that the electronic spectrum of bigraphene is sufficiently different from that of monolayer graphene at the low energies, of the order of the interlayer hopping parameter $\gamma_1 \approx 0.3\,\text{eV}$ that describes the interlayer interaction. Thus, any optical peculiarities of bigraphene are expected to occur in the infrared range. Figure 5.19(a) illustrates possible phototransitions in bigraphene. For specificity, these phototransitions are shown for the case of the p-doped crystal (the Fermi level is in the valence band). We see that, instead of the single type of interband phototransition inherent for monolayer graphene, in bigraphene there are four types of transition between the valence and conduction bands and a transition between the lowest and upper valence bands.

However, the most interesting effect is the strong modification of the optical properties of bigraphene when this two-dimensional crystal is placed in an electric field perpendicular to the plane of the crystal. Indeed, according to the analysis

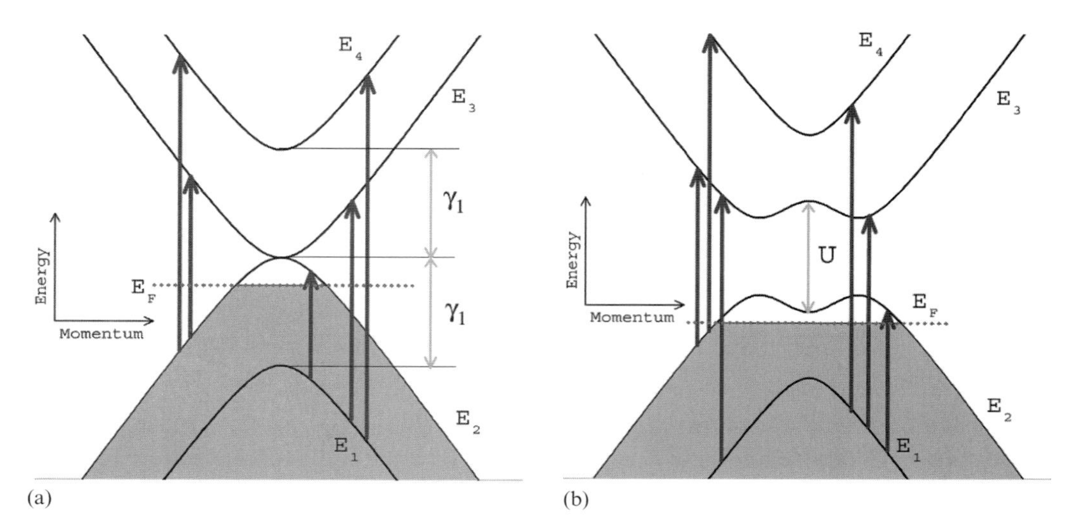

Figure 5.19 A schematic view of the band structure in a graphene bilayer, showing its modification via the bias voltage, U, applied across the two layers, for (a) $U = 0$ and (b) $U \neq 0$. The arrows show dipole-allowed transitions, which determine the optical response in the infrared spectral range. From Fig. 5(b) in M. Orlita and M. Potemski, Dirac electronic states in graphene systems: optical spectroscopy studies, *Semicond. Sci. Technol.* **25**, 063001 (2010). ©IOP Publishing. Reproduced with permission. All rights reserved.

of Subsection 3.6.2, such a field produces considerable changes in the spectrum; in particular, it gives rise to an energy bandgap between the valence band and the conduction band. Possible phototransitions in this case are shown in Fig. 5.19(b). Now for each of the four valence $\leftrightarrow$ conduction band transitions there is an onset of absorption/emission, i.e., a band edge. As a result, the spectra become quite complex. Indeed, in experiments, a very rich pattern was observed in the optical spectrum, as shown in Fig. 5.20 for p- and n-doped crystals (Figs. 5.20(a) and (b), respectively). One can see how the voltage applied across the bigraphene modified its optical conductivity. It is noteworthy that at high photon energies (>0.3 eV) the optical response of bigraphene is very close to twice the optical response of monolayer graphene (compare Figs. 5.18 and 5.20).

To conclude this short analysis of the optics of graphene and bigraphene, we mention that infrared optical methods have facilitated the direct observation of the Dirac-like electronic spectrum and the determination for two-dimensional crystals (graphene and bigraphene) of the band-structure parameters, such as the Fermi velocity, v_F, the interlayer hopping parameter γ_1, the band splitting, and the effective mass. The optical response of graphene and bigraphene in the terahertz and far-infrared frequency range can be used as the basis for photodetectors and emitters. The relatively easy tuning of the bandgap in graphene bilayers facilitates designing voltage-controlled optoelectronic devices.

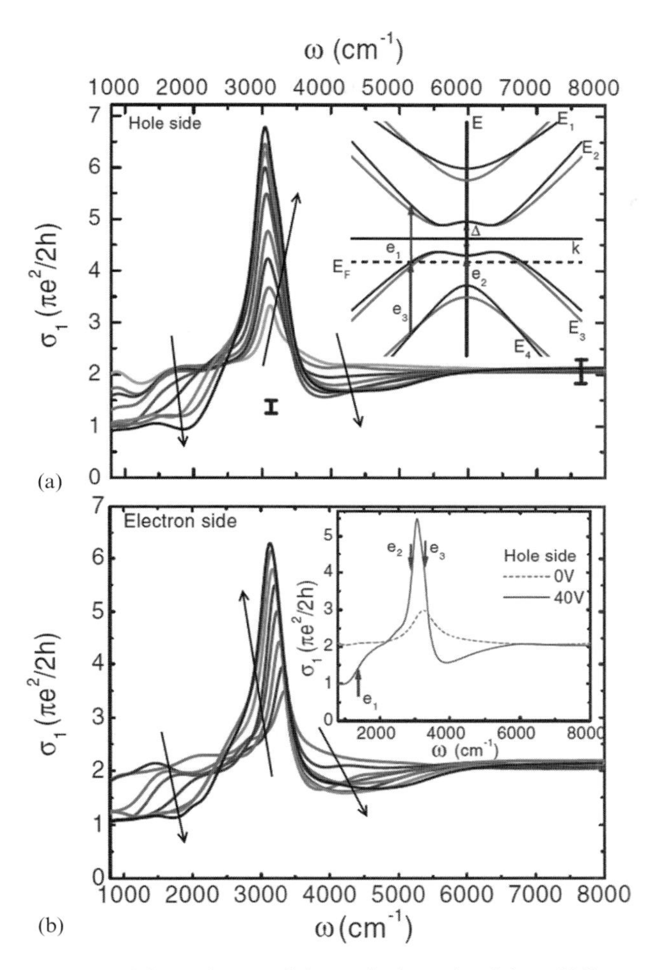

Figure 5.20 The real part of the optical conductivity of bilayer graphene at different applied voltages, i.e., at various hole (a) and electron (b) densities. The arrows show increases of voltage from 5 V to 82 V in (a) and from 10 V to 82 V in (b). The inset in (a) shows schematically the band structure of bigraphene calculated by different methods. The position of individual band-edge interband transitions in the optical conductivity is shown in the inset of (b). Reprinted figure with permission from Figs. 2(a), (b) of Z. Q. Li, *et al.*, *Phys. Rev. Lett.* **102**, 037403, 2009. Copyright (2009) by the American Physical Society.

5.4.2 The Optics of Transition-Metal Dichalcogenides

In contrast to graphene and bigraphene, the transition-metal dichalcogenides MoS_2, $MoSe_2$, WS_2, and WSe_2 have nonzero energy bandgaps. Remarkably, the bulk transition-metal dichalcogenide crystals are indirect-bandgap semiconductors, while these materials in the form of crystals of thickness a few layers show the evolution from indirect- to direct-bandgap materials. They become optically active in the visible and infrared portions of the electromagnetic spectrum. Naturally these

Table 5.2 Properties of the transition metal dichalcogenides

	MoS_2	$MoSe_2$	WS_2	WSe_2
Effective masses, m^*/m	0.5	0.6	0.4	0.4
Optical gap (eV)	2	1.7	2.1	1.75
Exciton binding energy (eV)	0.2–0.55	0.6	0.4–0.5	0.4–0.45
Conduction band spin–orbit splitting (meV)	−3	−20	30	35
Valence band spin–orbit splitting (meV)	150	180	430	470

Adapted by permission from table in Springer, *Nature Photonics*, "Photonics and optoelectronics of 2D semiconductor transition metal dichalcogenides," Kin Fai Mak and Jie Shan, Copyright 2016.

few-layer materials have direct bandgaps, a property well suited for photonics and optoelectronics applications.

Two other features of the optical properties of two-dimensional transition-metal dichalcogenides are as follows: these materials exhibit strong spin–orbit effects, as discussed in Chapter 3, Section 3.3.2; excitonic effects are pronounced in these materials. Some of their properties are summarized in Table 5.2, where, for the convenience of the reader, along with the bandgaps (optical gaps), the effective masses, exciton coupling energies, and spin–orbit splittings of the conduction and valence bands are presented.

As is evident from Table 5.2, the transition-metal dichalcogenides have extremely large exciton binding energies. It is instructive to compare the exciton binding energies of Table 5.2, which range from 0.2 to 0.6 eV, with those of Table 4.2, which range from 1 meV to tens of meV (with a few exceptional cases having larger exciton binding energies, i.e., group-III nitrides, discussed in Section 4.4.5, wide-bandgap ZnO [60 meV] and SnO_2 [130 meV]).

We will discuss experimental observations of the optical features of the transition-metal dichalcogenides for the example of MoS_2 materials. As in any indirect-bandgap material, the bandgap photoluminescence in bulk MoS_2 is a weak phonon-assisted interband process, which is known to have a negligible quantum yield. However, appreciable photoluminescence is observed from few-layer MoS_2 samples, and surprisingly bright photoluminescence is detected for monolayer samples. In Fig. 5.21, the measured room-temperature photoluminescence is presented for MoS_2 samples with different numbers of layers under identical excitation at 2.33 eV. One can see strikingly different luminescence near 1.9 eV: according to the result of the main panel of Fig. 5.21(a) and the inset, the quantum yield for monolayer MoS_2 exceeds that of the two-layer material by two orders of magnitude. The quantum yield progressively diminishes with increasing number of layers. The significant difference between the normalized photoluminescence spectra for monolayer and few-layer samples can clearly be seen in Fig. 5.21(b). The spectrum of the monolayer sample consists of a single narrow peak of width 50 meV,

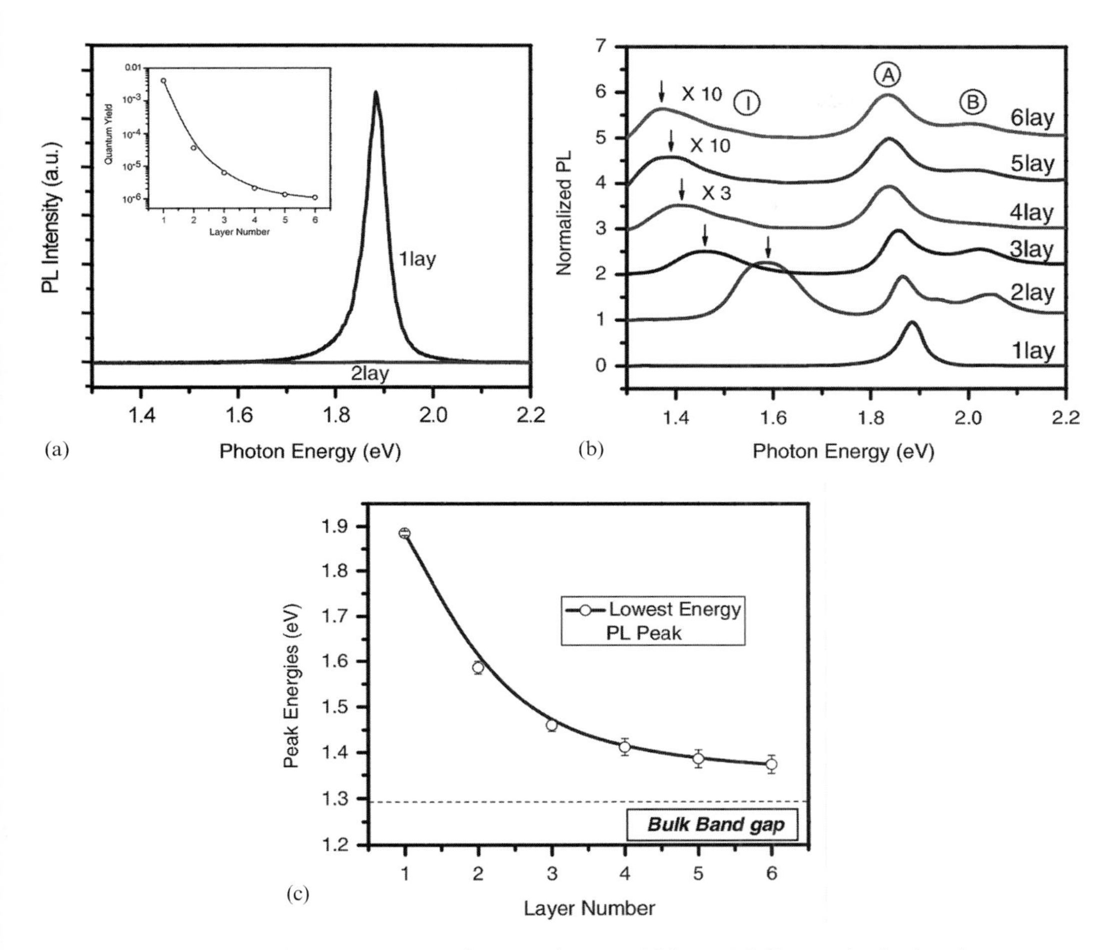

Figure 5.21 (a) Photoluminescence spectra for monolayer and bilayer MoS_2 samples in the photon energy range from 1.3 to 2.2 eV. The inset plots the quantum yield of photoluminescence for structures having one to six layers. (b) Photoluminescence spectra normalized by the intensity of peak A of thin layers of MoS_2. Feature I for four to six layers is magnified and the spectra are displaced for clarity. (c) The bandgap energy of thin layers of MoS_2, inferred from the energy of the feature I of the spectra for monolayer numbers from two to six and from the energy of peak A for the MoS_2 monolayer. The dashed line represents the (indirect) bandgap energy of bulk MoS_2. Reprinted figure with permission from Fig. 3 of Kin Fai Mak, *et al.*, *Phys. Rev. Lett.* **105**, 136805, 2010. Copyright (2010) by the American Physical Society.

centered at 1.90 eV. In contrast, the few-layer samples display multiple emission peaks (labeled A, B, and I). Peak A coincides with the monolayer emission peak. It shifts to the red and broadens slightly with increasing number of layers. Peak B is about 150 meV above peak A. The broad peak I, which lies below peak A, systematically shifts to lower energies, approaching the indirect-bandgap energy of the bulk material, 1.29 eV, and is less pronounced with increasing number of layers, as shown in Fig. 5.21(c).

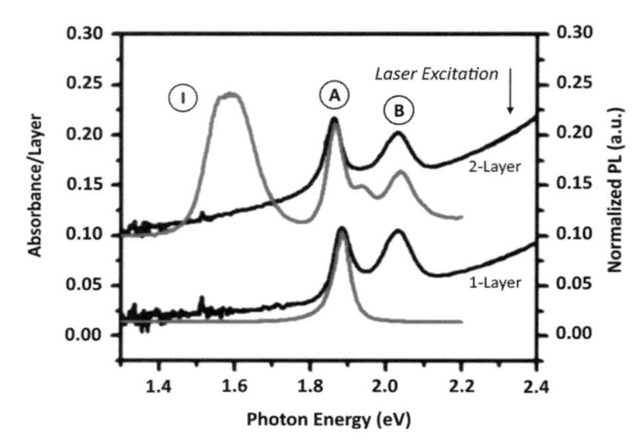

Figure 5.22 Absorption spectra (left axis, normalized by the number of the layers) and the corresponding photoluminescence spectra (right axis, normalized by the intensity of peak A). The spectra are displaced along the vertical axis for clarity. Reprinted figure with permission from Fig. 4 (a) of Kin Fai Mak, *et al.*, *Phys. Rev. Lett.* **105**, 136805, 2010. Copyright (2010) by the American Physical Society.

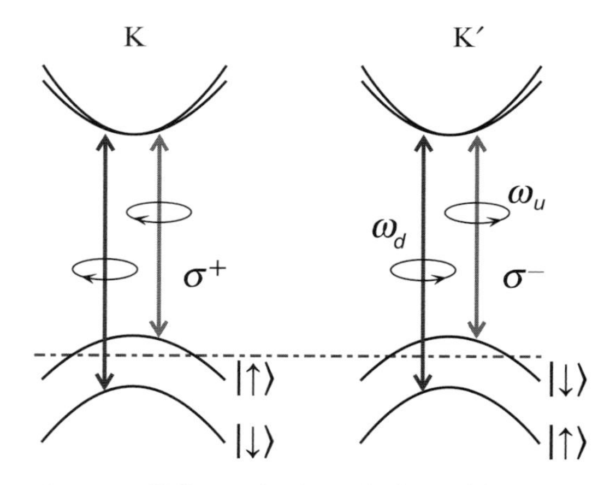

Figure 5.23 Valley and spin optical transition selection rules. Solid curves denote bands with spin down or up quantized along the out-of-plane direction. The splitting in the conduction band is exaggerated; ω_u and ω_d are the transition frequencies from the two split valence band tops to the conduction band bottom.

Additional information on the origin of these extraordinary phototransitions can be obtained from the absorption spectrum of MoS_2, which is presented in Fig. 5.22. In contrast with the luminescence spectrum, the absorption spectrum shows two peaks for monolayer samples, which can be interpreted as the consequence of the spin–orbit splitting of the valence bands near the K, K' points, as discussed previously in Section 3.3.2. The respective phototransition schemes are presented in Fig. 5.23. The interband transitions are coupled with the σ^+ circularly polarized optical wave at the K point (valley) and with the σ^- polarized wave at the K'

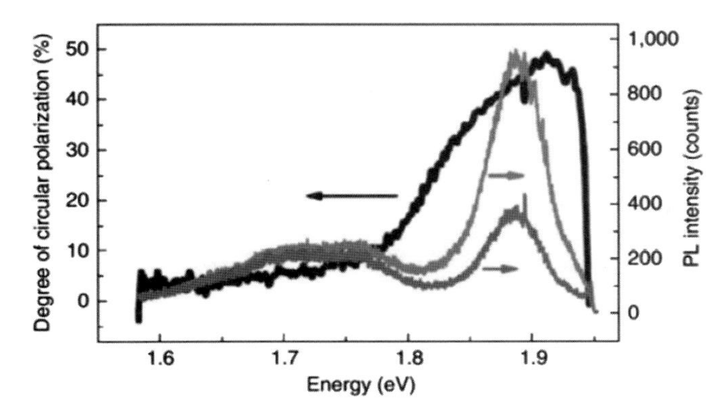

Figure 5.24 The circularly polarized photoluminescence of monolayer MoS_2 at 83 K along with the degree of circular polarization of the spectra. The two curves referred to the right-hand scale are the intensities corresponding to σ^+ (upper curve) and σ^- (lower curve) polarizations in the luminescence spectrum. The curve referred to the left-hand scale is the net degree of polarization. Reprinted by permission from figure 3(d) in Springer, *Nature Commun.*, "Valley-selective circular dichroism of monolayer molybdenum disulphide," Ting Cao, *et al.*, Copyright 2012.

point (valley). Because of the valley-dependent spin splitting of the valence bands, the spin-dependent selection rules give rise to *valley optical selection rules*, as illustrated in Fig. 5.23. In this figure, ω_u, ω_d denote frequencies corresponding to two phototransitions from the spin-split valence bands.

The circularly polarized photoluminescence of monolayer MoS_2 was observed at low temperatures. An example of such measurements is illustrated in Fig. 5.24. There the photoluminescence intensities with two polarizations are shown (the right vertical axis) and the respective degree of polarization is calculated (the left vertical axis) as a function of the photon energy. In this experiment, the degree of polarization reaches 50%.

Because of the spin–valley coupling, the selective excitation of carriers with various combinations of valley and spin indices becomes possible using optical waves of different circular polarizations and frequencies. Such spin- and valley-dependent selection rules can be used to generate long-lived electrons and holes with certain spin and valley indices. This can provide additional control of the electronic properties of this crystal, which can be used in new branches of electronics – *spintronics* and *valleytronics*.

It is worth mentioning that exciton-involved phototransitions contribute to the experimental spectra presented in Figs. 5.21, 5.22, and 5.24; as expected, the exciton states have a large effect on the optical properties of the transition-metal dichalcogenides. Moreover, the excitons in these materials include a three-particle charged exciton, a trion, as well as a four-particle bi-exciton. For example, in the case of a monolayer of WSe_2, where the exciton binding energy is especially large (see Table 5.2), the optical measurements directly confirm the presence of the trion in monolayer WSe_2.

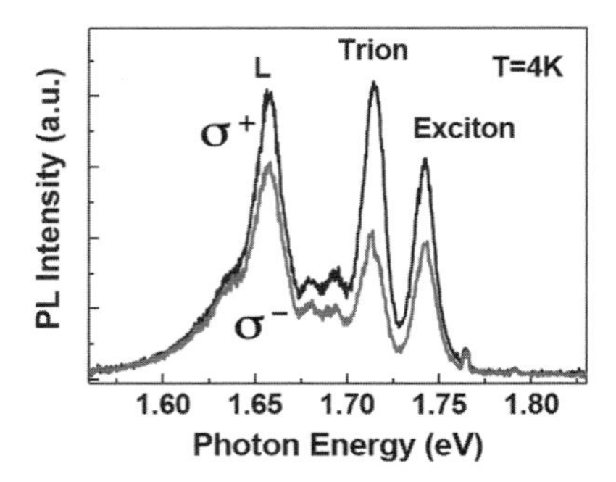

Figure 5.25 Exciton and trion photoluminescence spectra of WSe_2 at $T = 4$ K. Right (σ^+, higher intensity) and left (σ^-, lower intensity) circularly polarized components are observed under continuous-wave (cw) σ^+ polarized He–Ne laser excitation with photon energy 1.96 eV. Several emission peaks are observed below the trion emission peak. These peaks are assigned to localized exciton complexes. Reprinted figure with permission from Fig. 1(c) of Zhu C. R., *et al.*, *Phys. Rev.* **B 90**, 161302, 2014. Copyright (2014) by the American Physical Society.

Figure 5.25 displays the photoluminescence spectra at $T = 4$ K. The emission peaks at 1.742 eV and 1.714 eV correspond to the recombination emission of a neutral exciton and a charged exciton (i.e., a trion), respectively. The separation by 30 meV of the trion and the neutral exciton in a few-layer WSe_2 sample is a major advantage compared with MoS_2 samples, where these two lines are not resolved. The exciton and trion peaks are characterized by a significant degree of circular polarization (typically 33% and 23%, respectively, at $T = 4$ K), which demonstrates the optical initialization of valley polarization.

In summary, the study of two-dimensional materials for optoelectronics is advancing rapidly and this field holds promise for a wide variety of optoelectronic applications from the far-infrared to the ultraviolet range of electromagnetic spectra. In addition to graphene (as discussed previously), bigraphene, and two-dimensional transition-metal dichalcogenides, we can mention the very useful direct-bandgap semiconducting material black phosphorus (bandgap ≈ 2 eV) and the indirect-bandgap hexagonal boron nitride (bandgap ≈ 5.95 eV).

Importantly, the different two-dimensional structures discussed in this section are known to bind layers to each other by van der Waals interactions when they are stacked on top of each other. This type of binding is very different from that between materials that bind by ionic bonds such as GaAs and $Ga_xAl_{1-x}As$, where the strong ionic bonding forces lead to a lattice matching of the two materials at the interface. As we discussed in Chapters 2 and 3, in cases where the lattice constants of two materials do not match – such as in InAs and GaAs – the layer being grown on the substrate can adjust its lattice constant to match that of the substrate for some number of layers – only a few in some cases. This creates a so-called pseudomorphic structure, but after some growth layers have been grown, the energetics associated with the pseudomorphic layers becomes unfavorable and the strained growth layers degrade by forming defective layers containing misfit dislocations. This "lattice-matching" problem is absent in stacked two-dimensional layers since the van der Waals forces are thought to be too weak to perturb the band structure of each two-dimensional layer. The apparent lack of a "lattice-matching" problem for stacked two-dimensional layers is perceived as one of the major advantages of this field.

5.5 The Optics of Quantum Dots

In Sections 5.3 and 5.4 we studied the optical properties of two-dimensional electrons, i.e., electrons confined in one of three directions. Three-dimensional confinement of the electrons is possible in the quantum dot structures discussed in Section 3.5. This type of confinement changes the optical properties of the electrons, as shown in this section.

According to the analysis presented in Section 3.5, the energy spectra of electrons and holes in quantum dots are discrete (see Fig. 3.20). Accordingly, the interband optical spectrum of a quantum dot should consist of a series of very sharp lines. The positions of these lines depend on the particular quantum dot material, confining potential, etc.

The probabilities of absorption and emission of light by a quantum dot, $P(\omega)$, can be calculated with the use of Eqs. (4.45) and (4.49). Because of the three-dimensional confinement of the electrons in quantum dots, both the initial state, i, and the final state, f, now depend on a set of three quantum numbers. For example, in the case of the spherical quantum dots considered in Section 3.5.1 these numbers are the principal quantum number, n, the total orbital momentum, l, and its projection, m, while the energies of such a dot E_i depend only on the two numbers, n, l, i.e., they are $(2l+1)$-fold degenerate. The matrix element of the dipole, $e\langle i|\xi\vec{r}\,|f\rangle$, depends strongly on the quantum dot material and geometry; however, its order of magnitude can be estimated as $\approx eR$, with R the characteristic size of the quantum dot.

In practice, the spectral lines of a quantum dot have finite width. There are two contributions to the line broadening: homogeneous and inhomogeneous broadening. The homogeneous broadening arises due to the interaction of the electrons and holes with phonons and impurities, as well as due to the finite lifetime of the electrons and holes with respect to their recombination. Formally, this type of broadening can be accounted for in the phototransition probabilities by the following substitution in Eqs. (4.45) and (4.49):

$$\delta(E_f - E_i - \hbar\omega) \rightarrow \frac{1}{\pi}\frac{\Gamma_{i,f}}{(E_f - E_i - \hbar\omega)^2 + \Gamma_{i,f}^2},$$

where $\Gamma_{i,f}$ is the width of the spectral line corresponding to the $i \leftrightarrow f$ phototransition. This type of broadening is small; for example, for a typical radiation lifetime, τ_r, of about 10^{-10} s, the line width is $\Gamma \approx \hbar/\tau_r \approx 6.6\,\mu\text{eV}$.

It is useful to note that, because of the three-dimensional confinement, the homogeneous broadening of quantum dot optical spectra has a much weaker dependence on temperature than it does in bulk and two-dimensional systems. Indeed, in the latter systems the spectral width of band-to-band absorption/emission is proportional to the temperature, due to the thermal redistribution of the carriers over continuum energy states inherent to bulk and two-dimensional systems; see the discussion in Section 4.4.2 and Fig. 4.10.

Homogeneous broadening is characteristic for a single quantum dot. In the case of a quantum dot *ensemble*, each of the quantum dots has a slightly different size, shape, doping, etc. This results in variation among the energy spectra of individual dots and so-called inhomogeneous broadening of the spectral lines of the whole quantum dot ensemble. To describe such an ensemble, the probabilities of the phototransitions, $P(\omega)$, should be averaged over the ensemble of quantum dots. Let the quantum dots differ in their radius, R, and let the ensemble of quantum dots have a Gaussian distribution over radius, $\mathcal{W}(R)$, with mean radius R_0 and standard deviation $\sigma_R = \sqrt{\langle (R - R_0)^2 \rangle}$; then,

$$\mathcal{W}(R) = \frac{1}{\sqrt{2\pi}\sigma_R} \exp\left[-\frac{(R - R_0)^2}{2\sigma_R^2}\right]. \tag{5.54}$$

In Fig. 5.26 we present the resulting ensemble-integrated absorption of the spherical quantum dots calculated using Eq. (5.54), where the dimensionless deviation is introduced as $\xi = \sigma_R / R_0$. It can be seen that the spread in radius in the ensemble considerably affects the optical spectrum: the line structure of the spectrum at small

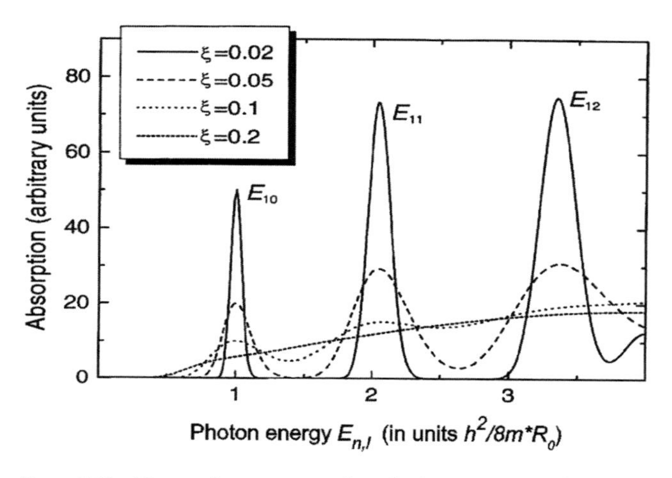

Figure 5.26 Absorption versus reduced photon energy for an ensemble of spherical quantum dots at different standard deviations σ_R/R_0. Republished with permission of John Wiley and Sons Inc., from Fig. 5.36 of *Quantum Dot Heterostructures*, D. Bimberg, M. Grundmann, and N. N. Ledentsov, 1st edn (1999); permission conveyed through Copyright Clearance Center, Inc.

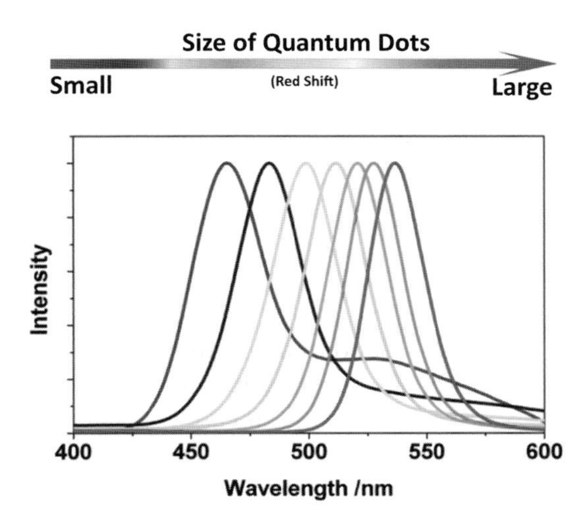

Figure 5.27 Emission spectra of CdSe quantum dots with radius ranging from 1 nm to 10 nm. After Fig. 1 in Debasis Bera, *et al.*, *Materials* **3**, 2260 (2010). Licensed under CC BY 3.0.

ξ is transformed to a wide structureless optical band at large ξ. We can conclude that the resonance properties of quantum dots can be preserved in their ensemble only if the spread of their parameters is small. Contemporary technologies provide multiple-quantum-dot systems which meet this condition.

For a given material the dependence of the optical spectra on quantum dot size is due mainly to the quantum confinement effect. The analysis presented in Section 3.5.1 showed that the energies of quantized levels in quantum dots increase when their characteristic size decreases. For interband transitions this leads to an increase in the energies of photons interacting with the quantum dots and a blue shift of the corresponding spectral lines. Figure 5.27 illustrates these changes in photoluminescence emission with size, for an ensemble of CdSe quantum dots. We see that by increasing the quantum dot size from 1 nm to 10 nm, one can cover the whole of the visible spectral range from violet to red light.

There are different methods of excitation of quantum dot emission. Here we consider briefly two methods of excitation of the *interband emission* of semiconductor quantum dots embedded in a semiconductor medium: photo-excitation and electric excitation. In the energy-coordinate diagram, a quantum dot embedded in a semiconductor can be represented by potential wells for both electrons and holes, as shown in Fig. 5.28. The depths of these wells are ΔE_c and ΔE_v for the electrons and holes, respectively. The energy levels for the electron and hole confined inside the dot are E_c and E_v, respectively. If an electron and a hole populate these energy levels, interband phototransision with emission of a photon becomes possible (this process is also called radiative recombination). The energy of the emitted photon is equal to the bandgap plus the bound-state energies for the electrons and holes measured from E_c and E_v, respectively.

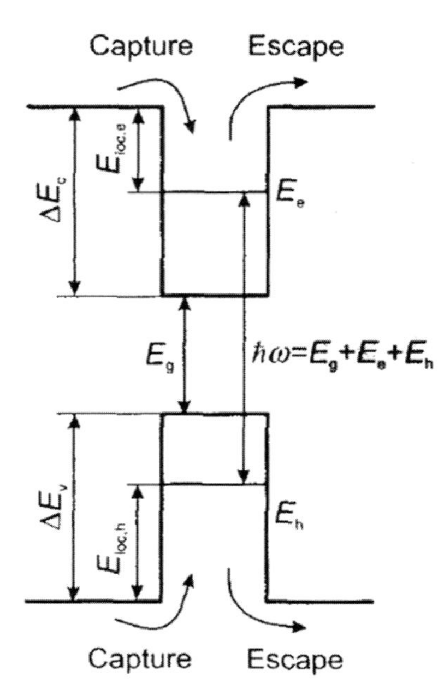

Figure 5.28 The energy-coordinate scheme of a quantum dot embedded in a semiconductor. Republished with permission of John Wiley and Sons Inc., from Fig. 5.38 of *Quantum Dot Heterostructures*, D. Bimberg, M. Grundmann, and N. N. Ledentsov, 1st Edition (1999); permission conveyed through Copyright Clearance Center, Inc.

For photo-excitation of such quantum dots, the semiconductor–dots system is illuminated by external light with the photon energy exceeding the energy bandgap of the material wherein the quantum dots are embedded, i.e., the material of the barrier. Absorption of such light generates electrons in the conduction band and holes in the valence band. Then these particles can be captured in the quantum dots, where they populate energy levels in the conduction band and in the valence band.

For the case of electric pumping, the quantum dots should be placed between n- and p-doped regions of a semiconductor; for example, one may use a p–i–n diode. On applying a direct electric bias to such a structure, electrons from the n-region and holes from the p-region are injected into the intermediate region with the quantum dots, where they are captured to the discrete levels of the dots. Furthermore, radiative recombination produces emission from quantum dots.

Both methods, photo-excitation and electric excitation, are widely used in emitters and in lasers based on quantum dots.

5.6 Closing Remarks to Chapter 5

In this chapter we studied optical interband transitions in quantum heterostructures. We derived basic formulae for optical effects in these systems and found that the optical spectra of quantum heterostructures have more complex forms and depend substantially on the materials and parameters of the structures. In particular, the optical density of states exhibits a sharper dependence on the frequency than does that of bulk materials. The excitonic absorption is much more pronounced in quantum heterostructures and is observable even at room temperature. All these results are important for optoelectronics, particularly for lasers and nonlinear optical devices based on quantum heterostructures.

We studied intraband phototransitions. In a bulk material these types of optical transitions are possible only if a "third" particle – a phonon, defect, interface, etc. – is involved in the process. In heterostructures, these optical transitions occur even for ideal structures. We calculated phototransitions between confined states of electrons and holes, and absorption from confined to extended states of carriers. Such intersubband optical spectra can be tailored for a desired spectral region. In many cases such tailoring techniques are applied in the infrared region of the electromagnetic spectrum. It is worth mentioning that intersubband phototransitions exist in heterostructures of any kind of material: both direct- and indirect-bandgap materials can be used. In Chapter 8, we will use these results to analyze photodetectors and lasers based on intersubband transitions.

In this chapter we considered only two-dimensional quantum structures. The extension of all equations to the case of quantum wires and quantum dots is straightforward: in addition to the quantization in the z-direction that was discussed in detail in this chapter, there is quantization in one more direction (say y) or correspondingly two more directions (x and y). As a result, carriers in quantum wires will interact with light of any polarization in the direction perpendicular to the longitudinal axis of the wire, x. The charge carriers in quantum dots interact with light of arbitrary polarization and consequently there are some advantages of quantum dot structures. The dispersion in size of the quantum dots that was discussed in Chapter 3 results in the broadening of absorption and emission spectra.

5.7 Control Questions

1. Explain why the decrement of light in a quantum well depends on the position of the quantum well in a Fabry–Pérot resonator.
2. What is the optical confinement factor?
3. For interband absorption, compare the frequency of light absorbed in a quantum well with that for absorption in the bulk of the same semiconductor.
4. Write down an equation for the lowest frequency for interband absorption of light in a quantum well, assuming very high confinement barriers for electrons and holes.
5. Using the same approach as was used to obtain Eq. (5.18), obtain an expression for the interband absorption coefficient in a quantum wire.
6. Write down the energy conservation equation for interband light absorption in a quantum wire (like Eq. (5.20)).

7. Using the same approach as was used to obtain Eq. (5.18), obtain an expression for the interband absorption coefficient in a quantum dot.
8. Write down the energy conservation equation for interband light absorption in a quantum dot (like Eq. (5.20)).
9. Obtain an expression for the optical density in a quantum wire (like Eq. (5.23)).
10. What are the reasons for the broadening of interband absorption in quantum wells?
11. Explain the difference between the dependence of the interband absorption coefficient on the polarization of light in quantum wells and that in the bulk material.
12. Explain the dependence of the intraband absorption coefficient in a quantum well on the light polarization.
13. Obtain an expression for the dependence of the intraband absorption coefficient in a quantum wire on the light polarization analogous to Eq. (5.37) and explain its difference from the quantum well case.
14. Obtain an expression for the dependence of the intraband absorption coefficient in a quantum dot on the light polarization analogous to Eq. (5.37) and explain its difference from the quantum well and quantum wire cases.
15. Discuss the difference between the absorption coefficients for an intraband transition and for transitions from subbands to extended states.
16. Explain how van der Waals bonding alleviates many of the difficulties caused by the lattice-mismatch problem in conventional heterostructures.

6 Electro-Optics and Nonlinear Optics

6.1 Introduction

The effect of an external electric field on the refractive index or the absorption coefficient, i.e., its effect on the propagation of light through a material or on the reflection of light, is known as *the electro-optical effect*. Owing to the high electric-field sensitivity of the electron energy spectra and the feasibility of controlling carrier concentrations in quantum heterostructures, the electro-optical behavior of heterostructure devices is unique.

In this chapter, we shall study the electro-optical effect for quantum heterostructures including quantum wells, double- and multiple-quantum-well structures, and superlattices. We shall consider the quantum confined Stark effect, the Burstein–Moss effect, and the effect of destroying excitons in gated heterostructures. In all three of these cases, the absorption edge is controllable over a wide range of external voltages. For double quantum wells, we shall study coherent oscillations of an electron wave packet and terahertz microwave emission controlled by an electric field. We also will briefly analyze the ultimate case of double quantum structures, namely bilayers of one-atom-thick two-dimensional crystals like bigraphene, etc.

We will show that there is a great potential for the optoelectronic applications of using a field to control light frequency and intensity and to realize the tunable generation of the emission of microwave, terahertz, infrared, and visible radiation.

The second part of this chapter is devoted to the consideration of *nonlinear optical effects* in quantum heterostructures. Such effects occur at high intensities of light, when the optical characteristics of a heterostructure become dependent on the amplitude of the light wave or the light intensity. Semiconductor heterostructures lead to enhancements of the nonlinear effects observed in bulk materials and they also exhibit the unique nonlinear optical behavior found in low-dimensional systems. This class of effects provides the possibility of controlling light by light and creates the basis for new devices with all-optical addresses.

6.2 Electro-Optics in Semiconductors

6.2.1 Electro-Optical Effects in Conventional Materials

We start with a brief review of the electro-optical effect in commonly used bulk materials.

The effect of an external electric field, $\vec{F}$, on the refractive index, $n(\omega)$, of a transparent dielectric medium can be introduced by use of a function $n(\omega, \vec{F})$. In fact, the field dependence is relatively weak and we can restrict ourselves to several terms of the expansion of $n(\omega, \vec{F})$ in a series with respect to the field, $\vec{F}$. Since n is a scalar value and $\vec{F}$ is a vector, the expansion depends on the symmetry of the medium under consideration.

In noncentrally-symmetric crystals, the first term in the expansion is linearly dependent on the magnitude of the field. Thus, for the linear electro-optical effect, we can write

$$n(F) = n - \frac{1}{2} a_{\mathrm{P}} n^3 F, \tag{6.1}$$

where n is the refractive index at $F = 0$ and a_{P} is the linear electro-optical coefficient. This case is known as the linear electro-optical effect, or the *Pockels mechanism.*

In centrally symmetric media, the quadratic electro-optical effect, or the *Kerr mechanism*, occurs. Changes in the refractive index are proportional to the square of the magnitude of the field:

$$n(F) = n - \frac{1}{2} b_{\mathrm{K}} n^3 F^2, \tag{6.2}$$

where b_{K} is the quadratic electro-optical coefficient. Generally, both coefficients, a_{P} and b_{K}, depend on the frequency of the light. Typical values of a_{P} lie in the range of 10^{-10} cm/V to 10^{-8} cm/V. For a field F of about 10^4 V/cm, such a material undergoes changes in refractive index from 10^{-6} to 10^{-4}. Typical values for b_{K} range from 10^{-14} cm^2/V^2 to 10^{-10} cm^2/V^2. That is, in a field of $F = 10^4$ V/cm, the changes in the refractive index are in the range from 10^{-6} to 10^{-2}.

From a practical point of view, the main operational function that can be provided by the electro-optical effect is the *modification of the optical properties of a material* as a result of an applied electric field. While acoustic waves and a magnetic field could be used for the same purposes via acousto-optical effects and magneto-optical effects, the electro-optical effect is of special importance because it is used widely for high-speed applications and it is compatible, in principle, with modern electronics.

In order to gain more insight, let us briefly sketch some applications of the effect under consideration. We know that a light wave can be characterized by three parameters: a frequency ω, an amplitude F_0 (or an intensity $\mathcal{I}$), and a phase ϕ. The electro-optical effect may be used to control two of these parameters – phase and intensity. For a plane wave, we can write

$$\vec{\mathcal{E}}(\vec{r}) = \vec{F}_0 \cos(\omega t - \phi(\vec{r})), \tag{6.3}$$

where the phase ϕ is position-dependent. For a homogeneous medium, $\phi = n\omega z/c$, where c is the velocity of light in free space and z is the direction of light propagation.

Consider first the phase modulation. If, by external means, one changes the refractive index n of a medium along the optical path L_z, the phase varies as

$$\phi = \phi_0 - \frac{\omega}{2c} a_{\mathrm{P}} n^3 F L_z \equiv \phi_0 - \pi \frac{\Phi}{\Phi_\pi}, \tag{6.4}$$

where ϕ_0 is the phase in the absence of an electric field, and we have introduced the voltage $\Phi = Fd$, the dimension, d, of the region with the field F, and a characteristic value $\Phi_\pi = 2\pi dc/a_{\mathrm{P}} n^3 \omega L_z$ which is appropriate to the case of the Pockels mechanism. According to Eq. (6.4), a variation of the field or, equivalently, the voltage, leads to a modulation of the wave's phase.

In Fig. 6.1, two schemes of phase modulators are shown. The case of Fig. 6.1(a) corresponds to the so-called longitudinal modulators, where the direction of the electric field coincides with the direction of light propagation (here the optical path, L_z, coincides with d). The case of Fig. 6.1(b) corresponds to a transverse modulator. Because the characteristic times of the nonstationary processes required to reach the refractive-index changes due both to the Pockels mechanism and the Kerr mechanism are extremely short (10^{-13} s or so), for both cases, shown in Figs. 6.1(a) and (b), the modulation speed is limited only by the transition time of the light through the crystal: $T = L_z n/c$. Thus, the transit-time-limited modulation bandwidth is $1/T$.

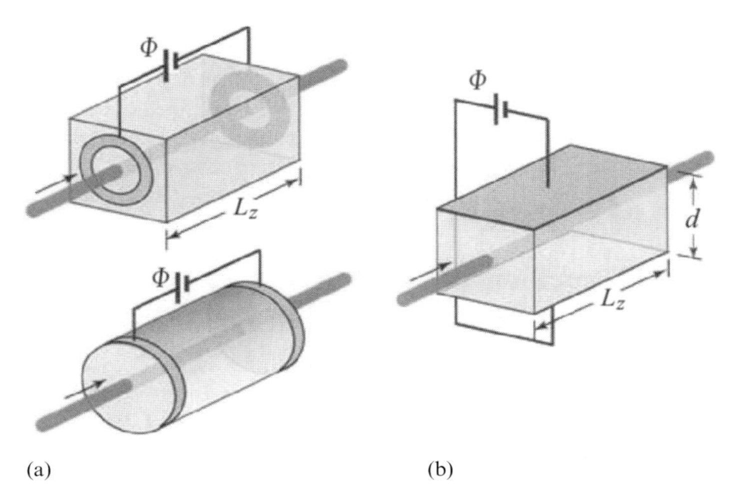

Figure 6.1 Electro-optical modulators (the arrows show the direction of propagation of light beams). (a) Two longitudinal modulators: an electric field is applied along the optical path. (b) A transverse modulator: the electric field is perpendicular to the optical path. Republished with permission of John Wiley and Sons Inc., from Fig. 20.1-2 (a,b) of *Fundamentals of Photonics*, B. E. A. Saleh and M. C. Teich, 2nd edn, Copyright 2007; permission conveyed through Copyright Clearance Center, Inc.

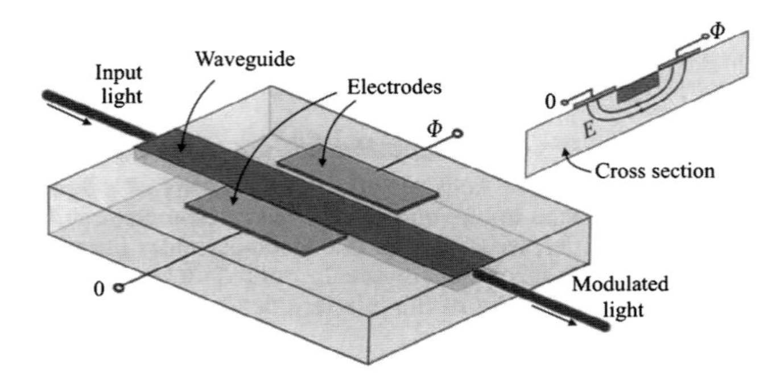

Figure 6.2 An integrated optical phase modulator. The waveguide and controlling electrodes are labeled. Input and output (modulated) light beams are shown. Republished with permission of John Wiley and Sons Inc., from Fig. 20.1-3 of *Fundamentals of Photonics*, B. E. A. Saleh and M. C. Teich, 2nd Edition, Copyright 2007; permission conveyed through Copyright Clearance Center, Inc.

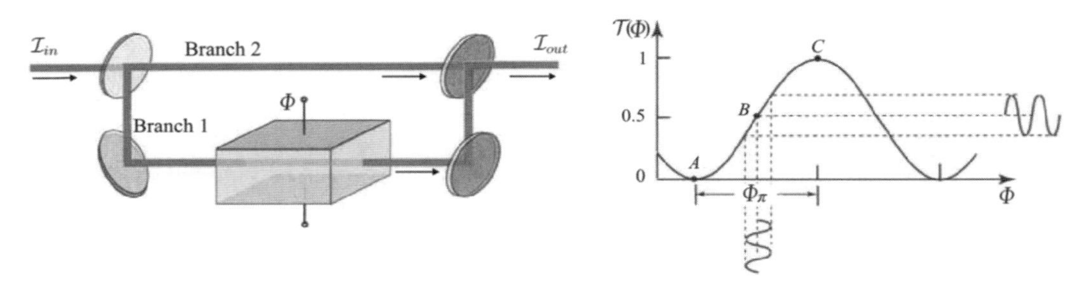

Figure 6.3 A Mach–Zehnder interferometer with a phase modulator placed in one branch. The right part of the figure shows the interferometer transmission, $\mathcal{T} = \mathcal{T}_{out}/\mathcal{T}_{in}$, as a function of the applied voltage, Φ. The dependence $\mathcal{T}(\Phi)$ is periodic with period equal to $2\Phi_\pi$. Near point B the device operates as an almost linear modulator of the light intensity, as illustrated by the figure. Voltage switching between points A and C provides optical switching between zero and total transmission. Republished with permission of John Wiley and Sons Inc., from Fig. 20.1-4 of *Fundamentals of Photonics*, B. E. A. Saleh and M. C. Teich, 2nd edn, Copyright 2007; permission conveyed through Copyright Clearance Center, Inc.

Integrated optical phase modulators working on the electro-optical effect have also been developed. Figure 6.2 presents such a modulator schematically. Light propagates through a waveguide which is placed between two electrodes. An applied voltage changes the refractive index of the waveguide and its surroundings and modulates the phase of the output light signal. The modulator can operate with a modulation speed of up to 100 GHz.

There are different methods for electric modulation of the intensity of light. For example, one can modulate the wave intensity if a phase modulator is placed in one branch of an interferometer. Such a case with a Mach–Zehnder interferometer is shown in Fig. 6.3. Here, the light wave with intensity $\mathcal{I}_{in}$ is split into two beams, say with equal intensity $\mathcal{I}_{in}/2$. As a result of the interference of the two beams, the output intensity, $\mathcal{I}_{out}$, is

$$\mathcal{I}_{out} = \frac{1}{2}\mathcal{I}_{in} + \frac{1}{2}\mathcal{I}_{in}\cos\phi = \mathcal{I}_{in}\cos^2\left(\frac{\phi}{2}\right), \quad (6.5)$$

where ϕ is the difference between the phase shifts in the two branches of the interferometer. Let this difference be ϕ_0 at zero electric field. Then, in accordance with Eq. (6.4), the transmission coefficient of the device is

$$\mathcal{T}(\Phi) = \frac{\mathcal{I}_{out}}{\mathcal{I}_{in}} = \cos^2\left(\frac{\phi_0}{2} - \frac{\pi}{2}\frac{\Phi}{\Phi_\pi}\right). \quad (6.6)$$

The phase ϕ_0 can be controlled, for example, by the length of the branches. Thus, different regimes of operation are possible. If $\phi_0 = \pi/2$, the light intensity can be modulated almost linearly around $\mathcal{I}_{out} = \mathcal{I}_{in}/2$. If $\phi_0 = 2\pi m$, where m is an integer, the device provides almost 100% modulation of the output intensity over the range of voltages from 0 to Φ_π.

The same principal idea can be used to realize an integrated optical intensity modulator. In such a device, an input signal is split and propagates through two waveguide branches; the refractive index for one of the branches is controlled by an applied voltage similarly to the case of Fig. 6.2. Modulation of the light intensity at a few GHz can be achieved for such devices.

The most common modulators are based on $LiNbO_3$ crystals. Unfortunately, this material is not directly compatible with modern optoelectronic and electronic devices based on heterostructures of III–V and group IV materials.

Recently, it has been verified that the electro-optical effect is very strong in quantum heterostructures. In addition to changing the refractive index, the electric field strongly affects light absorption in these structures. It is expected that quantum-based modulator structures – working on the basis of the principles just discussed – will play an increasingly important role in optical modulation as well as in other operations for controlling and processing light signals.

6.2.2 Electro-Optical Effects in Quantum Wells

The most well-known and important electro-optical effect in semiconductors is the Franz–Keldysh mechanism for modifying the fundamental edge of interband absorption in an electric field F. This mechanism can be understood as follows. In Fig. 6.4, the valence and conduction bands are presented for $F = 0$ and $F \neq 0$. At zero electric field, the bands are separated by an energy gap, E_g, and absorption starts at photon energies $\hbar\omega > E_g$ as discussed in Chapter 4. For a finite electric field F, the situation is quite different. In a strict sense, there is no longer any bandgap, since the wave functions of the electrons and holes overlap under the influence of the applied field and interband phototransitions become possible for $\hbar\omega < E_g$, as shown in Fig. 6.4. Thus, the Franz–Keldysh mechanism leads to electro-optical effects in absorption and refraction.

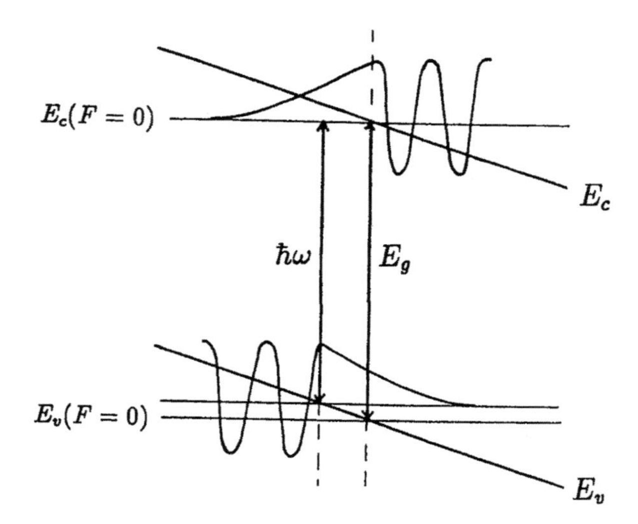

Figure 6.4 An illustration of the Franz–Keldysh effect.

As a result of the field, the fundamental absorption edge is modified and an absorption tail appears in the long-wavelength spectral region:

$$\alpha(\omega) \propto \exp\left[-\left(\frac{E_{\mathrm{g}} - \hbar\omega}{\hbar\omega_{\mathrm{F}}}\right)^{3/2}\right], \quad \hbar\omega < E_{\mathrm{g}},$$

where the factor ω_{F} depends on the electric field,

$$\omega_F^3 = \frac{e^2}{2\hbar}\left(\frac{m_{\mathrm{e}} + m_{\mathrm{h}}}{m_{\mathrm{e}} m_{\mathrm{h}}}\right) F^2,$$

and m_{e} and m_{h} are the effective masses of the electrons and holes, respectively. The presence of the reduced form of Planck's constant, $\hbar$, provides further insight into the quantum origin of the Franz–Keldysh mechanism.

Though the Franz–Keldysh effect is the most important electro-optical mechanism within the fundamental absorption edge, its use is quite limited in bulk crystals. Furthermore, this mechanism washes out excitonic effects, because relatively weak electric fields dissociate the excitons. For III–V compounds, the destruction of excitons begins in electric fields of the order of 10^3 V/cm.

Let us consider the case of a quantum well placed in an electric field perpendicular to the quantum well layer. Figure 6.5(a) depicts energy levels in quantum wells for both electrons and holes. These levels are modified by the field as shown in Fig. 6.5(b). One can see that a shift of energies occurs in the field; this leads to the so-called *quantum confined Franz–Keldysh* (or *Stark*) *effect*. For gas media, a shift of atomic levels caused by an electric field is simply known as the Stark effect. So the effect under consideration is also sometimes called the *quantum confined Stark effect*. In contrast to the case of bulk systems, in quantum well structures,

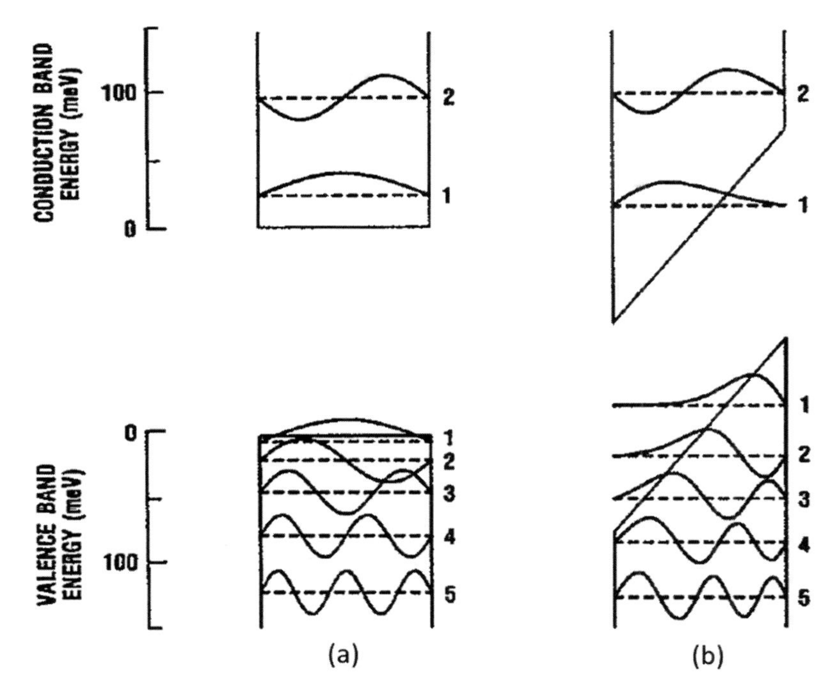

Figure 6.5 The modification of energy levels in a quantum well for electrons and holes, when an electric field is applied: (a) well and levels without an electric field, with the electron and hole subbands, and their wave functions shown; and (b) the same in an electric field. This field always leads to a decreasing energy separation of the electron and hole subbands and a red shift of interband optical spectra. Reprinted figure with permission from Fig. 3 of D. A. B. Miller, *et al.*, *Phys. Rev.* **B 33**, 6976, 1986. Copyright (1986) by the American Physical Society.

the gap between the conduction and valence bands still remains, but the separation between the lowest electron and hole subbands decreases in high electric fields. The upper subbands are also modified. This leads to a red shift of the interband optical spectra. In addition, excitonic effects are preserved up to very high fields of 10^5 V/cm and more, because the confining well potential suppresses dissociation processes. Indeed, the exciton level is shifted and follows the bottom of the subband.

A comparison of the bulk-like Franz–Keldysh effect with its confined analog can be performed using the same quantum well structure for two electric-field configurations: parallel and perpendicular to the layers of the structure. In Fig. 6.6, the effect of an electric field on the absorption coefficient is presented for fields parallel and perpendicular to the quantum well layer. The experiments were performed for a multiple-quantum-well structure with 95 Å-thick GaAs quantum wells separated by 98 Å-thick $Al_{0.32}Ga_{0.68}As$ barriers (60 periods). In accordance with the previous discussion, the parallel field configuration is rather similar to the Franz–Keldysh effect in a bulk crystal since quantum confinement is not of major importance. The

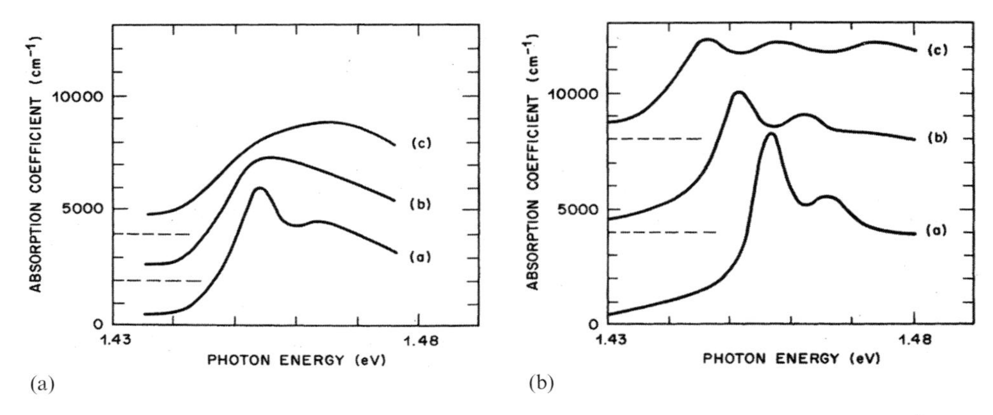

Figure 6.6 The effect of an electric field on the absorption coefficient of a sample with 95 Å-thick GaAs quantum wells for (a) parallel and (b) perpendicular field configurations. For the parallel configuration, curves labeled (a), (b), (c) correspond to fields of 0, 1.6×10^4, 4.8×10^4 V/cm, respectively. For the perpendicular configuration, curves labeled (a), (b), (c) correspond to fields 1×10^4, 4.7×10^4 and 7.3×10^4 V/cm, respectively. Reprinted figure with permission from Figs. 3 and 5 of D. A. B. Miller, *et al.*, *Phys. Rev.* **B 32**, 1043, 1985. Copyright (1985) by the American Physical Society.

field magnitude was varied from zero to 4.8×10^4 V/cm. The results presented for three magnitudes of the field show a red shift in the interband absorption tail and a suppression of the excitonic effect with increasing field, while the excitonic effect is clearly seen at zero field, with heavy- and light-hole excitons well resolved. In the perpendicular field configuration, the field is varied in a wider range, from zero to 2×10^5 V/cm. The results demonstrate a large red shift of light absorption and preservation of the excitonic effects up to high fields.

Note, that in contrast to the Pockels and Kerr mechanisms, electro-optical effects in quantum structures are neither linear nor quadratic functions of the electric field. The effects are, rather, characterized by complex field and frequency dependences. In order to enhance the electro-optical absorption effect, multiple-quantum-well (MQW) structures may be employed. The use of MQW structures facilitates reaching almost 100% modulation of the absorption. Thus, quantum structures can be employed for direct modulation of the light transmission (without an interferometer) due to the large electro-optical absorption effect.

As we discussed in Section 4.4.6, changes in the absorption coefficient are closely related to the electro-optical effect in the refractive index. The large magnitude of this effect in quantum wells leads to its use in controlling light by means of an electric field in devices based on the Mach–Zehnder interferometer. Accordingly, the sizes of the necessary heterostructures are much smaller than those of conventional light modulators based on the best available materials, such as $LiNbO_3$.

For quantum wells, there exist other natural mechanisms in addition to the electro-optical effect. The carrier concentration in gated heterostructures is changed by applying a voltage to the gate, as discussed in Section 3.9.1. One can deplete quantum wells or increase the carrier concentration up to sheet densities of about 10^{12} cm^{-2}. Variations of the carrier concentration induce significant changes in the absorption coefficient as a result of the following two dominant mechanisms.

The first is caused by the filling factor: increased population of the electron or hole subband blocks the absorption; for bulk crystals, this effect is known as the *Burstein–Moss effect*. Electrons and holes populate almost all energy states near the top of the valence band and the bottom of the conduction band, which makes phototransitions impossible in some spectral range near the fundamental absorption edge. If the quasi-Fermi levels for the two types of carriers are $E_{F,p}$ and $E_{F,n}$, the phototransitions are allowed for photon energies $\hbar\omega > E_g + E_{F,p} + E_{F,n}$. In other words, an increase in population of the bands leads to a blue shift in absorption. Since the density of states of two-dimensional subbands in quantum wells is quite small, the filling factor in quantum wells can be greater than that in bulk materials: the blue shift of the absorption coefficient is reached easily.

The second mechanism is related to the suppression of the excitonic effect: an increase in the carrier concentration provides more effective screening of the electron–hole Coulomb interaction, which leads to the destruction of excitons. This causes an additional blue shift in the absorption.

Both phenomena contribute to the electro-optical effect in quantum wells. They are illustrated in Fig. 6.7. In this figure, the absorption coefficient, α, and its change, $\Delta\alpha$, are presented as functions of the photon energy for a GaAs quantum well. The three curves in the left part of the figure correspond to the following cases: both the concentration per unit area and the field are zero ($N_s = 0, F = 0$); a finite electric field and zero concentration (F up); and a finite concentration and zero field (N_s up).

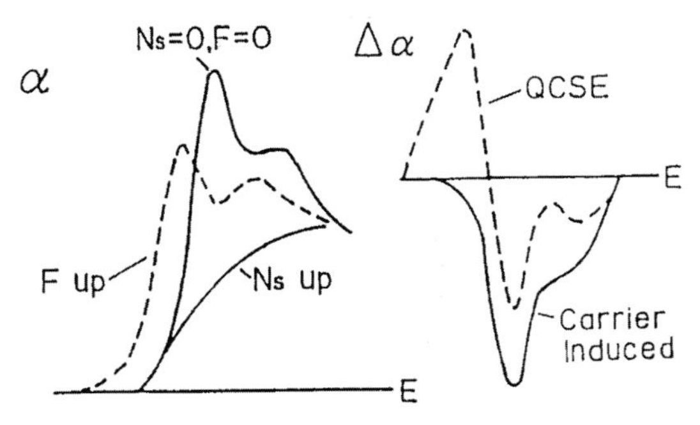

Figure 6.7 Schematic illustrations of changes in the absorption spectra of quantum wells. The electric-field effect or quantum confined Stark effect (QCSE) and the carrier-induced effect on absorption are shown. For comparison the absorption coefficient at zero electric field and in the absence of the carrier concentration is presented. The left part shows the interband absorption coefficient, α, as a function of the photon energy, E. The right part shows the partial contributions of the two effects. Reprinted by permission from figure 4 of Springer Nature: *Optical Switching in Low-Dimensional Systems* by H. Sakaki and H. Yoshimura, Copyright 1989.

In the right part of the figure, changes in the absorption coefficient are shown: one is due to the quantum confined Stark effect; the other one is the carrier-induced effect. A comparison shows that these two effects are of the same order of magnitude.

Within the absorption edge, carrier-induced changes in the refractive index can be large. The effect is strongly dependent on the light frequency and can reach values up to 0.01. Thus, controlling the carrier concentration by the gate voltage, one can realize effective electro-optical effects by varying either the absorption coefficient or the refractive index. Note, however, that the quantum confined Franz–Keldysh (Stark) effect and the carrier-induced effect are characterized by very different response times. The first effect involves the carrier energy states and their wave functions. So, the response time is limited by a value of the order of $\hbar/\epsilon_i$, where ϵ_i are the subband energies; this follows from the uncertainty relations. The carrier-induced effect is limited by the lifetime of the carriers, i.e., which is much larger and is on the nanosecond timescale or even longer. Figure 6.8(a) depicts schematically an integrated optical modulator based on both effects. The effective waveguide of an optical mode has a 5000 Å-long core of $Al_yGa_{1-y}As$ with $y = 0.24$. Beneath and on top of this core, there are cladding layers of $Al_xGa_{1-x}As$ with $x = 0.28$. Wave confinement in the lateral direction is achieved by side etching, for example.

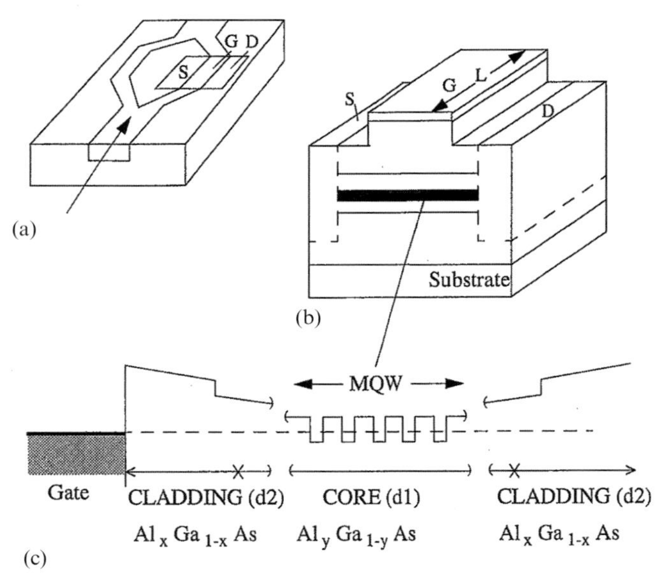

Figure 6.8 A schematic representation of a Mach–Zehnder interferometer with the quantum well modulator in one branch: (a) the principal scheme of the interferometer; (b) the MQW modulator with control of the electric field and the carrier concentration; and (c) the energy diagram of the heterostructure (the conduction band profile is shown). Reprinted by permission from figure 5 of Springer Nature: *Optical Switching in Low-Dimensional Systems* by H. Sakaki and H. Yoshimura, Copyright 1989.

A heterostructure modulator is embedded in one of the branches of the interferometer. The heterostructure consists of 90 Å GaAs MQWs separated by $Al_{0.24}Ga_{0.76}As$ barriers, as indicated in Fig. 6.8(c). The modulator length is about $L = 230\,\mu m$. To realize the quantum confined Stark effect and carrier-induced effect, a field-effect-transistor configuration is used as shown in Fig. 6.8(b). By applying a gate and source–drain voltages, one controls the electric field and carrier concentration in the region of the wells. Let us estimate the characteristic parameters of this modulator. Let both effects produce a change of the refractive index equal to Δn. If the confinement factor of the optical mode is Γ (this factor was introduced in Eq. (5.13)), then the effective change of the index of refraction of the guided mode is estimated as $\Gamma\,\Delta n$. For the above waveguide parameters, the confinement factor Γ is about 0.12. The maximum change in the refractive index for the case presented in Fig. 6.7 corresponds to the photon energy $\hbar\omega_{max} = 1.55\,eV$ and equals $\Delta n \approx 0.012$. To achieve a phase shift of π (i.e., 100% modulation according to Eq. (6.6)), the required length of the controlled part of the branch should be only 33 μm. Thus the device can effectively operate not only with the photon energy equal to $\hbar\omega_{max}$, but also over a wide range of energies.

It is important that the switching speed of such a modulator be essentially the same as that of the field-effect transistor. If the thickness of the electron channel of the transistor (the lateral thickness of the waveguide) is about 1 μm to 2 μm, the cut-off frequency for operation of the device is about 15 GHz or higher. In this example, the integrated optical modulator operates with light propagating parallel to the quantum well layers.

For the perpendicular geometry (the direction of the light propagation is normal to the layers composing a heterostructure), one uses transparent electrodes. In such a case, modulators based on MQW structures are efficient at thicknesses of roughly a few micrometers. Two improvements can be made. The first is to use an integrated multilayer reflector in a double-pass optical system. The second comes about by employing two such integrated reflectors to form a Fabry–Pérot resonator. These devices can increase the contrast ratio of the modulation by as much as a factor of 100. Modern technology facilitates the fabrication of arrays of these devices for electrically addressed spatial light modulators. Such modulators are a key component of any optical computing system which utilizes parallel processing of the signals.

6.2.3 Electro-Optical Effects in Superlattices

Systems with a large number of MQWs and with thin barrier layers manifest novel characteristics, including the formation of minibands with wave functions that extend throughout the structure. Such structures are known as superlattices (see Section 3.7). The electro-optics of superlattices is a more complex subject than that of isolated quantum wells. It involves reconstructions of the energy spectra in the electric field such as the Stark ladder.

To examine the electro-optical effect in superlattices, we sketch in Fig. 6.9(a) electron and hole MQW potentials and minibands for a type-I semiconductor material

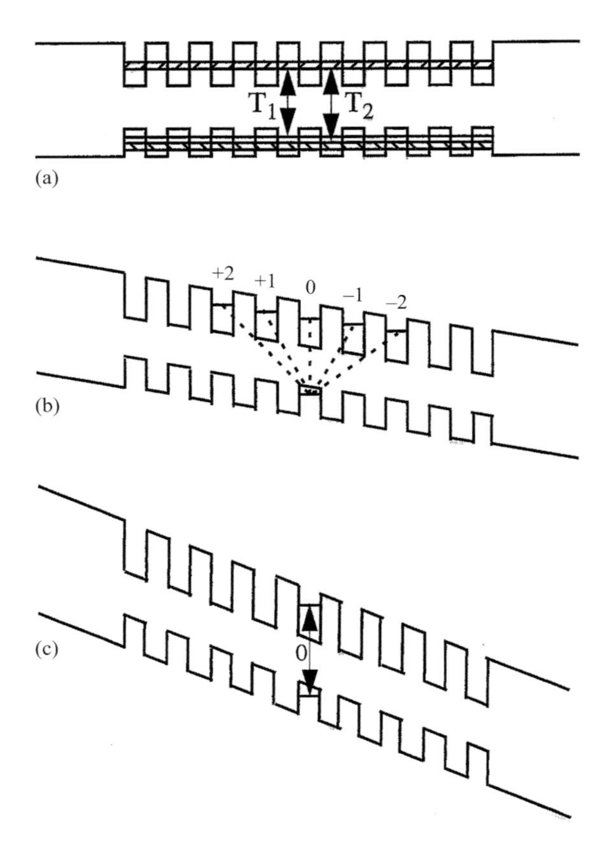

Figure 6.9 A diagram of the interband phototransitions in a superlattice. (a) The potential profile for electrons and holes in a type-I superlattice. One electron miniband and two hole minibands are shown. The transitions T_1 and T_2 originate from heavy- and light-hole minibands, respectively. (b) Interband phototransitions under a moderate electric field: 0 labels vertical transitions, ± 1 and ± 2 label transitions into neighboring wells. (c) Transitions at a high electric field. The miniband structure is destroyed, the carriers are localized in the wells, and dominant transitions are vertical. Adapted from Fig. 4 of *J. Luminesc.* **44**, E. E. Mendez and F. Agullo-Rueda, "Optical properties of quantum wells and superlattices under electric fields," 223, Copyright (1989), with permission from Elsevier.

without an electric field. The diagrams depict a $GaAs/Al_{0.35}Ga_{0.65}As$ superlattice with well and barrier thicknesses equal to 30 Å and 35 Å, respectively. The fields are of magnitude 2×10^4 V/cm for Fig. 6.9(b) and 1×10^5 V/cm for Fig. 6.9(c). In Fig. 6.9(a), only the lowest miniband is presented for the conduction band, while for the valence band, two minibands are plotted; one is associated with heavy holes and the other with light holes. In the absorption spectrum of such a superlattice, two types of interband phototransition should be observed: heavy-hole-miniband $\leftrightarrow$ electron-miniband (T_1) and light-hole-miniband $\leftrightarrow$ electron-miniband (T_2), as indicated in Fig. 6.9(a). Consistently with the calculations of the absorption spectra given in Chapters 4 and 5, each type of interband phototransition is proportional

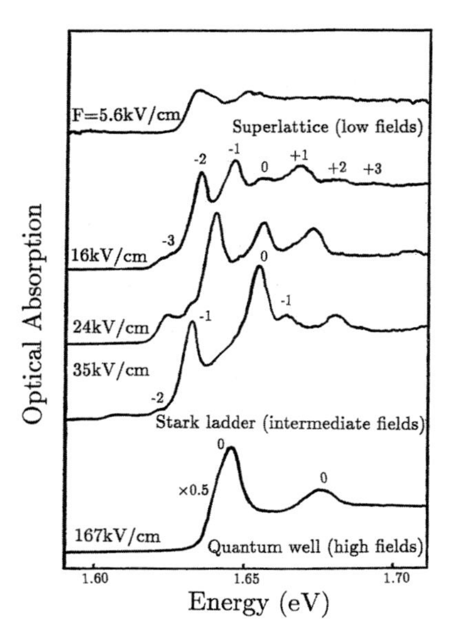

Figure 6.10 Absorption spectra of a superlattice with parameters discussed in the text for different electric fields. The magnitudes of the fields are indicated on the curves. At low fields the results correspond to absorption spectra of the superlattice and show two phototransitions, T_1 and T_2. At intermediate fields the peaks correspond to transitions from fully localized hole states to different states of the electron Stark ladder. At high fields the spectra correspond to interband phototransitions in a single-quantum-well structure. Reprinted from Fig. 5 of *J. Luminesc.* **44**, E. E. Mendez and F. Agullo-Rueda, "Optical properties of quantum wells and superlattices under electric fields," 223, Copyright (1989), with permission from Elsevier.

to the optical density of states of the superlattice. The curve corresponding to zero electric field has a shape very similar to that for the density of states of the superlattice. In realistic situations, both types of phototransition, T_1 and T_2, contribute to the spectrum; these contributions are overlapping and the spectrum becomes more complicated. One such example is shown in Fig. 6.10. This spectrum has been measured for a superlattice with 40 Å GaAs wells and 20 Å $Al_{0.35}Ga_{0.65}As$ barriers at $T = 5$ K. The upper curve is presented for a weak electric field of $F = 5.6$ kV/cm. For this case, contributions originating from both heavy- and light-hole minibands are resolved.

An electric field, F, applied to a superlattice with period d causes a localization of carriers and transforms the minibands into a set of equidistant energy levels,

$$\epsilon_n = \epsilon_0 + eFdn, \quad n = \pm 1, \pm 2, \pm 3, \ldots, \tag{6.7}$$

i.e., a Stark ladder of energy levels. In small fields, the localization of carriers is weak; that is, their wave function extends over several superlattice periods. As the field increases, the localization increases as well.

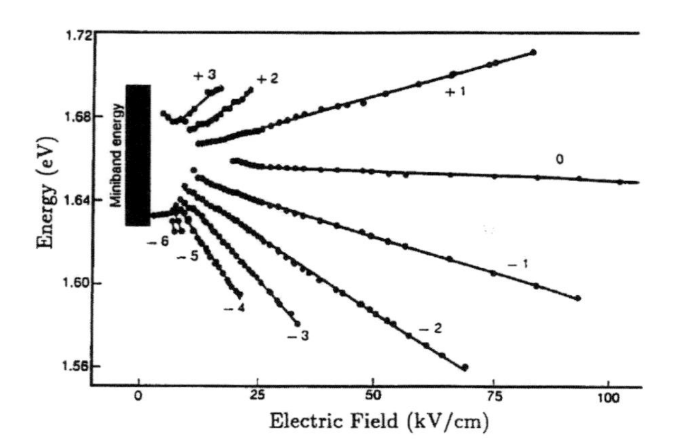

Figure 6.11 Interband transition energies as a function of the electric field for the superlattice with parameters discussed in Fig. 6.10. The numbers on the curves correspond to different energy states of the electron Stark ladder. Reprinted from Fig. 6 of *J. Luminesc.* **44**, E. E. Mendez and F. Agullo-Rueda, "Optical properties of quantum wells and superlattices under electric fields," 223, Copyright (1989), with permission from Elsevier.

The holes with larger effective masses are localized first. Thus, in intermediate fields, wave functions of both types of holes are confined in one of the identical quantum wells, while the electron wave functions are still delocalized over a few periods of the superlattice. As a result, besides strictly vertical interband transitions, which are labeled in Fig. 6.9 by 0, there are several strongly pronounced transitions, labeled as $\pm 1, \pm 2$, etc., into neighboring wells. That is, the transitions involve several energy levels of the electron Stark ladder of Eq. (6.7). The calculation for the simple model clearly demonstrates the appearance of these levels in the absorption spectrum at intermediate electric fields. For the superlattice under consideration, the experimental evidence of these levels is given in Fig. 6.10 for fields of 16 kV/cm to 35 kV/cm.

A further increase in the field causes the localization of electrons in the same manner as occurred for the holes. Figure 6.9(c) illustrates the interband transitions for this case. Now, the phototransitions and the absorption spectrum are those of a single quantum well. The absorption coefficient has a simple step-like shape. Experimental results are shown in Fig. 6.10 for $F = 167$ kV/cm, where both types of transition, hh $\leftrightarrow$ c and lh $\leftrightarrow$ c, are clearly seen. The evolution of the spectrum is dramatic: from two partially overlapping bands through the multi-band behavior and back again to two more peaked and highly resolved bands.

Plotting the positions of the observed peaks of the absorption coefficient as functions of the electric field, one can obtain the field dependence of the Stark ladder of Eq. (6.7). In this manner, the results obtained for the superlattice under consideration are collected in Fig. 6.11, which apparently shows the splitting of the electron miniband into a discrete level series and a linear displacement of the levels with increasing field (the width of the hole miniband is negligible). The numbers on the curves indicate different "steps" in the Stark ladder. Juxtaposing these results with those of Fig. 6.9(b), one can conclude that in Fig. 6.11 transitions "to the right" – as illustrated in Fig. 6.9(b) – over six quantum wells are present at low electric fields. The transitions "to the left" – as illustrated in Fig. 6.9(b) – are less observable for this particular heterostructure; however, three of them can clearly be seen in Fig. 6.11. (The transitions concerned are between the well labeled 0 in Fig. 6.9(b) and wells to the right – negative integers – and wells to the left – positive integers.) These results demonstrate that the electron wave functions have spatial coherence over no less than from 10 to 13 periods of the superlattice. With increasing electric field the high-order transitions disappear and only the vertical transition, 0, remains. The reasons for this behavior include the fact that there is a suppression of the coherent transverse motion as a result of a progressive localization of the electrons within several wells, and the fact there are low rates of tunneling and scattering processes. In superlattices grown by the most advanced technology, coherence lengths above 10 periods have been measured even at room temperature. The physical processes that we have just discussed result in a strong electro-optical effect. A typical blue shift of the absorption spectra is demonstrated in Fig. 6.10. The effect can be used in intensity modulators with a large contrast between "on" and "off" states.

6.3 Terahertz Coherent Oscillations of Electrons in an Electric Field

In this section we shall briefly discuss long-wavelength emission, which can be realized in quantum structures. For analysis of emission in the microwave and terahertz range one can use a semiclassical approach. Indeed, it is well known that classical vibrations of charged electrons emit microwave radiation. Semiclassical oscillations of the electric charge can be realized, for example, by the creation of an electron wave packet as a superposition of several states; this wave packet oscillates in real space.

Such oscillations and microwave emission have been studied for heterostructures which contain two quantum wells. Let us examine the case of coupled double quantum wells. Electron states in such structures were analyzed in Sections 3.6 and 3.9. In such a system of coupled wells, any electron state can be considered as a superposition of two states which originate from each of the wells. The coupling is due to the tunneling mechanism. For example, in the case of symmetric wells we have the splitting of initially degenerate states $\chi_{1,2}(z)$ to form the following two states:

$$\psi^{+}(t,z) = \frac{1}{\sqrt{2}}(\chi_1 + \chi_2)e^{it\epsilon_{+}/\hbar} \quad \text{and} \quad \psi^{-}(t,z) = \frac{1}{\sqrt{2}}(\chi_1 - \chi_2)e^{it\epsilon_{-}/\hbar}, \qquad (6.8)$$

with energy difference $\Delta\epsilon = \epsilon_{-} - \epsilon_{+} \equiv 4|T_{\mathrm{t}}|$, where T_{t} is the tunneling matrix element and ϵ_{+} and ϵ_{-} are the symmetric and antisymmetric electron states, respectively.

Now, let us suppose that, at $t=0$, we put the electrons inside the left well; that is, we populate an electron state with wave function

$$\Psi(t=0,z)=\chi_1(z).$$

Such a state is not an eigenstate of our problem and it should change in time. If we assume that other energy levels are situated quite far from ϵ_- and ϵ_+, we can write the time-dependent wave function $\Psi(t,z)$ as a superposition of the eigenstates of Eq. (6.8):

$$\Psi(t,z)=A\psi^+(t,z)+B\psi^-(t,z). \tag{6.9}$$

The coefficients A and B can be found from the initial condition at $t=0$ and the normalization condition. As a result of such a coherent superposition of states, the probability of finding the electrons at some point z, as determined by $|\Psi(t,z)|^2$, oscillates in space. That is, at some point z we get

$$|\Psi(t,z)|^2=\frac{1}{2}\left[|\chi_1(z)|^2\left(1+\cos\left(\frac{\Delta\epsilon}{\hbar}t\right)\right)+|\chi_2(z)|^2\left(1-\cos\left(\frac{\Delta\epsilon}{\hbar}t\right)\right)\right]. \tag{6.10}$$

Thus we see that the electric charge oscillates coherently in space with frequency $\Omega=\Delta\epsilon/\hbar$. For example, at $t=0$ we have $|\Psi(z)|^2=|\chi_1(z)|^2$, i.e., the electrons occupy the first quantum well, whereas at $t=\pi/2\Omega$ we find that $|\Psi(z)|^2=|\chi_2(z)|^2$, i.e., the electrons have transferred to the second well, and so on. These oscillations of the charged electrons should give rise to the emission of microwave radiation.

In the case of asymmetric quantum wells, by applying the same procedure again we can obtain two states $\psi^-(t,z)$ and $\psi^+(t,z)$, but, since an asymmetric structure is assumed, the total probabilities of finding the electrons in the left and right wells are different for these wave functions, and $\Delta\epsilon$ depends both on the tunneling matrix element and on the splitting due to the asymmetry. The latter is important because an applied electric field controls this splitting and, therefore, the frequency of the charge oscillations.

In order to detect these oscillations, asymmetric double-quantum-well structures, which manifest energy-level splitting in the absorption spectra, are used. One chooses light with a photon energy that is in the vicinity of the energies of the split levels. A short light pulse with a duration less than the inverse frequency of oscillations, Ω^{-1}, predominantly populates one of the wells. By this means, an initial coherent superposition of electron states – a wave packet – is created. This wave packet then oscillates between the two wells. The oscillations persist after the initial pulse for a timescale of the order of a relaxation time. In fact, the existence of the holes generated during the pulse complicates the whole picture since a hole wave packet can also oscillate, but with another frequency and another decay time.

A particular example of this effect is the mixing between heavy-hole and light-hole valence subbands after optical excitation in a single quantum well. In Fig. 6.12 we sketch the light-hole and heavy-hole states and their envelope wave functions for a quantum well in an electric field. From this figure it can be seen that in an electric field the centroids (i.e., average coordinates) of the holes of the two subbands are

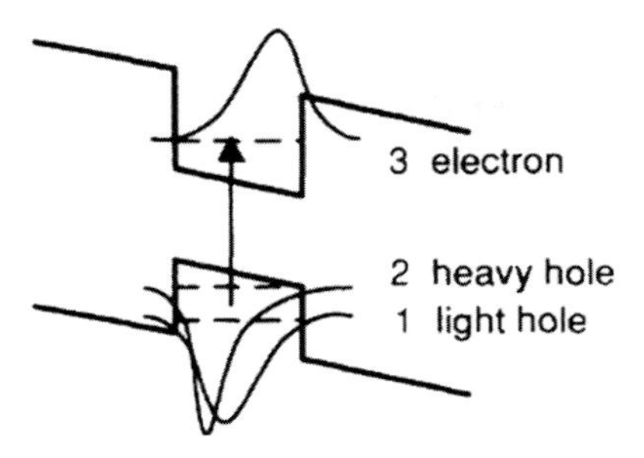

Figure 6.12 Wave function envelopes of the electron, light hole, and heavy hole in a quantum well biased with an electric field. Reprinted figure with permission from Fig. 1(a) of P. C. M. Planken, *et al.*, *Phys. Rev. Lett.* **69**, 3800, 1992. Copyright (1992) by the American Physical Society.

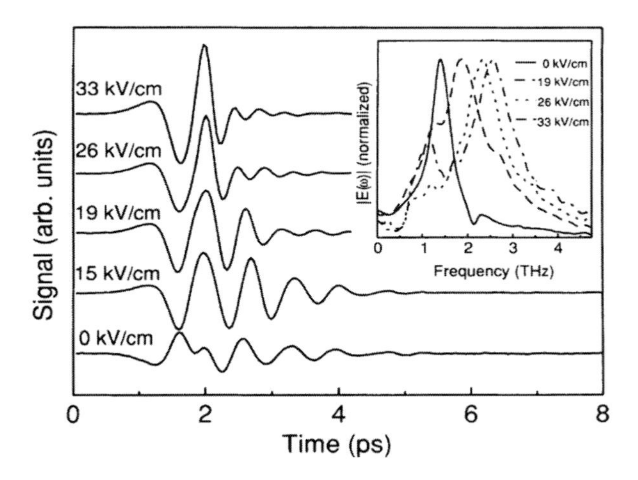

Figure 6.13 Measured terahertz wave forms for several electric fields in the MQW structure described in the text. The inset shows the Fourier transforms of only the oscillatory part of the signals. Reprinted figure with permission from Fig. 3 of P. C. M. Planken, *et al.*, *Phys. Rev. Lett.* **69**, 3800, 1992. Copyright (1992) by the American Physical Society.

different. Thus, as in the previous case, the time-dependent mixing of these subbands leads to a time-varying dipole moment.

In Fig. 6.13, we present measurements of the wave forms of terahertz emission of AlGaAs/GaAs MQWs placed in an electric field. The MQW structure consists of 15 periods of 175 Å GaAs wells separated by 150 Å $Al_{0.3}Ga_{0.7}As$ barriers. The MQW structure was incorporated into a Schottky diode, so that the light-hole–heavy-hole splitting was 6 meV, or 1.5 THz in the absence of an applied voltage. The experiment was conducted at low temperature ($\approx$10 K). The laser excitation was tuned to the light-hole and heavy-hole excitons. From Fig. 6.13 we can see that the amplitude of the oscillations increases when the applied electric field increases. The Fourier transforms presented in the inset of this figure show that the observed emission is in the terahertz range. The frequency of the oscillations changes from 1.4 THz at zero electric field to almost 2.6 THz at the highest electric field.

In conclusion, through the examples of a coupled double quantum well and a single quantum well, we have considered the effect of coherent oscillations of an electron wave packet, which generate electromagnetic radiation in the terahertz range. Similar oscillations can be excited in superlattices. Note that the terahertz frequency range is very interesting from a practical point of view.

6.4 Nonlinear Optics in Heterostructures

In this section, we consider optical effects of another class, which are related to the nonlinear propagation or reflection of light and constitute the so-called field of *nonlinear optics*.

We start with a study of the general nonlinear optical characteristics of materials. Then we consider the nonlinear optics of quantum heterostructures, focusing on the different physical mechanisms responsible for the effects. We consider both virtual and real populations of electron states in heterostructures. For the case of a real population, we introduce a hierarchy of characteristic times, which affords us the opportunity to classify a number of the processes leading to nonlinear effects and to understand their advantages and disadvantages in potential applications. Because the latter is an important subject in conventional optoelectronics, we provide a comparison of the two types of signal processing – all-optical and electronic. We show that the nonlinear optics of quantum heterostructures has great potential in signal-processing applications.

We begin by introducing basic ideas and results in this branch of optics, known as conventional nonlinear optics.

6.4.1 Linear and Nonlinear Optics

Throughout the previous sections, we considered linear optical phenomena. That is, previously we assumed that (1) the refractive index and absorption (amplification) coefficients are independent of the light intensity; (2) the light frequency is not altered during the passage of light through the medium; (3) the principle of superposition of waves holds; and (4) the light does not interact with any other light waves and, therefore, light cannot control light. These assumptions are valid and can be applied for a range of low-amplitude light waves.

In the opposite case, when the amplitude of the wave (and thus the light intensity) is large, light propagation through an optical medium can exhibit nonlinear behavior: (1) the refractive index and, consequently, the speed of light change with intensity; (2) light can alter its frequency; (3) the superposition principle fails; and (4) light can control light, i.e., photons interact with photons in a medium. Such a medium, which changes the behavior of light from linear to nonlinear, is known as *a nonlinear optical medium*.

In order to introduce parameters characterizing nonlinear optical properties, let us recall that the displacement vector, $\vec{\mathcal{D}}$, can be presented as

$$\vec{\mathcal{D}} = \epsilon_0 \vec{\mathcal{E}} + \vec{\Pi}, \tag{6.11}$$

where $\vec{\mathcal{E}}$ is the electric field of the light wave and $\vec{\Pi}$ is the polarization of a medium (see Section 4.2). In linear optical media, $\vec{\Pi}$, is simply proportional to $\vec{\mathcal{E}}$. For example, for an isotropic medium, we can write

$$\vec{\Pi} = \epsilon_0 \chi(\omega) \vec{\mathcal{E}} \quad \text{and} \quad \vec{\mathcal{D}} = \epsilon_0 \kappa(\omega) \vec{\mathcal{E}}, \tag{6.12}$$

where $\kappa(\omega) = 1 + \chi(\omega)$ is the dielectric constant and $\chi(\omega)$ is the electric susceptibility. These linear relations were employed widely in our previous discussions.

In the case when a material exhibits nonlinear properties, the latter can be accounted for with a nonlinear polarization, $\vec{\Pi}$, which, in general, is a functional of the electric field, $\vec{\mathcal{E}}$. The most general expansion of $\vec{\Pi}$ in a series with respect to $\vec{\mathcal{E}}$ is

$$\Pi_i = \epsilon_0 (\chi_{ij} \mathcal{E}_j + \chi^{(2)}_{ijl} \mathcal{E}_j \mathcal{E}_l + \chi^{(3)}_{ijlk} \mathcal{E}_j \mathcal{E}_l \mathcal{E}_k + \cdots), \tag{6.13}$$

where $\chi^{(2)}_{ijl}$ and $\chi^{(3)}_{ijlk}$ are coefficients describing second- and third-order nonlinear effects.

When the optical medium is nonlinear, the linear equation for light waves is no longer valid. For an isotropic medium we can rewrite $\vec{\Pi}$ as a sum of a linear part and a nonlinear contribution,

$$\vec{\Pi} = \epsilon_0 (\kappa - 1) \vec{\mathcal{E}} + \epsilon_0 \vec{\mathcal{P}}_{\mathrm{NL}}; \tag{6.14}$$

then the wave equation for $\mathcal{E}$ can be represented in the form

$$\vec{\nabla}^2 \vec{\mathcal{E}} - \frac{\kappa}{c^2} \frac{\partial^2 \vec{\mathcal{E}}}{\partial t^2} = -\vec{\mathcal{S}}, \tag{6.15}$$

where

$$\vec{\mathcal{S}} \equiv -\frac{1}{c^2} \frac{\partial^2 \vec{\mathcal{P}}_{\mathrm{NL}}}{\partial t^2}.$$

In this way, we keep the left-hand side of the wave equation linear, while the term $\vec{\mathcal{S}}$ is nonlinear. This term can be regarded as a source for the generation of electromagnetic fields different from the incident wave, including those with different frequencies, wavevectors, etc.

In contrast to the linear wave equation, Eq. (6.15) cannot be solved analytically in the general case. For each particular case it is necessary to develop a proper approach for analysis and solve the nonlinear Eq. (6.15). The most common method is the so-called *Born approximation*, which is similar to that used in perturbation theory in quantum mechanics. This approximation is an iterative procedure, wherein, for the first step, the incident wave is used for calculation of the right-hand side of Eq. (6.15), i.e., the "source" $\vec{\mathcal{S}}$. Thus, the first nonlinear correction to the incident wave is estimated as a solution of the linear wave equation with a given right-hand side, $\vec{\mathcal{S}}$, which is a nonlinear function of the amplitude of the incident wave. After finding this first-order correction the procedure should be iterated until one has achieved the accuracy necessary for the description of the nonlinear effect under consideration.

If this method cannot be applied, an analysis of the particular physical situation can sometimes lead to an approach for solving the specific problem at hand. For example, such an analysis can be based on a comparison of characteristic times. If a physical process in a material is described by a relaxation time, τ_{ch}, which is larger than the characteristic time for changes in the light wave's amplitude, t_{lw}, and the inverse of its frequency, ω,

$$\tau_{\mathrm{ch}} > t_{\mathrm{lw}}, \omega^{-1}, \tag{6.16}$$

then a set of susceptibilities of different orders can be introduced in accordance with Eq. (6.13), and the Born approximation can be used.

Another limiting case corresponds to the situation when the relaxation time is less than t_{lw}:

$$\omega^{-1} \ll \tau_{\mathrm{ch}} < t_{\mathrm{lw}}. \tag{6.17}$$

For this case, time-averaged parameters of the wave should specify the process. That is, the intensity of the wave governs the nonlinear effect. It can be taken into account by introducing a refractive index and an absorption coefficient that depend on the intensity of the wave, $\mathcal{I}$:

$$n = n(\omega, \mathcal{I}) \quad \text{and} \quad \alpha = \alpha(\omega, \mathcal{I}). \tag{6.18}$$

Since $\mathcal{I} \sim \overline{\mathcal{E}^2}$, Eq. (6.15) can be transformed into an equation for the wave amplitude with a dielectric function $\kappa \equiv n^2$ which depends on the square of the amplitude. Thus, one obtains another kind of nonlinear equation for $\mathcal{E}$; that is, $\kappa(\overline{\mathcal{E}^2})$ provides a new nonlinear relation, while in Eq. (6.15) one should use $\vec{\mathcal{S}} = 0$.

6.4.2 Optical Nonlinearities in Quantum Wells

Though there are a several different mechanisms which cause optical nonlinearities in semiconductors and semiconductor heterostructures, a nonlinear optical response depends on whether a *virtual* or *real* excited-state population is involved in the process.

If the light frequency is well below the absorption edge, the light wave induces *virtual excitations*: the electric field of the light brings about a coherent polarization of the material that persists only as long as the field is applied. The foregoing analysis is valid independently of whether the wave amplitude is large or small. But, if the wave amplitude is large, the induced polarization can couple various optical fields, providing a photon–photon interaction mediated via the material, etc. During the action of the field, the state of the crystal is a coherent superposition of its excited states. Such a coherence exists until relaxation processes cease to be active. In the opposite case, the quantum-mechanical phase of the excitation will be lost, i.e., there is absorption of a real photon. One can perform a very simple analysis based on the uncertainty principle. In a coupled system of an excited material and a light wave there is an oscillation of energy between these two parts. The frequency of the oscillation is determined by the detuning parameters,

$$\Delta\omega = (E - \hbar\omega)/\hbar,$$

where E is the excited energy level of the material. As long as this detuning is much greater than the inverse relaxation time, $1/\tau_{ch}$, the mechanism of the nonlinearity is predominantly virtual. Thus, large detuning below the band edge favors the virtual picture of the interaction of light and a semiconductor. According to this analysis, the virtual mechanism corresponds to the inequality of Eq. (6.16). In such a case, field-induced nonlinearities usually correspond to the slightly off-resonance case, but they are extremely fast; their characteristic times can be estimated as $(\Delta\omega)^{-1}$.

For a photon energy above the fundamental absorption edge, the predominant effect is the generation of a real population of excited states. The mechanisms change from virtual to real as a result of the fact that, within the detuning interval $\hbar\,\Delta\omega = \hbar\omega - E_g$, there are many electron states which absorb the energy of the light; then, the generated carriers scatter it. This strong absorption produces large populations, leading, in turn, to changes in optical properties. However, these changes are *incoherent* and depend only on the intensity of the wave as for the case of the inequality of Eq. (6.17). The populations of excited states have a finite lifetime and suffer various relaxation processes. Changes in optical spectra induced by such an excitation persist as long as the populations exist themselves; in particular, they can exist even when the light pulse is switched off rapidly. For the particular cases of semiconductors or their heterostructures, the indicated excitations are excitons and electron–hole pairs, which form an electron–hole plasma.

6.4.3 Virtual, Field-Induced Mechanism of Nonlinear Optical Effects

Let us consider light with a frequency well below the absorption edge so that we have the off-resonance case. As was discussed previously, in this case one can expect a coherent nonlinear response. Since the exciton levels also lie below the fundamental absorption gap, the contribution of the excitons in such an effect should be dominant. The simplest way to explain the nature of this virtual mechanism is to recall the close analogy between excitons and the hydrogen atom, which was emphasized in Chapter 3. From quantum mechanics, we know that a hydrogen atom placed in an ac electric field of amplitude F_0 manifests shifts of its energy levels known as *the ac Stark effect*. This effect is quadratic in the amplitude of the electric field. The shift and its sign depend strongly on the frequency of the light. The effect can be calculated to second order by quantum-mechanical perturbation theory. Thus, in the case of the exciton, one can also use second-order perturbation theory and obtain the shift of the exciton energy:

$$\delta E_{ex} = A \frac{|\mathcal{D} F_0|^2}{E_{ex} - \hbar\omega}, \tag{6.19}$$

where $\mathcal{D}$ is the interband matrix element (see Section 4.4), and F_0 is the amplitude of the light wave. The multiplier A is positive and dependent on the properties of the material, the heterostructure, etc. This result demonstrates that, if $\hbar\omega$ lies below the exciton energy, E_{ex}, the energy shift is positive, i.e., a blue shift appears. In order to examine this phenomenon further, the following experiment was performed. A short, high-power light beam (a pumping beam) with a frequency below the exciton line illuminates an AlGaAs/GaAs quantum well. Transient changes in the light transmission are measured with another relatively weak probe beam. The latter can be detuned either slightly below or above the exciton line. The results are presented in Figs. 6.14(a) and (b). The positions of two exciton lines associated with heavy holes, hh, and light holes, lh, and also the frequencies of two probe beams, $\omega_{probe} = \omega_a,\ \omega_b$, are identified in the insert. The main changes which can be seen in the transmission persist during the action of the pump; there is also a much weaker, long-time effect due to the photogeneration of an electron–hole plasma. In accordance with Eq. (6.19), in the region slightly below the exciton, as in Fig. 6.14(a), the transmission increases, while in the region slightly above the exciton, the transmission decreases as in Fig. 6.14(b). The results clearly demonstrate a blue shift of the exciton spectrum during illumination by the pump beam. In principle, these changes can be expressed in terms of a third-order susceptibility, $\chi^{(3)}$, but one can see that this quantity has very complex dependences on both of the frequencies, ω_{pump} and ω_{probe}. Figure 6.15(a) depicts the transmission as a function of the detuning energy, while Fig. 6.15(b) shows clearly that the nonlinear effect is of third-order and, accordingly, changes in the transmission are proportional to F_0^2 and they increase near the resonance condition $\hbar\omega = E_{ex}$ in accordance with Eq. (6.19).

6.4.4 Nonlinear Optical Effects Due to Generation of Excitons and Electron–Hole Plasma

If the light frequency is above the absorption edge, other mechanisms of nonlinear effects come into play. These mechanisms are due to the generation of real excitons and electron–hole pairs, which form a plasma. Before consideration of these mechanisms of a nonlinear response, let us classify the possible relaxation processes which can occur when real electrons and holes are generated optically.

The possible processes have very different timescales, which cover more than 10 decades. The real particles can be excited by a light pulse with a duration ranging from 10–100 fs to the continuous-wave regime. The lifetime of the excitations is of the order of 100 ps to 100 ns. Thus, there is a very wide time range, which should be analyzed. Let us suppose that the pulse is ultrashort. Since the time between electron-scattering events can be estimated as $\tau_{dph} = 100$–1000 fs, for $t < \tau_{dph}$ the behavior of the system is more like the virtual, or coherent, stage which was considered previously. During the coherent stage the carrier distribution is determined mainly by the spectrum of the pump beam. Then, a dephasing of the initial excitation occurs (for $t > \tau_{dph}$), the system absorbs photons, and we get real particles in both of the bands, with almost classical distribution functions centered around

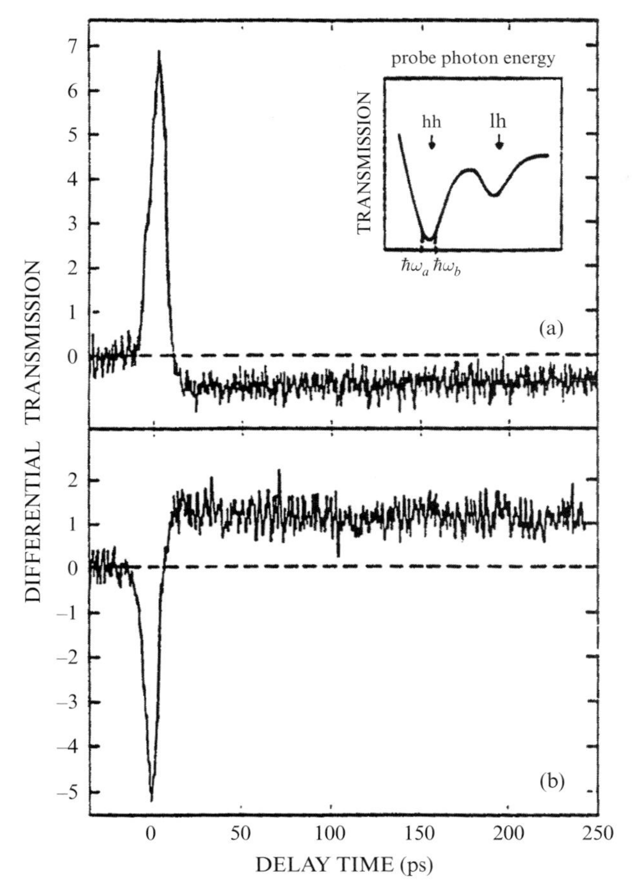

Figure 6.14 The differential transmission of an AlGaAs/GaAs quantum well as a function of the delay time between the pump and probe beams. (a) Transient changes in the light transmission measured with another relatively weak probe beam detuned slightly below the exciton line. (b) Transient changes in the light transmission measured with another relatively weak probe beam detuned slightly above the exciton line. The pump light has a photon energy approximately 25 meV below the hh-exciton resonance. The photon energies of two probe beams, a and b, of frequencies ω_a and ω_b are indicated in the insert, where the transmission under equilibrium is plotted. Both measurements show a blue shift of the exciton spectrum under such a virtual excitation. Reprinted with permission from figure 1 of *Opt. Lett.* **11**, 609 (1986), The Optical Society of America.

energies corresponding to those consistent with the conservation laws,

$$E_c(\vec{k}_c) - E_v(\vec{k}_v) = \hbar\omega, \quad k_c = k_v.$$

In the next stage, the particles exchange their momenta and energies and their distribution functions relax to Fermi functions, but their effective temperature, T_e, is far from that of the crystal; this situation defines the third relaxation stage.

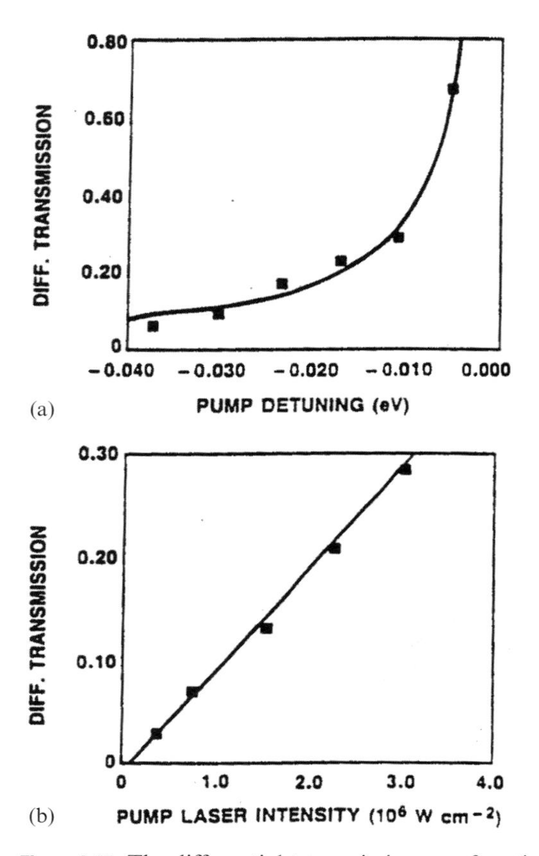

Figure 6.15 The differential transmission as a function of (a) the photon energy of the pump light and (b) the pump intensity. The latter case corresponds to fixed detuning 18 meV below the hh-exciton resonance. Reprinted with permission from figure 3 of *Opt. Lett.* **11**, 609 (1986), The Optical Society of America.

Both the second phase and the third phase correspond to a so-called nonthermalized plasma. At longer times, the electrons and holes lose excess energy and relax to distributions with the crystal temperature; at the end of this "cooling" stage one has a thermalized electron–hole plasma, and the electron–hole pairs begin to form excitons; this sequence defines the fourth stage. During the fifth, and last, stage, both the free electron–hole pairs and the excitons recombine toward the true equilibrium.

Now we consider the simplest case, which corresponds to the fifth stage discussed above. This case occurs when the excitons and the electron–hole plasma are *thermalized*, i.e., equilibrium within the conduction and valence bands is assumed to have been established. (In terms of the above brief analysis, this corresponds to the recombination stage.) In this case, only the concentrations of excitons, n_{ex}, and the electron–hole plasma, $n = n_c = n_v$, characterize the nonequilibrium state of the system. The optical parameters of the system can be regarded as functions of these concentrations which, in the turn, are dependent on the light intensity. Thus, one gets optical nonlinearity in the form of Eq. (6.18) via the generation of the concentrations of real electrons and holes.

In fact, the effect of the electrons and holes on the optical spectra was discussed previously, in Chapter 4 and Section 6.2, when we studied the electro-optical effect (see also Fig. 6.7, the curve marked N_s up). Optically generated carriers affect the spectra in the same manner. Both of the previously discussed phenomena, the filling factor and the screening of the electron–hole interaction, are important physical reasons for such an effect.

The filling factor reflects only the fact that the absorption coefficient is proportional to the difference in the distribution functions of generated holes and electrons: $\mathcal{F}(E_v) - \mathcal{F}(E_c)$. Populations of some states in the conduction and valence bands decrease this difference and lead to bleaching of the spectral regions corresponding to the transitions between populated states. When nonequilibrium carriers are thermalized within both of the bands, the mechanism leads to bleaching of the absorption edge and effectively increases the optical bandgap; this is the so-called *Burstein–Moss effect*.

The mechanism related to exciton screening has a more complex explanation. It depends on whether excitons or electron–hole pairs are excited. For the two-dimensional case we can assign an effective disk with radius a_{ex}, where a_{ex} coincides with the exciton radius. If exciton states are excited directly – by a resonant absorption of the pump beam – and there is no plasma, we can imagine that those disks are hard, so two disks cannot occupy the same space. If $\pi a_{ex}^2 n_{ex} \ll 1$, excitons do not interfere with one another. In the opposite case, the excitons are destroyed and the exciton peaks in the spectra are washed out. More accurate estimates give a threshold exciton concentration, above which the excitons disappear:

$$n_{ex,th} = \frac{0.117}{\pi a_{ex}^2}. \tag{6.20}$$

Thus, if one resonantly generates excitons, there is a threshold light intensity, $\mathcal{I}_{th} \approx \hbar\omega n_{ex,th}/\tau_{ex}$ (τ_{ex} is the lifetime of the exciton), beyond which the exciton absorption disappears as a result of the bleaching effect.

If free electron–hole pairs are generated, one can use for estimates a screening length that depends on the concentration. The meaning of this length is as follows. As the distance from an electric charge exceeds the screening length, the electric field of this charge is suppressed. Therefore, if the screening length becomes less than the exciton radius, the interaction between the electron and the hole decreases and that brings about the destruction of the exciton. For two-dimensional carriers and a low temperature, T, the screening length coincides with the exciton radius, a_{ex}, whereas at high temperature the screening length is proportional to $\sqrt{T}$. Calculations give the following criterion for destroying the excitons:

$$n\pi a_{ex}^2 \geq 0.056, \quad \text{if } k_B T \ll Ry_{ex}^*, \tag{6.21}$$

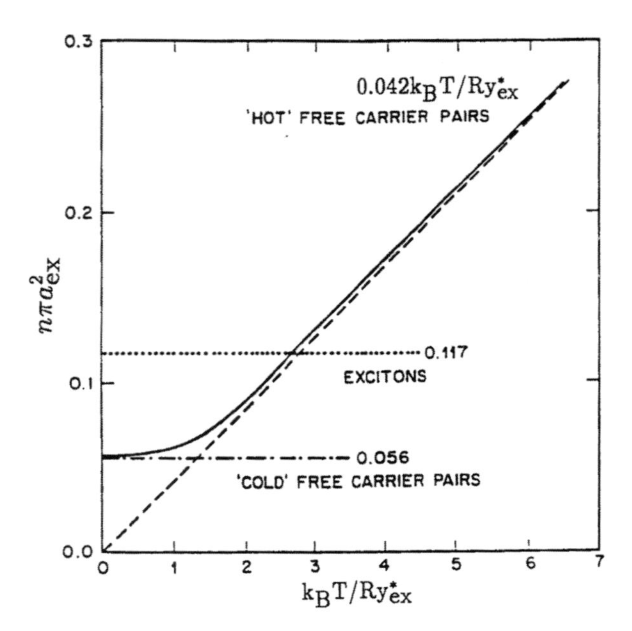

Figure 6.16 The concentration–temperature phase diagram for the existence of two-dimensional excitons. The concentration is given in units of πa_{ex}^2 and the temperature in units of Ry_{ex}^*. The dotted line shows the limitation associated with the criterion of Eq. (6.20). The solid line presents limitations due to plasma screening. The low-temperature and high-temperature limits correspond to Eqs. (6.21) and (6.22), respectively. Reprinted figure with permission from Fig. 1 of S. Schmitt-Rink, *et al.*, *Phys. Rev.* **B 32**, pp. 6601–6609, 1985. Copyright (1985) by the American Physical Society.

$$n\pi a_{\mathrm{ex}}^2 \geq 0.042\frac{k_{\mathrm{B}}T}{Ry_{\mathrm{ex}}^*}, \quad \text{if } k_{\mathrm{B}}T \gg Ry_{\mathrm{ex}}^*, \tag{6.22}$$

where Ry_{ex}^* is the coupling energy of the exciton. Using these results, one can obtain a "phase diagram" for the existence of an exciton. Because three essential parameters, n_{ex}, n, and T, control the effects, this phase diagram, in principle, should be a three-dimensional plot. We present this diagram as projections on an n, T plane as shown in Fig. 6.16. The results can be understood as follows. If n is the plasma concentration, the excitons exist only below the solid line. This line represents the screening effect, with the limiting cases of Eqs. (6.21) and (6.22) marked as contributions of "cold" and "hot" free-carrier pairs. But the exciton concentration, n_{ex}, cannot exceed the threshold of Eq. (6.20). The latter is represented by the dotted line in Fig. 6.16 for comparison.

Thus, the solid curve divides the n, T plane into two regions. Above the curve, the excitons and their features in the optical spectra vanish. Below the curve, there exist almost free excitons and exciton features in the optical spectra, if the exciton subsystem is weakly excited. At high excitation, the excitons exist only below the dotted line. A similar diagram can be plotted for the intensity–temperature variables; that is, there is a similar diagram in the $\mathcal{I}$, T plane.

In conclusion, in the case of thermalized photo-excitations, nonlinear optical effects manifest themselves as a vanishing of excitons in the spectra for high light intensities.

6.4.5 Nonlinear Effects Induced by Nonthermalized Electron–Hole Plasma

The processes of the thermalization of photo-excited plasmas and excitons take times ranging above tens of picoseconds. Thus, applications of these effects are restricted to sub-nanosecond or nanosecond timescales. Meanwhile, advanced optical techniques make it possible to generate light beams with temporal durations in the femtosecond range. Therefore, this technology facilitates the investigation and exploitation of femtosecond phenomena.

In the femtosecond regime, all physical processes, including nonlinear optical processes, differ greatly from those of the thermalized case. As discussed previously, the reason for such a difference is that the electron–hole pairs are generated initially in the continuum of the conduction band as well as that of the valence band; subbands serve as bands for the case of quantum well structures. In both cases, the states are populated in relatively narrow energy regions. The electrons and holes then begin to relax toward the bottom of the conduction band and the top of the valence band, respectively. The next stage of the relaxation process is the formation of coupled electron–hole pairs, i.e., excitons, which now can be screened dynamically by the relaxing plasma. As the electrons undergo transitions to lower energies and the holes make similar transitions as they are transferred in the direction of the band edge, there are significant time-dependent changes in the optical properties of the semiconductor medium. In sharp contrast with the other cases, there is a pronounced delay with respect to the initial, short femtosecond optical pulse. For a structure with multiple GaAs quantum wells, these fast changes have been examined on the basis of a pump–probe method employing two light beams. One such example is illustrated in Fig. 6.17. The first pulse – the higher-power pump pulse, which has an energy 20 meV above the ground state of the two-dimensional exciton but below the excited state associated with the next subband – has the spectrum shown in Fig. 6.17. A probe beam with a broad spectral band served for recording the light transmission with a controlled delay, Δt, between the two pulses. The differential transmission is plotted for several values of Δt. One can see that for $\Delta t \leq -50$ fs (just at the front of the pump) a "spectral hole" in the spectral region of the pump is formed in the transmission, i.e., there is an increase in the transmission. This spectral hole corresponds to the population of some states within the subbands. The created electron–hole pairs immediately screen the exciton lines belonging to the first and second subbands; for this time interval in the transmission there is no exciton absorption. In this phase, there are only small changes in the transmission. In the next time interval, the spectral hole shifts down to lower energies, which correspond to a progressive relaxation of the electron–hole pairs. Within about 200 fs, a transmission spectrum which exhibits bleaching near the fundamental absorption edge is formed. This corresponds to the case with the carriers occupying the lowest

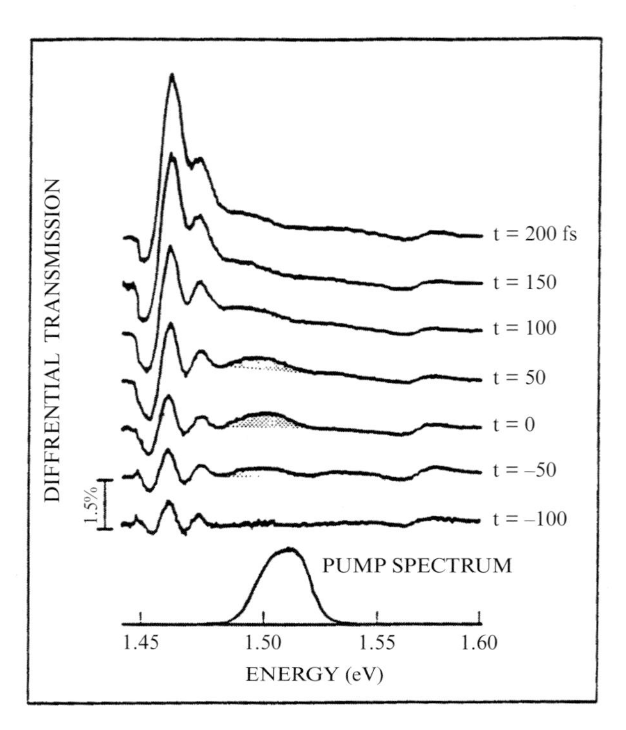

Figure 6.17 The differential transmission of excited AlGaAs/GaAs quantum wells as a function of photon energy. The transmission was measured with a broad spectral band probe pulse of 50 fs duration. The temporal dependence of the spectra was measured with 50 fs intervals before and after the excitation. At the bottom of the figure the spectrum of the pump light is shown. The bleaching of optical spectra near the fundamental absorption edge ($\hbar\omega = 1.455$ eV) demonstrates the dynamics of the thermalization of the photo-excited plasma. Reprinted from Fig. 4 of *Hot Carriers in Semiconductor Nanostructures: Physics and Applications*, J. Shah, *et al.*, "IV.2 – Optical studies of femtosecond carrier thermalization in GaAs," pp. 313–344, Copyright 1992, with permission from Elsevier.

energy states in both bands. The short excitation results in a strong suppression of the exciton absorption and a bleaching of the band-to-band absorption edge. Thus, this experiment supports the existence of a strong and fast nonlinear dynamic effect in the absorption spectra in the femtosecond time domain.

6.5 Plasmonics and its Peculiarities in Nanostructures

Plasmonics deals with ways to confine and control electromagnetic signals on dimensional scales smaller than or of the order of the wavelength of the electromagnetic field. Plasmonics exploits process where electromagnetic fields and conduction electrons interact at an *interface* between dielectric and conducting materials (metals, semiconductors), or in *nanostructures* to enhance optical fields in the subwavelength domain. In many instances such interactions are associated with surface localized plasmons and surface plasmon-polaritons. Accordingly, we shall describe both of these effects after a brief review of some of the basic properties of plasmons in bulk metals.

6.5.1 The Dielectric Permittivity of a Free-Electron Gas

Consider a system consisting of a free-electron gas and a fixed background of positive-ion cores. The total charge of the system is zero. Such a system is called a *plasma*. In a plasma, the electrons move with respect to the positive fixed charges. It is known that, in a plasma, various collective modes are possible. We consider long-wavelength modes in a three-dimensional plasma system. The motion of an electron in the presence of a time-dependent electric field, $\vec{\mathcal{E}}(t)$, is described by the Newton equation

$$m\frac{d^2\vec{r}}{dt^2} + m\gamma\frac{d\vec{r}}{dt} = -e\vec{\mathcal{E}}(t), \tag{6.23}$$

where $\vec{r}$ is the electron coordinate vector, and m is the mass of the electron. Damping processes are accounted for by the second term in Eq. (6.23), with a phenomenological damping constant, γ. Assuming harmonic time dependences, $\vec{\mathcal{E}}(t) = \vec{\mathcal{E}}_\omega \exp(-i\omega t)$ and $\vec{r}(t) = \vec{r}_\omega \exp(-i\omega t)$, with angular frequency ω, we obtain the following solution for $\vec{r}$:

$$\vec{r}_\omega = \frac{e}{m(\omega^2 + i\gamma\omega)}\vec{\mathcal{E}}_\omega.$$

The polarization of the system caused by n such electrons is $\Pi(t) = -e\vec{r}(t)n$, and it follows that

$$\vec{\Pi}_\omega = -\frac{e^2 n}{m(\omega^2 + i\gamma\omega)}\vec{\mathcal{E}}_\omega. \tag{6.24}$$

According to Eqs. (4.3) the dielectric displacement is $\vec{\mathcal{D}}(t) = \epsilon_0\vec{\mathcal{E}}(t) + \vec{\Pi}(t)$; this implies that

$$\vec{\mathcal{D}}_\omega = \epsilon_0\left[1 - \frac{e^2 n}{\epsilon_0 m(\omega^2 + i\gamma\omega)}\right]\vec{\mathcal{E}}_\omega = \epsilon_0\kappa_{\rm pl}(\omega)\vec{\mathcal{E}}_\omega,$$

where we have introduced the relative dielectric permittivity of the plasma system,

$$\kappa_{\rm pl}(\omega) = 1 - \frac{e^2 n}{\epsilon_0 m(\omega^2 + i\gamma\omega)} = 1 - \frac{\omega_{\rm pl}^2}{\omega^2 + i\gamma\omega}. \tag{6.25}$$

Here $\omega_{\rm pl} = \sqrt{ne^2/\epsilon_0 m}$.

For *longitudinal oscillations* of the plasma, a nontrivial solution to Eq. (4.1), $\vec{\nabla}\vec{\mathcal{D}} = 0$, is possible if

$$\vec{\mathcal{D}} = 0 \quad \text{at } \kappa_{\rm pl}(\omega) = 0. \tag{6.26}$$

This implies that $\omega^2 + i\gamma\omega - \omega_{\rm pl}^2 = 0$. Thus, oscillations occur with frequencies

$$\omega = -i\frac{\gamma}{2} \pm \sqrt{\omega_{\rm pl}^2 - \frac{\gamma^2}{4}}. \tag{6.27}$$

Usually, the damping constant γ is small and we obtain

$$\omega \approx \pm\omega_{\rm pl} - i\gamma/2 \quad \text{at } \omega_{\rm pl} \gg \gamma. \tag{6.28}$$

The frequency $\omega_{\rm pl}$ represents the frequency of the oscillation mode in the bulk plasma, where the electrons oscillate collectively relative to the positive charges of the ions. Indeed, from Eq. (6.26) it follows that for $\vec{\mathcal{D}} = \epsilon_0\vec{\mathcal{E}} + \vec{\Pi} = 0$ we can put $\vec{\mathcal{E}} = -\vec{\Pi}/\epsilon_0 = e\vec{r}n/\epsilon_0$. Then, Eq. (6.23) can be rewritten as an equation for a damped oscillator with frequency $\omega_{\rm pl}$:

$$\frac{d^2\vec{r}(t)}{dt^2} + \gamma\frac{d\vec{r}(t)}{dt} = -\frac{e\vec{\mathcal{E}}(t)}{m} = \frac{e\vec{\Pi}(t)}{m} = -\frac{e^2n}{\epsilon_0 m}\vec{r}(t) = -\omega_{\rm pl}^2\vec{r}(t).$$

When $\gamma = 0$, this is an undamped harmonic mode of frequency $\omega_{\rm pl}$. The quantum of these oscillations is known as a *plasmon*, and the plasmon energy is $\hbar\omega_{\rm pl}$. For plasmons, the oscillations of the electric field are potential, i.e., $\vec{\nabla} \times \vec{\mathcal{E}} = 0$.

6.5.2 Plasmons in Metals

With a small modification, the model presented for the free-electron plasma can be applied to describe charge oscillations in various metals. The modification takes into account that in an electromagnetic field the ion cores also become polarized. This effect can be described by introducing a dielectric constant κ_∞ to substitute the first term in Eq. (6.25):

$$\kappa(\omega) = \kappa_\infty - \frac{\omega_{\rm pl}^2}{\omega^2 + i\gamma\omega}. \tag{6.29}$$

Estimates of κ_∞ show that, typically, $1 < \kappa_\infty < 10$. Table 6.1 summarizes the plasmon energies for a variety of metals.

From the data of Table 6.1, one can conclude that the optical frequency range is below the plasmon frequency $\omega_{\rm pl}$ for all metals, i.e.,

$$\omega < \omega_{\rm pl}. \tag{6.30}$$

Table 6.1 Plasmon energies for a selection of metals

Metal	Ag	Au	Zn	Cr	Cu	In	Al	Ni
$\hbar\omega_{\rm pl}$ (eV)	9.01	9.03	10.01	10.75	10.83	12.8	14.98	15.92

Data were taken from A. D. Rakic, *et al.*, *Appl. Optics* **37**, 5271 (1998) and E. J. Zeman and G. C. Schatz, *J. Phys. Chem.* **91**, 634 (1987).

As can be seen from Eq. (6.25), the relative dielectric permittivity of the plasma is a complex function, $\kappa(\omega) = \kappa'(\omega) + i\kappa''(\omega)$, with

$$\kappa'(\omega) = \kappa_\infty - \frac{\omega_{\text{pl}}^2}{\omega^2 + \gamma^2}, \tag{6.31}$$

$$\kappa''(\omega) = \frac{\omega_{\text{pl}}^2 \gamma}{\omega(\omega^2 + \gamma^2)}. \tag{6.32}$$

According to Eq. (6.30), it follows that for optical frequencies the real part of the dielectric permittivity, $\kappa'(\omega)$, is typically negative. The imaginary part of κ is determined by the damping constant γ, related to the electron scattering time, τ, as $\gamma = 1/\tau$. The latter can be estimated by using the low-frequency conductivity, $\sigma = e^2\tau n/m$; i.e., finally we find $\gamma = \epsilon_0\omega_{\text{pl}}^2/\sigma$. For the noble metals the effective masses of the electrons are close to the free-electron mass and the conductivities are about $(1\text{–}5) \times 10^7$ S/m. These numbers give the following estimate for the damping constants: $\gamma \approx (0.4\text{–}2) \times 10^{14}\ \text{s}^{-1}$, i.e., $\hbar\gamma \approx 0.4\text{–}2$ meV.

Figure 6.18 presents a comparison of the dielectric permittivity given by the previously discussed model with measurements for the case of one particular metal – gold. One can see that both the real part and the imaginary part of the permittivity for the model correspond to the results of measurements for energies less than 2–3 eV. That is, the simple model is valid through the infrared and visible spectral ranges. At larger energies the model is not correct, because it does not take into account the interband transitions in metals.

In Section 4.4.6, it was shown that in a uniform medium the spatial dependence of an electromagnetic field has a simple exponential form, with the exponential factor defined by the refractive index, $n(\omega)$ (see Eq. (4.86)). For a complex dielectric permittivity $\kappa(\omega)$, the refractive index is also a complex function: $n(\omega) = n'(\omega) - in''(\omega)$. The real and imaginary parts of $n(\omega)$ are expressed via $\kappa'(\omega)$ and $\kappa''(\omega)$ by Eqs. (4.88) and (4.89). It is instructive to consider the following two limiting

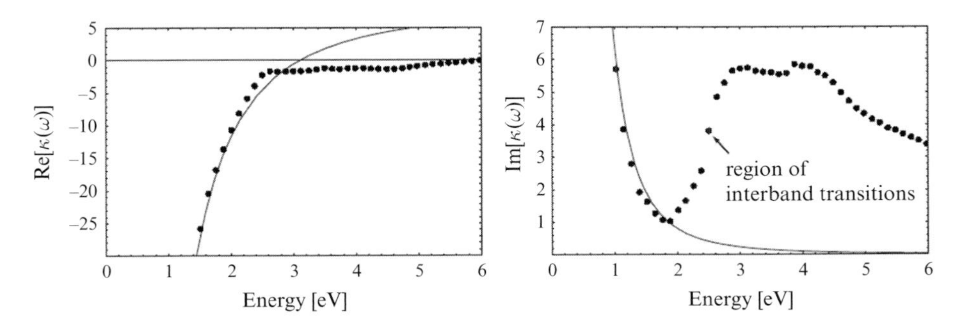

Figure 6.18 The relative dielectric permittivity of the electron gas fitted to the literature data for gold (solid line, calculations; dotted line, experiment). Interband transitions at high energy limit the validity of the presented model. Reprinted by permission from figure 1.1 of Springer Nature, *Plasmonics. Fundamentals and Applications* by S. A. Maier, Copyright 2007.

cases. The first is the case of low frequency, when $\omega \ll \gamma$. From Eqs. (6.31) and (6.32), it follows that $\kappa''(\omega) \gg |\kappa'(\omega)|$; thus $n'(\omega)$ and $n''(\omega)$ are of the same order of magnitude:

$$n' \sim n'' \approx \sqrt{\frac{\kappa''}{2}} = \sqrt{\frac{\omega_{\text{pl}}^2}{2\gamma\omega}}.$$

In this frequency region metals are mainly absorbing, and the text after Eq. (4.86) determines the absorption coefficient as

$$\alpha \approx \sqrt{\frac{2\omega_{\text{pl}}^2 \omega}{c^2 \gamma}}.$$

In the high-frequency range, $\omega \gg \gamma$, the imaginary part of the dielectric constant is much smaller than the real part, $|\kappa'(\omega)| \gg \kappa''(\omega)$, and

$$\kappa' = \kappa_\infty - \frac{\omega_{\text{pl}}^2}{\omega^2}. \tag{6.33}$$

In the complex refractive index, the real part, n', is large and dominating (see Eqs. (4.88) and (4.89)). Accordingly, reflection dominates over absorption and so the reflection coefficient of a metal is close to 1 (see Eq. (4.87)).

6.5.3 Surface Plasmon-Polaritons at Dielectric/Metal Interfaces

Because plasmonics deals with ways to confine and control electromagnetic waves on small-dimensional scales, we need to consider the propagation of electromagnetic waves in *nonuniform media*. We limit ourselves to the simplest case of a half-space dielectric ($z > 0$) adjacent to a half-space metal ($z < 0$), as illustrated in Fig. 6.19. Both media are supposed uniform. The dielectric has a positive and real dielectric permittivity κ_{d}, while the dielectric permittivity of the metal, $\kappa_{\text{m}}(\omega)$, is complex and $\kappa'_{\text{m}}(\omega) < 0$.

Analysis of the electromagnetic waves in such a nonuniform medium has to be based on the Maxwell equations (4.1)–(4.3). We will consider solutions of these equations describing waves propagating along the x-axis (see Fig. 6.19). Since the material system is uniform along the x coordinate, we will look for solutions for the electric field and the magnetic field strength in the following form:

$$\vec{\mathcal{E}} = \vec{\mathcal{E}}_0(z)e^{-i\omega t + i\beta x}, \quad \vec{\mathcal{H}} = \vec{\mathcal{H}}_0(z)e^{-i\omega t + i\beta x}. \tag{6.34}$$

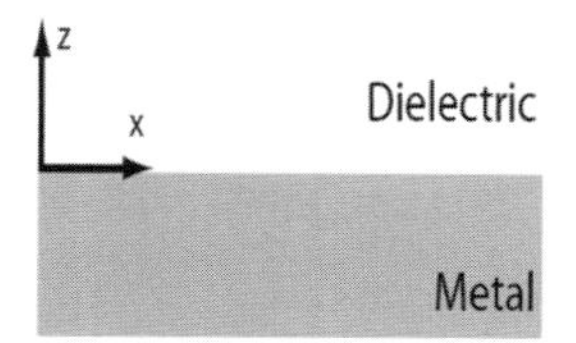

Figure 6.19 A half-space dielectric adjacent to a half-space metal.

Here $\vec{\mathcal{E}}_0(z)$ and $\vec{\mathcal{H}}_0(z)$ are unknown vector functions, and $\beta = \beta(\omega)$ is the propagation constant.

It can be shown that Maxwell's equations have two classes of solutions for $\vec{\mathcal{E}}_0(z)$ and $\vec{\mathcal{H}}_0(z)$ corresponding to two different polarizations of the waves. The first class consists of the transverse magnetic (TM) modes, where only two electric-field components, $\mathcal{E}_x$ and $\mathcal{E}_z$, and one magnetic-field component, $\mathcal{H}_y$, are nonzero. The second class consists of the transverse electric (TE) modes, with the following nonzero components: $\mathcal{H}_x$, $\mathcal{H}_z$, and $\mathcal{E}_y$.

It turns out that a confinement of electromagnetic fields is possible only for the TM class of modes. For the three unknown functions $\mathcal{E}_{0,x}(z)$, $\mathcal{E}_{0,z}(z)$, and $\mathcal{H}_{0,y}(z)$, Maxwell's equations read

$$\mathcal{E}_{0,x} = -\frac{i}{\epsilon_0 \kappa \omega}\frac{\partial \mathcal{H}_{0,y}}{\partial z}, \quad \mathcal{E}_{0,z} = -\frac{\beta}{\epsilon_0 \kappa \omega}\mathcal{H}_{0,y}, \tag{6.35}$$

$$\frac{\partial^2 \mathcal{H}_{0,y}}{\partial z^2} + \left(\frac{\omega^2 \kappa}{c^2} - \beta^2\right)\mathcal{H}_{0,y} = 0, \tag{6.36}$$

where $\kappa = \kappa_\mathrm{d}$ for $z > 0$ and $\kappa = \kappa_\mathrm{m}$ for $z < 0$.

Equation (6.36) has two exponential solutions, $\mathcal{H}_{0,y} \propto e^{\pm Kz}$, with

$$K \equiv \sqrt{\beta^2 - \frac{\omega^2 \kappa}{c^2}}. \tag{6.37}$$

Assuming that $\mathrm{Re}[K] > 0$, we may construct wave solutions confined to the interface, i.e., decaying far away from the interface, as follows:

$$\begin{gathered}\mathcal{E}_{0,x}(z) = \frac{iA_\mathrm{d}K_\mathrm{d}}{\epsilon_0 \kappa_\mathrm{d}\omega}e^{-K_\mathrm{d}z}, \quad \mathcal{E}_{0,z}(z) = -\frac{A_\mathrm{d}\beta}{\epsilon_0 \kappa_\mathrm{d}\omega}e^{-K_\mathrm{d}z}, \\ \mathcal{H}_{0,y}(z) = A_\mathrm{d}e^{-K_\mathrm{d}z}, \quad \text{at } z > 0;\end{gathered} \tag{6.38}$$

$$\begin{gathered}\mathcal{E}_{0,x}(z) = \frac{iA_\mathrm{m}K_\mathrm{m}}{\epsilon_0 \kappa_\mathrm{m}\omega}e^{K_\mathrm{m}z}, \quad \mathcal{E}_{0,z}(z) = -\frac{A_\mathrm{m}\beta}{\epsilon_0 \kappa_\mathrm{m}\omega}e^{K_\mathrm{m}z}, \\ \mathcal{H}_{0,y}(z) = A_\mathrm{m}e^{K_\mathrm{m}z}, \quad \text{at } z < 0.\end{gathered} \tag{6.39}$$

Here, A_d, A_m are still unknown constants, and K_d, K_m are the parameters given by Eq. (6.37) calculated for the dielectric and metallic media, respectively.

The constants A_d and A_m and the propagation constant (wavevector) β can be found from the boundary conditions at the interface at $z = 0$. Indeed, continuity of the magnetic field strength, $\mathcal{H}_y$, through the interface requires that $A_\mathrm{d} = A_\mathrm{m}$, and continuity of the electric-field component, $\mathcal{E}_x$, implies that

$$\frac{K_\mathrm{d}}{\kappa_\mathrm{d}} = -\frac{K_\mathrm{m}}{\kappa_\mathrm{m}}. \tag{6.40}$$

The latter relationship can be satisfied only if simultaneously $\mathrm{Re}[\kappa_\mathrm{d}] > 0$ and $\mathrm{Re}[\kappa_\mathrm{m}] < 0$; i.e., waves confined at an interface can exist only if the adjacent materials have the real parts of their dielectric permittivities of opposite signs, as realized here for the dielectric/metal system.

By substituting the expressions for K_d and K_m into Eq. (6.40), we find the propagation constant β as a function of the frequency:

$$\beta(\omega) = \frac{\omega}{c}\sqrt{\frac{\kappa_d(\omega)\kappa_m(\omega)}{\kappa_d(\omega)+\kappa_m(\omega)}}. \tag{6.41}$$

Here and below, we assume that ω is a real value. However, β is a complex value, $\beta = \beta' + i\beta''$, because $\kappa_m(\omega)$ at least is a complex function. The solution obtained is called the *surface plasmon-polariton*.

To understand better the dispersion relation of Eq. (6.41), let us analyze $\beta(\omega)$ for the actual case, $\kappa''_m \ll |\kappa'_m|$:

$$\beta(\omega) \approx \beta'(\omega) \approx \frac{\omega}{c}\sqrt{\frac{\kappa_d(\omega)\kappa'_m(\omega)}{\kappa_d(\omega)+\kappa'_m(\omega)}} = \frac{\omega\sqrt{\kappa_d(\omega)}}{c}\sqrt{\frac{1}{1-\kappa'_d(\omega)/|\kappa'_m(\omega)|}}. \tag{6.42}$$

This simplified equation facilitates drawing some qualitative conclusions concerning the obtained solution. Indeed, from Eqs. (6.41) and (6.42) it follows that at small ω/c the propagation constant approaches the wavevector of light in the dielectric medium, remaining, however, slightly larger. This latter fact means that the surface plasmon-polariton cannot transform into light; i.e, it is a form of *nonradiative* excitation. The largest value of β' is reached at

$$|\kappa'_m| \to \kappa'_d. \tag{6.43}$$

According to Eq. (6.33), the condition (6.43) is equivalent to

$$\omega \to \omega_{spl} = \frac{\omega_{pl}}{\sqrt{\kappa_\infty + \kappa'_d}}. \tag{6.44}$$

The characteristic frequency ω_{spl} has a clear physical interpretation. Indeed, consider the limiting case $\omega/c \to 0$, when the Maxwell equations reduce to the Poisson equation (see the first equation in the set of Eqs. (4.1)). As discussed in Section 6.5.1, this limit corresponds to longitudinal charge oscillations. From Eqs. (6.41) and (6.42), it can be is seen that, as $\omega/c \to 0$, the propagation constant, i.e., the wavevector of the excitation (the oscillations), can be finite, if the condition (6.43) is satisfied, which is possible at $\omega = \omega_{spl}$. Thus, in the limit of a large propagation constant (wavevector), the solution corresponds to electric-field and charge oscillations localized at the dielectric/metal interface, and ω_{spl} is the frequency of these surface plasma oscillations. Such longitudinal and purely potential oscillations are known as *surface plasmons*.

It is worth emphasizing that a surface plasmon-polariton is an electromagnetic excitation with oscillations of the charges, and of the electric and magnetic fields, while a surface plasmon consists of self-consistent oscillations of the charge and the electrostatic field.

The electric potential oscillations localized near the dielectric/metal interface can be obtained directly from the first Maxwell equation, $\vec{\nabla}\vec{\mathcal{D}} = 0$ (see Eq. (4.1) for

$\rho = 0$). Indeed, in each half-space, $z < 0$ and $z > 0$, presented in Fig. 6.19, in terms of the electrostatic potential, ϕ, this equation has the form of the Laplace equation:

$$\nabla^2 \phi \equiv \frac{\partial^2 \phi}{\partial x^2} + \frac{\partial^2 \phi}{\partial z^2} = 0. \tag{6.45}$$

For a wave propagating along the x-axis, as in Eq. (6.34), the solutions of this equation are

$$\phi_{\rm d}(x,z) = A_{\rm d} e^{i\beta x} e^{-\beta z} \text{ at } z > 0 \quad \text{and} \quad \phi_{\rm m}(x,z) = A_{\rm m} e^{i\beta x} e^{\beta z} \text{ at } z < 0,$$

with $\text{Re}[\beta] > 0$. The continuity of the potential through the interface and the tangential component of the electric displacement, $\mathcal{D}_z = -\kappa(\omega)\partial\phi/\partial z$, require $A_{\rm d} = A_{\rm m}$ and

$$\kappa_{\rm d}(\omega) + \kappa_{\rm m}(\omega) = 0. \tag{6.46}$$

For $\kappa_{\rm m}(\omega)$ given by Eq. (6.33), the latter equation determines the frequency of the surface plasmons given by Eq. (6.44). Note that, in the model being considered, the frequency of the surface plasmons does not depend on the propagation constant (wavevector), β. This means that surface plasmons with arbitrarily high wavevectors oscillate with the same frequency.

Now, we return to the surface plasmon-polaritons. Such an excitation at the interface between a metal and a dielectric material has the combined character of an electromagnetic wave and a surface charge, as illustrated in Fig. 6.20(a). The waves are a transverse magnetic field ($\vec{\mathcal{H}}_0$ is in the y-direction) and an electric field, $\vec{\mathcal{E}}_0$,

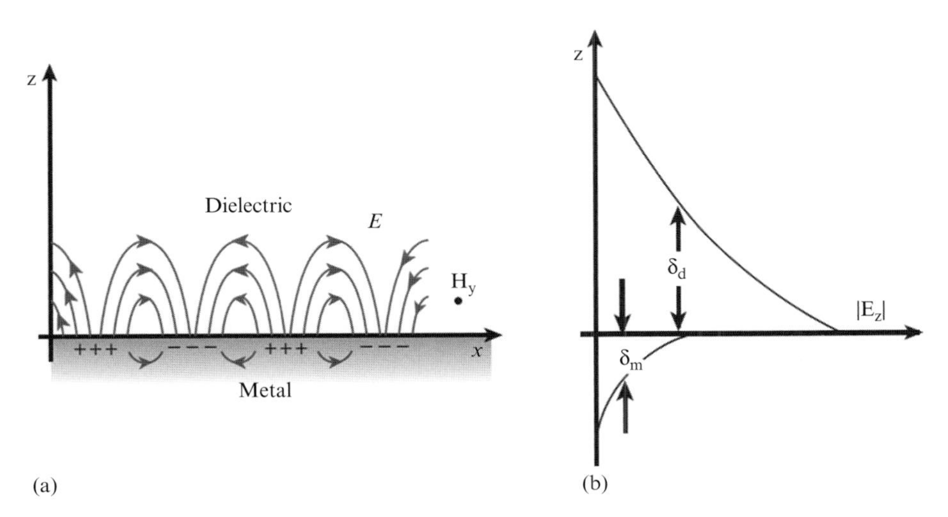

Figure 6.20 (a) The electric and magnetic fields, and surface charge in a surface plasmon-polariton. (b) The evanescent character of such an excitation illustrated for the normal component of the electric field. For the fields, the decay lengths in the dielectric, $\delta_{\rm d}$, and the metal, $\delta_{\rm m}$, are different. Reprinted by permission from Figs. (*a*) and (*b*) of Box 1 in Springer, *Nature*, "Surface plasmon subwavelength optics," W. L. Barnes, *et al.*, Copyright 2003.

with two components, $\mathcal{E}_{0x}$, $\mathcal{E}_{0z}$. The surface charge is localized near the surface, providing different amplitudes of the $\mathcal{E}_{0z}$ component in the dielectric and the metal. The electric and magnetic fields decay exponentially with distance away from the surface (see Fig. 6.20(b)), i.e., the fields are evanescent, corresponding to the bound nature of surface plasmon-polaritons.

Figure 6.21(a) demonstrates the general behavior of the dispersion relation, ω versus β', of Eq. (6.42) for a metal with negligible damping. To show that the dispersion relation of the plasmon-polariton is dependent also on the properties of the dielectric, results are presented for two interfaces: air/metal ($\kappa_d = 1$, $\kappa_\infty = 1$, 1–3 interface) and silica/metal ($\kappa_d = 2.2$, $\kappa_\infty = 1$, 2–3 interface). This figure also facilitates determination of the imaginary part of β at a given frequency (dashed curves). In the limit $\gamma \to 0$, β'' is nonzero only for frequencies from the interval $\omega_{spl} < \omega < \omega_{pl}$.

We see that the dispersion curves $\omega(\beta')$ never cross the dispersion curves $\omega = c\beta/\sqrt{\kappa_d}$ – the dispersion of a light wave in the dielectric medium – which corresponds to the localization of the surface plasmon-polariton at the interface. Special phase-matching techniques such as corrugation of the surface (to be discussed in Section 6.5.4, which deals extensively with graphene), or prism-coupling are required in order to provide the excitation by light and radiation of the surface plasmon-polariton.

The prism method can be implemented in two geometries: Otto's geometry and Kretschmann's geometry. Figure 6.21(b) depicts Kretschmann's geometry. In this case, the thin film, 3, is placed on the bottom face of a prism and the film serves as the active medium. As a result of the strong absorption in the active medium, the film thickness has to be rather small so that most of the wave energy reaches

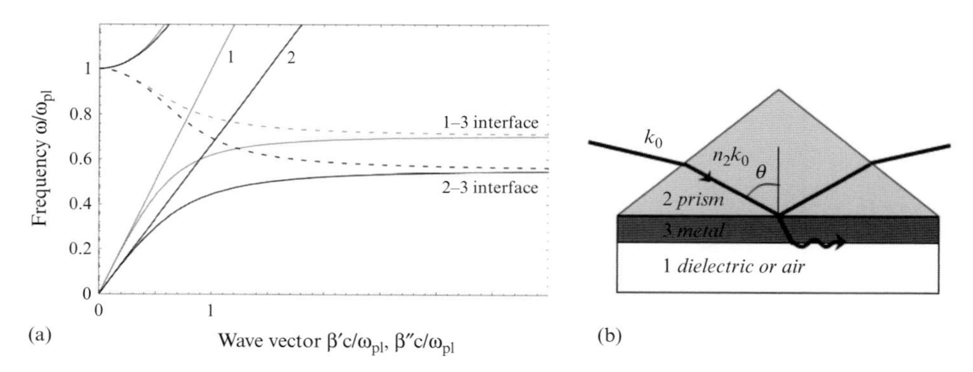

Figure 6.21 (a) The dispersion relation $\omega(\beta')$ for surface plasmon-polaritons at the interface between a metal with κ_m given by Eq. (6.33) and air (line 1 for air and the lighter gray solid curve for the 1–3 interface), and silica (line 2 for silica and the darker gray solid curve for the 2–3 interface). Relationships between ω and β'' are presented as dashed curves. Here ω is normalized with respect to the plasmon frequency, ω_{pl}, while β' and β'' are normalized with respect to the wavevector of the light wave in free space, ω_{pl}/c. (b) Prism coupling of light and surface plasmon-polaritons in Kretschmann's configuration. Figure part (a) reprinted by permission from figure 2.3 of Springer Nature, *Plasmonics. Fundamentals and Applications* by S. A. Maier, Copyright 2007.

the metal/dielectric (1–3) interface. As can be seen from Fig. 6.21(a), the dispersion relation for surface plasmon-polaritons at the 1–3 interface intersects with line 2, which corresponds to the dispersion curve in the prism, so the matching of these two waves is possible and surface plasmon-polaritons at the 1–3 interface can be excited. It is necessary to take into account that the projection of the wavevector n_2k_0 on the axis of wave propagation along the prism/metal interface is $n_2k_0 \sin\theta$, so angle θ should be close to $\pi/2$. Also, a substantial difference between the refractive indices, n, of media 1 and 2 will facilitate matching and, if the refractive index n_2 is large enough, medium 1 can be a dielectric with small n_1, rather than air as indicated in Fig. 6.21(b).

The dispersion relations presented in Fig. 6.21(a), with "quasi-saturation" of the $\omega(\beta')$ dependences, clearly illustrate one of the main ideas for using plasmonics in nanostructures. Indeed, if a coupling between a free electromagnetic wave and a surface plasmon-polariton of the same frequency is realized, one can transform the wave into a plasmon-polariton excitation with much larger wavevector and much smaller wavelength, and then operate with the electromagnetic energy at the subwavelength scale.

In the previous discussion we analyzed the surface plasmon-polariton for an ideal metal, neglecting the damping effects which are due to electron scattering in metals (see Eq. (6.23)) and interband transitions at high frequency (see Fig. 6.18(b)). These damping processes determine the imaginary part of the dielectric permittivity of metals and result in reconstruction of the dispersion relation $\omega(\beta)$ in comparison with that presented in Fig. 6.21. Indeed, Fig. 6.22 shows the dispersion relation of the surface plasmon-polaritons for the cases of air/silver and silica/silver interfaces, when the damping effects in silver are taken into account. We see that the most significant reconstruction of the dispersion occurs near the frequency of the surface plasmon, ω_{spl}. The dependence $\beta'(\omega)$ has a clear resonant character, and

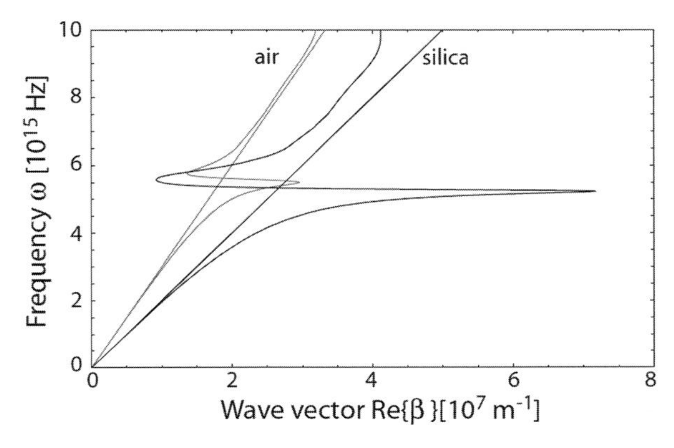

Figure 6.22 Dispersion curves $\omega(\beta')$ for surface plasmon-polaritons at air/silver (gray solid curves) and silica/silver (black solid curves) interfaces. The normalizations of ω and β' are as in Fig. 6.21. Reprinted by permission from figure 2.4 of Springer Nature, *Plasmonics. Fundamentals and Applications*, by S. A. Maier, Copyright 2007.

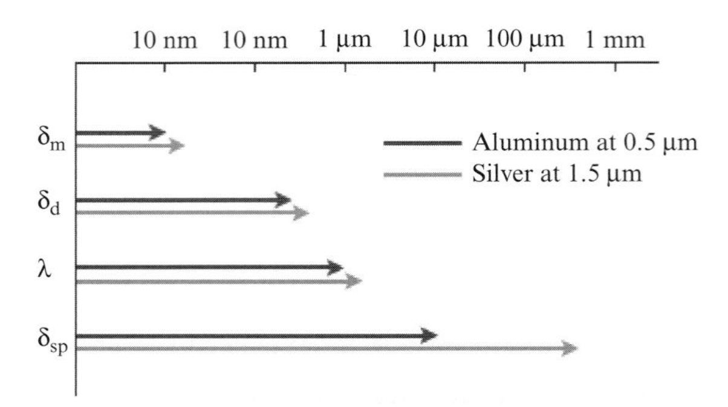

Figure 6.23 Comparison of the characteristic length scales of the surface plasmon-polaritons for air/aluminum and air/silver structures at frequencies corresponding to light wavelengths 0.5 μm and 1.5 μm. Reprinted by permission from the figure in Box 2 of Springer Nature, "Surface plasmon subwavelength optics," W. L. Barnes, *et al.*, Copyright 2003.

there appears to be a limitation (maximum) for the propagation constant and, thus, a lower bound for the wavelength of the excitation, $\lambda_{pl} > 2\pi/\max(\beta')$. According to Eq. (6.42), the limitation of β' also leads to a limitation of the degree of confinement of the surface plasmon-polaritons. Besides, in Fig. 6.22 there is an intersection of the plasmon-polariton dispersion curve with the free electromagnetic wave dispersion (straight lines). This means that, near this intersection, the surface plasmon-polariton becomes a *quasi-bound* (*leaky*) excitation.

Consider the spatial extensions of the electromagnetic fields and the electric charges of the surface plasmon-polariton. There are three length scales characterizing this type of excitations, namely the propagation length, $\delta_{spl} = 1/\mathrm{Im}[\beta]$, the decay length in the dielectric, $\delta_d = 1/K_d$, and the decay length in the metal, $\delta_m = 1/K_m$. All three are dependent on the frequency. The propagation length is determined mainly by loss in the metal and, for frequencies corresponding to the visible spectrum, varies from 2 μm for a relatively absorbing metal such as aluminum to 20 μm for low-absorbing metals such as silver. For silver, an increase in the wavelength to 1.55 μm gives rise to a propagation length of up to 1 mm. The decay length in the dielectric material, δ_d, is typically of the order of half the wavelength of light involved, while the decay length in the metal, δ_m, is approximately tens of nanometers, i.e., one to two orders of magnitude smaller than the wavelength of light. In Fig. 6.23, we present a diagram of the characteristic lengths for the surface plasmon-polaritons propagating over interfaces of air/aluminum and air/silver structures at the frequencies corresponding to light with wavelengths 0.5 μm and 1.55 μm.

6.5.4 Plasmons in Low-Dimensional Systems

Now we briefly consider collective oscillations of the electrons in low-dimensional conductors, such as semiconductor quantum wells and quantum wires, and two-dimensional crystals. As shown in previous chapters, for the systems considered here it is characteristic to have a quantization of electron motion in one or two directions, which radically changes the electronic properties. For the collective charge oscillations in low-dimensional systems, the electrostatic effects are more important, and they depend significantly on the geometry of the system. This makes the character of charge oscillations in low-dimensional conductors different from that in bulk-like samples.

Consider the following simple model for the electrons confined to a thin plane layer of width d (a quantum well, or a two-dimensional crystal). The wavelength of the excitations, λ, is assumed to be much greater than d. Then, the layer with the electrons can be thought of as infinitesimally thin and the electron concentration can be defined as $n = n_s\delta[z]$, with n_s the surface electron concentration, where $\delta[z]$ is the Dirac function. The layer is surrounded by a dielectric with dielectric permittivity κ_d. The electron motion is still modeled by Eq. (6.23), which can be rewritten in terms of the electron velocity, $\vec{v}$:

$$m\frac{d\vec{v}}{dt} + m\gamma\vec{v} = -e\vec{\mathcal{E}}_{||}(t). \tag{6.47}$$

Here $\vec{v}$ is the two-dimensional vector in the x, y plane, and $\vec{\mathcal{E}}_{||} = \{\mathcal{E}_x, \mathcal{E}_y\}$ is the in-plane component of the electric field. The surface electron concentration obeys the continuity equation

$$\frac{\partial n_s}{\partial t} - \frac{1}{e}\vec{\nabla}_2\vec{\mathcal{J}} = 0, \quad \vec{\mathcal{J}} = -en_s\vec{v}, \tag{6.48}$$

where $\vec{\mathcal{J}}$ is the density of the surface current and $\vec{\nabla}_2 \equiv \{\partial/\partial x, \partial/\partial y\}$. For the electrostatic potential ϕ, we can write the Poisson equation

$$\frac{\partial^2\phi}{\partial x^2} + \frac{\partial^2\phi}{\partial y^2} + \frac{\partial^2\phi}{\partial z^2} = \frac{e}{\epsilon_0\kappa_d}(n_s - n_{s0})\delta[z], \tag{6.49}$$

where n_{s0} is the uniform concentration of the compensating positive charge. The electric field in Eq. (6.47) is $\vec{\mathcal{E}}_{||} = -\vec{\nabla}_2\phi$.

Assuming as in the previous subsection that the excitation propagates along the x-axis, and restricting ourselves to consideration of solutions independent of the y coordinate, i.e.,

$$v_x = v_{x,\omega}e^{i(kx-\omega t)}, \quad \phi = \phi_\omega(z)e^{i(kx-\omega t)},$$

$$\mathcal{E}_x = \mathcal{E}_{x,\omega}e^{i(kx-\omega t)}, \quad \tilde{n}_s \equiv (n_s - n_{s0}) = \tilde{n}_\omega e^{i(kx-\omega t)},$$

we obtain the following solutions of the system of Eqs. (6.47)–(6.49):

$$v_{x,\omega} = -\frac{e\,\mathcal{E}_{x,\omega}}{m\gamma[1 - i\omega/\gamma]}, \quad \tilde{n}_\omega = \frac{ekn_{s0}\,\mathcal{E}_{x,\omega}}{m\omega\gamma[1 - i\omega/\gamma]},$$

$$\phi_\omega(z) = A_d^+e^{-kz} \text{ at } z > 0 \quad \text{and} \quad \phi_\omega(z) = A_d^-e^{kz} \text{ at } z < 0.$$

To be more specific, here and below we set $k > 0$. From the condition of continuity of the potential at $z = 0$, we obtain $A_d^+ = A_d^-$. From Eq. (6.49) we find the boundary condition for $\partial\phi/\partial z$ at $z=0$:

$$\left(\frac{\partial\phi_\omega(z)}{\partial z}\right)_{z\to+0}-\left(\frac{\partial\phi_\omega(z)}{\partial z}\right)_{z\to-0}=\frac{e\tilde{n}_\omega}{\epsilon_0\kappa_d}. \tag{6.50}$$

Using Eq. (6.50) and the above presented solutions of the system of Eqs. (6.47)–(6.49), we obtain the condition for the solvability of our system of equations:

$$\omega\gamma\left[1-i\frac{\omega}{\gamma}\right]=-i\frac{e^2 n_{s0}k}{2\epsilon_0\kappa_d m}, \tag{6.51}$$

which is the dispersion relation for the plasmons in the electron gas confined in a narrow (two-dimensional) layer. For a small damping parameter γ, we get the following simple expression for the plasmon frequency:

$$\omega'_{2\text{Dpl}}\approx\sqrt{\frac{e^2 n_{s0}k}{2\epsilon_0\kappa_d m}},\quad \omega''_{2\text{Dpl}}=-\frac{\gamma}{2}. \tag{6.52}$$

In contrast to the case of bulk plasmons, we have obtained that the plasmon frequency depends significantly on the wavevector: $\omega'_{2\text{Dpl}}\propto\sqrt{k}$.

Equation (6.52) was obtained for electrons with an effective mass m. In the case of graphene the effective mass equals zero (see Section 3.3.1) and this result is not applicable. However, for doped graphene there are also collective charge oscillations with frequency

$$\omega_{\text{Gpl}}=\sqrt{\frac{e^2 v_F\sqrt{n_{s0}}\,k}{2\sqrt{\pi}\epsilon_0\kappa_d\hbar}}. \tag{6.53}$$

Here $k_F=\sqrt{\pi n_{s0}}$ is the Fermi wavevector of the electrons in graphene. These oscillations are weakly damped due to electron scattering, if the plasmon energy, $\hbar\omega_{\text{Gpl}}$, is less than $2E_F$ ($E_F=\hbar v_F k_F$ is the Fermi energy of doped graphene), when interband transitions are suppressed. As in Eq. (6.52), for the long-wavelength oscillations ω_{Gpl} is proportional to $\sqrt{k}$, but it is more weakly dependent on the electron concentration, being proportional to $n_{s0}^{1/4}$. The dispersion relation for the plasmon in graphene is shown in Fig. 6.24(a). For comparison, the dispersion relation of the two-dimensional plasmon in GaAs quantum well with the electron concentration $n_s=5\times10^{11}$ cm^{-1} is shown in Fig. 6.24(b).

For semiconductor structures with confined electrons, as well as for two-dimensional crystals, the concentration of electrons, n_{s0}, can be changed over a wide range, from 10^{11} cm^{-2} to 10^{13} cm^{-2}, and the frequency of plasmons can be varied from the terahertz range to the far-infrared range.

It is necessary to add the following comment related to plasmon damping processes. In the equations of the electron motion (6.23) and (6.47), the damping parameters, γ, were introduced to account for electron scattering on lattice imperfections (defects, lattice vibrations, etc.). These processes obviously lead to plasmon damping in bulk conductors, in low-dimensional semiconductor heterostructures, and in two-dimensional crystals. Besides the scattering on imperfections there is

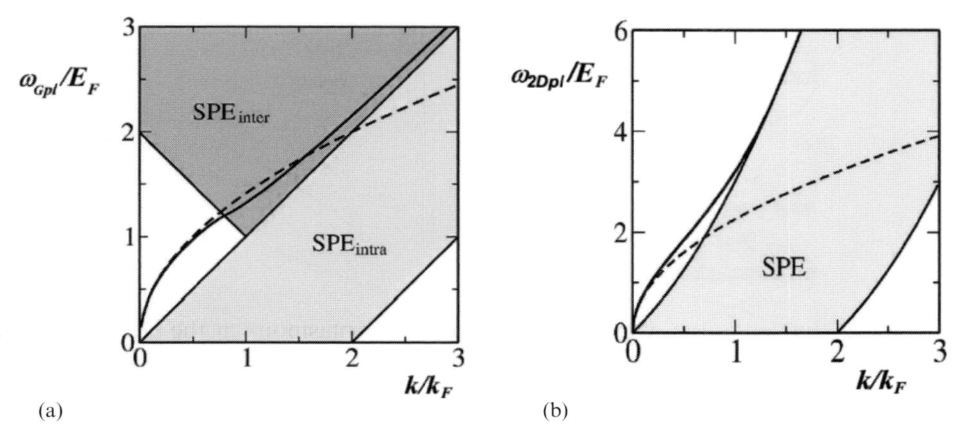

Figure 6.24 The dispersion relations of the plasmons in graphene (a) and in a two-dimensional electron gas (b). The plasmon energy and the wavevector are normalized with respect to the Fermi energy and the Fermi wavevector, respectively. The regions of the plasmon energies and momenta where decay processes involve intraband single-particle excitations (SPE_{intra}) and interband single-particle excitations (SPE_{inter}) are indicated. The dashed lines show the square-root dependences inherent for small wavevectors. Reprinted figure with permission from Fig. 1 of E. H. Hwang and S. Das Sarma, *Phys. Rev.* **B 75**, 205418, 2007. Copyright (2007) by the American Physical Society.

another channel of plasmon damping – decay of the *collective plasma excitations* into *one-particle excitations*. This decay process is possible if the energy and momentum conservation laws are valid. For a plasmon with wavevector $\vec{k}$ and energy $\hbar\omega_{\text{pl}}$, these conservation laws tell us that:

$$\hbar\omega_{\text{pl}} = E(\vec{K}+\vec{k}) - E(\vec{K}), \tag{6.54}$$

where $\vec{K}$ and $\vec{K}+\vec{k}$ are the initial and final wavevectors of the electron to which the plasmon energy is transfered. If condition (6.54) is met, the decay process takes place even in a *collisionless* plasma and is called *collisionless Landau damping*. The Landau damping of the plasmons can be important at low temperatures, when scattering processes in the electron gas are suppressed. At low temperatures the electron gas is, typically, degenerate, and the electrons can be characterized by the Fermi energy, E_{F}, and the wavevector K_{F}. In Fig. 6.24, the regions of the plasmon energies and momenta where decay processes involve intraband single-particle excitations (SPE_{intra}) and interband single-particle excitations (SPE_{inter}) for graphene are shown. For massless electrons in graphene and for an electron gas with a finite mass in the usual bulk materials, these regions are quite different. From Fig. 6.24 we can see that for a plasmon in a two-dimensional gas collisionless Landau damping is an important decay mechanism, whereas for plasmons in graphene Landau damping is absent, which is a consequence of the Dirac spectrum of the electrons.

Equations (6.52) and (6.53) define the plasmon frequency of an ungated two-dimensional semiconductor layer and graphene, respectively. As is shown in

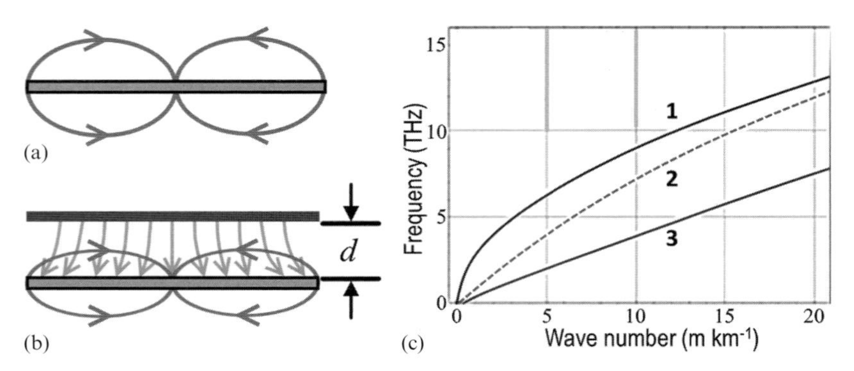

Figure 6.25 The electric-field distribution in (a) an ungated and (b) a gated two-dimensional semiconductor or graphene. (c) Changes in the dispersion relation on passing from the ungated structure (curve 1) to the gated one with decrease of the dielectric thickness, *d*, as shown by curve 2 for an intermediate thickness and by line 3 for a thin dielectric, $k^{-1} \gg d$.

Fig. 6.25(a), the electric field is spread widely in the surrounding dielectric or air in the ungated structure, and the plasmon frequency is proportional to the square root of the wavevector, $\omega'_{2\mathrm{Dpl}} \propto \sqrt{k}$, as is shown schematically by curve 1 in Fig. 6.25(c). In the case of the gate structure shown in Fig. 6.25(b), the electric field equals zero on the equipotential gate. As a result, the plasmon frequency goes down with decreasing thickness of dielectric, *d*, between the gate and the two-dimensional structure, as is shown schematically by curve 2 in Fig. 6.25(c). If *d* is smaller than the inverse wavevector, $d < 1/|k|$, the dispersion relations (6.52) and (6.53) transform into a linear relation as shown by curve 3 in Fig. 6.25(c):

$$\omega_{\mathrm{pl}} = k s_{\mathrm{pl}}. \tag{6.55}$$

When the dispersion relation becomes like the dispersion for light, $\omega = kc$, this has a very special implication for gated graphene. In gated graphene the plasmon velocity, s_{pl}, is substantially smaller than the speed of light, *c*, but it is larger than the Fermi velocity, v_{F}, in graphene (see Eq. (3.34)), and the process of interband absorption of plasmons in graphene is as depicted in Fig. 5.16 for the interband absorption of photons. Moreover, the dispersion relations for photons and plasmons are both linear, and the absorption of plasmons can be derived from the absorption coefficient of photons, $\pi e^2/c\hbar$ (see Fig. 5.15(b) and Section 5.4.1) by replacing *c* by s_{pl}. As a result, we have for the interband absorption coefficient of the plasmons in graphene, α_{pl}:

$$\alpha_{\mathrm{pl}} = \frac{\pi e^2}{s_{\mathrm{pl}}\hbar}. \tag{6.56}$$

Owing to the strong inequality $s_{\mathrm{pl}} \ll c$, the value of α_{pl} is several orders of magnitude larger than the interband absorption coefficient of photons in graphene,

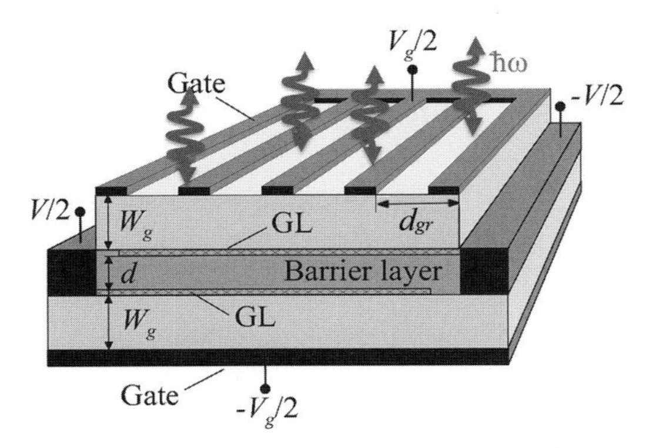

Figure 6.26 A grated upper gate is used to control the electron concentration in the upper graphene layer as well as to match radiation to plasmons in graphene. Reprinted from Fig. 1 of V. Ryzhii, *et al.*, *Appl. Phys. Lett.* **104**, 163505 (2014), with the permission of AIP Publishing.

$\alpha_{\rm ph} = \pi e^2/c\hbar = 2.3\%$. The plasmon velocity in graphene is equal to

$$s_{\rm pl} = \sqrt{\frac{4\ln 2\, e^2 d k_{\rm B} T}{\kappa \hbar^2}} \propto d^{1/2} T^{1/2} \quad \text{for } E_{\rm F} \ll k_{\rm B} T \tag{6.57}$$

and

$$s_{\rm pl} \propto v_{\rm F} d^{1/4} \Phi_{\rm g}^{1/4} \quad \text{for } E_{\rm F} \gg k_{\rm B} T. \tag{6.58}$$

Here $\Phi_{\rm g}$ is the voltage applied between the gate and graphene and $E_{\rm F}$ is the Fermi energy, which is controlled by the gate voltage and the thickness of the dielectric:

$$E_{\rm F} = \hbar v_{\rm F} \sqrt{\frac{\kappa \Phi_{\rm g}}{2ed}}. \tag{6.59}$$

Plasmon enhancement is very important for the interaction of electrons with light in graphene. Figure 6.21(b) shows an example of prism matching, but that method is convenient only for characterization of structures. In devices, matching of electromagnetic waves to plasmons can be carried out by means of a diffraction grating which diffracts radiation under particular angles. An example is shown in Fig. 6.26 for the graphene structure that was discussed in Section 3.9.1 (see Fig. 3.40(b)). The period of the grating, $d_{\rm gr}$, is chosen to match the wavevector of the plasmons, k, to the wavevector, k_0, of an electromagnetic wave of the same frequency in free space. The matching condition is simplified compared with the one that was discussed for the case of matching with a prism (see Fig. 6.21(b) and the text explaining the figure) and is practically defined by the grating period rather than just by the angle of incidence, θ, being

$$k = k_0 \sin\theta \pm l \frac{2\pi}{d_{\rm gr}}, \tag{6.60}$$

where l is an integer: $l = 1, 2, 3, \ldots$

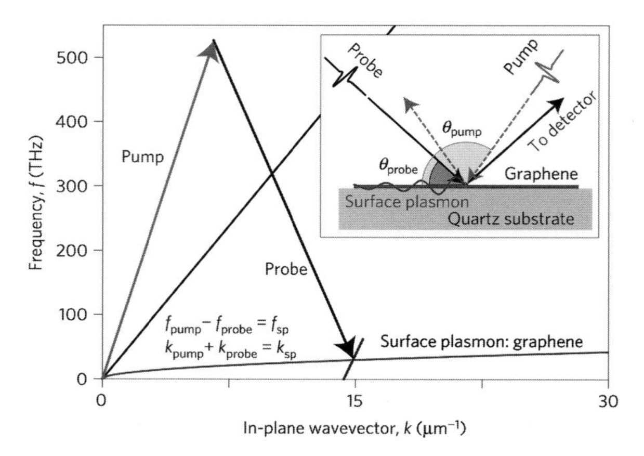

Figure 6.27 Illustrated here is a difference-frequency-generation technique where a graphene surface plasmon is created using a pump wavelength ranging from 615 nm to 545 nm with the probe wavelength fixed at 615 nm. The inset shows the experimental arrangement used to excite surface plasmons on graphene. Reprinted by permission from Fig. 1 of Springer, *Nature Physics*, "All-optical generation of surface plasmons in graphene," T. J. Constant, *et al.*, Copyright 2016.

Recently, a promising all-optical plasmon coupling technique for the production of plasmons in graphene has been demonstrated. In this technique, visible light pulses are used to generate surface plasmons in graphene using wave mixing and the control of phase matching. It was demonstrated that graphene surface plasmons can be generated over a wide range of frequencies. This matching of wavevectors and frequencies is illustrated in Fig. 6.27, where the difference-frequency-generation technique is illustrated for a pump wavelength ranging from 615 nm to 545 nm, with the probe wavelength fixed at 615 nm. As can be seen in Fig. 6.27, this combination of visible laser wavelengths allows conservation of energy (laser frequency) and momentum (wavevector) such that coupling to the graphene dispersion curve is possible, with the result that graphene surface plasmons are created through this all-optical technique.

6.5.5 Localized Surface Plasmons and Field-Enhancement Effects

The previous discussion was focused on *propagating* plasmons and plasmon-polaritons in a bulk material, over a thin conductive layer or two-dimensional conducting crystals, etc. Now we turn to *localized* plasmons in nanostructures, one of the commonly encountered excitations in the field of plasmonics.

We present an analysis of the localized plasmons for the example of a conducting sphere. Figure 6.28 illustrates the alternate displacement of an electron cloud relative to the positive-ion charges in a metallic sphere, as well as the electric field. The left and right halves of the drawing correspond to different phases of the

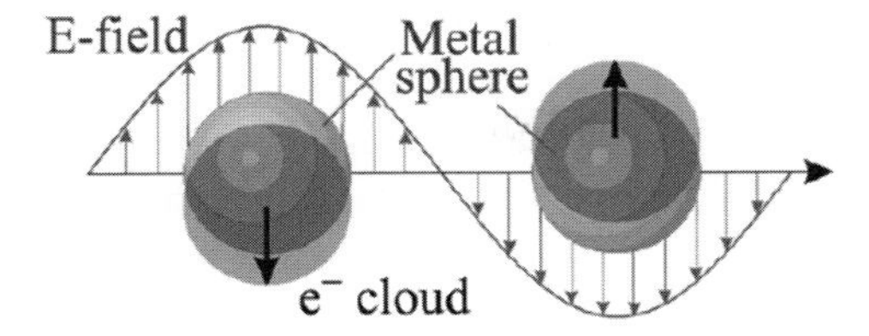

Figure 6.28 Displacement of electrons relative to the positive charges in a conducting nanosphere. Reprinted with permission from Fig. 1 of K. Lance Kelly, *et al.*, *J. Phys. Chem.* **B 107**, 668 (2003). Copyright 2003 American Chemical Society.

self-consistent oscillations of the charges and the field. These self-consistent oscillations represent a *localized* plasmon in the conducting sphere.

A description of the scattering of an electromagnetic wave from such a spherical particle has been available for many years, and the scattering process is known as Mie scattering. One of the Mie results is the resonance-like increase of the scattering cross-section at some discrete frequencies dependent on the size and on the dielectric permittivities of the sphere and of the medium in which it is embedded. Excitation of these resonances by an incident electromagnetic field produces a significant enhancement of the field in the vicinity of the sphere. The simplest way to study these resonances is to apply the electrostatic approximation (as was used above for the plasmons on a planar dielectric/metal interface). This approximation can be applied when the diameter of the sphere is much smaller than the wavelengths corresponding to the frequency of the resonances being studied.

We assume that an incident electromagnetic wave illuminates a sphere with dielectric permittivity $\kappa_\mathrm{m}(\omega)$. If the sphere's diameter, $2a$, is much smaller than the wavelength, λ, the instantaneous electric field of the wave, $\mathcal{E}_0$, is approximately constant on scales of the order of $2a$. The electrostatic field and the potential of the sphere subjected to the incident field, $\mathcal{E}_0$, can be found as solutions of the three-dimensional Laplace equation

$$\nabla^2\phi = 0. \tag{6.61}$$

The potential inside and outside of the sphere may be represented in terms of Legendre polynomials:

$$\phi_\mathrm{m} = \sum_{l=0}^{\infty} A_l r^l P_l(\cos\theta), \quad \phi_\mathrm{d} = \sum_{l=0}^{\infty} [B_l r^l + C_l r^{-(l+1)}] P_l(\cos\theta), \tag{6.62}$$

where A_l, B_l, C_l are constants and the functions $P_l(\cos\theta)$ define a set of orthonormal Legendre functions, used widely in electromagnetic theory as well as in other branches of physics and engineering. Taking $\mu = \cos\theta$, the first few Legendre polynomials are

$$P_0(\mu) = 1, \quad P_1(\mu) = \mu, \quad P_2(\mu) = \frac{1}{2}(3\mu^2 - 1), \quad P_3(\mu) = \frac{1}{2}(5\mu^3 - 3\mu).$$

It can be shown that for an ideal sphere a solution for ϕ can be constructed for any given l, which corresponds to a series of resonances.

Here we present the calculations for the lowest resonance with $l=1$. As follows from the first of Eqs. (6.62), at $l=1$ the electric field inside the sphere is constant. At large distances from the sphere, the potential must be the potential associated with the incident electric field, i.e., $\phi(r\to\infty)=-\mathcal{E}_0 r\cos\theta$, so that $B_1=-\mathcal{E}_0$, and all other $B_l=0$ at $l\neq 1$. The remaining coefficients may be determined from the continuity of the tangential components of the electric field and continuity of the normal component of the electric displacement:

$$-\frac{\partial\phi_{\mathrm{m}}}{\partial\theta}=-\frac{\partial\phi_{\mathrm{d}}}{\partial\theta},\qquad -\kappa_{\mathrm{m}}\frac{\partial\phi_{\mathrm{m}}}{\partial r}=-\kappa_{\mathrm{d}}\frac{\partial\phi_{\mathrm{d}}}{\partial r}\quad \text{at } r=a.$$

As a result, one can define the coefficients, $A_1,\ B_1,\ C_1$:

$$A_1=-\frac{3\kappa_{\mathrm{d}}}{2\kappa_{\mathrm{m}}+\kappa_{\mathrm{d}}}\mathcal{E}_0,\qquad C_1=\frac{\kappa_{\mathrm{m}}-\kappa_{\mathrm{d}}}{2\kappa_{\mathrm{d}}+\kappa_{m}}a^3\,\mathcal{E}_0,$$

and $A_l=C_l=0$ at $l\neq 1$. Finally, we obtain the potential in the form

$$\begin{aligned}\phi_{\mathrm{m}}&=-\frac{3\kappa_{\mathrm{d}}}{2\kappa_{\mathrm{m}}+\kappa_{\mathrm{d}}}\mathcal{E}_0 r\cos\theta,\\ \phi_{\mathrm{d}}&=-\mathcal{E}_0 r\cos\theta+\frac{\kappa_{\mathrm{m}}-\kappa_{\mathrm{d}}}{2\kappa_{\mathrm{d}}+\kappa_{\mathrm{m}}}\mathcal{E}_0\frac{a^3}{r^2}\cos\theta.\end{aligned}\tag{6.63}$$

Here the first term is the potential of the incident wave and the second term represents the scattered wave. The total electric field inside the sphere is

$$E_{\mathrm{m}}=\frac{3\kappa_{\mathrm{d}}}{2\kappa_{\mathrm{m}}+\kappa_{\mathrm{d}}}\mathcal{E}_0.$$

The potential of Eq. (6.63) can be rewritten in the form

$$\phi_{\mathrm{d}}=-\mathcal{E}_0 r\cos\theta+\frac{\vec{p}\vec{r}}{4\pi\epsilon_0\kappa_{\mathrm{d}}r^3},\tag{6.64}$$

where we have introduced the dipole moment of the sphere,

$$\vec{p}=4\pi\epsilon_0\kappa_{\mathrm{d}}\frac{\kappa_{\mathrm{m}}-\kappa_{\mathrm{d}}}{2\kappa_{\mathrm{d}}+\kappa_{\mathrm{m}}}a^3\vec{\mathcal{E}}_0.$$

This dipole moment may be interpreted as the volume integral of the polarization of the sphere. Now we have found the complex polarizability of a small sphere of subwavelength diameter:

$$\alpha=4\pi a^3\frac{\kappa_{\mathrm{m}}-\kappa_{\mathrm{d}}}{2\kappa_{\mathrm{d}}+\kappa_{\mathrm{m}}}.\tag{6.65}$$

From Eqs. (6.63)–(6.65) one can see that the polarization, the surface charge, and the scattered field (relating to the second term in Eq. (6.63)) have maxima, if

$$\mathrm{Re}[2\kappa_{\mathrm{d}}(\omega)+\kappa_{\mathrm{m}}(\omega)]\to 0.$$

The corresponding frequency, $\omega_{\mathrm{lpl}}^{(1)}$, is the frequency of the lowest localized plasmon in the sphere (compare this with the condition for the surface plasmon given by

Eq. (6.46)). For the simplest dielectric permittivity for a conducting medium given by Eq. (6.25), we obtain

$$\omega_{\mathrm{lpl}}^{(1)} = \omega_{\mathrm{pl}} \sqrt{\frac{1}{1 + 2\kappa_{\mathrm{d}}}}.$$

The resonances with arbitrary l can be obtained from the equation

$$\frac{\kappa_{\mathrm{m}}(\omega)}{\kappa_{\mathrm{m}}(\omega) - \kappa_{\mathrm{d}}} = \frac{1 + l}{1 + 2l}. \tag{6.66}$$

In particular, for $\kappa_{\mathrm{m}}(\omega)$ given by Eq. (6.25) the frequencies of the localized plasmons are

$$\omega_{\mathrm{lpl}}^{(l)} = \omega_{\mathrm{pl}} \sqrt{\frac{l}{\kappa_{\mathrm{d}} + (1 + \kappa_{\mathrm{d}})l}}. \tag{6.67}$$

For any given l, similarly to the case with $l = 1$, the scattering rate, the absorption rate, and the polarizability of a spherical particle all contain the factor $1/[(1 + l)\kappa_{\mathrm{d}} + l\kappa_{\mathrm{m}}(\omega)]$, which produces a strong resonance when $\omega = \omega_{\mathrm{lpl}}^{(l)}$. The strength of these resonances is reduced by absorption in the conducting sphere and by higher-order corrections not restricted by the approximation $a \ll \lambda$. However, when the resonance condition for excitation of the localized plasmon, $\omega = \omega_{\mathrm{lpl}}^{(l)}$, is satisfied there is a large enhancement in the polarizability and in the magnitude of the electromagnetic field near the sphere.

To illustrate the effect of the field enhancement, let us consider a metal sphere in vacuum. Then the lowest resonance ($l = 1$) takes place at $\kappa'_{\mathrm{m}} = 2$, and the field enhancement equals

$$T_{\mathrm{enh}} = \left| \frac{\mathcal{E}(a)}{\mathcal{E}_0} \right|^2 = \left| \frac{3\kappa'_{\mathrm{m}}(\omega_{\mathrm{lpl}}^{(1)})}{\kappa''_{\mathrm{m}}(\omega_{\mathrm{lpl}}^{(1)})} \right|^2.$$

For silver the lowest resonance is at $\omega_{\mathrm{lpl}}^{(1)} \approx 350\,\mathrm{nm}$ ($\approx 3.5\,\mathrm{eV}$) and the corresponding $\kappa'' = 0.28$, so that $T_{\mathrm{enh}} \approx 480$. Such a field enhancement at the plasmon resonances facilitates many applications of metal nanoparticles in optical devices and sensors. In particular, this resonant enhancement is responsible for many enhanced scattering process such as in surface-enhanced Raman scattering (SERS) where a small particle, in close proximity to the nanostructure being characterized by Raman scattering, produces a plasmon that causes a many-fold enhancement in the electric field associated with the incident radiation in the Raman scattering.

6.6 Closing Remarks to Chapter 6

In this chapter, we have studied two classes of optical phenomena: electro-optical effects and nonlinear effects. These effects are well known for conventional bulk-like materials and they manifest remarkable features for the case of quantum heterostructures.

Large electro-optical effects in quantum structures occur as a result of two major mechanisms. The first is an analog of the Franz–Keldysh or Stark effect, i.e., a reconstruction of the energy spectra of two-dimensional free carriers and excitons in an electric field. This spectrum reconstruction results in large changes in optical absorption near the fundamental absorption edge. The second mechanism arises because the filling factor is controlled by an external field, i.e., there is a field-dependent population of energy states in quantum structures. An electric field affects both the absorption and the refractive index of the structures; it also provides a way to achieve effective control of optical properties.

We have considered electrically biased superlattices and double-quantum-well structures, and found that high-frequency microwave emission (up to the terahertz range) can be generated by such structures.

A hierarchy of different mechanisms is responsible for the nonlinear optical properties of quantum structures. The fastest mechanisms are due to the virtual processes used for excitation of the structures. In this case, an off-resonance short light pulse creates a coherent superposition of excited states – the polarization of a quantum structure – providing optical nonlinearity (photon–photon interaction). This mechanism occurs for light frequencies below the lowest exciton energy. As discussed in Chapters 2, 3, and 5, in low-dimensional systems, exciton features are much more pronounced in the optical spectra; thus, the virtual mechanism leads to larger optical nonlinearities. At higher, resonant frequencies, when real excitons and electron–hole pairs are generated, the nonlinear mechanisms and their magnitudes depend significantly on the timescale of the light pulse. The fastest nonlinearity (the subpicosecond range) arises as a result of the filling factor and the screening of the Coulomb interaction. This occurs in a nonthermalized photo-excited plasma. On a longer timescale (the nanosecond range), the electron–hole plasma is thermalized, and the screening effect and bleaching near the absorption edge determine the nonlinearity. The required intensities of light are sufficiently lower than those for bulk-like crystals.

Plasmonics was discussed, as well as its focus on achieving ways to confine and control electromagnetic signals on dimensional scales smaller than, or of the order of, the wavelength of the electromagnetic field. It was explained how plasmonics exploits process whereby electromagnetic fields and conduction electrons interact at interfaces or in nanostructures to enhance optical fields in the subwavelength domain.

6.7 Control Questions

1. What is the electro-optic energy, $\hbar\omega_F$, associated with the Franz–Keldysh effect for an $In_{0.15}Ga_{0.85}As$ layer grown on an unstrained GaAs [111] surface? Assume that the strain-induced piezoelectric field in the $In_{0.15}Ga_{0.85}As$ layer is equal to 2.1×10^5 V/cm and is directed along the [111] direction. Assume that the

dominant Franz–Keldysh effect occurs as a result of the interaction of the electron wave function and the heavy hole wave function; take $m_e[111] = 0.067m$ and $m_{hh}[111] = 0.65m$. Further assume that the InGaAs may be treated as a region of bulk material and that the only nonzero electric field in this material is the piezoelectric field. (Readers interested in learning more about the use of the Franz–Keldysh effect to measure the strain-induced piezoelectric field in InGaAs may refer to "Direct measurement of piezoelectric field in a [111] B grown InGaAs/GaAs heterostructure by Franz–Keldysh oscillations," H. Shen, M. Dutta, W. Chang, R. Moerkirk, D. M. Kim, K. W. Chung, P. P. Ruden, M. I. Nathan, and M. A. Stroscio, *Appl. Phys. Lett.*, **60**, 2400, 1992.)

2. Consider the two-well system of Section 6.3, which has been used to observe coherent oscillations of electrons. Derive a relationship between the energy separation, $\Delta\epsilon$, and the period of oscillation, T. Is this result consistent with the Heisenberg uncertainty relation?
3. Suppose that the exciton radius is 50 Å for a system with $k_B T / Ry_{ex}^* = 4$. What value of the exciton density, n, corresponds to the density at which these excitons will be "washed out" from exciton–exciton interactions?
4. It is known that Ti_2O_3 is an indirect narrow-bandgap semiconductor with a bandgap of only about 0.1 eV at 300 K. At 600 K the bandgap closes completely and the Ti_2O_3 becomes a semi-metal. It is known further that coherent optical phonons with a frequency of 7 THz may be excited in Ti_2O_3 by an appropriate laser field. Explain why it is possible to pick a suitable operating temperature such that the laser excitation of Ti_2O_3 produces semiconductor-to-semi-metal transitions at a frequency of 7 THz. (See, if necessary, "Modulation of a semiconductor-to-semimetal transition at 7 THz via coherent lattice vibrations," T. K. Cheng, L. H. Acioli, J. Vidal, H. J. Zeiger, G. Dresselhaus, M. S. Dresselhaus, and E. P. Ippen, *Appl. Phys. Lett.*, **62**, 1901, 1998; https://doi.org/10.1063/1.109537.)
5. Give a detailed derivation of the relationships between A_l and C_l for $l = 1$, as well as for $l > 1$, for each of the boundary conditions – continuity of the tangential components of the electric field and continuity of the normal component of the electric displacement – and show the steps needed to obtain the final expressions given above for ϕ_m and ϕ_d.

7 Light-Emitting Devices Based on Interband Phototransitions in Quantum Structures

7.1 Introduction

In Chapters 4, 5, and 6, we studied the optical properties of quantum heterostructures. We found that these properties are frequently significantly different from those of bulk materials and that they depend on the parameters of the heterostructure. The novel optical properties of heterostructures open new possibilities for the application of quantum heterostructures in various optoelectronic devices. Among the different optoelectronic devices, the light-emitting devices are the most important. In this chapter, we analyze the applications of quantum heterostructures to devices emitting near-infrared, visible, and ultraviolet light. These semiconductor devices exploit the phototransitions between the valence and conduction bands, i.e., the interband phototransitions studied in Chapters 4 and 5.

We start with a study of population inversion and light amplification due to interband phototransitions. As the major method of pumping such systems, we consider the injection of electrons and holes into the active region of a p–n junction with embedded quantum structures. Since the width of such an active region is small, specially designed heterostructure waveguides are necessary for the confinement of a light wave in this region. Accordingly, our discussion includes an analysis of the effect of optical confinement. Then, we show that the amplification coefficient of the confined optical mode can be large enough to produce laser oscillations in an optical cavity. Our discussion focuses on heterostructures which facilitate the simultaneous formation of optical modes and the realization of laser oscillations which convert electric power into laser emission. A variety of different designs of light-emitting diodes and laser diodes are analyzed. Indeed, we consider a set of particular quantum well, quantum wire, and quantum dot lasers and light-emitting diodes, including devices operating in the near-infrared, visible, and ultraviolet optical regions.

7.2 Light Amplification in Semiconductors

This section is devoted to a subject of prime importance: the phenomenon of light amplification in bulk semiconductors and semiconductor heterostructures. In Section 4.3, we established the existence of two fundamental processes: light absorption and the spontaneous and stimulated emission of light. Stimulated emission can dominate over absorption when there is a population inversion for two energy levels separated by the energy of the light, $\hbar\omega$. Such a physical system can serve as an optical amplifier. We shall study the criteria for the formation of a population inversion between the conduction and valence bands as well as how these criteria depend on the light frequency, pumping rate, temperature, etc. In addition, we calculate the gain and discuss methods to create population inversion. Then we briefly consider double heterostructures from the point of view of light amplification.

7.2.1 Criteria for Light Amplification

In order to examine light amplification in the case of interband phototransitions in semiconductors and semiconductor structures, let us return to the results obtained in Chapter 4 for the light absorption coefficients; see Eqs. (4.50), (4.55), (4.68), and (4.70). Note that analysis of the population inversion does not depend on the dimensionality of an electron system. We already know that the absorption rate is proportional to

$$\mathcal{F}(E_{\rm v}(\vec{k})) - \mathcal{F}(E_{\rm c}(\vec{k})),$$

where $\mathcal{F}(E_{\rm v}(\vec{k})) \equiv \mathcal{F}_{\rm v}$ and $\mathcal{F}(E_{\rm c}(\vec{k})) \equiv \mathcal{F}_{\rm c}$ are the distribution functions of the electrons in the valence and conduction bands, respectively. Hence, the condition for "negative absorption," or light amplification, is

$$\mathcal{F}(E_{\rm c}) - \mathcal{F}(E_{\rm v}) > 0, \tag{7.1}$$

where the energies $E_{\rm v}(\vec{k})$ and $E_{\rm c}(\vec{k})$ satisfy the condition

$$E_{\rm c}(\vec{k}) - E_{\rm v}(\vec{k}) = \hbar\omega. \tag{7.2}$$

Here ω is the light frequency. Obviously, the inequality of Eq. (7.1) imposes the population inversion of the energy states $E_{\rm v}(\vec{k})$ and $E_{\rm c}(\vec{k})$, i.e., that the states with the higher energy $E_{\rm c}(\vec{k})$ are more populated than the states with the lower energy $E_{\rm v}(\vec{k})$.

A population inversion can be obtained only under strongly nonequilibrium conditions. In general, the functions $\mathcal{F}_{\rm c}$ and $\mathcal{F}_{\rm v}$ may take a variety of forms depending upon the conditions. Two types of processes – intraband and interband – have special importance in establishing the forms of these functions. As discussed in Section 6.4.4, intraband processes are much faster than interband processes. In the most interesting case, that of a high carrier concentration, the carrier–carrier interaction defines the shape of the functions $\mathcal{F}_{\rm c}$ and $\mathcal{F}_{\rm v}$; this interaction is also responsible for the evolution of these functions to forms close to Fermi distributions. As a consequence of electron–electron interactions, quasi-equilibrium distribution functions can be assumed for each band. Thus

$$\mathcal{F}_{\rm v,c} = \frac{1}{\exp\left((E - E_{\rm F_{v,c}})/k_{\rm B}T\right) + 1}, \tag{7.3}$$

where E_{F_v} and E_{F_c} are *quasi-Fermi levels*, which depend on the nonequilibrium electron concentrations in the valence and conduction bands. Within this approximation, the population numbers depend on two parameters: the lattice temperature T and the quasi-Fermi levels $E_{F_{c,v}}$; in other words, they depend on the concentrations of electrons in the conduction and valence bands.

Now, one can easily deduce that the inequality of Eq. (7.1) is equivalent to

$$\hbar\omega < E_{Fc} - E_{Fv}.$$

Thus, to achieve light amplification the photon energy must be smaller than the separation between the quasi-Fermi levels of the electrons and the holes. Taking into account the fact that interband transitions are possible only for $\hbar\omega > E_g$, where E_g is the bandgap, one can obtain the criteria for light amplification for interband phototransitions:

$$E_g < \hbar\omega < E_{Fc} - E_{Fv}. \tag{7.4}$$

The two-band semiconductor system under consideration is transparent for light in the frequency range $\omega < E_g/\hbar$, amplifies light in the frequency range given by Eq. (7.4), and absorbs light in the range $\hbar\omega > E_{Fc} - E_{Fv}$. Under equilibrium conditions, $E_{Fc} = E_{Fv}$, i.e., absorption is possible if $\hbar\omega > E_g$.

Both quasi-Fermi levels are functions of concentration and, therefore, functions of the pumping rates of the electrons and holes. The difference on the right-hand side of Eq. (7.4) increases as the rate of pumping increases, i.e., as the rates of both carrier injection and optical generation of the electrons and holes increase. The inequalities of Eq. (7.4) indicate the existence of a *threshold level of pumping*, which corresponds to the very appearance of the population inversion for a fixed light frequency, ω. For the case when population inversion is driven by an injection current, there should be a threshold current which leads to a population inversion and to light amplification at frequency ω. When the pumping rate increases, the spectral bandwidth over which the amplification exists expands as well. In this respect, a semiconductor laser amplifier is unlike an atomic or molecular laser amplifier, where the bandwidth is relatively independent of the pumping parameters.

Let us consider the spectral dependence of the amplification coefficient for interband transitions in bulk semiconductors. According to Eq. (4.70), the attenuation coefficient is

$$\alpha^{\text{attn}} = \frac{4\pi^2\omega|(\vec{\xi}\vec{\mathcal{D}}_{\text{c,v}})|^2}{c\sqrt{\kappa}\hbar}\varrho(\omega)\left[\mathcal{F}(E_{\text{v}}) - \mathcal{F}(E_{\text{c}})\right]. \tag{7.5}$$

For the condition of Eq. (7.3), this attenuation coefficient is negative. Instead of it we introduce the amplification coefficient

$$\alpha^{\text{ampl}} = -\alpha^{\text{attn}}.$$

The prefactor in Eq. (7.5) is a slowly varying linear function of frequency and can be ignored; thus, the main frequency dependence comes from both the optical density of states, $\varrho(\omega)$, and the population inversion, $\mathcal{F}_c - \mathcal{F}_v$. At low temperatures, we set $\mathcal{F}_c(E_c) = 1$ for $E_c < E_{Fc}$ and 0 otherwise; as well, we have $\mathcal{F}_v(E_v) = 0$ for $E_v > E_{Fv}$ and 1 otherwise. Thus,

$$\mathcal{F}_v(E_v) - \mathcal{F}_c(E_c) = \begin{cases} +1, & \hbar\omega < E_{Fc} - E_{Fv}, \\ -1, & \hbar\omega > E_{Fc} - E_{Fv}. \end{cases}$$

For the case of a bulk crystal, both frequency factors are presented in Figs. 7.1(a) and (b). As a result, the frequency dependence of $\alpha^{\text{ampl}}(\omega)$ almost coincides with the optical density of states and changes its sign at $\hbar\omega = E_{Fc} - E_{Fv}$; see Fig. 7.1(c). Finite temperatures results in a flattening of the functions $\mathcal{F}_v(E_v)$, $\mathcal{F}_c(E_c)$, and $\alpha^{\text{ampl}}(\omega)$.

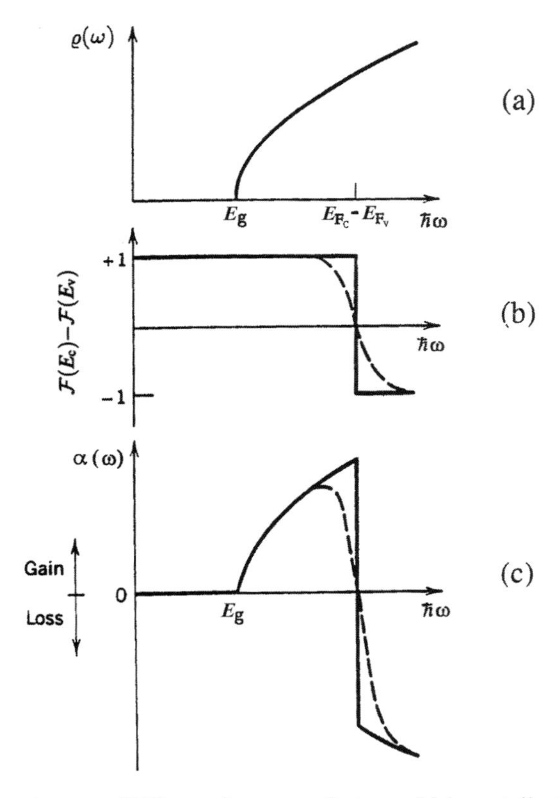

Figure 7.1 Different frequency factors which contribute to the amplification coefficient, α: (a) the energy dependence of the optical density of states, $\varrho(\omega)$, as a function of the photon energy $\hbar\omega$; (b) the population factor, $\mathcal{F}(E_c) - \mathcal{F}(E_v)$ (the solid line corresponds to $T = 0$, the dashed line corresponds to a finite temperature); and (c) the amplification coefficient. Light amplification and absorption correspond to $\alpha > 0$ and $\alpha < 0$, respectively. Reprinted with permission from Fig. 16.2-2 of *Fundamentals of Photonics*, B. E. A. Saleh and M. C. Teich, 1st Edition. Copyright ©1991 John Wiley & Sons, Inc.

7.2.2 Estimates of Light Gain

As the pumping rate grows, we see that the amplification spectrum broadens and the amplification coefficient increases. At low temperatures, when electrons and holes can be considered as completely degenerate gases, one can easily calculate the gain bandwidth, $\Delta\omega$, and the maximum gain, α_{max}^{ampl}, as functions of the carrier concentration, $n = n_e = n_h$. Using equations for E_{Fc} and E_{Fv} in the three-dimensional case, we find

$$\Delta\hbar\omega = (3\pi^2)^{2/3}\frac{\hbar^2}{2m_r}n^{2/3}$$

and

$$\alpha_{max}^{ampl} = P(3\pi^2)^{1/3}\sqrt{\frac{\hbar^2}{2m_r}}n^{1/3},$$

where the factor P is determined by Eq. (4.74), and m_r is the reduced electron–hole mass. The concentrations of electrons and holes are assumed to be equal. The maximum of α^{ampl} is reached at the edge of the spectral gain band, as shown in Fig. 7.1. Because the carrier concentration n is proportional to the pumping rate, the above equations show how the gain spectral bandwidth and the gain coefficient increase with increased pumping. More accurate calculations of α for $T = 297$ K are shown in Fig. 7.2(a) for an InGaAsP alloy for different electron and hole concentrations. This compound supports light amplification in the near-infrared spectral range from 1.1 μm to 1.7 μm. The maximum values of the gain are collected in Fig. 7.2(b), which clearly demonstrates the threshold character of the population inversion. It is worth noting that the absolute values of the amplification coefficients are large, typically of the order of several hundred cm^{-1}.

In accordance with the inequality of Eq. (7.4), at a finite temperature there is a minimum threshold value of the carrier concentration, n_{th}, below which there is no amplification in terms of this threshold concentration. Near the threshold it is possible to approximate the gain by

$$\alpha = \gamma\left(\frac{n}{n_{th}} - 1\right). \tag{7.6}$$

This equation is useful for making rough estimates of the gain coefficient. For example, in the case of InGaP, the parameters needed to estimate the gain α in Fig. 7.2 are $\gamma = 600\ cm^{-1}$ and $n_{th} = 1.8 \times 10^{18}\ cm^{-3}$; thus, for $n = 1.3n_{th}$, one obtains $\alpha = 180\ cm^{-1}$. If we choose the length of the active region of the amplifier in the direction of light propagation to be $L_y = 300$ μm, we find that the total gain,

$$G \equiv \frac{\mathcal{I}_{out}}{\mathcal{I}_{in}} = e^{\alpha L_y},$$

is equal to $e^{5.4} = 221 = 2.3$ dB. As a result of facet reflections from facet structure imperfections as well as other loss mechanisms, the losses of light in optical cavities are typically of the order of 3 dB to 5 dB per path. Such losses must,

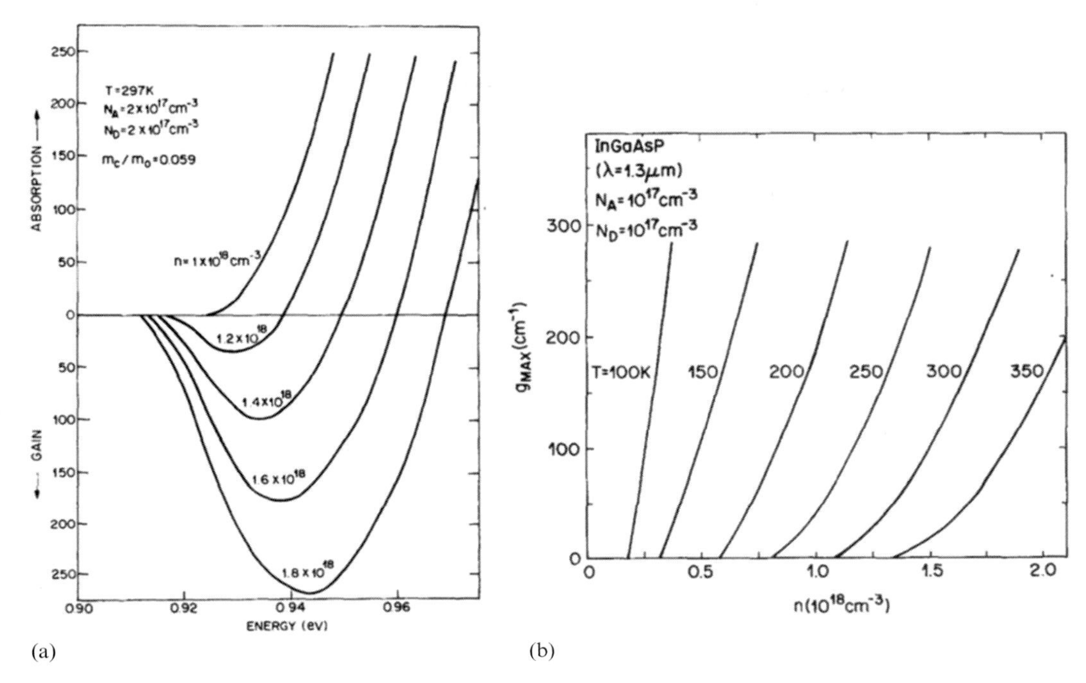

Figure 7.2 Characteristics of an InGaAsP light amplifier. (a) The calculated light-amplification coefficient as a function of the photon energy. Different curves correspond to different injected carrier concentrations, indicated on the curves. The threshold injected concentration is about 1×10^{18} cm^{-3}. The maximum light amplification is centered near the photon energy corresponding to 1.3 μm. (b) The maximum amplification coefficient as a function of the carrier concentration at different temperatures. Reprinted (a) from Fig. 8 of N. K. Dutta, *et al.*, *J. Appl. Phys.* **51**, 6095 (1980), and (b) from Fig. 17 of N. K. Dutta, *et al.*, *J. Appl. Phys.* **53**, 74 (1982) with the permission of AIP Publishing.

of course, be overcome by the gain if the medium is to function as an amplifier. Figure 7.3 gives another example of a calculation of $\alpha(\hbar\omega)$ for GaAs at $T = 300$ K and for different carrier concentrations. Despite the different photon energy range (1.42 eV to 1.48 eV), the order of magnitude of the concentrations required to obtain population inversion remains almost the same.

7.2.3 Methods of Pumping

From the previous numerical examples, one can see that to create population inversion for a bulk crystal one should generate a density of nonequilibrium carriers as high as 10^{18} cm^{-3}. A light gain comparable to the optical losses requires even more intensive pumping. In the active region of a device, where the electron and hole concentrations are highly nonequilibrium, the characteristic lifetime of these excess carriers is small. Figure 7.4 shows the radiative lifetime for electrons and holes with different concentrations in direct-bandgap GaAs at 300 K. For the concentrations of about 10^{18} cm^{-3} which are encountered in practice, this time is less than 10 ns.

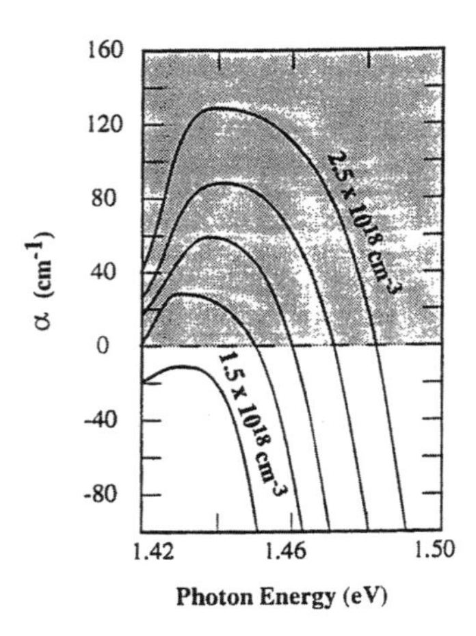

Figure 7.3 The amplification coefficient versus the photon energy in a GaAs light amplifier for different carrier concentrations. The temperature is 300 K. The results are given for five values of the concentration with equal steps of 0.25×10^{18} cm^{-3}. Reprinted with permission from Fig. 11.16 in *Physics of Semiconductors and Their Heterostructures* by J. Singh. Copyright McGraw-Hill Education (1993).

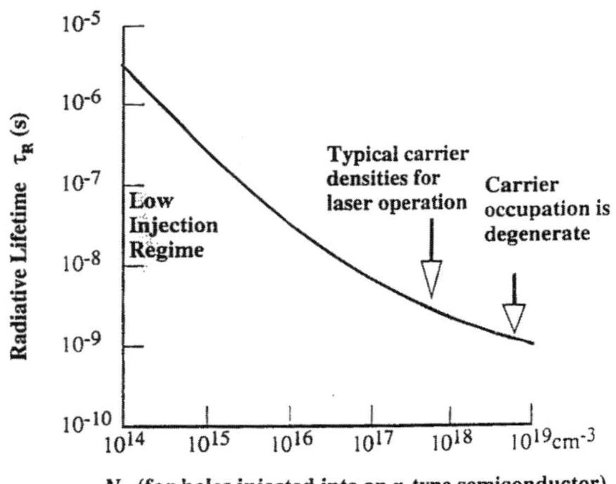

Figure 7.4 The radiative lifetime for GaAs as a function of electron and hole concentrations at room temperature. The low injection regime, the carrier concentrations necessary for the lasing, and the regime of strong degeneracy of the electron–hole gas are indicated. Reprinted with permission from Fig. 11.5 in *Physics of Semiconductors and Their Heterostructures* by J. Singh. Copyright McGraw-Hill Education (1993).

Let us introduce the pumping rate density, $\mathcal{R}_{\text{pump}}$, which represents the number of electron–hole pairs excited in a unit volume per unit time. For the case when the radiation mechanism is the decay of electron–hole pairs, one can write

$$\mathcal{R}_{\text{pump}} = Bn^2 = \frac{n}{\tau_R(n)}, \tag{7.7}$$

where B is a parameter, and $\tau_R = 1/(Bn)$ is the radiation lifetime. The previously discussed numerical values allow one to estimate the pumping rate density, $\mathcal{R}_{pump}$, necessary to induce a population inversion:

$$\mathcal{R}_{pump} \geq 10^{26}\ \mathrm{cm^{-3}\,s^{-1}}. \tag{7.8}$$

If the bandgap, E_g, is about 1 eV, the pumping rate density of Eq. (7.8) corresponding to the density of the *pumping power* given in the previous example is

$$E_g \mathcal{R}_{pump} \geq 16\ \mathrm{MW\,cm^{-3}}.$$

That is, a population inversion can be created only at an extremely high density of excitation power.

At very high levels of the excitation of electron–hole pairs, another mechanism competitive with radiative recombination comes into play; this is the *Auger mechanism*. Three carriers participate in Auger processes – two electrons and one hole, or one electron and two holes. The energy released by the recombining electron and hole is absorbed by the third particle and then dissipated by intraband relaxation processes. The rate of Auger recombination can be written as

$$\mathcal{R}_{Aug} = C_a n^3. \tag{7.9}$$

A typical order of magnitude for C_a is $5 \times 10^{-30}\ \mathrm{cm^6\,s^{-1}}$ for bulk GaAs at room temperature.

In the following discussion, we shall consider the conventional methods of pumping semiconductor materials to establish population inversion and light gain due to interband phototransitions.

The first method is *optical pumping*, where external light, often incoherent, with a photon energy larger than the bandgap, is absorbed and creates nonequilibrium electron–hole pairs. Then, because of the short time associated with the intraband relaxation, the electrons and holes relax to the bottom of the conduction band and the top of the valence band, respectively, where they accumulate. If the rate of optical pumping is sufficiently large, population inversion can be induced between the bands, and light amplification becomes possible as illustrated in Fig. 7.5.

The optical pumping corresponds to a conversion of one kind of radiation, not necessarily coherent, into coherent radiation with a lower photon frequency. One employs optical pumping in cases where electric current pumping either is not possible or is ineffective. This pumping method is often used to test prototype laser structures before designing the current-pumping system.

A much more convenient method for pumping semiconductor lasers is electron and hole injection into the depletion region of a p–n junction. The formation of such a junction with two adjacent regions doped by donors and acceptors is illustrated in Fig. 7.6. Let us show that such a junction can be used to obtain population inversion. The p–n junction shown in Fig. 7.6 corresponds to an unbiased state where electrons and holes have the same equilibrium Fermi level, E_F, and are separated in space. If a voltage bias is applied, the electrons are injected into the region with the existing hole population and vice versa. Nonequilibrium electron and hole concentrations are

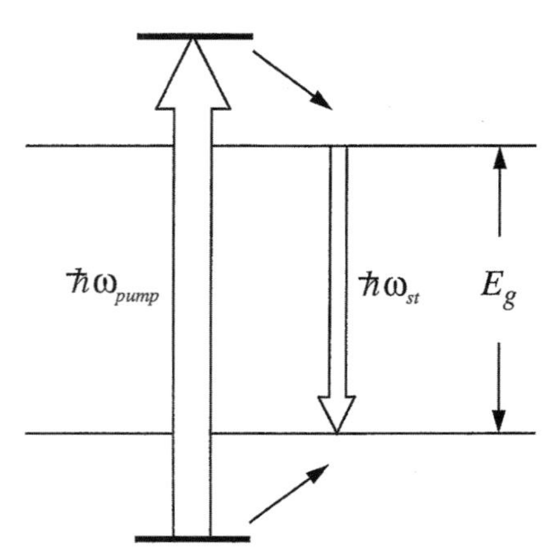

Figure 7.5 The optical pumping process. A pump light with photon energy $\hbar\omega_{\rm pump}$ creates electron–hole pairs with an excess energy. The electrons relax to the conduction band bottom, while the holes relax to the top of the valence band. Stimulated emission occurs with energy $\hbar\omega_{\rm st}$, near the bandgap, $E_{\rm g}$.

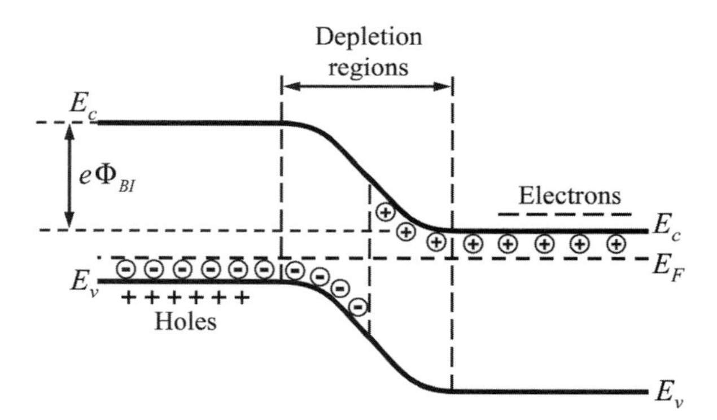

Figure 7.6 The formation of a p–n junction. The energy scheme of a semiconductor with p and n doping (left and right parts, respectively). $\Phi_{\rm BI}$ is the built-in potential arising due to the charge redistribution.

determined by the quasi-Fermi levels designated as $E_{\rm Fn}$ and $E_{\rm Fp}$ shown in Fig. 7.7. As a result of the carrier injection, a space region is formed where both electrons and holes exist simultaneously and interband phototransitions occur. An increase in the voltage bias leads to separation of the quasi-Fermi levels, as shown in Fig. 7.7. When these levels are well separated, light amplification becomes possible. In accordance with Eq. (7.4), the applied voltage should exceed $\Phi_{\rm th}$, which is given by

$$e\Phi_{\rm th} \geq E_{\rm g} - E_{\rm Fn} + E_{\rm Fp}.$$

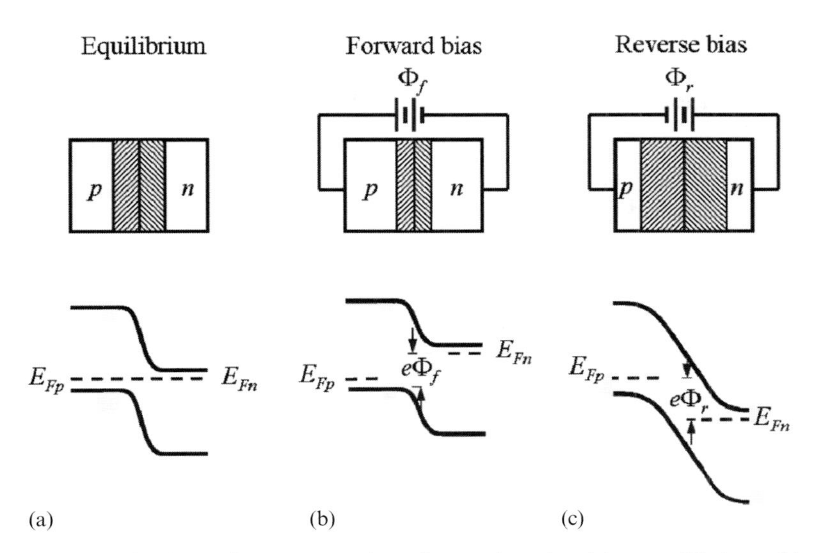

Figure 7.7 A schematic representation of a p–n junction (a) at equilibrium, (b) under forward bias, Φ_{f}, and (c) under reverse bias, Φ_{r}. For nonequilibrium cases, the two quasi-Fermi levels for the holes and the electrons, E_{Fp} and E_{Fn}, are shown to differ.

Note that the region of population inversion – the so-called *active region* – is quite narrow. Its typical thickness, d, for GaAs and InGaAsP is 1–3 μm.

Let us introduce the cross-sectional area $A = L_x \times L_y$, through which the current I is injected into the p–n region of a diode; L_x and L_y are the linear sizes of this area. Then, in steady state, the rate of injection of electrons and holes into a unit volume per unit time can be expressed as

$$\mathcal{R} = \frac{I}{eAd} = \frac{j}{ed}, \tag{7.10}$$

where j is the injection current density. Injection leads to the accumulation of nonequilibrium electron–hole pairs with concentration

$$n = \tau \mathcal{R} = \frac{\tau}{ed} j, \tag{7.11}$$

where τ is the total lifetime of the nonequilibrium pairs in the active region. Because the injected concentration is proportional to the current density, we can rewrite Eq. (7.6) for the light gain in another form:

$$\alpha^{\mathrm{ampl}}(\omega) = \gamma \left(\frac{j}{j_{\mathrm{th}}} - 1 \right). \tag{7.12}$$

Here we introduce the threshold current, j_{th}, given by

$$j_{\text{th}} = \frac{ed}{\eta_{\text{i}} \tau_{\text{R}}} n_{\text{th}}. \tag{7.13}$$

The ratio

$$\eta_{\text{i}} \equiv \frac{\tau}{\tau_{\text{R}}}, \tag{7.14}$$

characterizes the *internal efficiency of emission* in the p–n junction. Note that at $j = j_{\text{th}}$ the absorption (amplification) coefficient $\alpha = 0$; this condition corresponds to the case where there is no absorption or amplification. One can say that the threshold current corresponds to *transparent conditions*.

From Eq. (7.13), we can draw some important conclusions. First of all, one can see that j_{th} is proportional to the thickness of the active region of the p–n junction, d. This means that a lower threshold current can be achieved in a narrow p–n junction. The current density, j_{th}, is inversely proportional to the internal efficiency; therefore, a material with high internal efficiency (large probability of radiative recombination) is preferable.

Let us make some numerical estimates of the characteristic parameters for a semiconductor diode light amplifier. An InGaAsP system can function at room temperature and has the following typical parameters: $\tau_{\text{R}} = 2.5$ ns, $\eta_{\text{i}} = 0.5$, $n_{\text{th}} = 1.8 \times 10^{18}$ cm^{-3}, and $\gamma = 600$ cm^{-1}. Let the junction thickness and transverse sizes of the diode be $d = 2$ μm, $L_x = 10$ μm, and $L_y = 200$ μm; then the threshold current density is $j_{\text{th}} = 4.6 \times 10^4$ A/cm^2. The current density $j = 5.4 \times 10^4$ A/cm^2 results in a light gain coefficient, α, of 100 cm^{-1}, or a total gain (in the y-direction), G, of magnitude $G = e^{\alpha L_y} = 7.3$. The total current flowing through the area $A = 2 \times 10^{-5}$ cm^2 is about 1 A. These numbers show that laser amplifiers usually exploit very high electric currents and electric powers, in accordance with our previous estimates.

So far we have considered that the electron–hole concentrations are fixed and that the amplification coefficient does not depend on the light intensity. However, if the light intensity, $\mathcal{I}$, is large, the phototransitions induced by the light wave change the balance between generation and recombination of the carriers; in turn, this results in a decrease both in the carrier concentration and in the population inversion. Accordingly, the amplification coefficient decreases. The concentration balance is given by

$$\frac{dn}{dt} = \mathcal{R}_{\text{pump}} - \frac{n}{\tau} - \mathcal{R}_{\text{st}}, \quad n = p, \tag{7.15}$$

where $\mathcal{R}_{\text{pump}}$ is the rate of electron–hole generation in the active region, and $\mathcal{R}_{\text{st}}$ is the rate of stimulated emission. According to Section 4.4, we have

$$\mathcal{R}_{\text{st}} = \alpha \frac{c}{\sqrt{\kappa}} \mathcal{I}. \tag{7.16}$$

Here, we take into account that the generation of one photon corresponds to one event of radiative electron–hole recombination. Thus, the amplification becomes a function of the light intensity. In the limit of very high intensity, the population inversion goes to zero,

$$\mathcal{F}_c - \mathcal{F}_v \to +0,$$

and the light amplification saturates. This *saturation effect in the light amplification* is very important for laser operation because it is responsible for steady-state laser performance.

7.2.4 Motivations for Using Heterostructures for Light Amplification

From the previous considerations, one can see that, if the thickness of the active region decreases, the electric current required to obtain population inversion and light amplification also decreases. Thus, the length of the active region of the p–n junction is one of the critical parameters for injection pumping of the population inversion. As we already know, an excess extrinsic carrier concentration in some spatial region always leads to carrier diffusion out of this region. This "leakage" of the carriers is prevented by their finite lifetime; for the case under consideration, this lifetime is the recombination time τ. If the diffusion coefficient is D, the characteristic width of the region with the excess concentration cannot be less than the diffusion length $L_D = \sqrt{D\tau}$, which is the average distance of electron (hole) transfer before recombination. That is,

$$d > L_D.$$

Since the diffusion length of electrons and holes in III–V compounds is of the order of a few micrometers, it is not possible to make the active region of a size less than the diffusion length in typical homostructure p–n junctions.

In order to localize nonequilibrium electrons and holes in a smaller active region, one can employ double heterojunctions. The basic idea of using the double heterostructure is to design potential barriers on both sides of the p–n junction – this prevents electrons and holes from undergoing diffusion. The potential profile of a double heterostructure is sketched in Fig. 7.8(a). The heterostructure consists of three materials with the bandgaps E_{g1}, E_{g2}, and E_{g3}. The band offsets are chosen so as to design a structure with a barrier for electrons in the left part of the structure before the p region and a barrier for holes in the right part before the n region. The middle region, with bandgap E_{g2}, is accessible by both types of carriers and it serves as the active region. Figure 7.8(a) corresponds to the case of a flat-band condition of a p–n junction with a double heterostructure embedded in the depletion region. In this case, it is not the diffusion length but the distance between the barriers that determines the size of the active region. As a result, the size can be as small as 0.1 μm and the critical electric current decreases by one order of magnitude or more compared with that of a conventional homostructure p–n junction.

Another reason for employing heterostructures for light amplification arises as a result of the electrodynamic peculiarities of light propagation in a narrow active

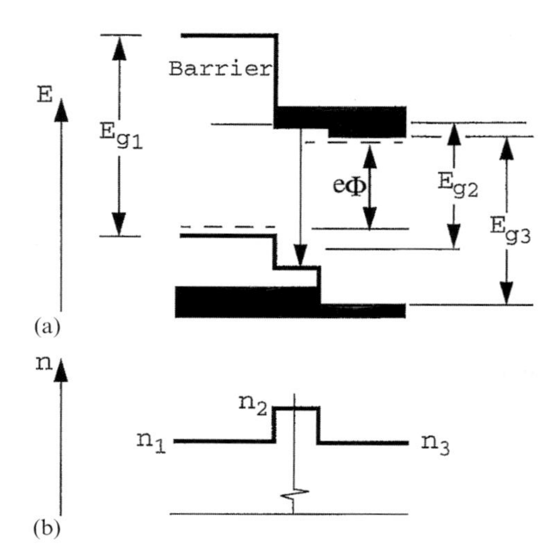

Figure 7.8 (a) The energy band diagram of a double heterostructure for the light amplifier. The applied voltage, Φ, induces the flat-band condition; E_{g1}, E_{g2}, and E_{g3} are the bandgaps in different regions of the structure. (b) The refractive index corresponding to the double heterostructure; n_1, n_2, and n_3 are the values of the refractive index in the regions with different bandgaps.

region. In Section 4.2.2, we discussed such peculiarities. We came to the conclusion that, for spatially inhomogeneous media, the concept of a simple plane light wave is no longer valid. We found that propagation and amplification are strongly dependent both on the magnitude and geometry of the inhomogeneity and on the wave direction. We showed that, in nonuniform media, light modes are formed due to the inhomogeneity of the refractive index. On the basis of these modes, one can introduce confined photons in these media and then consider the amplification of these modes. If the volume of the active region is less than the volume confining the light mode, the amplification of the mode decreases roughly by the ratio of these volumes. This reduction is described in terms of the so-called optical confinement factor (see Eq. (5.13)). For the case where the active region is in a plane layer perpendicular to the z-axis, the confinement factor is

$$\Gamma \approx \frac{\int_{\text{over active region}} dz\, \kappa(z)|F(z)|^2}{\int_{\text{over sample}} dz\, \kappa(z)|F(z)|^2}, \tag{7.17}$$

where $F(z)$ is the distribution of the wave amplitude as a function of z. For plane waves, we can easily approximate this factor by $\Gamma \approx d/L_z$, where L_z represents the size of the sample in the z-direction. The factor Γ is, in fact, the portion of the electromagnetic energy of the wave which propagates through the active region. One can easily understand that, for a narrow active region, the factor Γ is small and a decrease of the amplification by this factor can be crucial.

The second disadvantage associated with the use of a narrow active layer is that there is absorption of the light in the non-amplifying regions adjacent to the active layer of the device; indeed, some optical losses always exist there. Even if these losses are small, they can result in a significant reduction or even elimination of the total optical gain if the passive region is wide. Thus, there are potential problems associated with the *localization of light within the active region.*

With double heterostructures, it is possible to avoid these disadvantages. Let us return to the simplest double-heterostructure design, which requires three basic layers of different materials. Figures 7.8(a) and (b) depict schematically a double heterostructure and the associated profiles of bands and the refractive index. Layer 1 is p-doped and it has a bandgap E_{g1} and a refractive index n_1. The second layer has a bandgap E_{g2} and a refractive index n_2; usually this layer is undoped. The third layer is n-doped and it has a bandgap E_{g3} and a refractive index n_3. If the bandgaps satisfy the conditions

$$E_{g2} < E_{g1}, E_{g3},$$

the refractive index of the middle layer, n_2, is greater than that of the surrounding layers. This structure is designed for two goals.

1. The band discontinuities in the conduction and valence bands lead to confinement of the nonequilibrium electrons and holes within the active layer. In comparison with a conventional homogeneous p–n junction, this double heterostructure efficiently accumulates both types of carrier in a thinner region.
2. The larger refractive index in the second, *narrow-bandgap*, layer causes optical confinement within this layer. When solving the wave equation for light, we obtain solutions localized within the layer. The lowest mode, $\vec{F}_0(z)$, that has no nodal points is presented in Fig. 7.9. There are other solutions with transverse distributions of the electric field $\vec{F}_1(z)$, $\vec{F}_2(z)$, ..., which have 1, 2, ..., nodal points and weaker confinement. These waves are the *transverse modes*. In Fig. 7.9, the lowest modes with wavelength $\lambda_L = 0.9$ μm are presented for three different active-layer thicknesses in the range from 0.1 μm to 0.3 μm; the discontinuity in n equals 0.1. This discontinuity leads to localization of the modes with characteristic extents of about 0.5 μm to 1 μm.

Thus, in a double heterostructure, one can assume the proper localization of photon modes, for which the factor Γ is much greater than that for unconfined, extended waves. For the confined modes, the factor Γ can be of the order of 1.

A complication can arise from the mechanism of electron–hole recombination at the interfaces of the heterostructure. Within the context of a phenomenological model, we can describe this mechanism by introducing the *interface recombination velocity*, s_{if}, so that the rate of this additional recombination is

$$R_{\text{if}} = s_{\text{if}} n, \tag{7.18}$$

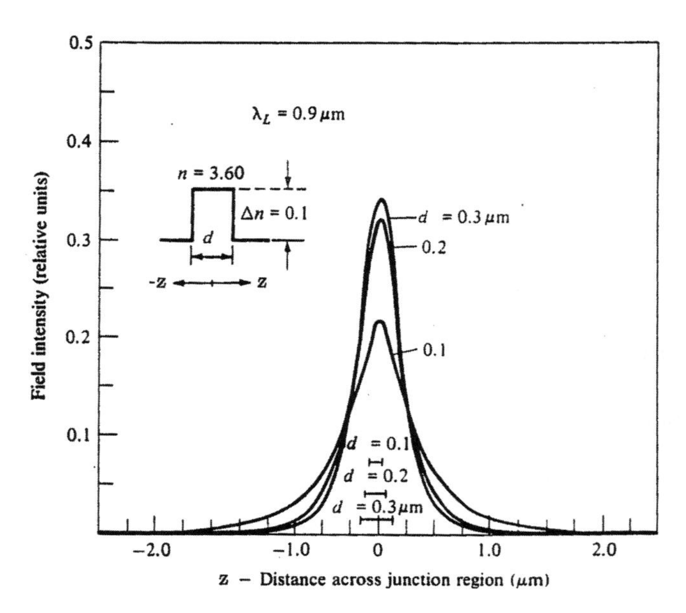

Figure 7.9 Distributions of the intensity in the lowest optical modes confined by step discontinuities in the refractive index. The insert shows the refractive index across the heterostructure. The results are given for the wavelength $\lambda = 0.9$ μm and three widths of the dielectric waveguide, *d*. Reprinted from figure 7.2.3a of *Semiconductor Lasers and Heterojunction LEDs*, 1st Edition, H. Kressel and J. K. Butler, "Chapter 7 – Modes in laser structures: mainly experimental," pp. 205–247, Copyright 1977, with permission from Elsevier.

where n is the three-dimensional carrier concentration in the barriers of the heterostructure. Typical values of s_{if} are of the order of 100 cm/s to 1000 cm/s. At modest levels of pumping, this channel of carrier losses can be very important. At higher injection levels, radiative recombination dominates over interface recombination because the former is proportional to the square of the carrier concentration; see Eq. (7.7).

As a particular example of light localization in a double heterostructure, we consider a GaAs/AlGaAs diode with an active layer fabricated from GaAs with $E_{g2} = 1.42$ eV and $n_2 = 3.6$. The surrounding layers can be made of $Al_xGa_{1-x}As$ with x in the range from 0.35 to 0.5, and $E_{g1,3} > 1.43$ eV; such layers have refractive indices $n_{1,3}$ that are less than 3.6 by 5% to 10%. This diode can amplify light within the optical band from 0.82 μm to 0.88 μm.

Another example is given by the InGaAsP/InP double heterostructure. In this case, the $In_{1-x}Ga_xAs_{1-y}P_y$ layer is the narrow-bandgap active region and it simultaneously localizes the light wave, while the InP layers have a wider bandgap and serve as barriers for the carriers. The alloy fractions, x and y, are chosen so that all materials are lattice-matched. Chapter 2 contains a detailed discussion on the lattice matching of such material layers. As a result, light amplification can be obtained in the spectral range from 1.1 μm to 1.7 μm.

The double heterostructures of our previous examples have typical sizes of the active region, d, of about 0.2 μm to 0.3 μm. According to Eq. (7.13), this means that the threshold currents for double heterostructures are one order of magnitude smaller than those for homogeneous p–n junctions.

7.2.5 Light Amplification in Quantum Wells, Quantum Wires, and Quantum Dots

If the thickness of the double heterostructure decreases further, the influence of quantum effects on carrier motion becomes important. For the structure shown in Fig. 7.8, quantum effects do not lead to any advantages because the same advantages can be obtained if heterostructures are designed so that quantum confinement applies both to electrons and to holes. Here we consider only the case of light propagation along the active layer. In the next section, we will study particular structures in more detail. The simplest case of quantum confinement can be achieved if a quantum well layer is embedded in an active region of a type-I heterostructure. Three possible designs of the active regions exhibiting quantum confinement are sketched in Fig. 7.10. Figures 7.10(a) and (b) correspond to the resultant confinement in a single quantum well, while Fig. 7.10(c) corresponds to confinement in multiple quantum wells. For these designs, electrons and holes, which are either generated by external light or injected from p and n regions, move in barrier layers and then are captured into the active region and quantum wells. The characteristic time of this capture is less than 1 ps. Escape processes require additional energy and have low relative probabilities of occurrence. Carriers in the quantum wells relax to the lowest energy states available. This results in the accumulation of both types of carrier in an extremely narrow active region, which is typically 100 Å wide or even narrower. A similar situation can be realized if quantum wires or quantum dots are used as the active regions.

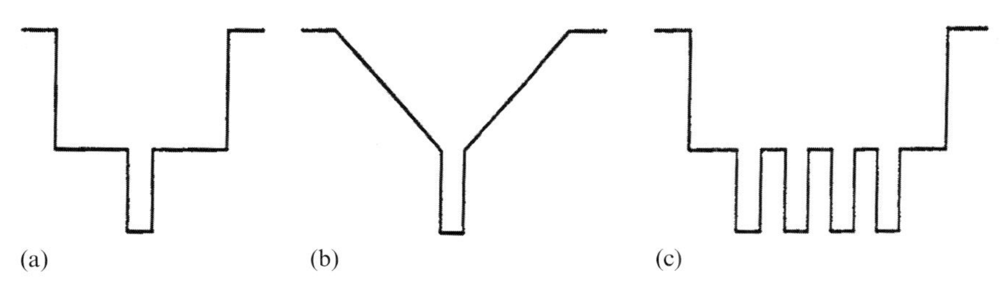

Figure 7.10 Composition profiles of $(Al_xGa_{1-x})_{0.5}In_{0.5}P$ heterostructures providing simultaneously a quantum confinement of the carriers and an optical confinement: (a) a single quantum well and a step-like-refractive-index heterostructure, (b) a single quantum well and graded-index optical confinement structure, and (c) a multiple quantum well and a step-like-index structure. Reprinted from figure 13 of *Quantum Well Lasers*, 1st Edition, P. S. Zory Jr., *et al.*, "Chapter 9 – AlGaInP quantum well lasers," pp. 415–460, Copyright 1993, with permission from Elsevier.

From the analysis of interband phototransitions of confined electrons and holes given in Section 5.2.2, it is clear that the criterion for population inversion given by Eq. (7.1) can be presented in the form of the inequality of Eq. (7.4), with only one modification: the bandgap of the bulk material, E_g, should be replaced by the optical bandgap of the quantum confining layer,

$$E_g + \epsilon_{c1} + \epsilon_{v1},$$

where ϵ_{c1} and ϵ_{v1} are the bottom and top of the lowest electron and hole energy subbands, respectively. For low-dimensional cases, the difference between the quasi-Fermi levels determining population inversion increases with increasing carrier concentration faster than in a bulk crystal. The Fermi level E_F is proportional to n_{2D} in quantum wells and to n_{1D}^2 in quantum wires, where n_{2D} and n_{1D} are the surface and linear carrier concentrations, respectively. These scaling results imply that it is possible to satisfy the criterion of Eq. (7.4) and create population inversion at lower pumping rates.

As discussed previously, two factors – the optical density of states, $\varrho(\omega)$, and the value of the population inversion, $\mathcal{F}_c - \mathcal{F}_v$, determine the frequency dependence of light amplification as described by Eqs. (5.23) and (5.24). As shown in Chapter 3, quantum confinement modifies the density of states significantly. In turn, this causes changes of the spectral dependence of the amplification coefficient, α^{ampl}. Figure 7.11 illustrates both factors for a quantum well layer. A comparison with the bulk case of Fig. 7.1 shows the differences in the frequency dependences of light amplification. The results shown in Fig. 7.11 are obtained for a pumping rate such that only the lowest electron and hole subbands exhibit population inversion. From Fig. 7.11, it can be seen that the step-like density of states causes a narrower spectral range of light amplification. This fact is used in quantum well lasers for the generation of a smaller number of optical modes. Thus, using quantum wells is in many cases advantageous for improving the spectral characteristics of light amplification. For light amplification in quantum wells, the optical confinement factor Γ is a critical parameter. It is convenient to represent the amplification coefficient of Eq. (5.23) by isolating the factor Γ as follows:

$$\alpha^{ampl}(\omega) = \Gamma \alpha_0^{ampl}(\omega). \tag{7.19}$$

The narrow-bandgap quantum well layer can serve simultaneously as a waveguide for the light. From Fig. 7.9 it can be seen, however, that the degree of optical localization is lower in thinner confining layers. Thus, a single quantum well does not lead to good localization of the light wave. Accordingly, the factor Γ is usually small.

In order to localize the light within a quantum well and to realize simultaneously a high rate of capture into the well, one designs special heterostructure waveguides in the active region. Two such designs of the active region with variable refractive index are shown in Fig. 7.12 for InGaAsP/InGaP systems. These designs correspond to the structure shown in Figs. 7.10(a) and (b). The structure provides a step-like or

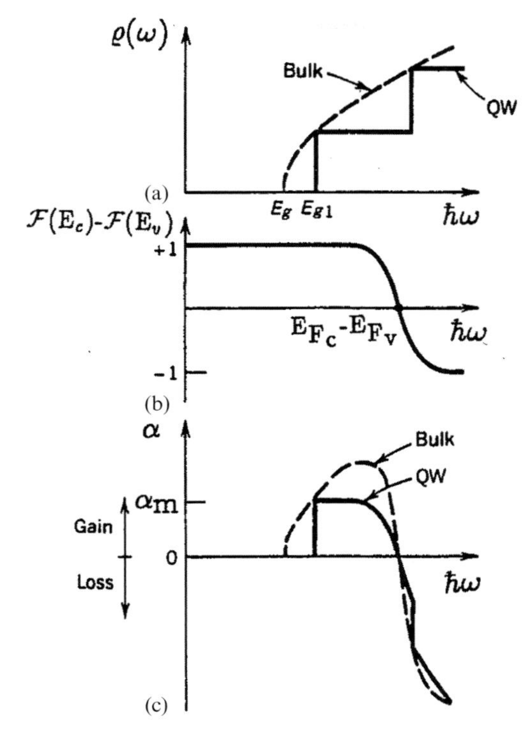

Figure 7.11 Different frequency factors, which contribute to the amplification coefficient in a quantum well layer (for comparison these factors are also shown for a bulk material, by dashed lines). (a) A schematic representation of the energy dependence of the optical density of states, ϱ. (b) The population factor, $\mathcal{F}(E_c) - \mathcal{F}(E_v)$ (a finite temperature is supposed). (c) The amplification/absorption coefficient. Reprinted with permission from Fig. 16.3-13 of *Fundamentals of Photonics*, B. E. A. Saleh and M. C. Teich, 1st Edition. Copyright ©1991 John Wiley & Sons, Inc.

gradual change in the refractive index and confines light modes within the active region. The dependence of the confinement factor Γ on the half-thickness, $d/2$, of this structure is presented in Fig. 7.12 for the same (100 Å-wide) quantum well. The nonmonotonic dependences $\Gamma(d/2)$ can be understood as follows. The limits $d \to 0$ and $d \to \infty$ both correspond to confinement of the mode in a single quantum well layer. In the case of the first limit, this layer provides a step in the refractive index corresponding to the composition with $x = 1$; for the case of the second limit, the step corresponds to the composition with $x = 0.6$. Both quantum well layers lead to relatively weak localization of light modes, with Γ about 1%. If d is finite, the modes are more localized. Figure 7.12 shows that, if the half-thickness of the structure changes from 0 to 0.25 μm, the factor Γ changes from 1% to 3.7%. That is, specially designed heterostructures improve the confinement factor and the gain coefficient in accordance with Eq. (7.19).

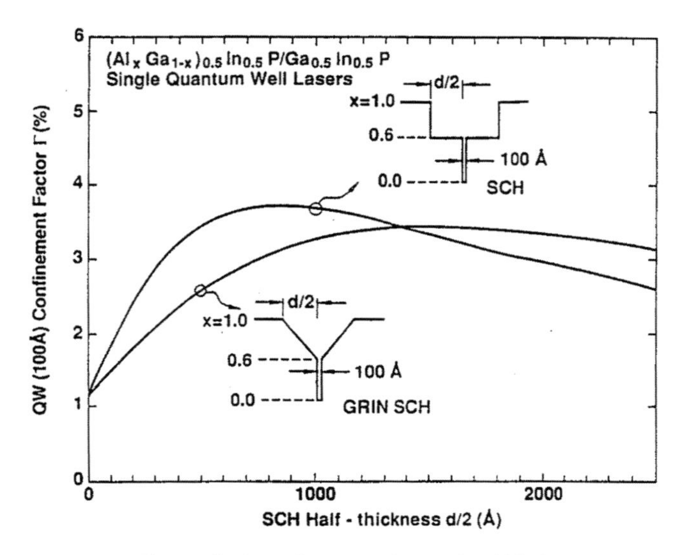

Figure 7.12 The optical confinement factor for $(Al_xGa_{1-x})_{0.5}In_{0.5}P/\ Ga_{0.5}In_{0.5}P$ single-quantum-well lasers. Results are given for the two designs of separate-confinement heterostructure (SCH) shown in the insets. Reprinted from figure 14 of *Quantum Well Lasers*, 1st Edition, P. S. Zory Jr., *et al.*, "Chapter 9 – AlGaInP quantum well lasers," pp. 415–460, Copyright 1993, with permission from Elsevier.

In the case of injection current pumping, the quantum well layer and the light-confining waveguide structure should be embedded in the region of the p–n junction. For such a case, results of calculations of the parameter $\alpha_0(\omega)$ in Eq. (7.19) are shown in Fig. 7.13(a) as a function of the current density, j. The calculations are given for InGaAsP/InP and AlGaAs/GaAs heterostructures with 50 Å quantum wells. The results demonstrate that the zero-gain (threshold) currents for both heterostructures are low: 20 A/cm^2 and 60 A/cm^2, for InGaAs/InP and AlGaAs/GaAs, respectively. Both $\alpha_0(j)$ curves increase sharply with the injection-current density. For comparison, we also present in Fig. 7.13(b) the theoretical and experimental results for double-heterostructure diodes based on the same materials and for a bulk-like active region where the carriers are not quantized. As one can see, the threshold currents are approximately 100 times less in structures with quantum wells than those in double heterostructures with bulk-like carriers.

Taking into account the small value of the confinement factor (see Fig. 7.12), one can estimate that for optimally designed quantum structures it is possible to obtain a threshold current for laser operation 5–10 times less than that of classical double heterostructures. The threshold can be lowered further if instead of a single quantum well one employs a multiple-quantum-well structure embedded in the active region; see Fig. 7.10(c). In a multiple-quantum-well amplifier, the capture of carriers into the wells is more effective; moreover, all the wells interact with the same optical

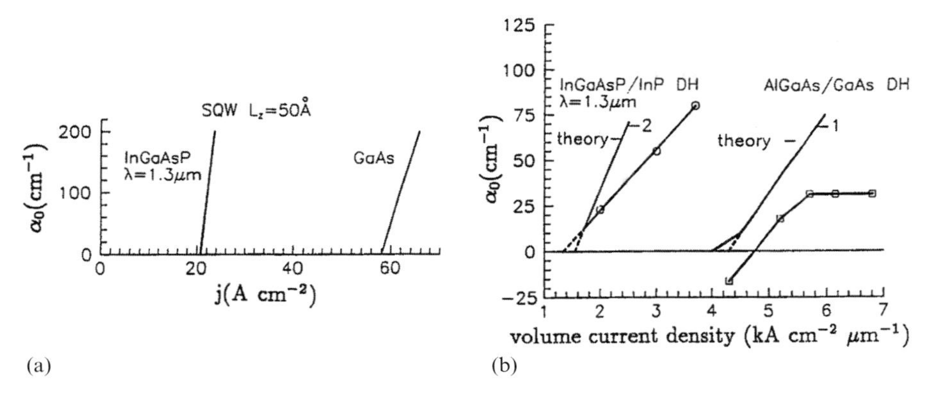

Figure 7.13 (a) The amplification coefficient, α_0, as a function of the pumping current density for single InGaAsP and GaAs quantum wells of width 50 Å. (b) The amplification coefficient, α_0, versus the current density for InGaAsP/InP and AlGaAs/GaAs double heterostructures (DM) with bulk-like carriers. Theoretical and experimental results are presented. Dashed lines show the zero-amplification current density. Reprinted from figures 10 and 9 of *Quantum Well Lasers*, 1st Edition, P. S. Zory Jr., *et al.*, "Chapter 6 – Single quantum well InGaAsP and AlGaAs lasers: a study of some peculiarities," pp. 227–327, Copyright 1993, with permission from Elsevier.

mode and emit the same photons. As a result, both the optical confinement factor and the light amplification increase.

7.3 Light-Emitting Diodes and Lasers

7.3.1 Light-Emitting Diodes

In the previous section, we stressed that carrier injection into the active region of a semiconductor diode is the most applicable method for the generation of nonequilibrium electrons and holes to bring about the creation of population inversion and stimulated emission. Although diode stimulated emission is very important, subthreshold operation of the diode – when only spontaneous light is emitted – is in many cases advantageous. It does not require feedback to control the power output, it facilitates operation over a wide range of temperatures, and it is reliable. Such diodes operating on spontaneous light emission are called *light-emitting diodes.* Light-emitting diodes are less expensive than laser diodes.

Let us consider these light-emitting diodes. We base our considerations on the results obtained in the previous section. For a forward-biased p–n diode, we can write the rate of spontaneous emission, or the total photon flux, as

$$\Upsilon = \frac{n}{\tau_R}\mathcal{V}, \tag{7.20}$$

where n is the electron–hole concentration per unit area in the active region of the diode, $\mathcal{V}$ is the volume of this region, and τ_R is the radiative lifetime. Using

Eqs. (7.11) and (7.14) we obtain

$$\Upsilon = \eta_{\mathrm{i}} \frac{I}{e}, \tag{7.21}$$

where η_{i} is the internal quantum efficiency of the diode. In fact, not all photons emitted in the active region contribute to the output light flux. This is so because photons are lost as a result of internal absorption in the n and p regions, reflections from air/semiconductor boundaries, etc. In order to evaluate these losses, one introduces the overall transmission efficiency, η_{t}, so that the product $\eta_{\mathrm{ext}} = \eta_{\mathrm{t}}\eta_{\mathrm{i}}$ characterizes the *external efficiency of the diode* and the output photon flux can be written as

$$\Upsilon_{\mathrm{out}} = \eta_{\mathrm{ext}} \frac{I}{e}. \tag{7.22}$$

The quantity η_{ext} is the ratio of the externally produced photon flux to the injected electron flux. Obviously, η_{ext} depends not only on the physical mechanisms but also on the design of the diode. The optical power of the light-emitting diode can be calculated from Eq. (7.22) as

$$\mathcal{P} = \hbar\omega \Upsilon_{\mathrm{out}} = \eta_{\mathrm{ext}} \hbar\omega \frac{I}{e}, \tag{7.23}$$

where $\hbar\omega$ is the photon energy. Practically, the linear dependence of $\mathcal{P}(I)$ is valid only within a limited current range. At high injection currents, the light output starts to saturate for a variety of reasons: a decrease of the radiative lifetime, heating of the device, etc. Figure 7.14 illustrates a typical *light–current characteristic* of the diode.

The spectral distribution of diode emission is one of the important characteristics. For a homogeneous p–n junction, the spectral density of spontaneous emission is

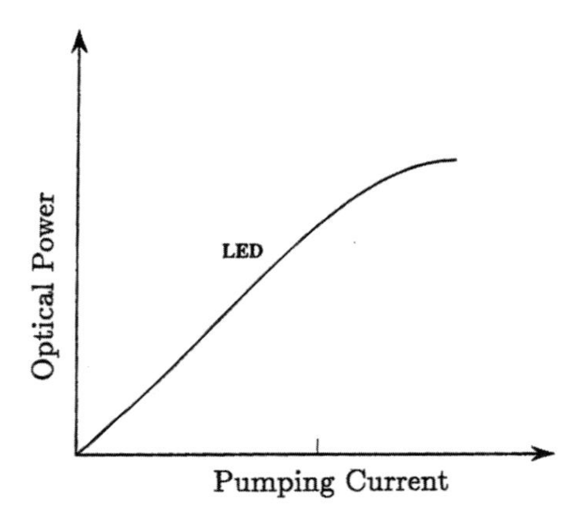

Figure 7.14 The output of a light-emitting diode as a function of the pumping current.

given by Eq. (4.77); see also Fig. 4.10. For relatively weak pumping, when the quasi-Fermi level is within the bandgap, the peak value of the spectral distribution is

$$\hbar\omega_{\mathrm{m}} = E_{\mathrm{g}} + \frac{1}{2}k_{\mathrm{B}}T.$$

The full width at half maximum of the distribution is $\Delta\omega \approx 2k_{\mathrm{B}}T/\hbar$ and it is independent of ω. In terms of the wavelength, λ, we obtain $\Delta\lambda = (\lambda_{\mathrm{m}}^2/2\pi c)\Delta\omega$, or

$$\Delta\lambda = 1.45\lambda_{\mathrm{m}}^2 k_{\mathrm{B}}T, \tag{7.24}$$

where λ_{m} corresponds to the maximum of the spectral distribution, $\Delta\lambda$ and λ_{m} are expressed in micrometers, and $k_{\mathrm{B}}T$ is expressed in electron-volts. Figure 7.15 shows the spectral density as a function of the wavelength of light-emitting diodes based on various materials. The spectral density is normalized so that its maximum equals 1 for all samples. For different materials, the spectral linewidths increase in proportion to λ^2, in accordance with Eq. (4.77). From Fig. 7.15 one can see that light-emitting diodes cover a wide spectral region from the infrared – about 8 μm for InGaAsP alloys – to near ultraviolet – 0.4 μm for GaN. They are, indeed, very universal light sources.

Light-emitting diodes may be designed either in a *surface-emitting* configuration or in an *edge-emitting configuration*. These configurations are illustrated in Figs. 7.16(a) and (b), respectively. Surface-emitting diodes radiate from the face parallel to the plane of the p–n junction. The light emitted in the opposite direction is either absorbed by a substrate or reflected by metallic contacts. Edge-emitting diodes radiate from the edge of the junction region. Usually, surface-emitting diodes are more efficient.

Since the diodes under consideration radiate through spontaneous emission, the spatial patterns of the emitted light depend only on the geometries of the

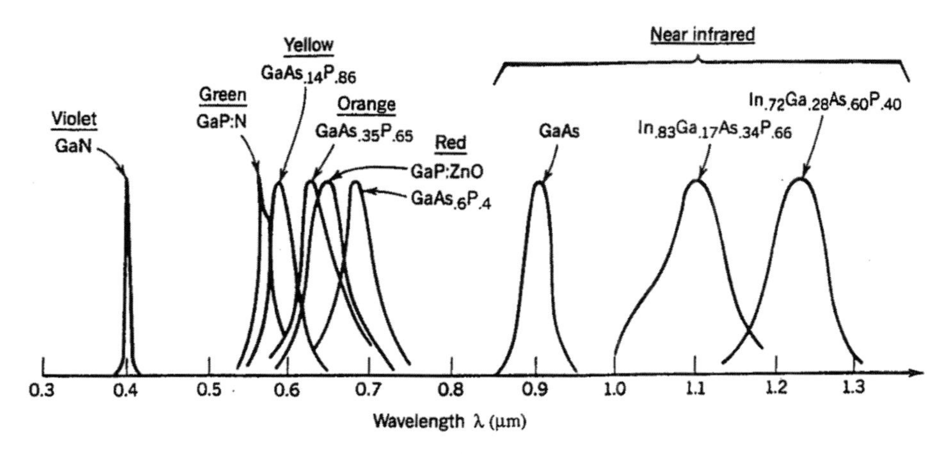

Figure 7.15 The spectra of light-emitting semiconductor diodes with different bandgaps. Reprinted with permission from Fig. 13 in chapter 12 of *Physics of Semiconductor Devices*, S. M. Sze, 2nd Edition. Copyright ©1981 Wiley-Interscience.

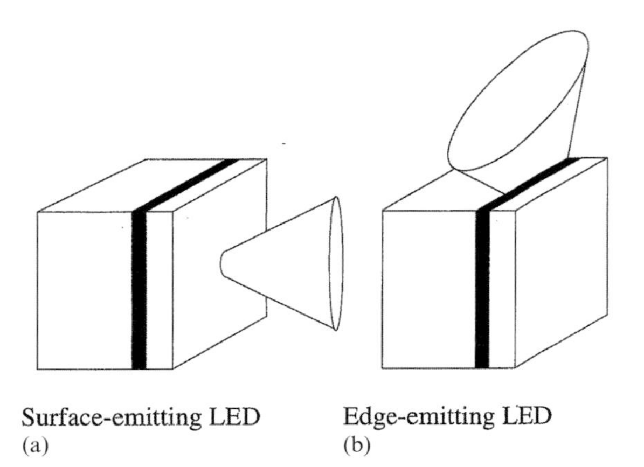

Figure 7.16 Two possible outputs of light-emitting diodes: (a) a surface-emitting diode and (b) an edge-emitting diode.

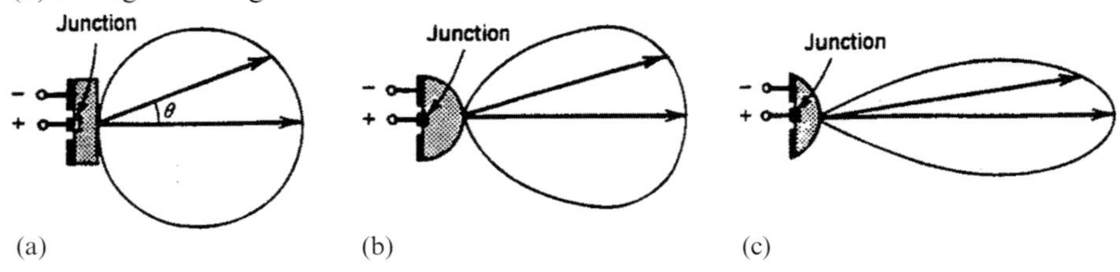

Figure 7.17 Patterns of output emission for surface-emitting diodes: (a) the pattern in the absence of a lens, (b) the pattern of a diode with a hemispherical lens, and (c) the pattern of a diode with a parabolic lens. Reprinted with permission from Fig. 16.1-12 of *Fundamentals of Photonics*, B. E. A. Saleh and M. C. Teich, 1st Edition. Copyright ©1991 John Wiley & Sons, Inc.

devices. Figure 7.17(a) illustrates emission patterns for a surface-emitting diode in the absence of a lens, when the intensity varies as $\cos\theta$, where θ is the angle from the normal to the face. Different lenses can improve the emission pattern, as shown in Figs. 7.17(b) and (c) for hemispherical and parabolic lenses. Usually, edge-emitting diodes have narrower emission patterns.

Light-emitting diodes find many applications in the field of signal processing and communications. Therefore, an important figure of merit is their response time. To estimate this time, one can consider a time-dependent low injection level, $I_1 \cos(\Omega t)$, so that all processes are described by linear differential equations. Then, one can assume an injection current in the form

$$I = I_0 + I_1 \cos(\Omega t), \quad I_1 \ll I_0. \tag{7.25}$$

Let us use this expression to calculate the injection rate of Eq. (7.10). Then, from Eq. (7.15) one can find the electron and hole concentrations and calculate the optical power of Eq. (7.23). The result is

$$\mathcal{P} = \mathcal{P}_0 + \mathcal{P}_1 \cos(\Omega t + \phi),$$

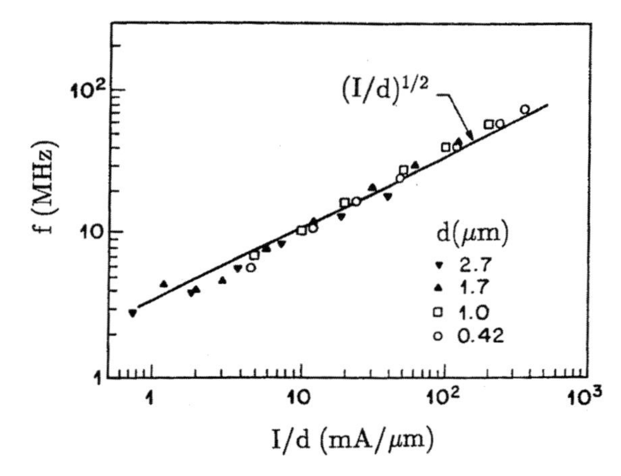

Figure 7.18 The modulation bandwidth as a function of the pumping-current-to-active-region-width ratio, I/d, for InGaAsP light-emitting diodes. ©1981 IEEE. Reprinted, with permission, from Fig. 6 of O. Wada, *et al.*, "High radiance InGaAsP/InP lensed LED's for optical communication systems at 1.2–1.3 μm," *IEEE J. Quantum Electron.* **QE-17**, pp. 174–178 (1981).

where ϕ is a phase shift. For the modulated output $\mathcal{P}_1$, one gets

$$\mathcal{P}_1 \sim \frac{I_1}{\sqrt{1 + (\Omega\tau)^2}}. \tag{7.26}$$

Here τ is the lifetime of injected carriers in the active region. This time limits the device speed. As a modulation characteristic, one introduces the *modulation bandwidth*:

$$f \equiv \frac{1}{2\pi\tau}. \tag{7.27}$$

The values of the bandwidth are presented in Fig. 7.18 for different injected concentrations. We can consider that $\tau \approx \tau_R \approx 1/(Bn)$ (see Eq. (7.7)). Then, using Eq. (7.27), we obtain an expression for the modulation bandwidth as a function of the current:

$$f = \frac{1}{2\pi}\sqrt{\frac{BI}{eAd}}. \tag{7.28}$$

This square-root dependence on the current agrees with the experimental data presented in Fig. 7.18 for InGaAsP light-emitting diodes. Results are given for diodes with different thicknesses of the active region, d, and cross-sections, A. The typical bandwidth of such devices is of the order of hundreds of MHz, though a bandwidth above 1 GHz has also been demonstrated.

7.3.2 Amplification, Feedback, and Laser Oscillations

So far we have considered only one effect related to stimulated emission, namely light amplification. Lasing can be achieved if a light amplifier is supplied with a

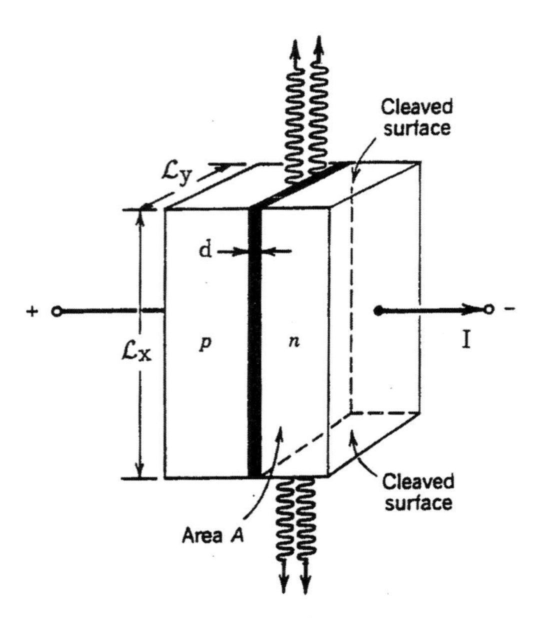

Figure 7.19 A schematic representation of an injection laser with two cleaved facets that act as reflectors. Reprinted with permission from Fig. 16.3-1 of *Fundamentals of Photonics*, B. E. A. Saleh and M. C. Teich, 1st Edition. Copyright ©1991 John Wiley & Sons, Inc.

path for optical feedback. A straightforward way to provide the optical feedback is to place an amplifier between the two mirrors of an optical resonator. Upon passing through the amplifier, the light wave gains energy; then it is reflected back with some energy losses and, on passing through the amplifier again, it gains energy once again, and so forth. If the amplification exceeds the optical losses in the system, the light energy inside the resonator increases. This increase may not be infinite because of the saturation effect; specifically, an increase in the light energy leads to additional electron–hole recombination through the radiative channel, and consequently to decreasing population inversion and amplification. Steady-state laser operation will be reached when the amplification becomes equal to the optical losses. This criterion defines the steady-state amplitude of a light wave in the optical resonator. These considerations are valid for any type of laser.

Let us consider a semiconductor injection laser. Our goal is to find the threshold current for lasing. For an injection laser, the feedback is usually obtained by cleaving the crystal planes normal to the plane of the p–n junction. In Fig. 7.19, a device with two cleaved surfaces forming an optical resonator is shown. For the light reflected from the crystal boundaries, we define the reflection coefficient as

$$r = \frac{\mathcal{I}_{\mathrm{r}}}{\mathcal{I}_{\mathrm{in}}},$$

where $\mathcal{I}_{\text{in}}$ and $\mathcal{I}_{\text{r}}$ are the intensities of incident and reflected light, respectively. The reflection coefficient for an air/semiconductor boundary is

$$r = \left(\frac{n_{\text{ri}} - 1}{n_{\text{ri}} + 1}\right)^2.$$

Since semiconductor materials usually have larger refractive indices, n_{ri}, than that of air, the coefficients r are large enough. The intensity of the light transmitted through this mirror is

$$\mathcal{I}_{\text{out}} = (1 - r)\mathcal{I}_{\text{in}}.$$

Let two cleaved surfaces be characterized by two coefficients, r_1 and r_2. After two passages through the device, the light intensity is attenuated by the factor $r_1 r_2$. We can define an effective overall distributed coefficient of optical losses:

$$\alpha_{\text{r}} \equiv \frac{1}{2\mathcal{L}_x} \ln\left(\frac{1}{r_1 r_2}\right).$$

Here, $\mathcal{L}_x$ is the distance between the cleaved surfaces. Note that the optical cavity area, $\mathcal{L}_x\mathcal{L}_y$, may be larger than the active area, $L_x L_y$, of the device. In principle, there can be other sources of optical losses in the resonator. Let them be characterized by the absorption coefficient, α_{s}. Then the total loss coefficient is

$$\alpha_{\text{ls}} = \alpha_{\text{r}} + \alpha_{\text{s}}.$$

If α^{ampl} is the amplification coefficient of some light mode in this resonator, we can write the *criterion for laser oscillations* as

$$\alpha^{\text{ampl}} \geq \alpha_{\text{ls}} = \alpha_{\text{r}} + \alpha_{\text{s}}. \tag{7.29}$$

For injection lasers, the criterion of Eq. (7.29) is a condition imposed on the magnitude of the injection current density, j. Using Eq. (7.12) for $\alpha(j)$ we find the threshold injection current density necessary for laser oscillations:

$$j_{\text{th}}^{\text{l}} = \frac{\alpha_{\text{ls}} + \gamma}{\gamma} j_{\text{th}}, \tag{7.30}$$

where j_{th} is the threshold current density for population inversion. If the loss coefficient is small, $\alpha_{\text{ls}} \ll \gamma$, the threshold currents nearly coincide.

The threshold *total* current $I_{\text{th}}^{\text{l}} = j_{\text{th}} L_x L_y$ is the key parameter characterizing laser diodes. Equation (7.30) shows that in order to minimize j_{th}^{l} one needs to minimize the optical losses α_{ls} and optimize the structure parameters. For numerical estimates, we can use the example of the InGaAsP amplifier with a homogeneous p–n junction which was considered in Section 7.2.3. We have $j_{\text{th}} = 4.6 \times 10^4$ A/cm^2 for a diode with sizes $d \times L_x \times L_y = 2$ μm $\times$ 10 μm $\times$ 200 μm and $\gamma = 600$ cm^{-1} at $T =$ 300 K. The reflection coefficient is $r = 0.3$ and the corresponding effective absorption coefficient is $\alpha_{\text{r}} = 60$ cm^{-1} for $\mathcal{L}_x = L_x$. We suppose that other intracavity losses give the same absorption $\alpha_{\text{s}} = \alpha_{\text{r}}$ and thus, $\alpha_{\text{ls}} = 120$ cm^{-1}. According to Eq. (7.30), we find the laser threshold current density $j_{\text{th}}^{\text{l}} = 1.2 \times j_{\text{th}} = 5.6 \times 10^4$ A/cm^2 and the total current $I_{\text{th}}^{\text{l}} = 1.1$ A. The current through a device with a homogeneous p–n junction is so high that for continuous-wave operation of this laser diode it is necessary to provide effective heat removal from the device. In contrast, for a double-heterostructure device, where the characteristic currents are one order of magnitude lower, laser operation at room temperature becomes possible. For example, if we assume that the size of the active region of the double heterostructure is $d = 0.1$ μm and the optical confinement factor is $\Gamma = 1$, we obtain the threshold current density $j^{1}_{\rm th} = 2.3 \times 10^{3}$ A/cm^2 and the total threshold current of the laser operation is $I^{1}_{\rm th} = 55$ mA.

7.3.3 Laser Output Power and Emission Spectra

According to Eq. (7.29), laser oscillations are possible for those resonator modes which have the lowest optical losses. As we have seen in Chapter 4, the optical resonator strongly discriminates different optical modes, and a relatively small number of them enjoy low losses as a result of their high quality factors. Below the threshold, $j < j^{1}_{\rm th}$ and there is no stimulated emission, so the intracavity photon flux Υ of these high-quality modes is almost zero. Above threshold, this flux grows with time. According to Eqs. (4.51), (4.52), and (7.16), we can rewrite the differential equation for the photon flux taking into account the total loss coefficient, $\alpha_{\rm ls}$:

$$\frac{d\Upsilon}{dt} = \frac{c}{\sqrt{\kappa}} \Upsilon(\alpha(\Upsilon) - \alpha_{\rm ls}) \,. \tag{7.31}$$

It is important that the amplification coefficient is now a function of Υ. Above threshold, when $\Upsilon \neq 0$, in the steady state the amplification coefficient is equal to the total loss coefficient:

$$\alpha(\Upsilon_0) = \alpha_{\rm ls}, \tag{7.32}$$

where Υ_0 is the *photon flux in the steady state*. In order to calculate $\alpha(\Upsilon_0)$ at fixed injection current density, j, we should return to the balance equation for the carrier concentration; see Eq. (7.15). Now, we can rewrite this equation as

$$\frac{dn}{dt} = \mathcal{R} - \frac{n}{\tau} - \alpha(\Upsilon)\frac{\Upsilon}{\mathcal{L}_y d} = 0, \tag{7.33}$$

where $\mathcal{R} = j/ed$ is the injection rate density, and τ is the electron lifetime. According to Eq. (4.43), the last term in Eq. (7.33) represents the stimulated emission rate per unit volume; see also Eq. (7.16). Thus, instead of Eq. (7.11), we obtain

$$n_0 = \frac{j}{ed}\tau - \frac{\alpha_{\rm ls}\tau}{\mathcal{L}_y d}\Upsilon_0. \tag{7.34}$$

Let us use Eq. (7.6) for the calculation of $\alpha(n)$. Now we have three algebraic equations, (7.32), (7.6), and (7.34), for the three variables α, n_0, and Υ_0, which can be easily solved. For the intracavity photon flux we obtain

$$\Upsilon_0 = \begin{cases} 0, & j < j^{1}_{\rm th}, \\ \eta_{\rm i}(I - I^{1}_{\rm th})/e, & j > j^{1}_{\rm th}. \end{cases} \tag{7.35}$$

Here we have introduced the internal efficiency of the stimulated emission as $\eta_{\rm i} \equiv \mathcal{L}_y/(\alpha_{\rm ls}A) \equiv 1/(\alpha_{\rm ls}\mathcal{L}_x)$ and the current $I = j\mathcal{L}_x\mathcal{L}_y$.

Since the light transmission through the mirrors is $(1 - r_{1,2})$, the laser output flux has the form

$$\Upsilon_{\rm out} = (1 - r)\eta_{\rm i}\frac{I - I_{\rm th}^{\rm l}}{e} \equiv \eta_{\rm ext}\frac{I - I_{\rm th}^{\rm l}}{e}, \tag{7.36}$$

where we have assumed that $r_1 = r_2 = r$ and have introduced the external quantum efficiency of the laser diode,

$$\eta_{\rm ext} = (1 - r)\eta_{\rm i}.$$

Finally, above the laser threshold the laser output power equals

$$\mathcal{P}_{\rm out} = \hbar\omega\eta_{\rm ext}\frac{I - I_{\rm th}^{\rm l}}{e}. \tag{7.37}$$

In Fig. 7.20, the light–current characteristic of Eq. (7.37) is shown for an ideal laser diode and is compared with that of a real InGaAsP injection laser emitting light at 1.3 μm. For the latter, the characteristic saturates at currents greater than 75 mA; this is not shown in the figure.

In order to illustrate a typical laser output power, let us use our numerical example of the InGaAsP laser. For the double-heterostructure design, when the laser threshold current is 55 mA and when $\eta_{\rm ext} = 0.25$, one obtains an output power $\mathcal{P}_{\rm out} = 4.9$ mW.

The special distribution of the output light is one of the important characteristics which can be improved considerably by stimulated emission. In general, several

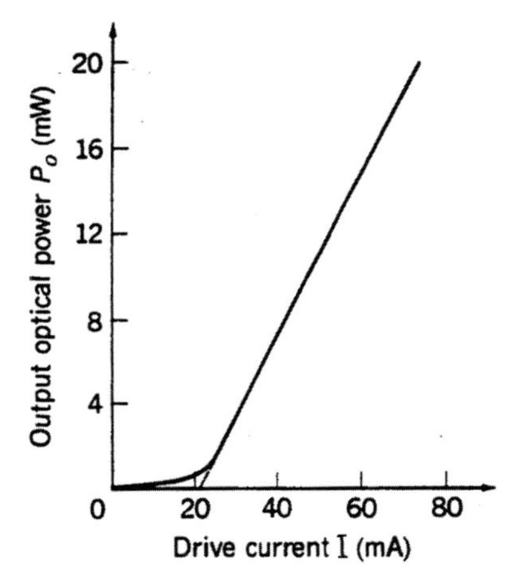

Figure 7.20 The light–current characteristic for an ideal (straight line) and real (solid curve) InGaAsP injection laser operating at $\lambda = 1.3$ μm. The laser uses an index-guided heterostructure. Reprinted with permission from Fig. 16.3-4 of *Fundamentals of Photonics*, B. E. A. Saleh and M. C. Teich, 1st edn. Copyright ©1991 John Wiley & Sons, Inc.

factors govern the spectral distribution. One such factor is that stimulated emission is possible only in the spectral region where the amplification exceeds the optical losses: $\alpha(\omega) > \alpha_{ls}$. Then, the oscillations are possible only for resonator modes with high quality factors. In the case of a resonator with plane mirrors, these modes are separated by a spacing $\Delta\omega = \pi c/(\mathcal{L}_x\sqrt{\kappa})$; see Section 4.2.3. For the InGaAsP laser with a central wavelength $\lambda = 1.3$ μm and $\mathcal{L}_x = 300$ μm, the intermode spacing is $\Delta\lambda = 0.9$ nm while the bandwidth of the amplification coefficient approximately equals that of the spontaneous emission of Eq. (7.26) and is about 7 nm. Thus, approximately eight longitudinal modes can be generated. Figure 7.21 shows a specific example of the spectral distribution of a 1.3 μm InGaAsP index-guided heterostructure laser, where light confinement is provided by variation of the refractive index. Comparing this figure with Fig. 7.15, where the spontaneous spectra are presented, one can conclude that the laser spectra are much narrower. By reducing

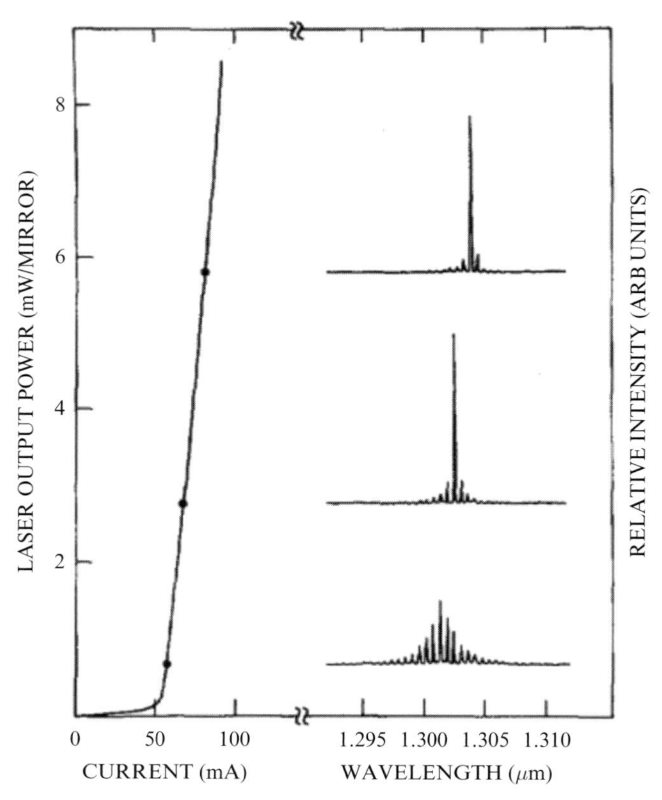

Figure 7.21 The spectral distribution of the emission from a 1.3 μm InGaAsP index-guided heterostructure laser, at three different output powers that are indicated on the left curve, of the power dependence on the post-threshold current. ©1981 IEEE. Reprinted, with permission, from Fig. 6 of R. J. Nelson, *et al.*, "CW electrooptical properties of InGaAsP ($\lambda = 1.3$ μm) buried-heterostructure lasers," *IEEE J. Quantum Electron.* **QE-17**, pp. 202–207 (1981).

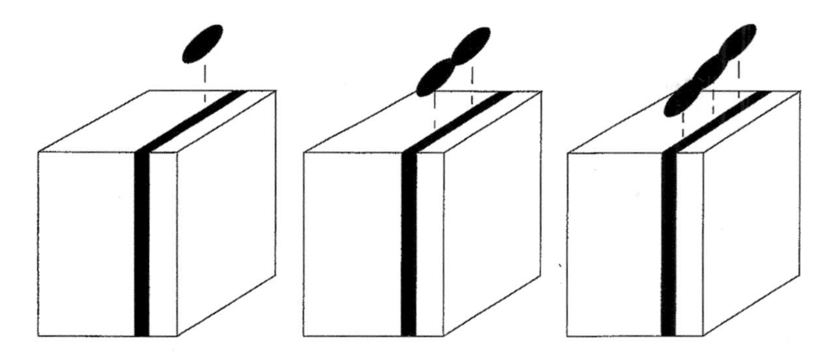

Figure 7.22 Schematic illustration of the spatial distributions of the light output for the lowest waveguide modes.

the resonator length to about 30 μm to 40 μm, one can obtain only one longitudinal mode within the amplification bandwidth; in this case, single-mode generation is possible, the linewidth of the generated mode being approximately equal to 0.01 nm.

Another advantage of stimulated emission is the sharp spatial distribution of the output power. Let us consider an injection-current laser. The spatial distribution of the laser emission is determined by the number and type of the resonator modes which are excited under lasing conditions. Suppose we have an index-guided double-heterostructure laser. The dielectric waveguide has a cross-sectional area $d\mathcal{L}_y$. For simplicity we can treat this waveguide as a dielectric slab with the same rectangular cross-section. This waveguide has modes with different transverse structures. Since the ratio d/λ is small, the waveguide maintains only a single mode in the transverse direction perpendicular to the plane of the p–n junction. But, since $\mathcal{L}_y$ is much greater than λ, the waveguide maintains several modes having spatial structure in the direction parallel to the junction; these modes are called *lateral modes*. Fortunately, only a few of these lateral modes enjoy low optical losses and can be excited during laser operation. Note that the number of excited lateral modes can be reduced by reducing the transverse device size $\mathcal{L}_y$. Figure 7.22 illustrates the generation of several transverse modes. We can estimate the far-field angular divergence of the laser emission. In the plane perpendicular to the junction, the angular divergence is approximately λ/d, while in the plane parallel to the junction it is approximately $\lambda/\mathcal{L}_y$. For example, if the cross-section of the waveguide is 2×10 μm^2, the two angles are 40° and 8.5°. Thus, the laser diode has a significantly sharper spatial distribution of emission than do light-emitting diodes operating in the spontaneous regime; see Fig. 7.16.

7.3.4 Modulation of the Laser Output

As was emphasized previously, a major application of advanced emitters is optical communication and processing systems. This application requires high-speed modulation of the light output. Unlike other lasers that are modulated externally via changing the mirror transmission, semiconductor lasers can be modulated internally by modulating the injection current. In Section 7.3.1, we considered the modulation of light-emitting diodes. This is an easy task because the number of radiatively recombining electron–hole pairs is directly governed by the injection current. In estimating the laser modulation one should take into account that, in general, the lasing process is nonlinear. Besides pumping, it involves stimulated emission. The steady state is obtained as a result of a self-consistent solution of the nonlinear time-dependent equations (7.31)–(7.33) and (7.6). Now we again suppose that the current consists of dc and ac components as in Eq. (7.25). Then we define the small-signal modulation response of the electron–hole concentration and the photon flux:

$$n - n_0 = n_1 = \delta n\, e^{i\Omega t} + \delta n^*\, e^{-i\Omega t},$$

$$\Upsilon - \Upsilon_0 = \Upsilon_1 = \delta\,\Upsilon\, e^{i\Omega t} + \delta\Upsilon^*\, e^{-i\Omega t}.$$

We rewrite Eqs. (7.31) and (7.33) as

$$\frac{d\Upsilon_1}{dt} = \frac{c}{\sqrt{\kappa}}\Upsilon_0 \frac{\gamma n_1}{n_{\rm th}}$$

and

$$\frac{dn_1}{dt} = \frac{I_1 \cos(\Omega t)}{ed\mathcal{L}_x\mathcal{L}_y} - n_1\left(\frac{1}{\tau} + \Upsilon_0 \frac{\gamma}{n_{\rm th}\mathcal{L}_y d}\right) - \Upsilon_1 \frac{\alpha_{\rm ls}}{\mathcal{L}_y d}.$$

The last two terms in the second equation are related to the self-consistent action of the pumping current and the stimulated emission flux. This system of equations can be easily solved. Since our main interest is in the modulation response, we present here the ratio of the complex amplitude $\delta\Upsilon$ and the amplitude of the current variation, I_1:

$$\frac{\delta\Upsilon}{I_1} = -\frac{c\gamma}{2e\sqrt{\kappa} n_{\rm th}\mathcal{L}_x\mathcal{L}_y d} \times \frac{\Upsilon_0}{\Omega^2 - i\Omega\left[1/\tau + \Upsilon_0\gamma/(\mathcal{L}_y d n_{\rm th})\right] - c\Upsilon_0\gamma\alpha_{\rm ls}/(\sqrt{\kappa} n_{\rm th}\mathcal{L}_y d)}. \tag{7.38}$$

A typical response,

$$\frac{\Upsilon_1}{I_1} \equiv \frac{|\delta\Upsilon|}{I_1},$$

is shown in Fig. 7.23 for different voltage biases. The curves labeled 1 to 5 correspond to an increase in the bias current. The response curve is flat at low frequencies, as in the case of light-emitting diodes described by Eq. (7.26), but unlike in that case it has a peak at the frequency

$$\Omega_{\rm m} = \sqrt{\frac{\gamma\Upsilon_0\alpha_{\rm ls}c}{n_{\rm th}\mathcal{L}_y d\sqrt{\kappa}} - \frac{1}{2}\left(\frac{1}{\tau} + \frac{\gamma\Upsilon_0}{n_{\rm th}\mathcal{L}_y d}\right)^2}. \tag{7.39}$$

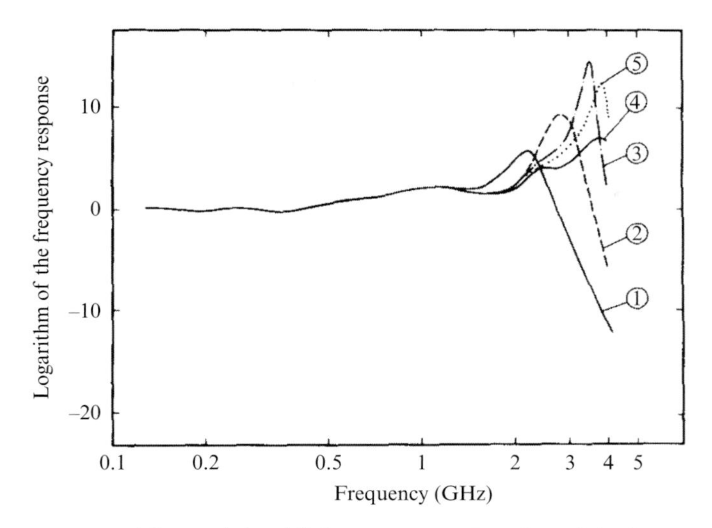

Figure 7.23 The modulated light output as a function of the frequency of the current modulation at different dc currents, whose magnitudes increase with the numbers at the curves. Reprinted from K. Y. Lau, *et al.*, "Superluminescent damping of relaxation resonance in the modulation response of GaAs lasers," *Appl. Phys. Lett.* **43**, pp. 329–331 (1983), with the permission of AIP Publishing.

For a typical semiconductor laser, only the first term under the square root in Eq. (7.39) is important so that, to a good degree of accuracy,

$$\Omega_{\mathrm{m}} = \sqrt{\frac{\gamma \Upsilon_0 \alpha_{\mathrm{ls}} c}{n_{\mathrm{th}} \mathcal{L}_y d \sqrt{\kappa}}}. \tag{7.40}$$

If Ω exceeds the value of Eq. (7.40), the response decreases very fast. We can introduce the modulation bandwidth as

$$f = \frac{1}{2\pi} \Omega_{\mathrm{m}} = D\sqrt{\mathcal{P}_{\mathrm{out}}}, \tag{7.41}$$

where D is some constant. (Compare this with Eq. (7.27) for light-emitting diodes.) Equations (7.40) and (7.41) show that, to increase the modulation response, one needs to increase the light amplification, to decrease the optical losses, and to operate at as high a photon flux as possible. For advanced laser structures, modulation bandwidths as high as hundreds of GHz are achieved.

In closing this section, it is worth stressing that, like light-emitting diodes, semiconductor lasers operate over a very wide spectral region: from the middle infrared to the near ultraviolet region. In Fig. 7.24, data on spectral regions of different semiconductor lasers are presented. Most of these laser systems operate with injection-current pumping. Some of them require optical excitation. Semiconductor lasers operating at $\lambda > 3\ \mu\mathrm{m}$ usually require cooling below room temperature.

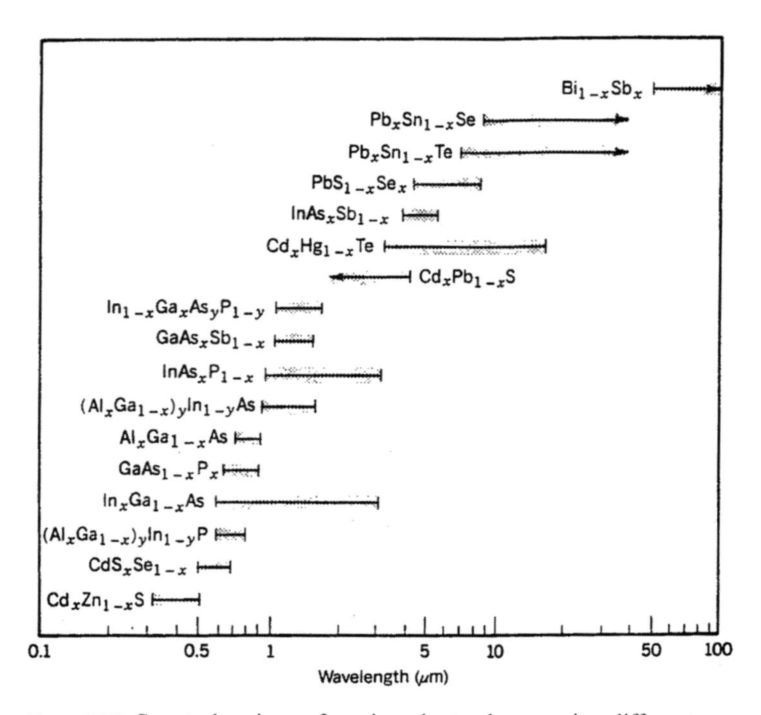

Figure 7.24 Spectral regions of semiconductor lasers using different compounds. The lasers emitting in the region $\lambda > 3$ μm usually operate at low temperatures. Reprinted with permission from Fig. 16.3-11 of *Fundamentals of Photonics*, B. E. A. Saleh and M. C. Teich, 1st Edition. Copyright ©1991 John Wiley & Sons, Inc.

7.3.5 Quantum Well Lasers

In Section 7.2.5, we analyzed light amplification in heterostructures with quantum wells. Using the results of Section 7.2.5, we now study lasers that are based on light amplification in quantum wells. We start with a consideration of a single quantum well and rewrite Eq. (7.19) in terms of the optical confinement factor of this well, Γ_0, and the coefficient, α_0, calculated for a single well:

$$\alpha(\omega) = \Gamma_0 \alpha_0(\omega).$$

The threshold condition for laser generation as given by Eq. (7.29) now takes the form

$$\Gamma_0 \alpha_0(\omega) > \alpha_s + \frac{1}{2\mathcal{L}_x} \ln\left(\frac{1}{r_1 r_2}\right). \tag{7.42}$$

The intracavity optical losses, α_s, include light scattering out of the waveguide, α_{sc}, and losses in the passive region of the guide, α_{pp}, as well as those of the active region in the quantum well, α_{qw}. Thus, one can write these losses as $\alpha_s = \alpha_{sc} + \Gamma_0 \alpha_{qw} + (1 - \Gamma_0)\alpha_{pp}$. Note that one of the main mechanisms of loss in the passive region of the guide is absorption by free carriers. This mechanism becomes important at high injection levels, when a large number of carriers move over the barrier layers in the guiding region and contribute to α_{pp}.

Since the electric current governs the injection and pumping processes, let us rewrite the criterion of Eq. (7.42) as a condition on the current. Following Eq. (7.12), we can represent $\alpha_0(\omega)$ as

$$\alpha_0 = \gamma\left(\frac{j}{j_{th}} - 1\right), \tag{7.43}$$

where j_{th} is the threshold current density for the establishment of population inversion in a quantum well. Then, the criterion of Eq. (7.42) can be rewritten as

$$j > j_{th}\left(1 + \frac{\alpha_{sc}}{\gamma\Gamma_0} + \frac{1}{2\gamma\Gamma_0\mathcal{L}_x}\ln\left(\frac{1}{r_1 r_2}\right)\right). \tag{7.44}$$

We see the critical role of the optical confinement factor in determining the injection current needed for laser action.

In order to estimate the parameters j_{th} and γ we can use, for example, the data of Fig. 7.13(b) for the InGaAsP quantum well laser. According to these data, one can set $j_{th} \approx 20$ A/cm^2 and $\gamma = 10^3$ cm^{-1}. The optical confinement factor can be estimated by using the results of Fig. 7.12, obtaining $\Gamma_0 = 0.03$. A typical value of α_{sc} is 3 cm^{-1}. Thus, the criterion is

$$j\,(\text{A/cm}^2) > 20 + 2 + \frac{1}{3\mathcal{L}_x}\ln\left(\frac{1}{r_1 r_2}\right).$$

Let the length of the optical cavity, $\mathcal{L}_x$, be 200 μm. For cleaved surfaces we can estimate the reflection coefficients as $r_1 = r_2 = 0.31$. Finally, we obtain

$$j\,(\text{A/cm}^2) > 20 + 2 + 38 = 60,$$

where we intentionally present all three contributions in the inequality of Eq. (7.44) in explicit form. Since the first contribution corresponds to the threshold for population inversion, we can conclude that for this typical example the lasing threshold current is as much as three times higher than that of population inversion. In order to lower the laser threshold one can either increase the mirror reflectivity or increase the cavity length, $\mathcal{L}_x$.

Because of the inherently small transverse size (width) of quantum wells, the optical confinement factor, Γ_0, is always small and can be increased only by adding more quantum wells to the guiding region. If ν_{QW} is the number of quantum wells in the region, for estimates of the threshold current one can use Eqs. (7.42) and (7.44) with

$$\Gamma = \Gamma_0 \nu_{QW}. \tag{7.45}$$

It is easy to see that for values of ν_{QW} of 10 to 30, the laser threshold current can be decreased to practically the value of the current for population inversion, j_{th}, which means that the confinement factor is of the order of 1.

Modulation bandwidth. Important applications of light-emitting diodes and conventional laser diodes in communication systems, optical signal processing, etc., require high-speed modulation of the light output. Hence, the modulation bandwidth is a crucial parameter of these devices. Considering the modulation speed,

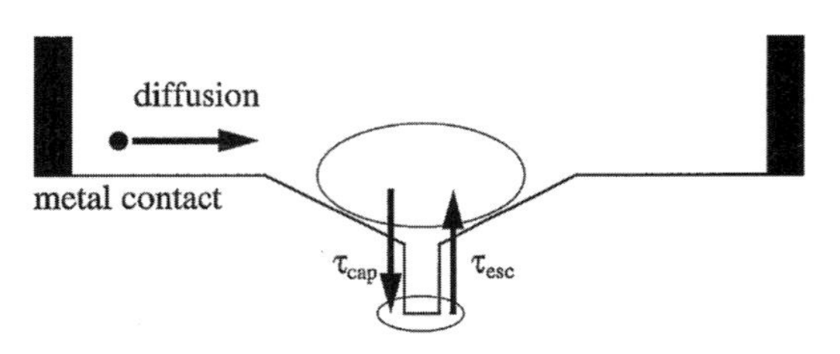

Figure 7.25 A schematic illustration of the main processes which determine the rate of population of the quantum well states.

one must take into account two additional relaxation stages in the process of quantum well pumping. In Fig. 7.25, these two processes are shown schematically: the diffusion (or drift) of charge carriers through the region of the optical confinement and their subsequent capture into the wells. In the optical confinement region, the charge carriers are in extended states. In a quantum well, they are in confined states. The transition between these two types of states can be characterized by a time which usually exceeds the intraband relaxation times. One can introduce the surface concentration of carriers in the extended states, n, and the surface concentration of carriers in the confined states, n_{QW}. Instead of a single equation for the carriers, as in Eq. (7.33), we now write two equations:

$$\frac{dn}{dt} = \mathcal{R} - \frac{n}{\tau} - \left(\frac{n}{\tau_{cap}} - \frac{n_{QW}}{\tau_{esc}}\right), \tag{7.46}$$

$$\frac{dn_{QW}}{dt} = \left(\frac{n}{\tau_{cap}} - \frac{n_{QW}}{\tau_{esc}}\right) - \frac{n_{QW}}{\tau^*} - \alpha\frac{\Upsilon}{\mathcal{L}_y d}, \tag{7.47}$$

where $\mathcal{R}$ is the injection rate density, τ and τ^* are the lifetimes due to interband recombination in extended and confined states, respectively, and τ_{cap} and τ_{esc} are the *effective times* of the capture and escape processes. The last term in Eq. (7.47) is the stimulated emission rate; see Eq. (7.33). The effective time τ_{cap} includes the processes of delivery of carriers to the quantum well layer and subsequent quantum transitions of carriers from extended to confined states. If the delivery process is carrier diffusion, one can estimate these parameters as

$$\tau_{cap} = \frac{d^2}{2D} + \tau_{cap}^{q}\frac{d}{L}, \tag{7.48}$$

where D is the diffusion coefficient and τ_{cap}^{q} is the average lifetime of the carriers in the region of the quantum well due to quantum transitions into the well. The first term is equal to the time of carrier diffusion over a distance d. The intrinsic quantum capture time, τ_{cap}^{q}, is scaled up by a factor d/L. Recall that L is the width of the well. For typical values of $\tau_{cap}^{q} \approx 0.5$ ps and $d/L = 10$, the maximum modulation bandwidth due to quantum capture is $1/\tau_{cap} \approx 30$ GHz. The diffusion contribution in Eq. (7.48) dominates for $d \geq 0.2$ μm; it increases τ_{cap} and limits the modulation width. Structures with graded optical confinement regions, as in Fig. 7.12, have built-in forces which can produce drift-like carrier motion and considerably decrease the delivery time. (Such a drift effect in a built-up effective field was discussed in detail in Section 2.5.2.) In this case, the intrinsic quantum capture restricts τ_{cap}. Therefore, we can conclude that, with proper design of the narrow optical confinement region, quantum well lasers can be modulated by an injection current with frequencies up to 30 GHz and higher.

Strained-layer quantum well lasers. So far we have considered examples of AlGaAs and InGaAsP quantum well lasers which are based on lattice-matched heterostructures. In some cases, lasers based on *lattice-mismatched heterostructures* demonstrate superior properties. These heterostructures were briefly analyzed in Chapters 2 and 3. Lasers using lattice-mismatched heterostructures can operate in spectral regions different from those of lasers based on lattice-matched structures, and they can provide lower threshold currents or higher amplification. Let us consider briefly a strained-layer GaAs/InGaAs/GaAs laser. The lattice constants of both materials are significantly different: $a = 6.0585$ Å (InAs) and $a = 5.6535$ Å (GaAs). If a thin layer of InGaAs is embedded between two thick GaAs layers, the InGaAs layer experiences a considerable in-plane compression. This compressive strain changes the band structure of the layer in three ways: (1) it increases the bandgap, as shown in Fig. 7.26(a); (2) it removes the degeneracy at $\vec{k} = 0$ between the light and heavy holes, as illustrated in Fig. 7.26(b); and (3) it makes the valence band strongly anisotropic. In Fig. 7.26(b), the negative and positive wavevectors are presented for different directions. As can be seen from Fig. 7.26(b), in the directions parallel to the layer plane, the topmost band has a light effective mass, while in the direction perpendicular to the plane the same band has a heavy effective mass. These changes in the energy spectrum of the valence band improve the performance of strained-layer lasers. First, the peak emission of the laser is shifted to higher energies due to an

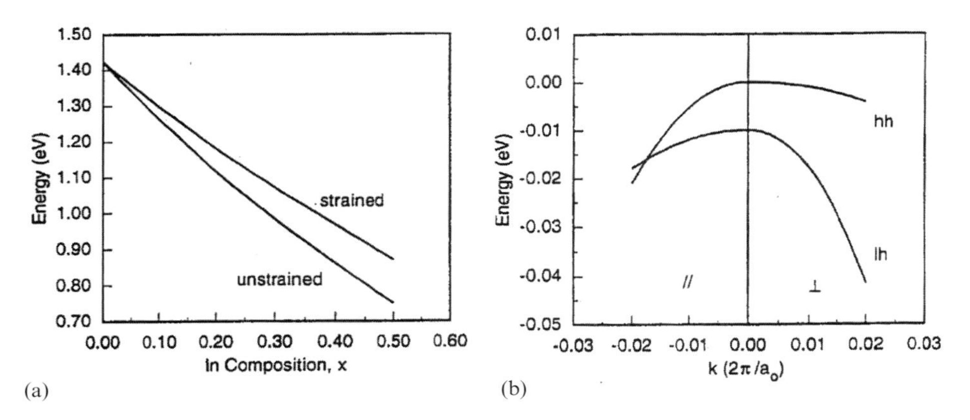

Figure 7.26 The bandgap and energy dispersion in hole bands under strain: (a) the bandgaps for strained and unstrained $In_xGa_{1-x}As$ as a function of the indium component, and (b) the heavy-hole and light-hole valence band structure resulting from a biaxial compression for $In_xGa_{1-x}As$. Reprinted from figures 9 and 11 of *Quantum Well Lasers*, 1st Edition, P. S. Zory Jr., *et al.*, "Chapter 8 – Strained layer quantum well heterostructure lasers," pp. 367–413, Copyright 1993, with permission from Elsevier.

increase in the energy of the lowest hole subband of the strained quantum well layer. Second, the decrease of in-plane effective mass leads to a reduction of the threshold current. This result follows from the criterion for population inversion expressed by Eqs. (7.1) and (7.4). Since in the strained layer the hole effective mass decreases, the quasi-Fermi level of the holes increases at the same carrier concentration. Thus the difference in the quasi-Fermi levels increases under strain and the threshold injection current decreases.

As an example of the family of strained-layer lasers, we mention InGaAsP devices operating within the broad wavelength region from 0.9 μm to 1.55 μm. One particular laser fabricated from a multiple-quantum-well structure with 20 Å-thick $In_{0.78}Ga_{0.22}As$ quantum wells and 200 Å barriers operates at $\lambda = 1.55$ μm with a sub-milliampere threshold current. Another example is a GaInP/InGaAlP strained-layer quantum well laser which emits light with $\lambda = 634$ nm and achieves an output power in excess of 0.5 W.

7.3.6 Surface-Emitting Lasers

Thus far, we have studied the amplification of light propagating along quantum well layers. In Chapter 5, we mentioned another possible design where light propagates perpendicular to the layers in a so-called vertical geometry. The amplification of light passing through a quantum well layer can be defined as

$$\frac{\mathcal{I}_{\text{out}} - \mathcal{I}_{\text{in}}}{\mathcal{I}_{\text{in}}} \equiv \beta.$$

Here $\mathcal{I}_{\text{in}}$ and $\mathcal{I}_{\text{out}}$ are the input and output light intensities, respectively. The quantity β can be expressed through $\alpha_0(\omega)$ of Eq. (7.19) as

$$\beta(\omega) = \alpha_0(\omega)L,$$

where L is the quantum well width. It is easy to see that β is typically very small. For example, if $\alpha_0 = 100\ \text{cm}^{-1}$ and $L = 100$ Å, we get $\beta = 10^{-4}$. In order to obtain laser oscillations in a vertical-geometry structure, one should employ a multiple-quantum-well structure and provide near-perfect mirrors with extremely high reflection. Figure 7.27(a) shows the scheme of a *surface-emitting laser*. The laser design includes an active region providing high light gain, dielectric multilayers, metallic contacts, and implanted regions which form the light output. Layered dielectric mirrors give very high reflection, while the active region contains a multiple-quantum-well structure. The lateral sizes of this laser can be reduced to the 1–10 μm range. A decrease in the surface area of the diode leads to a considerable decrease in the magnitude of the threshold current. In the case of the quantum well structures just considered, we can assume a characteristic current density of about 100 A/cm^2; accordingly, if the lateral sizes are each 5 μm, the pumping surface area is 2.5×10^{-7} cm^2 and the threshold current equals 25 μA.

Surface-emitting quantum well lasers offer the advantage of a high packing density. Nowadays, technology allows one to fabricate an array of about 10^6

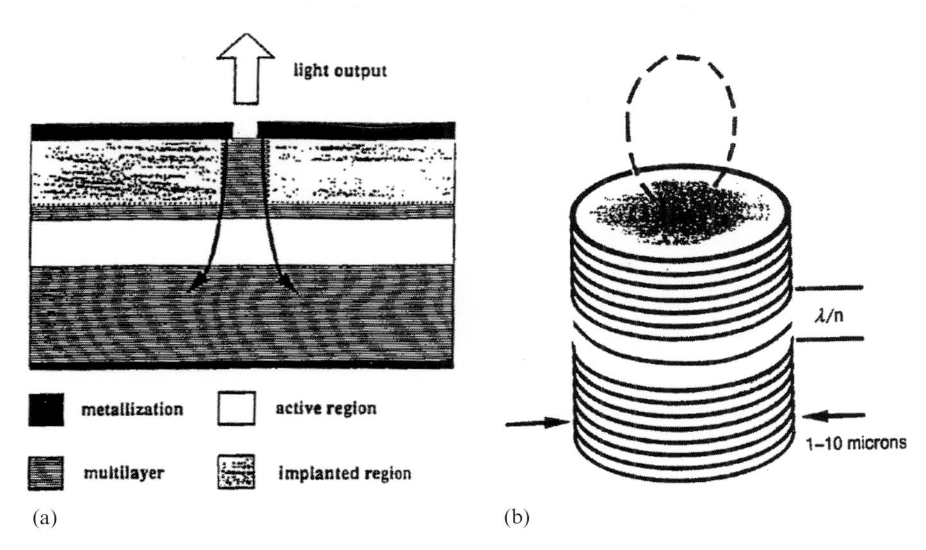

Figure 7.27 Surface-emitting lasers: (a) a schematic diagram of a surface-emitting laser and (b) a surface-emitting microlaser. Dielectric mirrors, the active region (with a width of λ/n), and the output pattern (dashed line) of the lowest mode are indicated.

surface-emitting electrically pumped lasers. The structure of such a single microlaser is shown in Fig. 7.27(b). The cylindrical boundary is formed by an etching procedure that results in lateral dimensions of less than 10 μm. The dominant mode pattern emitted from this microlaser is shown schematically. Microlasers can operate at room temperature and have threshold currents of the order of 0.1 mA.

7.4 Blue and Ultraviolet Light-Emitting Diodes

Among light-emitting diodes of different spectral regions, the devices operating in the blue and ultraviolet regions (wavelengths below 400 nm) are especially important because of their numerous applications, which include white lighting, high-density optical storage, ultraviolet sensing systems, uses in the medical and biochemical fields (sterilization and decontamination), etc. In the diagram in Fig. 7.28, different applications of the short-wavelength emitters are indicated, with an acceptable spectral range for every particular application.

7.4.1 Nitride-Based Light-Emitting Diodes

It is clear that efficient short-wavelength-emitting devices may be realized with the use of wide-bandgap semiconductor materials and their heterostructures. The direct-bandgap group-III nitrides studied in Section 4.4.5 represent a suitable class of materials for such emitting devices. For an extended period of time, researchers tried unsuccessfully to develop techniques for the growth of high-quality

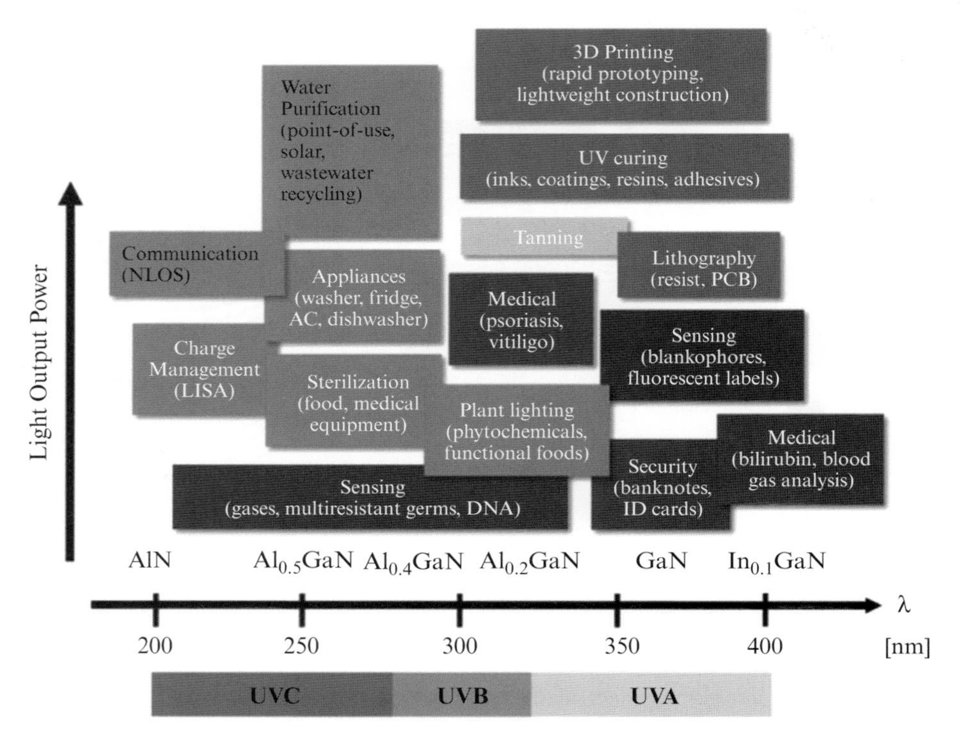

Figure 7.28 Applications of ultraviolet emitters. The different spectral regions are 400–320 nm, UVA; 320–280 nm, UVB; and 280–200 nm, UVC. Adapted by permission from Fig. 1.2 of Springer Nature: Springer, *III-Nitride Ultraviolet Emitters* by M. Kneissl and J. Rass ©2016.

group-III nitride materials and heterostructures. Additionally, p-doping, which is necessary for light-emitting devices, has proven to be a difficult problem. Recently, however, progress in the fabrication of high-quality single-crystal GaN and of ternary and quaternary alloys such as AlGaN, InGaN, and AlGaInN, and the success of the p-type doping technologies for these wide-bandgap materials, have facilitated the realization of short-wavelength-emitting devices.

The electronic structure of AlGaInN-based materials is very similar to that of the GaN crystal studied in Section 4.4.5. The edge optical absorption and emission of AlGaInN alloys are illustrated by the diagram shown in Fig. 7.29, where two important parameters – the energy bandgaps, E_g, and the lattice constants, a_0, are presented. For the binary compounds AlN, GaN, and InN, these parameters are indicated by the open circles (see also Table 4.2 and Fig. 7.15). The full lines connecting pairs of open circles represent the $E_g(a_0)$ dependences for the ternary compounds. For example, the full line connecting the GaN and AlN parameters presents the $E_g(a_0)$ dependence for the ternary AlGaN compound. The full lines restrict the region of the parameters for the quaternary AlGaInN compounds. The region of the parameters suitable for short-wavelength emitters (allowed wavelength

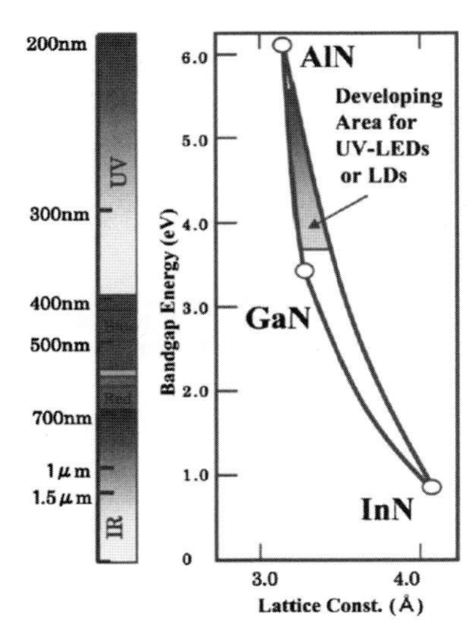

Figure 7.29 A diagram of the direct-transition energy bandgap and corresponding light wavelength versus lattice constant of AlGaInN materials. The region of the parameter space that is suitable for short-wavelength emitters is indicated by shading. Reprinted from Fig. 1 of H. Hirayama, *J. Appl. Phys.* **97**, 091101 (2005), with the permission of AIP Publishing.

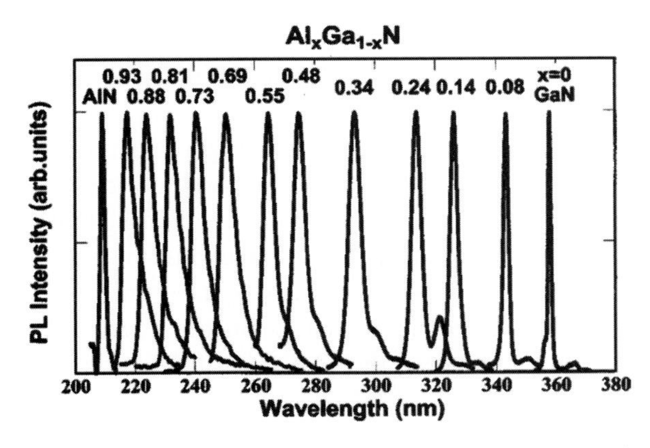

Figure 7.30 Spectra of the photoluminescence observed from $Al_xGa_{1-x}N$ alloys over the compositional range with x from 0 to 0.98, measured at 77 K. Reprinted from Fig. 3(a) of H. Hirayama, *J. Appl. Phys.* **97**, 091101 (2005), with the permission of AIP Publishing.

less than 350 nm) is shaded. This diagram facilitates the design and fabrication of heterostructure devices emitting specific short-wavelength light.

According to the diagram in Fig. 7.29, the ternary AlGaN compounds facilitate the coverage of the short-wavelength optical region. This is supported by direct measurements of the photoluminescence from different $Al_xGa_{1-x}N$ compositions with x from 0 to 0.98, which are presented in Fig. 7.30 for 77 K. The corresponding emitting devices include a series of multi-quantum $Al_xGa_{1-x}N$ wells separated by

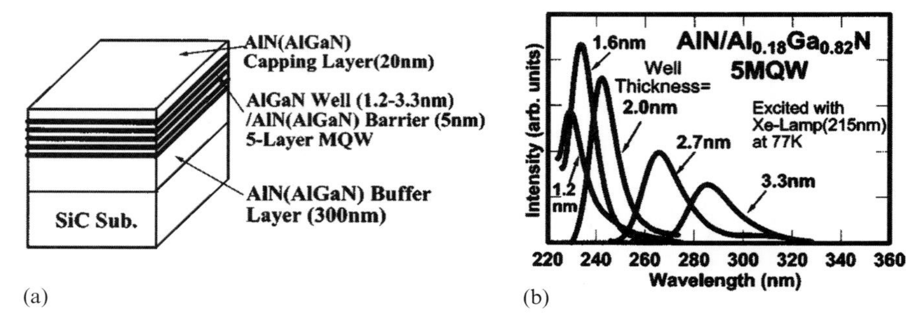

Figure 7.31 (a) An $AlN/Al_{0.18}Ga_{0.82}N$ heterostructure with five quantum well layers grown on a 300 nm AlN buffer over the SiC substrate. (b) Photoluminescence spectra of heterostructure devices of various well thicknesses at $T = 77$ K. Reprinted from Figs. 4 and 5(a) of H. Hirayama, *J. Appl. Phys.* **97**, 091101 (2005), with the permission of AIP Publishing.

AlN barriers. In Fig. 7.31(a) a sketch of such a device is presented for the case of five quantum well layers. It is supposed that the emitting device is grown on an SiC substrate. The emission spectra of such devices are dependent on the material of the quantum well layers, as well as on their thicknesses. Figure 7.31(b) illustrates the significant shift of the emission spectra for $AlN/Al_{0.18}Ga_{0.82}N$ heterostructures when the quantum well thickness is changed from 1.2 nm to 3.3 nm. This shift is obviously due to the quantization of the electron and hole energies in the quantum wells. The most efficient emission was observed for wavelength 234 nm and well thickness ≈ 1.6 nm. As discussed for the multiple-quantum-well device, at $T = 77$ K the intensity of the emission was 20 to 30 times larger than that of the bulk $Al_xGa_{1-x}N$ materials. However, the emissions from the AlGaN- and GaN-based quantum wells are much weaker at room temperature. Enhancement of the room temperature emission of such devices requires further improvement in the quality of the AlGaN heterostructures (for example, reducing the threading-dislocation density on the AlGaN buffer).

Another approach for the realization of short-wavelength-emitting devices is the use of quaternary AlGaInN compounds. Indeed, Fig. 7.29 illustrates that short-wavelength emission is possible also with a small proportion of In added to AlGaN alloys. From studies of emission from GaAs and GaN quantum well layers comprising In, it is known that the incorporation of a small amount of In leads to efficient emission even at room temperature. The reason for this phenomenon is related to the formation of In segregates in quantum well layers. The In segregates produce additional potential wells localizing both electrons and holes, which results in the enhancement of radiative electron–hole recombination. These In-segregation effects can also be applicable to InAlGaN-based quantum wells: the room-temperature short-wavelength emission of AlGaN-based quantum wells may be significantly increased by the incorporation of a small amount of In. In Fig. 7.32(a), the design

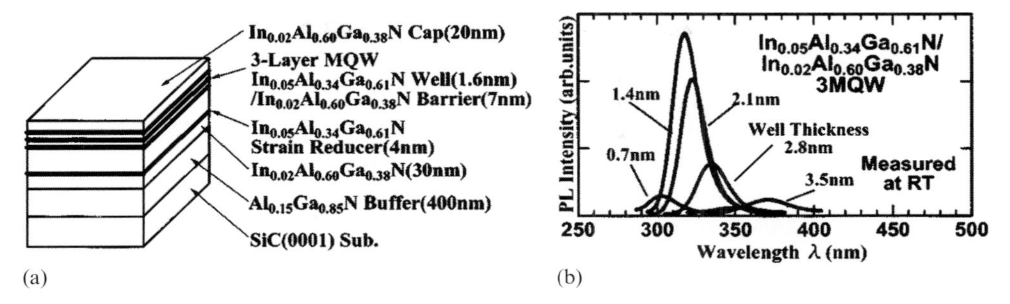

Figure 7.32 (a) The $In_{0.05}Al_{0.034}Ga_{0.61}N/In_{0.02}Al_{0.6}Ga_{0.32}N$ heterostructure with three quantum well layers grown on the SiC substrate. The thicknesses and sequence of the layers are indicated. (b) Photoluminescence spectra of heterostructure devices of various well thicknesses at room temperature. Reprinted from Figs. 12 and 13 (b) of H. Hirayama, *J. Appl. Phys.* **97**, 091101 (2005), with the permission of AIP Publishing.

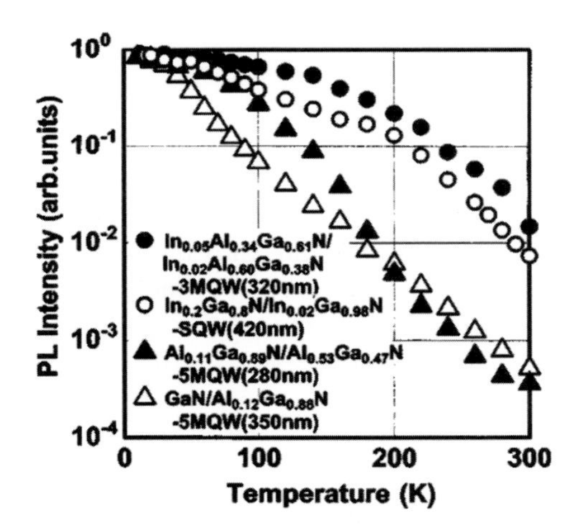

Figure 7.33 Temperature dependences of photoluminescence intensity of the InAlGaN devices (see Fig. 7.32(a)) and the AlGaN devices (see Fig. 7.31(a)) with optimized quantum well layer thicknesses. Reprinted from Fig. 15 of H. Hirayama, *J. Appl. Phys.* **97**, 091101 (2005), with the permission of AIP Publishing.

of a particular InAlGaN heterostructure with three $In_{0.05}Al_{0.034}Ga_{0.61}N$ quantum well layers and $In_{0.02}Al_{0.6}Ga_{0.32}N$ barrier layers is shown. The emission spectra of such devices with different quantum well thicknesses are presented in Fig. 7.32(b). The most efficient emission was obtained at wavelength 318 nm and with optimal thickness ≈ 1.4 nm.

Measured temperature dependences of the photoluminescence intensity of InAlGaN and AlGaN devices with optimized quantum well thicknesses are collected in Fig. 7.33. The intensity of each sample is normalized with respect to the intensity obtained at 20 K. It can be seen that the temperature dependence of the emission

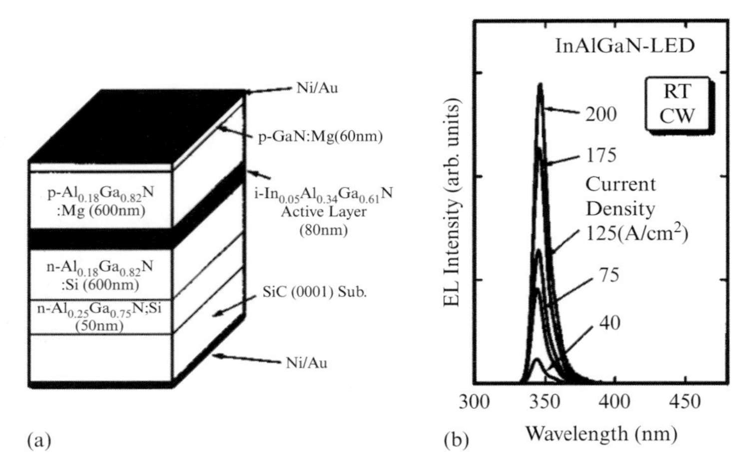

Figure 7.34 (a) A schematic representation of the structure of an electrically pumped quaternary InAlGaN light-emitting diode. (b) Room-temperature emission with wavelength 345 nm at different electric currents. Reprinted from Fig. 17 of H. Hirayama, *J. Appl. Phys.* **97**, 091101 (2005), with the permission of AIP Publishing.

for the InAlGaN devices is much better (one to two orders of magnitude higher) than that of the AlGaN devices.

Summarizing, intense ultraviolet emission at room temperature in the wavelength range 350 nm to 290 nm has been achieved for InAlGaN multiple-quantum-well heterostructures. These results indicate that quaternary InAlGaN is very promising for use as the active medium of bright ultraviolet-light-emitting devices and laser diodes.

As discussed above, to realize electrically pumped light-emitting devices it is necessary to create p-doped and n-doped regions, i.e., to fabricate p–n or p–i–n structures. Note that, for group-III nitrides, as well as other wide-bandgap semiconductors, obtaining highly doped p-type regions is a difficult problem. The difficulties in achieving high hole concentrations originate from (i) the relatively low solubility of the typical acceptor elements (Mg, Zn, C, etc.), (ii) neutralization of these acceptors by the formation of complexes with hydrogen and other material defects, and (iii), most importantly, the deep energy levels of the known acceptors (for example, the ionization energy of Mg in GaN is about 250 meV). The fabrication of heavily p-doped device regions is even more difficult to realize for AlGaN materials, in which the energies of the acceptor levels are found to be larger.

The structure of a short-wavelength-emitting diode based on a quaternary InAlGaN heterostructure is presented schematically in Fig. 7.34(a). For this structure the active region, which consists of multiple quantum wells, as shown in Figs. 7.31(a) and 7.32(a), is placed between the p- and n-doped regions. Both the p and n regions have metallic contacts.

In Fig. 7.34(b), the electroluminescence intensity is shown for an InAlGaN structure consisting of a 50 nm Si-doped AlGaN buffer layer, a 600 nm Si-doped

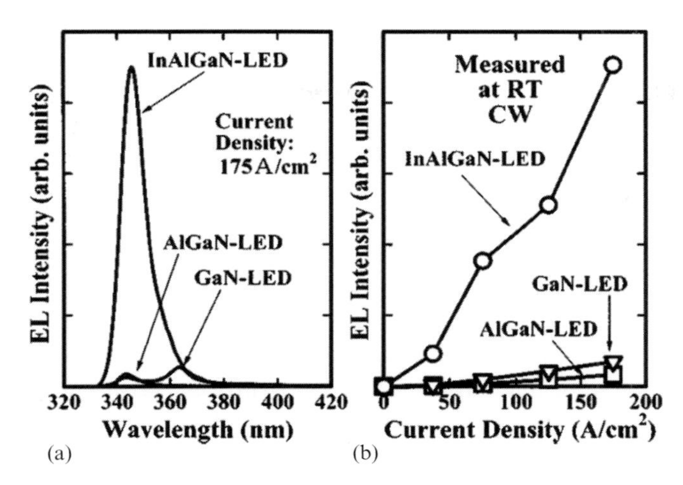

Figure 7.35 A comparison of the emission spectra (a) and emission intensities (b) as functions of the injection current density for emitting devices with three types of group-III-nitride active region. Reprinted from Fig. 18 of H. Hirayama, *J. Appl. Phys.* **97**, 091101 (2005), with the permission of AIP Publishing.

$Al_{0.17}Ga_{0.83}N$ layer, an 80 nm undoped quaternary $In_{0.05}Al_{0.34}Ga_{0.61}N$ active layer, a 600 nm Mg-doped $Al_{0.18}Ga_{0.82}N$ layer, and a thin Mg-doped GaN capping layer. The amount of Mg incorporated into the Mg-doped AlGaN layer was approximately $10^{20}\,cm^{-3}$, while the hole and electron concentrations in the Mg-doped and Si-doped layers were $\approx 10^{17}\,cm^{-3}$. The emission spectra were measured at room temperature under continuous-wave (cw) operation. A single-peaked emission at wavelength 345 nm was observed for this device, with intensity increasing for greater injection current.

In Fig. 7.35(a), a comparison of the emission spectra of binary (GaN), ternary (AlGaN), and quaternary (InAlGaN) devices is presented. The layer structures of the GaN, AlGaN, and InAlGaN were the same except for the active regions. The dependences of the emission intensity on the electric current are given in Fig. 7.35(b). The conditions of the measurements were the same for all three devices (room temperature and cw operation). From these results it can be seen that the short-wavelength emission in the case of the optimized (as discussed above) quaternary devices was more than one order of magnitude larger than that from the ternary and binary group-III-nitride devices. Thus, the use of quaternary InAlGaN active layers as the active emitting regions is significantly advantageous.

It is worth noting that technologies involving short-wavelength-emitting devices based on group-III nitrides are at the beginning phase of being introduced into practice. Further improvements of the quality of these materials are expected, and it is also anticipated that the search for methods of enhancement of hole-injection currents will lead to enhanced performance.

7.4.2 Methods of Enhancement of Hole-Injection Currents in Group-III-Nitride Heterostructures

The example presented in Fig. 7.34 clearly indicates the problem in creating high hole-injection currents – indeed, for p-doping at the level of 10^{20} cm^{-3} incorporated Mg atoms, the concentration of free holes is about 10^{17} cm^{-3}. To overcome the low-acceptor-activation problem, it has been suggested that a p-doped ternary compound material with a spatially modulated chemical composition, e.g., a multiple-quantum-well structure or a superlattice, could enhance the average hole concentration. Direct measurements support this idea: the average hole concentration in Mg-doped ternary superlattices can be increased by up to one order of magnitude. However, the main drawback of this approach is that most of the holes ionized from the acceptors are localized inside the quantum wells, which have potential barriers as high as 100 meV to 400 meV. These barriers hinder the participation of the holes in the vertical transport required in typical light-emitting devices. In this subsection we consider two methods of injecting hole currents.

To increase the overbarrier hole concentration and the vertical hole current, a *two-terminal hole injector* has been proposed, as illustrated schematically in Fig. 7.36. The injector consists of a p-doped superlattice base and two anode contacts A_1 and A_2. The injector is separated from the rest of the device by an undoped i region. A bias voltage applied between the A_1 and A_2 contacts provides lateral hole acceleration and increases the effective temperature of the holes, T_h, resulting in an enhancement of overbarrier hot-hole concentration. Such an effect is known as the *real space transfer effect*. In general, a light-emitting device can be thought of as a three-terminal device or a light-emitting triode, as shown in Fig. 7.36(b), with a hole-injector region, an intrinsic barrier i-layer, and an n-doped region, having a contact C. With the A_2 contact as ground and a voltage V_G applied to the A_1 contact, the injector would yield a lateral hole current and heating of the holes. Assuming that a negative bias V_C is applied to the cathode C, both the hot holes from the superlattice and the electrons from the n region would be injected into the

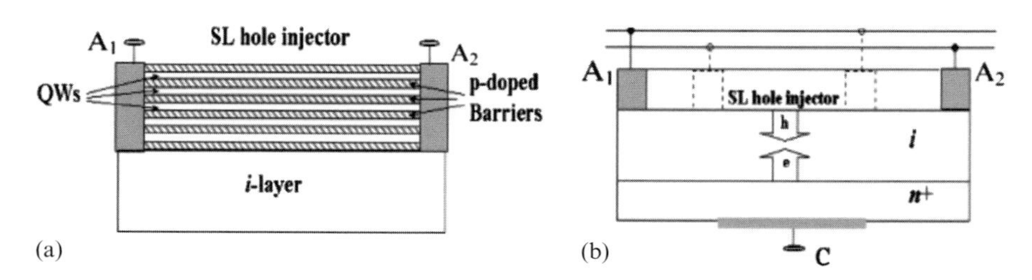

Figure 7.36 A schematic illustration of a two-terminal superlattice hole injector (a), and a light-emitting triode with hole injector (b). Reprinted from Fig. 1 of *Solid-State Electronics*, **47**, S. M. Komirenko, *et al.*, "Enhancement of hole injection for nitride-based light-emitting devices," pp. 169–171, Copyright (2003), with permission from Elsevier.

i-layer, as illustrated in Fig. 7.36(b). Emission of light would occur in the i-layer as a result of electron–hole recombination.

Assume that the p-doped superlattice consists of quantum well layers of thickness L_{QW}, separated by barrier layers of thickness L_B. The height of the energy barrier is U_B. The average over the barrier hole concentration in the superlattice under equilibrium is $N_{3D\text{-av}} = n_s/(L_B + L_{QW})\exp(-U_b/k_BT)$, with n_s the areal hole concentration in the quantum wells. The concentration of holes moving over the barrier increases exponentially with the hole temperature. Thus, for the case of holes in quantum wells heated up to temperature T_h by an electric field, the concentration of the overbarrier holes in the superlattice is

$$N_{3D\text{-hot}} \approx N_{3D-av} e^{-(U_b/k_BT - U_b/k_BT_h)},$$

where T is the lattice temperature. The hole temperature T_h can be found from the energy-balance equation $e\mu_h F^2 = k_B(T_h - T)/\tau_\epsilon$, where e is the elementary charge, μ_h is the lateral mobility of the holes in the quantum wells, F is the lateral electric field, and τ_ϵ is the energy relaxation time for the holes. This equation gives us an estimate of the increase of the hole concentration in the p region of the device.

Figure 7.37(a) shows the changes both in the hole temperature, $T_h = T + \Delta T_h$, and in the hole concentration, $N_{3D\text{-hot}}/N_{3D\text{-av}}$, as functions of the distance between

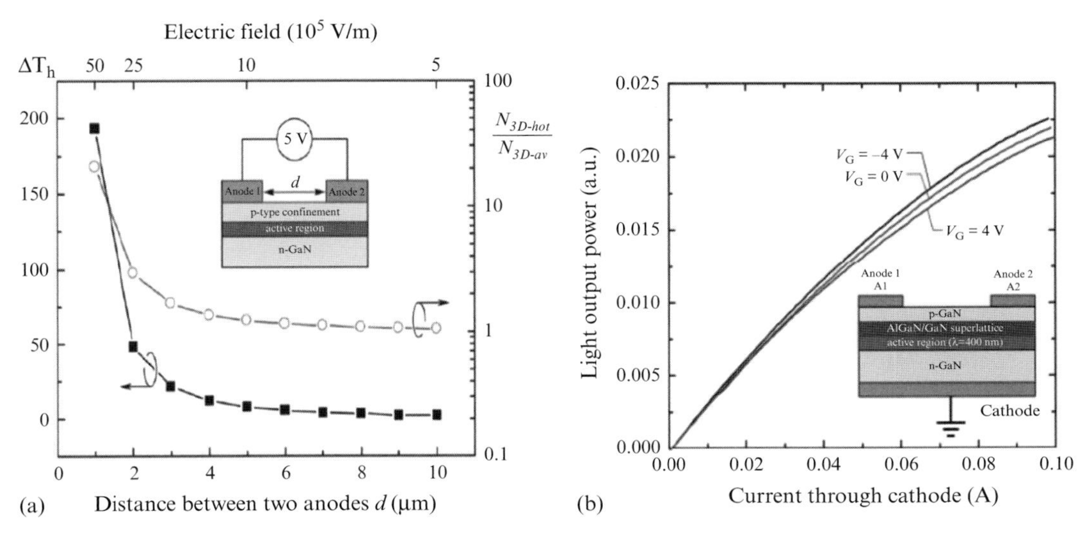

Figure 7.37 (a) Variation in the hole temperature, ΔT_h (the left axis) and the normalized hole concentration, $N_{3D\text{-hot}}/N_{3D\text{-av}}$ (the right axis) as functions of the inter-anode distance, d, at inter-anode bias $V_G = 5$ V. The upper axis shows the lateral electric fields, E, corresponding to the given bias and the inter-anode distances. (b) The light-output-power-versus-current characteristics of the ultraviolet triode, with and without hole heating by the lateral electric field. Republished with permission of The Electrochemical Society, from Figs. 3 and 7b of "GaN light-emitting triodes for high-efficiency hole injection," J. K. Kim, *et al.*, **153**, G734–G737 (2006); permission conveyed through Copyright Clearance Center, Inc.

the two anodes, A_1 and A_2. In the calculations, it is assumed that the energy scattering time $\tau_\epsilon = 10^{-12}$ s, the mobility $\mu_h = 10\,\text{cm}^2/\text{V s}$, $T = 300$ K, and the bias voltage $V_G = 5$ V. The upper axis shows the electric field, E, corresponding to the inter-anode distances at given bias voltage V_G. The values of both characteristics, ΔT_h and $N_{3\text{D-hot}}/N_{3\text{D-av}}$, increase with decreasing distance between the anodes, i.e., with increasing electric field. Significant improvement in the hole concentration can be expected for inter-anode distances less than 4 μm. In Fig. 7.37(b) the measured light-output-power-versus-current characteristics of the ultraviolet light-emitting triode are shown with and without heating of the holes in an AlGaN/GaN superlattice. The results indicate an enhancement of the light output with increasing negative bias to the additional anode A_2, despite the fact that the inter-anode distance was above 4 μm. In conclusion, the light-emitting device in triode configuration demonstrates enhanced hole-injection efficiency due to heating of the holes by the lateral electric field. Optimized device geometry, including the use of small distances between the anode contacts, can enhance the hole injection by one order of magnitude.

Another approach to achieve a high-density nonequilibrium electron–hole plasma in group-III nitrides for efficient light emission is based on the suggestion that one could exploit the *lateral electric current* for pumping, instead of the vertical currents discussed previously for all devices. Indeed, contemporary technologies facilitate the fabrication of planar p–i–n structures with high concentrations of two-dimensional electrons and holes. In such structures, the highly efficient double injection of electrons and holes can occur when a bias is applied along the quantum wells.

The main element of such lateral nonuniform structures in a single quantum well is illustrated in Fig. 7.38. The quantum well layer is confined by selectively doped barriers. Each barrier is doped laterally, so that an initial region doped with acceptors is followed (in the lateral direction) by an undoped (intrinsic) region and, finally, by a region doped with donors. The dopant in the barrier supplies carriers to the quantum well layer. Since these nitrides form type-I heterostructures, this layer can accumulate both types of free carrier, which leads to the formation of the *lateral p–i–n structure*. The contacts are to be made to the p and n regions as illustrated in Fig. 7.38(a).

The energy band diagram of the unbiased lateral p–i–n structure is illustrated in Fig. 7.38(b). The energy barriers separating the p, i, and n regions arise due to the formation of space charges near the p–i and i–n junctions. The band bending is similar to that of a p–i–n homostructure; see Figs. 7.6 and 7.7. In a forward-biased structure the potential barriers decrease, providing for an injection of confined holes and electrons into the i region. This *planar double injection* gives rise to a nonequilibrium two-dimensional electron–hole plasma in the i region. Radiative recombination of the plasma in the active region results in light emission. A structure arranged from the quantum wells considered here can form an efficient source of radiation – a lateral-current-pumped emitter.

For the case of vertical transport, the scenario of double injection in p–i–n homostructures and heterostructures has been well studied. It is based on a few

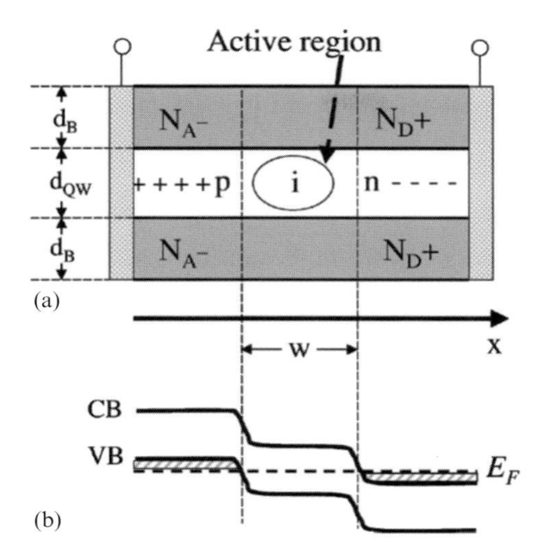

Figure 7.38 Schematic illustration of (a) the multilayered lateral p–i–n structure and (b) the energy band diagram in the lateral direction with no bias. CB and VB indicate the lowest populated energy levels in the conduction and valence bands, respectively. Reprinted from Fig. 1 of S. M. Komirenko, *Appl. Phys. Lett.* **81**, 4616 (2001), with the permission of AIP Publishing.

well-proven assumptions: (i) in electrically biased structures, the p, i, and n regions are mostly quasi-neutral; and (ii) the charged (depleted) regions remain very narrow. These assumptions allow one to avoid a detailed description of the processes in the depleted regions. To expand this scenario to lateral double injection, we need to compare the length scales characterizing the structure under consideration. These include different groups of scales: geometric scales – the quantum-well-layer thickness, d_{QW}, the barrier-layer thickness, d_{B}, and the size of the i region, w; the kinetic lengths – the diffusion lengths of the electrons, L_{n}, and holes, L_{p}; and the length of the screening of an electric charge by carriers, l_{sc}. In light-emitting devices the i-region should be extended: $w \geq L_{\mathrm{n}}, L_{\mathrm{p}}$, where the macroscopic diffusion lengths are of the order of a few micrometers, while the screening lengths, l_{sc}, are less than 10 nm. Thus, we obtain the following inequalities: $w \geq L_{\mathrm{n}}, L_{\mathrm{p}} \gg d_{\mathrm{QW}}, d_{\mathrm{B}} \geq l_{\mathrm{sc}}$. The lateral extents of the p–i and i–n junctions are estimated to be less than, or of the order of, $d_{\mathrm{QW}}, d_{\mathrm{B}}$. Thus, the charged layers are very narrow and electron–hole recombination there can be neglected just as it can for vertical structures. The extended i region can be considered to be quasi-neutral, with the carrier transport described by the usual bipolar drift–diffusion equations. Finally, we conclude that the well-known scenario of double injection can also be applied to lateral p–i–n structures.

Consider, for example, a GaN/AlGaN lateral-current-pumped emitter. Assuming a linear recombination mechanism with recombination time τ_{R} that is different in the different device regions, we set $\tau_{\mathrm{R}}^{(\mathrm{p})} = \tau_{\mathrm{R}}^{(\mathrm{n})} = 0.1$ ns, $\tau_{\mathrm{R}}^{(\mathrm{i})} = 1$ ns, $\mu_{\mathrm{n}}/\mu_{\mathrm{p}} = 20$,

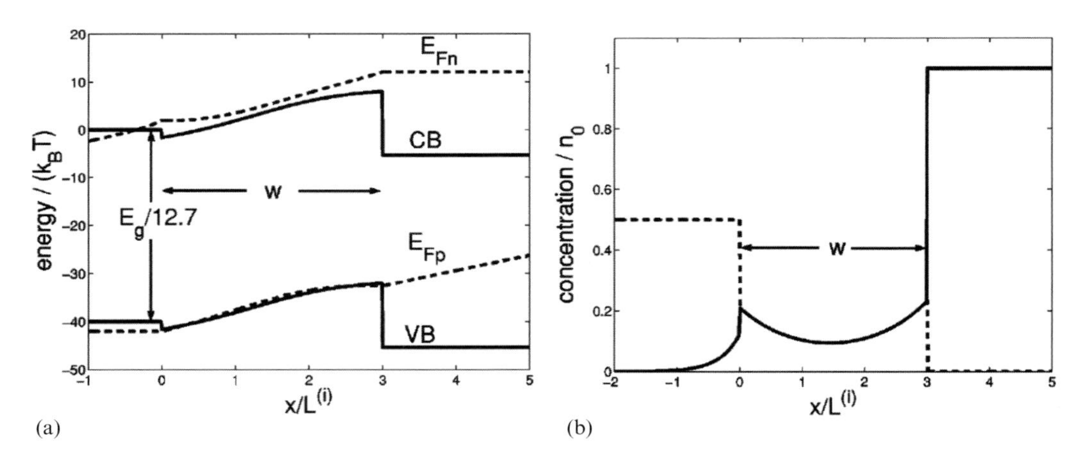

Figure 7.39 Characteristics of the lateral p–i–n structure with GaN quantum wells at $T = 80$ K and the current through a single quantum well, $j = 16$ mA/mm (corresponding to the case GaN (a) in Table 7.1). (a) The energy band bending. CB and VB indicate the lowest populated energy levels in the conduction and valence bands, respectively. (b) Concentrations of the injected electrons and holes normalized with respect to the initial doping n_0 of the n region. The electron (hole) density is represented by the solid (dashed) line. Reprinted from Fig. 2 of S. M. Komirenko, *Appl. Phys. Lett.* **81**, 4616 (2001), with the permission of AIP Publishing.

$\mu_n = 500$ cm^2/s V. For $T = 80$ K, these parameters result in ambipolar diffusion length $L^{(i)} = 1.2$ μm and diffusion lengths $L_n^{(p)} = 1.7$ μm and $L_p^{(n)} = 0.4$ μm.

These parameters facilitate calculations of the lateral distribution of the potential energy across the p–i–n structure, of the energy band bending, and of the concentrations of injected electrons, n, and holes, p, for a given current in the quantum well, j.

In Fig. 7.39 we present results of these calculations for the p–i–n structure with $w = 3.6$ μm at $J = 16$ mA/mm. The corresponding energy diagram is shown in Fig. 7.39(a). The potential barriers at the junctions, Δ_p, Δ_n, are finite and decrease with increasing current. For example, the p–i junction barrier vanishes first at $J = 188$ mA/mm. The spatially nonuniform potential in the i region facilitates the spreading of "slow" holes through the extended i region. For this current, the total voltage drop across the p–i–n structure is 4.22 V for quantum wells of thickness $L_{QW} = 5$ nm. In the quasi-neutral region $n \approx p$, the concentrations are nonmonotonic (see Fig. 7.39(b)), with a maximum p_M at the n–i junction and a minimum p_m in the middle of this region.

In Table 7.1, the following parameters characterizing planar double injection in nitride-based structures are presented: the current density through a single quantum well, J; the energy barriers Δ_p and Δ_n at the p–i and i–n junctions, respectively; the total lateral bias voltage, U (in energy units), corresponding to the given lateral current, and the minimum, p_m, and maximum, p_M, area concentrations in the active region. The data were calculated for different temperatures. The results of Table 7.1 for GaN and InN quantum wells show that, under planar double

Table 7.1 Parameters characterizing lateral double injection in nitride-based p–i–n structures

Quantum well material	T (K)	J (mA/mm)	Δ_p (meV)	Δ_n (meV)	U (meV)	p_m (10^{12} cm^{-2})	p_M (10^{12} cm^{-2})
GaN							
(a)	80	16	11	92	66.6	0.9	2.3
(b)	250	80	20	42	284	2	6
InN							
(a)	80	37	12	153	74	1	2.9
(b)	300	110	27	73	285	2.4	6.6

(a) $p_0 = n_0/2 = 5 \times 10^{12}$ cm^{-2}; (b) $p_0 = n_0 = 10^{13}$ cm^{-2}. U/e is the voltage drop across the i region. Reprinted from Table 1 of S. M. Komirenko, *Appl. Phys. Lett.* **81**, 4616 (2001), with the permission of AIP Publishing.

injection, high densities of the electron–hole plasma in the active region – 10^{12} cm^{-2} to 5×10^{12} cm^{-2} – can be achieved (for a 5 nm quantum well, these densities are equivalent to bulk concentrations above 2×10^{18} cm^{-3} to 10^{19} cm^{-3}, respectively).

For 5 nm quantum wells, the radiative recombination of the electron–hole plasma produces light emission centered at ≈ 344 nm and ≈ 587 nm for GaN and InN quantum wells, respectively. The wavelength can be scaled readily to the deep-ultraviolet range by using AlGaN or AlGaInN for the lateral-current-pumped emitter.

This analysis shows that laterally selective doped quantum wells and superlattices can be used for the light emitters based on group-III nitrides. In such devices, planar p–i–n regions with high concentrations of two-dimensional electrons and holes are formed, and highly efficient double injection occurs when a bias is applied along the quantum well layers. This results in high densities of two-dimensional electron–hole plasma in an extended i region. Planar double injection can be used as an efficient method for the electrical pumping of short-wavelength lasers that are based on a wide-bandgap material, when high p doping is complicated to achieve.

The above analysis of planar p–i–n structures for a short-wavelength-emitting device is based on an idealized model. The practical realization of such structures involves complicated technological methods. In Fig. 7.40, an example of a structure with lateral p–i–n junctions is sketched. The structure consists of an active region of i-GaN or InGaN/GaN multiple quantum wells with laterally overgrown GaN layers, e.g., with selective area regrowth. The narrow n- and p-doped regions are fabricated side by side. The double injection mechanism is clear: the electrons and holes flow to the continuous active layer through bipolar diffusion. Strong electron–hole radiative recombination is achieved in the multiple i-GaN or InGaN/GaN quantum wells. Such structures can be realized using either selective-area growth or ion-implantation techniques. Figure 7.41 presents typical current-dependent electroluminescence spectra of a light-emitting planar device, where the n region was fabricated by Si implantation. Only one peak, corresponding to the interband transitions in the $In_{0.2}Ga_{0.8}N$ quantum wells (at around 450 nm), was observed in this

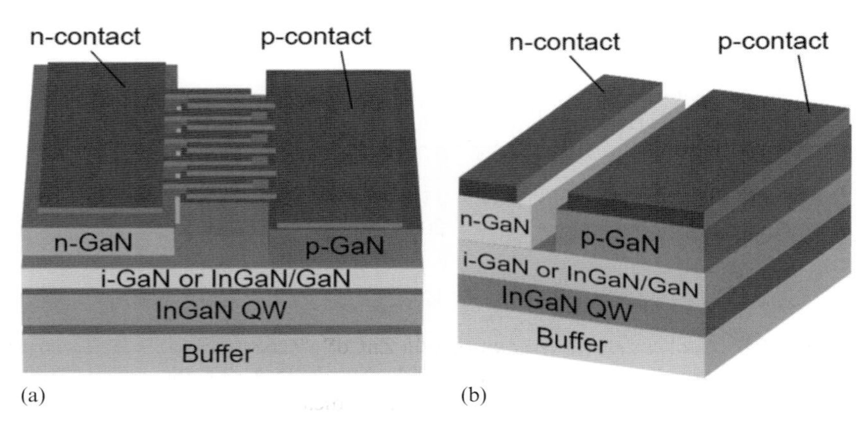

Figure 7.40 (a) A schematic illustration of a lateral heterojunction with a finger structure, which is used to form the lateral p–i–n junctions and for the double injection of the carrier into the quantum wells. (b) A perspective image of a side view between two fingers. After Fig. 5 in I. Kim, *et al.*, *Materials* **10**, 1421 (2017). Licensed under CC BY 4.0.

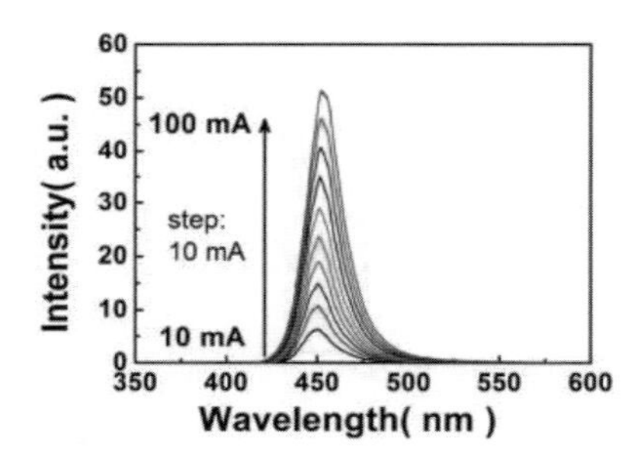

Figure 7.41 Electroluminescence spectra taken from an Si-implanted planar light-emitting device. ©1981 IEEE. Reprinted, with permission, from Fig. 4(b) of M.-L. Lee, *et al.*, *IEEE Trans. Electron Dev.* **64**, 4156 (2017).

device. When the device was driven with the current increasing from 10 mA to 100 mA, the emission intensity was increased by one order of magnitude.

The reported demonstrations of planar p–i–n structures for short-wavelength emitters could have great potential for further development and for opening new horizons for electrically pumped devices that cannot be reached by using conventional vertical-current-injection heterostructure systems.

7.4.3 Short-Wavelength Laser Diodes

In this subsection we consider short-wavelength lasers with electrical pumping. As indicated above, the realization of short-wavelength light-emitting devices is of considerable interest for a number of applications, some of which are shown in Fig. 7.28. From that diagram it can be seen that many applications require short-wavelength emission of increased power. A high light power can be reached by laser devices.

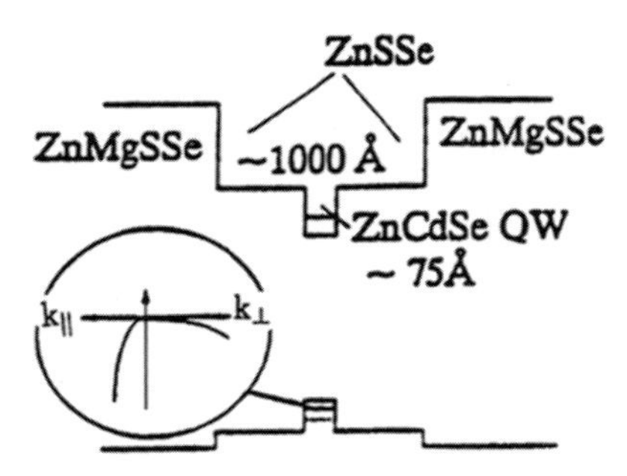

Figure 7.42 A schematic representation of a ZnCdSe/ZnSSe/ZnMgSSe heterostructure for a blue laser. In the energy diagram of the device, the 75 Å ZnCdSe quantum well and 1000 Å ZnSSe optical confinement region are indicated. The lower part of the figure illustrates the anisotropy of the lowest hole band in the ZnCdSe strained layer. From Fig. 1(a) in *Proceedings of 22nd Int. Conf. on the Physics of Semiconductors*, vol. 1, A. V. Nurmikko and R. L. Gunshor, ed. D. J. Lockwood, pp. 27–34, Copyright ©1995 World Scientific Publishing.

Historically, the first family of short-wavelength injection lasers was based on II–VI materials. For this family, all of the necessary elements – quantum wells and a waveguide layer embedded in a p–n junction – have been developed. Figure 7.42 depicts a particular example of such a ZnCdSe/ZnSSe/ZnMgSSe structure. The quantum well layer is made of narrow-bandgap $Zn_{1-x}Cd_xSe$ material with a thickness of about 75 Å. The optical confinement layer is ZnSSe with a thickness of about 1000 Å. The left and right parts of the structure are made of $Zn_{1-z}Mg_zS_ySe_{1-y}$ and are p- and n-doped, respectively. The quantum well layer is pseudomorphic, with lattice mismatch greater than 1%. This produces strain and a deformation of the valence bands. The anisotropy of the lowest hole subband is shown in the inset. This effect was discussed in the previous section; see also Fig. 7.26(b). The depths of the quantum wells are 180 meV for the electrons and 70 meV for the holes.

Unlike in the III–V compounds, in II–VI-based quantum wells the exciton binding energy is large. Such binding energies were discussed in Section 3.7.2. For ZnCdSe and ZnSe quantum wells, this energy is about 40 meV and exceeds k_BT, even at room temperature. These factors make electron–hole correlations and exciton effects even more important than those of the III–V materials. The exciton is stable against thermal dissociation, screening, and other many-body effects. In accordance with the general considerations of Section 5.2.4, exciton effects enhance the strength of the radiative transitions and, as a result, they dominate in the optical spectra of these structures. Another important consequence of excitonic effects is that radiative recombination through two-dimensional excitons dominates over possible nonradiative paths, including the recombination of free electron–hole pairs. These dominant radiative transitions result in increased internal emission efficiency for devices fabricated from these materials. The stability of the excitons against screening facilitates the realization of high exciton concentrations. All of these factors result in a peak amplification coefficient below the edge of the

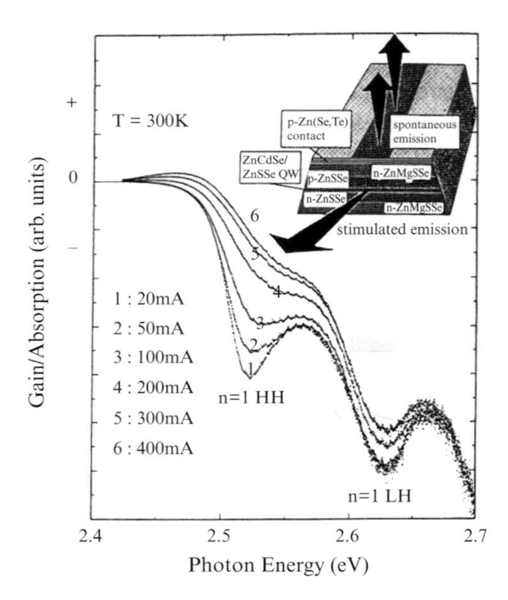

Figure 7.43 The amplification (absorption) coefficient as a function of the photon energy for the heterostructure presented in Fig. 7.42. The curves correspond to different pumping currents. Reprinted figure with permission from figure 1 of J. Ding, *et al.*, *Phys. Rev.* **B 50**, 5787 (1994), Copyright 1994 by the American Physical Society.

band-to-band transitions. This is in strong contrast to the case of III–V compound lasers, where, under conditions of population inversion, the exciton peaks completely disappear. In Fig. 7.43, the amplification/absorption spectra are shown for the structure under consideration at different injection currents and at room temperature. The transparency current – corresponding to the threshold for population inversion – is equal to approximately 300 μA. At higher currents, positive amplification occurs at energies below the $n = 1$ heavy-hole resonance, i.e., in the exciton region for $\hbar\omega < 2.5$ eV. For ZnCdSe/ZnMgSSe structures, an amplification coefficient of the order of 1500 cm^{-1} is reached at room temperature. The inset of Figure 7.43 shows a sketch of this laser and its outputs of stimulated and spontaneous emission. In Fig. 7.44, the current–voltage and light-power–current characteristics are shown for such a blue laser diode. The light-power–current characteristic clearly demonstrates the threshold behavior of the stimulated emission. The light output is roughly several milliwatts for pumping currents above 10 mA.

The major problems associated with quantum well lasers of this class are device degradation and short device lifetimes. The electric power dissipated in present-day II–VI diodes is still high and causes rapid degradation as a result of the generation of intrinsic defects. Much improvement must be realized in this field in order for there to be wide practical application of these devices.

In contrast to the remaining problems with II–VI light-emitting devices, great progress has been achieved for current-pumped emitters that are based on the

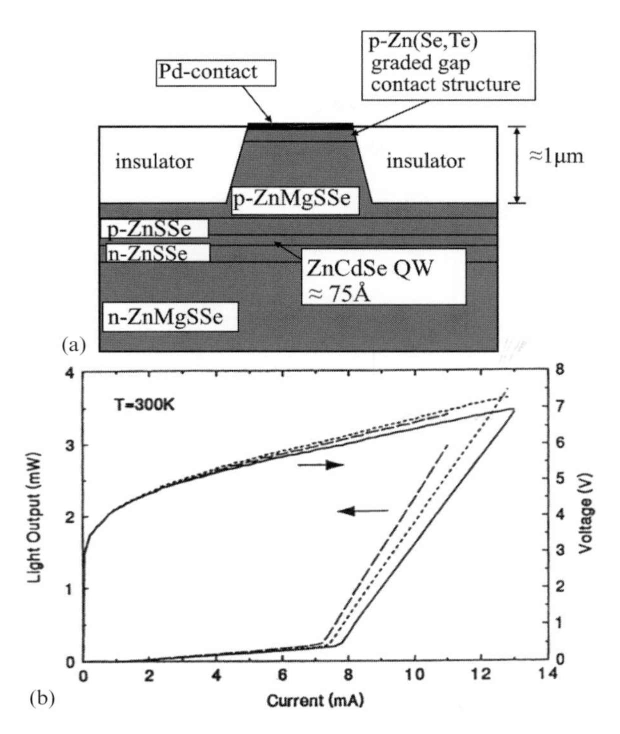

Figure 7.44 The current–voltage and light-output–current characteristics of the blue laser presented in Figs. 7.42 and 7.43. From Fig. 2 in *Proceedings of 22nd Int. Conf. on the Physics of Semiconductors*, vol. 1, A. V. Nurmikko and R. L. Gunshor, ed. D. J. Lockwood, pp. 27–34, Copyright ©1995 World Scientific Publishing.

group-III-nitride materials, as was demonstrated in Section 7.4.1 for different examples among the efficient light-emitting diodes fabricated from these materials.

However, achieving a population inversion and lasing in a short-wavelength region under electrical pumping is a much more difficult problem. Indeed, it requires more complex structures, thicker layers, lower dislocation densities, stronger optical and electrical confinement, higher currents, and solutions of many other more difficult technological tasks in comparison with the case of devices with spontaneous emission. Nevertheless, solutions are being found for this problem and ultraviolet nitride-based laser diodes are being successfully created. Below we present a few examples of these lasers.

In Fig. 7.45, the example shown is a InGaN multiple-quantum-well laser diode. The active region, consisting of a 26-period $In_{0.2}Ga_{0.8}N/In_{0.05}Ga_{0.85}N$ superlattice, was grown above a thick n-type $Al_{0.15}Ga_{0.85}N$ and 0.1 μm-thick GaN layers. The thick AlGaN layer was grown to prevent cracking of the adjacent layers during their fabrication. The 2.5 nm-thick $In_{0.2}Ga_{0.8}N$ quantum wells were separated by 5 nm-thick $In_{0.05}Ga_{0.85}N$ barrier layers. On top of this active layer, p-type $Al_{0.2}Ga_{0.8}N$ and GaN layers with Mg doping were grown. A 20 nm-thick layer of $Al_{0.2}Ga_{0.8}N$ was used to prevent the dislocation degradation of the InGaN

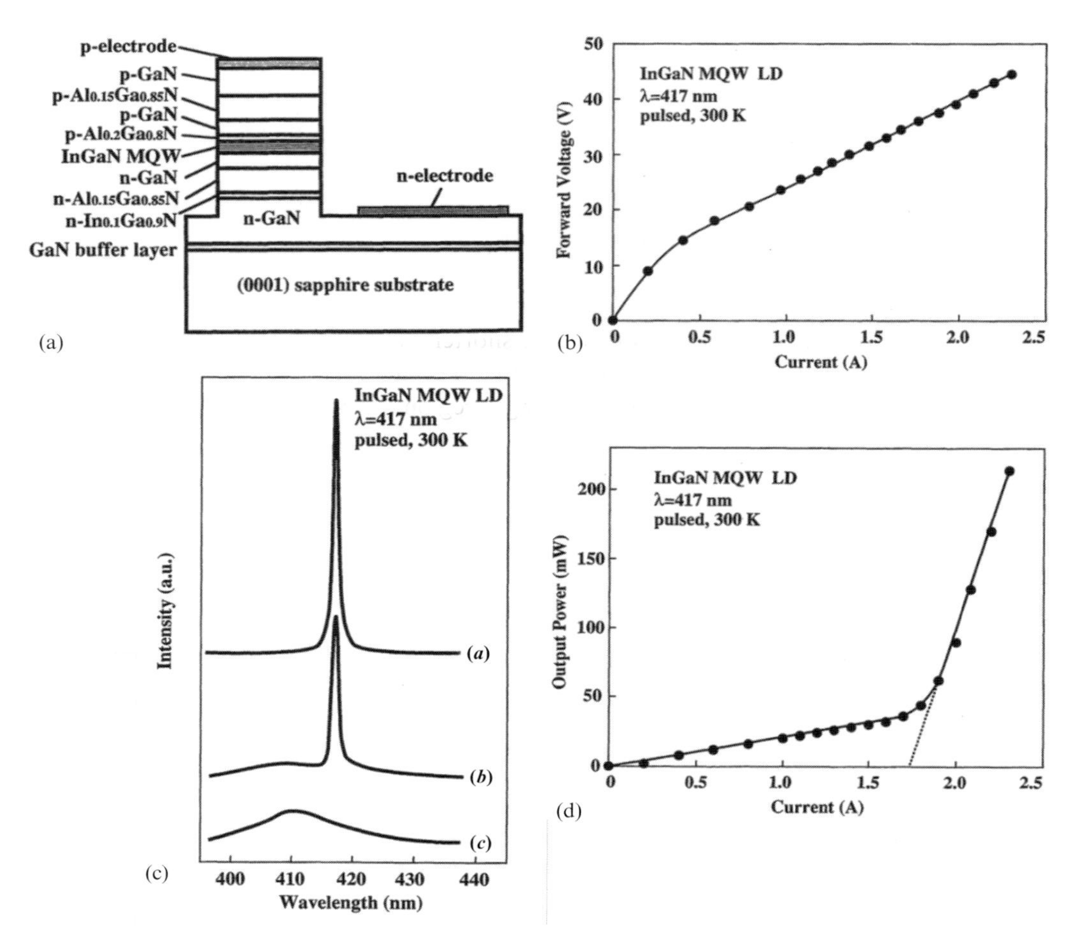

Figure 7.45 An InAlGaN multiple-quantum-well ultraviolet laser diode. (a) The schematic diagram shows the layer structure of the diode. (b) The electrical characteristics of the laser diode. (c) Optical spectra of the laser diode at currents 2.1 A (*a*), 1.7 A (*b*), and 1.3 A (*c*). (d) The light output power at 417 nm as a function of the diode current. From Figs. 1, 6, 4, and 3 of S. Nakamura *et al.*, *Jpn J. Appl. Phys.*, **35**, L74 (1996). Copyright 1996 The Japanese Society of Applied Physics.

p-layers. The thick n-type and p-type $Al_{0.15}Ga_{0.85}N$ layers served also as cladding layers for the confinement of light in the active multiple-quantum-well region. Other details of the diode are shown in Fig. 7.45(a). The active area of the device was 30 μm × 1500 μm. Special technological efforts were applied to produce high-quality facets for the cavity mirrors to make a strip laser diode (in which generated light propagates along the active region).

The electric characteristics of the diode were measured at room temperature in the pulse mode (the pulse width and period were 2 μs and 2 ms, respectively). The dependence of the voltage on the current for this device is presented in Fig. 7.45(b).

The emission spectrum of the device is shown in Fig. 7.45(c) for several currents. It can be seen that a narrow spectrum line was formed at currents exceeding 1.3 A.

Below the laser threshold, an emission maximum was observed at 410 nm with a full width at half maximum of 20 nm. The dependence of the emission intensity on the current shown in Fig. 7.45(d) allows one to estimate the threshold current as ≈ 1.7 A, which corresponds to a threshold current density of 4 kA/cm^2. Above the laser threshold, the emission maximum was at 417 nm and the full width at half maximum was estimated to be 1.6 nm. A pulsed output power as high as 215 mW was obtained at a device current of 2.3 A.

It is important to emphasize that the laser diode discussed above includes a certain amount of indium in the active layers. The inclusion of indium in the active layers inhibits stimulated emission at shorter wavelengths because of the smaller bandgap of the InN-containing layers (see the diagram in Fig. 7.29). A further shift of the emission to the shorter-wavelength region can be achieved by increasing the AlN mole fraction in the active layers. However, as discussed in Section 7.4.1, a lack of indium in the AlGaN active layers increases the probability of nonradiative recombination. Thus, progression to shorter-wavelength laser devices with AlGaN requires a significant reduction in the density of dislocations, which act as nonradiative recombination centers. In the following discussion, we present an example of a laser diode with a low-dislocation-density AlGaN active layer obtained by use of an improved fabrication technology.

The low-dislocation-density AlGaN layers, including relatively high AlN mole fractions up to 0.3, were fabricated by metal–organic vapor-phase epitaxy. Special testing showed that these AlGaN layers had a low dislocation density and were suitable for producing a laser structure. Details of the structure and the doping of this device are shown in Fig. 7.46. Because of the low refractive index of the $Al_{0.3}Ga_{0.7}N$ cladding layers, this design provides suitable optical confinement in the waveguide, of length 900 μm.

The emission spectra of this device were measured at room temperature in the pulsed-current mode with a pulse duration of 10 ns and a repetition frequency of 5 kHz. Spontaneous emission was observed with a peak wavelength of approximately 345 nm with a full width at half maximum of only ≈ 6 nm at small currents. The narrow spectrum of the spontaneous emission implies a low fluctuation of the quantum well characteristics such as the width and the AlN compositions in the active layer. The peak of the spontaneous emission shifts to a shorter wavelength and the width becomes narrower on increasing the injection current. Relatively broad lasing emission with a peak wavelength of 342.7 nm was observed at a current of 415 mA, as can be seen in Fig. 7.46(b). With increasing injection current the peak of the emission slightly shifts to a shorter wavelength and the width becomes narrower, as shown in Fig. 7.46(b). At a current of 512 mA a sharp lasing emission with a wavelength of 342.3 nm and full width at half-maximum of only 0.3 nm were observed.

The measured peak wavelength and the full width at half-maximum of spectrum as a function of the current are shown in Fig. 7.46(c). By decreasing the current-pulse duration it was checked that the thermal bandgap narrowing is negligible. The observed shift of the emission wavelength can be attributed to the effects of screening

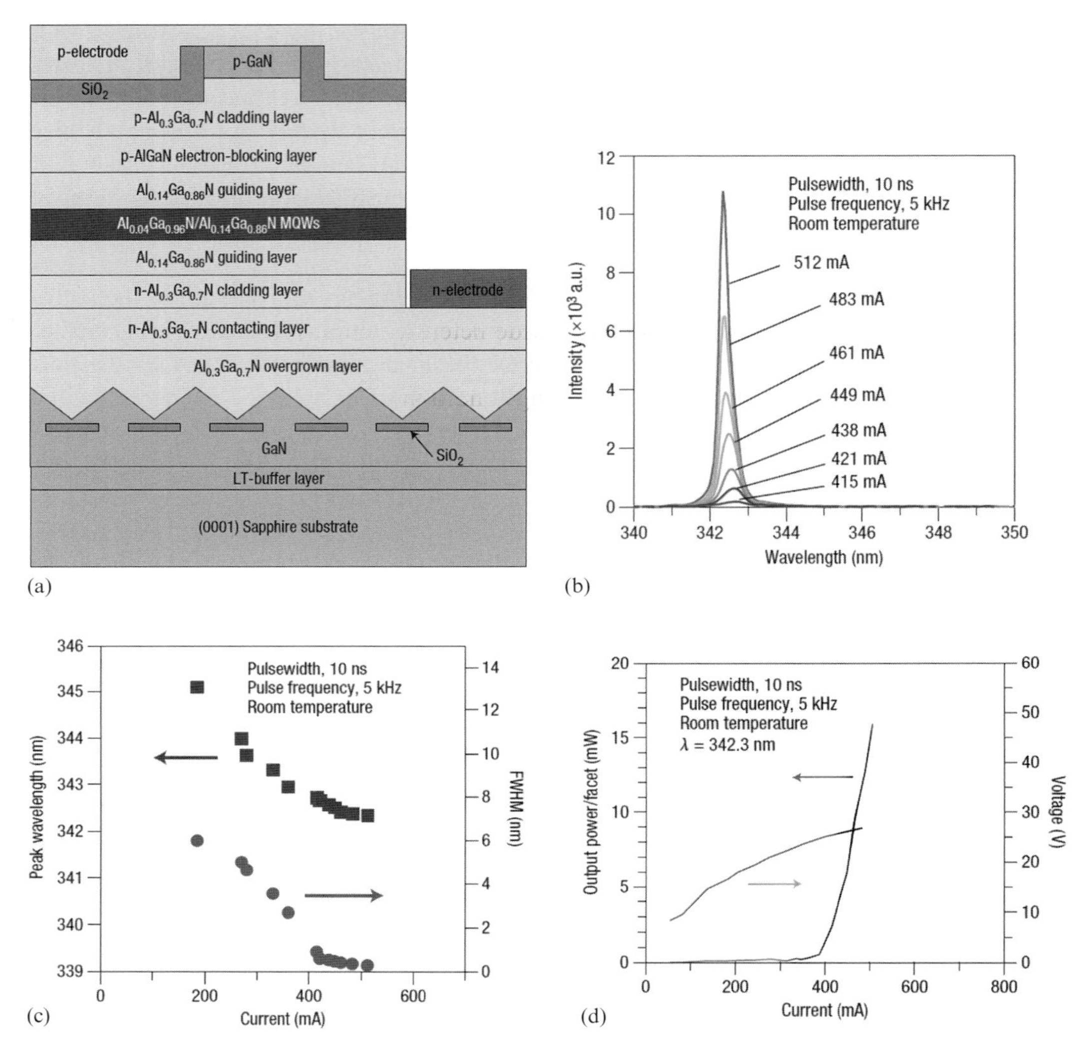

Figure 7.46 AlGaN multiple-quantum-well ultraviolet laser diode. (a) The schematic shows the layer structure of the diode. (b) Room-temperature emission spectra for different injection currents above the threshold. (c) The dependences of the peak wavelength and full width at half maximum of the emission spectra on the current. (d) The light-output power at 342.3 nm and electrical characteristics of the laser diode. Reprinted by permission from Figs. 1, 2(b), (c), and 4 in Springer Nature: Macmillan Publishers Ltd., *Nature Photonics*, "A 342-nm ultraviolet AlGaN multiple-quantum-well laser diode," H. Yoshida, *et al.*, Copyright 2008.

of the internal electric fields and the change in the quantum confined Stark effect. These fields arise due to spontaneous and piezoelectric polarizations, which are characteristic for structures with a high AlN fraction. Also, the screening effect can change the overlap of the electron and hole wave functions, affecting electron–hole recombination in the quantum wells, as well as the optical gain necessary for lasing.

Note that the internal fields are responsible for the observed strong polarization of the emitted laser beam.

The light output versus current and the current–voltage characteristics of the laser are shown in Fig. 7.46(d). The threshold current was estimated to be 390 mA, which corresponds to a threshold current density of 8.7 kA/cm^2; the threshold voltage was 25 V. The pulse output power reached 16 mW. These output characteristics are comparable to those for the previously discussed 417 nm lasing in the InAlGaN structure.

Summarizing this subsection, in spite of the great challenges faced in working to improve the quality of group-III-nitride heterostructures, significant progress has been realized toward achieving ultraviolet laser diodes, in particular through the optimization of growth and processing conditions.

7.5 Quantum Wire and Quantum Dot Emitters and Lasers

As we learned from Chapter 3, additional carrier confinement leads to modifications of the energy spectra and the density of states. These modifications may significantly improve laser performance as the structures under consideration change from quantum well structures to quantum wire and quantum dot structures. Let us recall the density of states per unit volume for electron systems with different dimensionalities:

$$\varrho_{3\mathrm{D}} = \frac{(2m^*/\hbar^2)^{3/2}}{2\pi^2}\sqrt{E},$$

$$\varrho_{2\mathrm{D}} = \frac{m^*}{\pi\hbar^2 L_z}\sum_i \Theta(E - \epsilon_i),$$

$$\varrho_{1\mathrm{D}} = \frac{(2m^*)^{1/2}}{\pi\hbar L_y L_z}\sum_{i,j} \frac{1}{\sqrt{E - \epsilon_{i,j}}}\Theta(E - \epsilon_{i,j}),$$

$$\varrho_{0\mathrm{D}} = \frac{2}{L_x L_y L_z}\sum_{i,j,k} \delta(E - \epsilon_{i,j,k}),$$

where L_x, L_y, L_z are the sizes of the structures in the confined dimensions, E is the kinetic energy, and ϵ_i, $\epsilon_{i,j}$, $\epsilon_{i,j,k}$ are the energy levels in quantum wells, quantum wires, and quantum dots, respectively. A comparison of the densities of states for different dimensionalities is presented in Fig. 7.47. The density of states acquires sharper features as the carrier dimensionality decreases. By comparing these results one can obtain several general conclusions related to the transition from a quantum well (confinement in one direction) to a quantum wire and quantum dot (multi-dimensional confinement): a sharper density of states leads to a narrower spectral linewidth of emission and to an increase in the quasi-Fermi levels E_{Fn} and E_{Fp}. According to the criterion for population inversion as expressed by Eq. (7.4), an increase of E_{F} leads to a reduction of the threshold carrier concentration and, hence, to a lowering of the threshold current. Other advantages of multi-dimensional

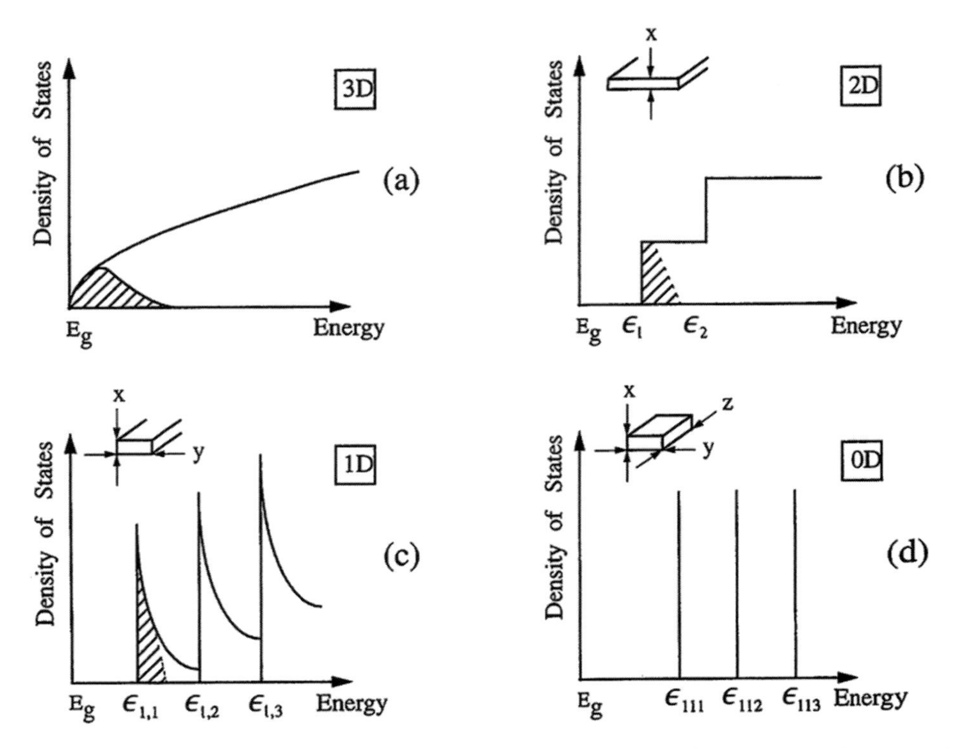

Figure 7.47 Schematic representations of the energy dependences of the density of states and occupations for systems with different dimensionality. The hatched areas indicate occupied states: (a) a three-dimensional system, (b) a quantum well, (c) a quantum wire, and (d) a quantum dot. ©1992 IEEE. Reprinted, with permission, from Fig. 1 of E. Kapon, *Proc. IEEE* **80**, 398, (1992).

confinement are related to the small sizes of these structures, the possibility of densely packed laser arrays, the potential for monolithic integration with low-power electronics, etc.

7.5.1 Quantum Wire Lasers

First, let us consider briefly some results obtained for quantum wire lasers. For the parallel laser configuration, when the light wave propagates along the wires, we can use Eq. (7.19), where the optical confinement factor is now the ratio of the area within which the carriers are confined and the area of optical confinement:

$$\Gamma \approx \frac{S_{\mathrm{QWR}}}{S_{\mathrm{mode}}}.$$

Here S_{QWR} and S_{mode} are the areal cross-sections of the quantum wire and waveguide region, respectively. We assume a linear dependence of the amplification coefficient on the carrier concentration, n:

$$\alpha_0 = \gamma^{1D}\left(\frac{n}{n_{th}} - 1\right),$$

where γ^{1D} is to be calculated for phototransitions between one-dimensional subbands of the electrons and holes. In order to pump the quantum wire one needs to apply the following electric current per unit length of the wire,

$$I = \frac{eS_{QWR}}{\eta_i \tau} n,$$

where η_i is the internal efficiency, and τ is the recombination time. Hence, the threshold transparency current is

$$I_{th} = \frac{eS_{QWR}}{\eta_i \tau} n_{th}.$$

One can rewrite the local amplification coefficient as

$$\alpha_0 = \gamma^{1D}\left(\frac{I}{I_{th}} - 1\right).$$

Now the lasing threshold current can be written as

$$I > I_{th} + \frac{I_{th}}{\gamma^{1D}\Gamma}\left(\alpha_s + \frac{1}{2\mathcal{L}_x}\ln\left(\frac{1}{r_1 r_2}\right)\right) \equiv I_{th} + I_{cav}, \tag{7.49}$$

where the first term is related to the appearance of population inversion; the second depends on cavity losses and the cavity design. Let us examine the various terms in Eq. (7.49) in order to gain an understanding of the ultimate limit of the laser threshold current. The transparency current density, I_{th}, is proportional to the wire cross-section and can be extremely small for quantum wire structures. For example, assume a quantum wire fabricated with a cross-section of 100 Å × 100 Å and a length of 100 μm. The threshold concentration for creating population inversion (the transparent condition) is $n_{th} \approx 10^{18}$ cm^{-3} and $\tau = 3$ ns; see Fig. 7.4. Then, for $\eta_i = 0.5$, we obtain a total transparency current, $I_{th} \times \mathcal{L}_x$, as small as 1 μA. To estimate the last term in Eq. (7.49) we set $\gamma^{1D}/n_{th} \approx 10^{-15}$ cm^2 and $S_{mode} = 1$ μm × 1 μm; then $I_{th}/(\gamma^{1D}\Gamma) = 1.1$ mA. If $r_1 = r_2 = 0.9$ and $\alpha_s = 10$ cm^{-1}, we obtain $I_{cav} \times \mathcal{L}_x =$ 110 μA. For highly reflective mirrors with $r_1 = r_2 \approx 0.99$, we find the threshold current to be in the range of tens of microamperes. These simple estimates illustrate the dominant trends affecting device design and the achievement of the lowest current thresholds.

In Fig. 7.48, a calculation of the local amplification coefficient α_0 is presented as a function of the wavelength for a GaInAs/GaInAsP rectangular quantum wire with dimensions 120 Å × 200 Å at different carrier densities, n. The calculation takes into account the broadening of the subbands due to intrasubband relaxation, with characteristic time $\tau_{in} = 0.1$ ps; see Section 6.4.5. The carrier density changes

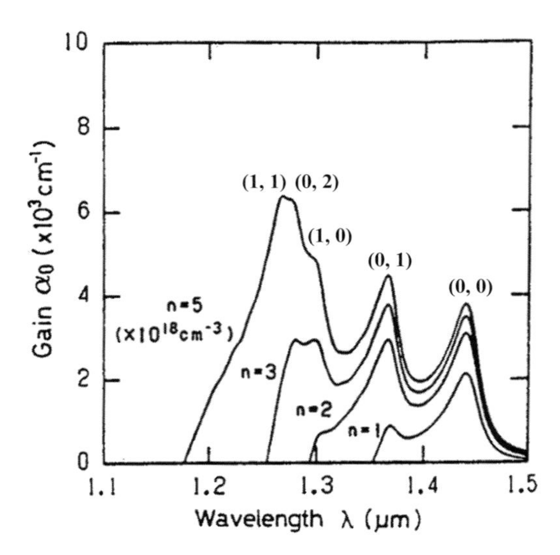

Figure 7.48 The amplification coefficient as a function of the light wavelength for the GaInAs/GaInAsP quantum-wire laser with a 120 Å × 200 Å cross-section. The results are calculated for the set of carrier concentrations, n, and $T = 77$ K. The indices over the curves indicate the contributions of different one-dimensional subbands, (n_1, n_2). After Figure 2 of M. Cao, *et al.*, *Trans. IEICE* **E73**, 63 (1990).

from 1×10^{18} cm^{-3} to 5×10^{18} cm^{-3}. The complex structure of $\alpha_0(\lambda)$ is due to phototransitions from the upper subbands. According to Section 5.2, phototransitions between different electron and hole subbands contribute to the spectra. Each subband is characterized by two quantum numbers (n_1, n_2). The contributions of electron and hole subbands with the same pair (n_1, n_2) are the most important; they are marked in Fig. 7.48. This figure shows that the local amplification coefficient reaches a value of the order of 10^3 cm^{-1}. The dependence of these results on the concentration supports the previously assumed estimate for γ/n_{th}.

There are different technological approaches for the fabrication of quantum wire lasers. The most direct approach is fabrication by the etching and regrowth technique. In Fig. 7.49, such a device is presented schematically. The structure is fabricated by first growing a conventional quantum well structure. Next, a wire pattern is formed by etching the quantum well active region using an electron-beam-written grating mask. Then, the upper part of the structure is formed in the second growth step. The wires are of dimensions 100 Å × 300 Å. The distance between the wires is 400 Å. The lower part of the structure is n-doped InP, the upper part is p-doped InP. They are accessed by electrical contacts. The optical confinement region is made of a GaInAsP layer with a width of about 5000 Å. The cleaved facets serve as light reflectors. Other details of the structure are given in Fig. 7.49. A similar procedure is used for the formation of an array of quantum dots with lateral dimensions less than 1000 Å.

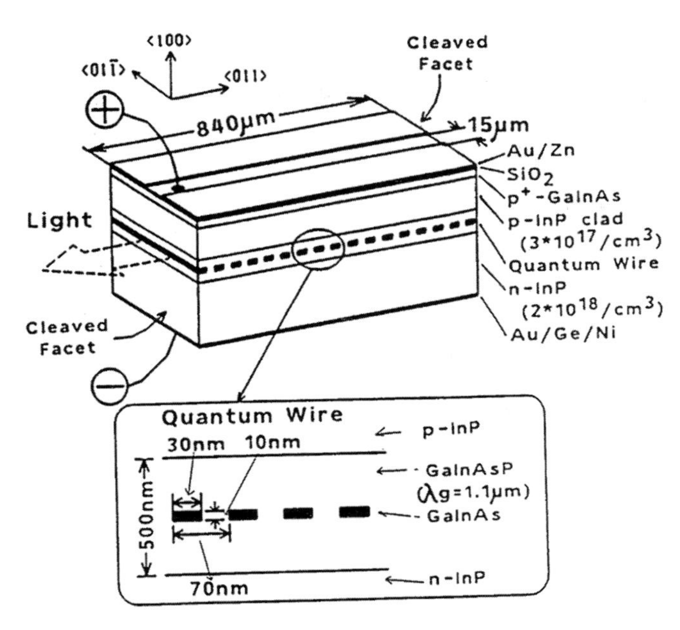

Figure 7.49 A schematic representation of the GaInAs/GaInAsP quantum wire laser fabricated by etching and regrowth. The lower part shows the optical confinement GaInAsP region with embedded quantum wires. After figure 3 of M. Cao, *et al.*, *Trans. IEICE* **E73**, 63 (1990).

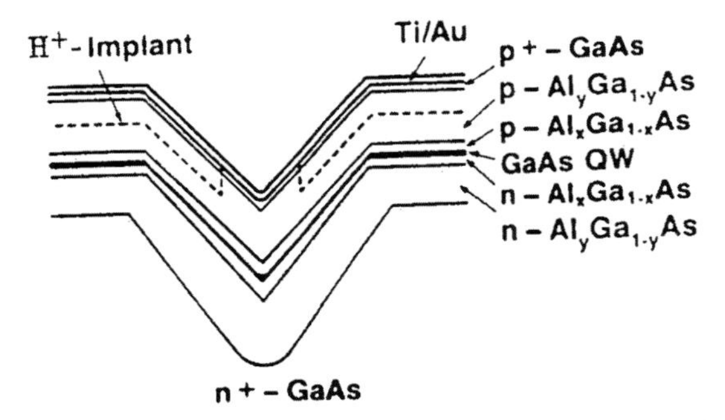

Figure 7.50 The cross-section of an AlGaAs/GaAs single-quantum-well laser grown on a V-grooved substrate. Reprinted from Fig. 1(a) of E. Kapon, *et al.*, *Appl. Phys. Lett.* **55**, 2715 (1989), with the permission of AIP Publishing.

Another method employs growth on a patterned nonplanar substrate. Cross-sections of a single AlGaAs/GaAs quantum-wire laser grown on a V-grooved GaAs substrate are shown in Fig. 7.50. The figure shows the sequence of layers and their doping. The quantum well layer is situated in an undoped i region. The lower and upper parts are heavily p- and n-doped GaAs layers, respectively. The H^+ implantation provides electric insulation of the side parts of the structure. It is worth emphasizing that, for this structure, the waveguide layer of $Al_xGa_{1-x}As$

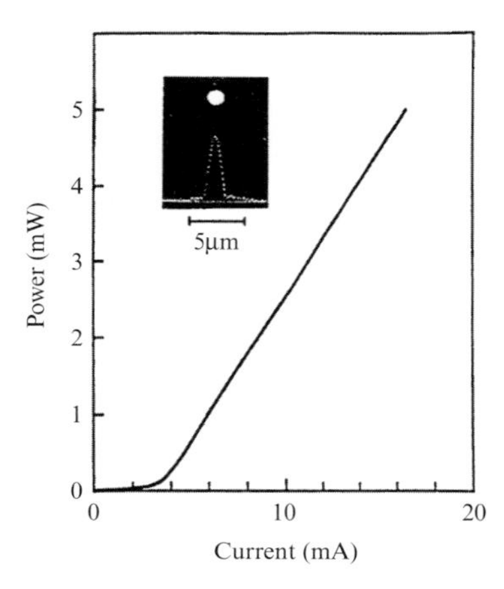

Figure 7.51 The light output versus the pumping current of the single-quantum-wire laser with the structure presented in Fig. 7.50. The inset shows the lateral distribution of the output light in the near-field zone. Reprinted from Fig. 2 of E. Kapon, *et al.*, *Appl. Phys. Lett.* **55**, 2715 (1989), with the permission of AIP Publishing.

between $Al_yGa_{1-y}As$, with $x < y$, gives rise to *two-dimensional optical confinement*. Analogously, "lateral" confinement for the carriers occurs due to the bent quantum well layer. The size of this confinement region is about 200 Å. The output power–current characteristic of this device is presented in Fig. 7.51. This particular single-quantum-wire device has a cavity with a length of 350 μm and uncoated facets, and operates in a pulsed regime at room temperature with a laser threshold of about 4 mA. The insert in Fig. 7.51 shows the lateral distribution of the emission in the near-field zone (when one can neglect the diffraction of the light beam). This distribution is almost circular with roughly a 1 μm full width at half maximum.

Figure 7.52 depicts the evolution of the emission spectra for a single-quantum-wire laser with a cavity length of 270 μm and a laser threshold current of 4.3 mA. The design of this laser corresponds to that of Fig. 7.50. Near the threshold many longitudinal modes oscillate. It is instructive to compare this behavior with the spectral dependence of the light-amplification coefficient presented in Fig. 7.48. The envelope of these modes clearly demonstrates that two different subbands are involved in the lasing. At high levels of the pumping current, the oscillations shift to shorter wavelengths as a result of the filling of the upper subbands, in agreement with the results of Fig. 7.48. As the current increases, the intensity of stimulated emission increases, while the number of oscillating modes decreases.

7.5.2 Quantum Dot Emitters

We now consider emitters and lasers based on quantum dot heterostructures. As discussed in detail in Section 3.5, the three-dimensional confinement and the quantization of the carriers in quantum dots lead to a further increase of the density of energy states, while the interband optical spectra take the form of sharp lines

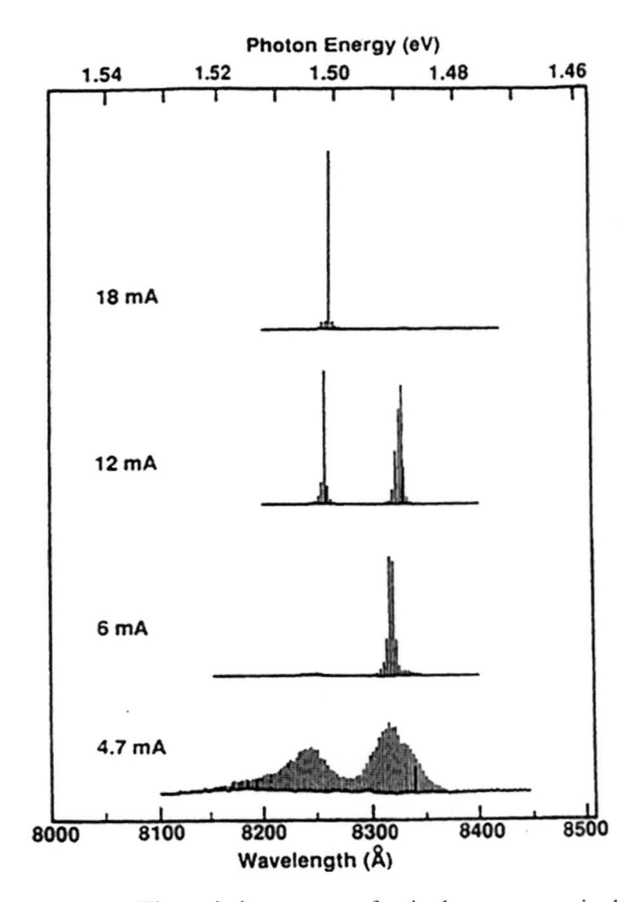

Figure 7.52 The emission spectra of a single quantum wire laser with the structure presented in Fig. 7.50 for different pumping currents. Reprinted from Fig. 4 of E. Kapon, *et al.*, *Appl. Phys. Lett.*, **55**, 2715 (1989), with the permission of AIP Publishing.

(see Section 5.5). Let some pumping method provide carriers in an active layer with quantum dots. Particular mechanisms for the population of the quantum dots are illustrated in Fig. 5.28. Then, the injected carriers concentrate in the discrete levels, i.e., in a narrow energy range near the band edge. Consequently, the maximum amplification increases, while the temperature dependence of the emission parameters is reduced.

These effects present the main advantages of quantum well and quantum wire heterostructures in comparison with bulk materials. Indeed, for the latter case the carriers are continuously distributed over energy and, when the temperature is raised, a redistribution of the carriers toward higher energies occurs, which leads to a shift and broadening of the spectra, as well as a decrease of the light gain. In zero-dimensional structures like quantum dots such factors as the spectral degradation with increasing temperature are minimized. In addition, for electrical pumping,

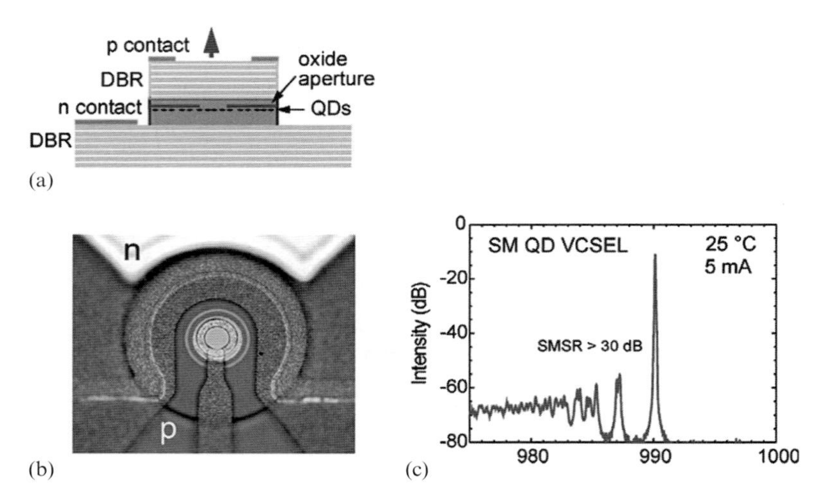

Figure 7.53 A vertical-cavity surface-emitting laser (VCSEL) based on a GaAs/AlAs quantum dot structure. (a) Schematic view in cross-section. The optically active medium comprising the quantum dots (QDs) is clad by two distributed Bragg reflectors (DBRs), which form an optical cavity. The radiation is vertically emitted, as indicated by the arrow. (b) Top view of a typical device showing the aperture and the contacts. (c) The emission spectrum presented on a logarithmic scale. Reprinted from Figure 4 of *Materials Today* **14**, D. Bimberg and U. W. Pohl, "Quantum dots: promises and accomplishments," pp. 388–397, Copyright 2011, with permission from Elsevier.

an increase in the light amplification leads to a decrease of the threshold current density.

As follows from the discussions presented in Chapters 2, 3, and 5, a great number of quantum dot structures are fabricated from direct-bandgap materials, which are optically active from the infrared to the ultraviolet optical region. Correspondingly, there are numerous applications of these quantum dot emitting structures. Below we present only a few examples of optoelectronic devices which are based on these structures.

Vertical-cavity surface-emitting lasers. As discussed above, semiconductor lasers with a vertical cavity have a number of favorable properties.

The configuration of such a GaAs/AlAs laser with active layer(s) containing quantum GaAs dots is shown in Fig. 7.53(a). This active layer is placed in a vertical cavity formed by distributed-Bragg-reflector mirrors. Low losses at each reflection are crucial, since the cavity is very short (one layer, or just a few layers) and the amplification in a single passage through the system is low. In this device, the large output aperture provides a low divergence of the generated beam. The top view of the processed GaAs-based quantum dot vertical cavity laser is shown in Fig. 7.53(b), where the p and n contacts and the circular output aperture can be seen.

Recall that the bandwidths of emission and amplification of a quantum dot ensemble are relatively broad, as discussed in Section 5.5 (see Fig. 5.26). The spectrum of laser generation (on a logarithmic scale!) is presented in Fig. 7.53(c)

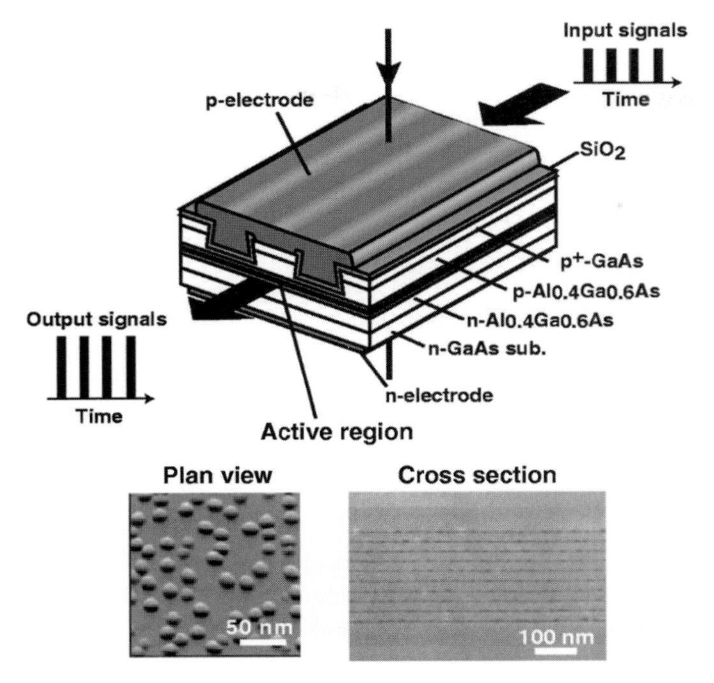

Figure 7.54 The scheme of a linear amplifier based on a quantum dot active region. Reprinted figure with permission from figure 1 of M. Sugawara, *et al.*, *Phys. Rev.* **B 69**, 235332 (2004), Copyright 2004 by the American Physical Society.

for a pumping current of 5 mA at room temperature. The wavelength of the generated light is fixed by the high-quality resonator. The observed very narrow spectrum corresponds to single-mode generation.

The device presented in Fig. 7.53 generates a circular beam of high quality, which can be easily coupled to a fiber, and operates at low threshold currents, enabling high-speed modulation. These GaAs/AlAs injection lasers emit at the standard wavelengths for the data communication industry and facilitate large signal data rates at 40 Gb/s.

Other applications of the stimulated emission of quantum dots. The broad bandwidth and ultrafast gain recovery of the quantum dot active media allow one to use them for linear amplifiers, modulators, switches, and other optoelectronic devices. For example, Fig. 7.54 shows a linear amplifier based on a quantum dot heterostructure. Such a device consists of a waveguide structure with a quantum dot active layer and has nonreflecting cavity facets to prevent self-lasing. The active region is pumped by the vertical current, as in the surface-emitting laser presented in Fig. 7.53.

In Sections 7.3.4 and 7.3.5 it was indicated that the high-frequency modulation of laser emission has many important applications. Quantum dot lasers also are

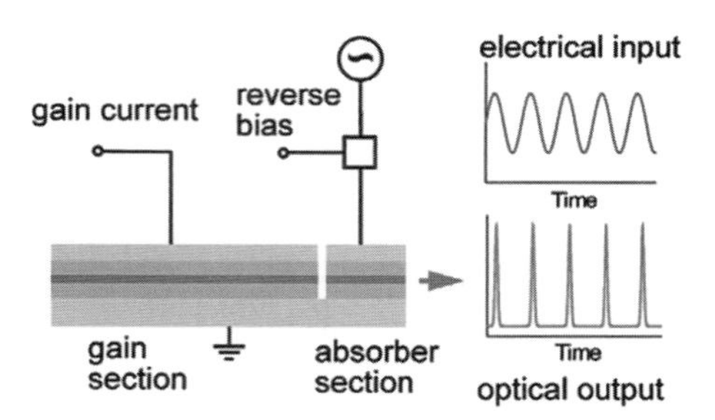

Figure 7.55 A schematic diagram of a mode-locked laser, with hybrid mode locking and with the quantum dot active region pumped by the vertical current. The waveguide laser structure consists of two sections, gain and absorber. The electrically modulated absorption generates a synchronous optical pulse train. Reprinted from Figure 2a of *Materials Today* **14**, D. Bimberg and U. W. Pohl, "Quantum dots: promises and accomplishments," pp. 388–397, Copyright 2011, with permission from Elsevier.

used for the high-speed modulation of optical signals and the generation of high-frequency optical pulse trains for data communication, optical clocks in electronic circuits, optical sampling of high-speed signals, etc.

Figure 7.55(a) depicts schematically a mode-locked laser, in which the quantum dot active layer is pumped by the vertical current as in Figs. 7.53 and 7.54. The laser cavity is formed in the lateral direction as a waveguide structure. This waveguide is divided into two sections: a gain section and a reverse-biased absorber section. Pulse trains or generated light modulation are produced by modulation of the absorption in the passive section of the device. Figure 7.55(b) illustrates the electrically modulated light output of this laser. The presented device demonstrates data transmission at 40 Gb/s by an external synchronized modulation of the laser pulse-train. It is expected that an improved version of such a device can be suitable for generating optical data streams at 160 Gb/s.

In Section 6.2 we analyzed electro-optical effects – changing the refractive index of materials and heterostructures by the application of external electric fields. These effects, in particular the quantum confined Stark effect, can occur in the active layers with the quantum dots. Because the spectra of the quantum dots are of a resonant character, the quantum confined Stark effect, which is responsible for the shift of the optical resonance, is large in such systems and can be exploited for high-frequency modulation of the emission. Figure 7.56(a) depicts the refractive index near resonance of a quantum-state excitation and the shift of the resonance induced by an external electric field, which produces a corresponding change, Δn, in the refractive index. If the voltage-controlled medium is placed in a resonant cavity, the modulation of the light output can be especially efficient.

This effect is employed in a device with coupled cavities to change the transmittance of the modulator section. The design of the device with two sections, a vertical cavity section and a reverse-biased modulator section, is shown in Fig. 7.56(b). For this design, the device has three distributed-Bragg-reflector mirrors (black and

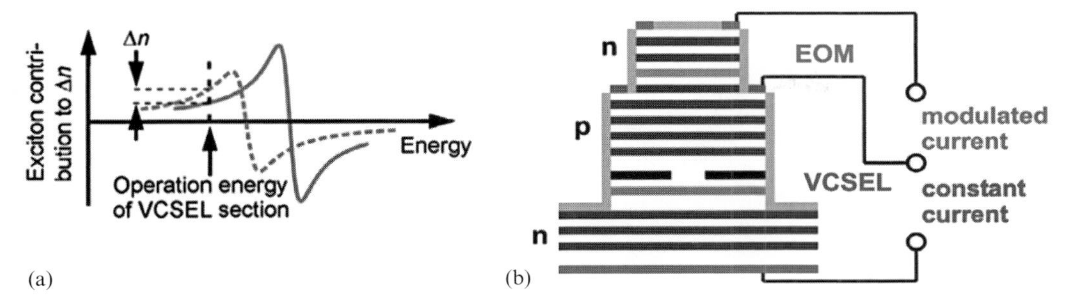

Figure 7.56 A monolithic electro-optically modulated vertical-cavity surface-emitting laser on quantum dots. (a) A schematic diagram of the electro-optic effect used for modulation. The refractive index changes by Δn at the operating wavelength of the device due to the field-induced energy shift of the resonant absorption. (b) A schematic diagram of the electro-optically modulated laser. The device consists of two cavities, a vertical laser cavity and the vertical electro-optical modulation (EOM) cavity that modulates the optical output of the laser. Reprinted from Figure 5 of *Materials Today* **14**, D. Bimberg and U. W. Pohl, "Quantum dots: promises and accomplishments," pp. 388–397, Copyright 2011, with permission from Elsevier.

white lines). These three mirrors form two cavities, the vertical laser cavity operated in continuous-wave mode, and the electro-optically modulated cavity, which modulates the optical output of the laser. The device exhibits temperature-stable operation at 845 nm up to 85°C with low power consumption. This emitting device achieves an electrical bandwidth of 60 GHz and an optical bandwidth larger than 40 GHz.

Devices addressing individual quantum dots. The previously considered quantum dot devices use an *ensemble* of quantum dots. Research on individually addressed quantum dots is a subject of significant interest because of the potential applications in quantum information processing. A key element for quantum information processing is a *single-photon emitter*. Such an emitter must be able to emit just a single photon on demand. Obviously, this light source should exhibit nonclassical emission characteristics. Such emitting devices can be developed using individually addressed quantum dots.

An emitter with individually addressed quantum dots can be based on a p–i–n diode structure with a low-density quantum dot active layer and spatial filtering with the help of a small-area aperture. Individual addressing of just a single dot can be achieved by confining the current path in the p–i–n diode through the fabrication of an *oxide aperture* with a small opening in close proximity to a dot, as shown in Fig. 7.57(a).

Here, a specific device that uses a small $AlGaO_x$ aperture located 20 nm above a layer with InAs quantum dots is discussed. The emission spectrum of this device, consisting of a single line, is shown in Fig. 7.57(c). The emission originated from a single InAs quantum dot. The single-quantum-dot emission was enhanced by placing the quantum dot in a cavity formed by Bragg reflectors, as in other emitters considered in this subsection. The combination of the small aperture and the cavity

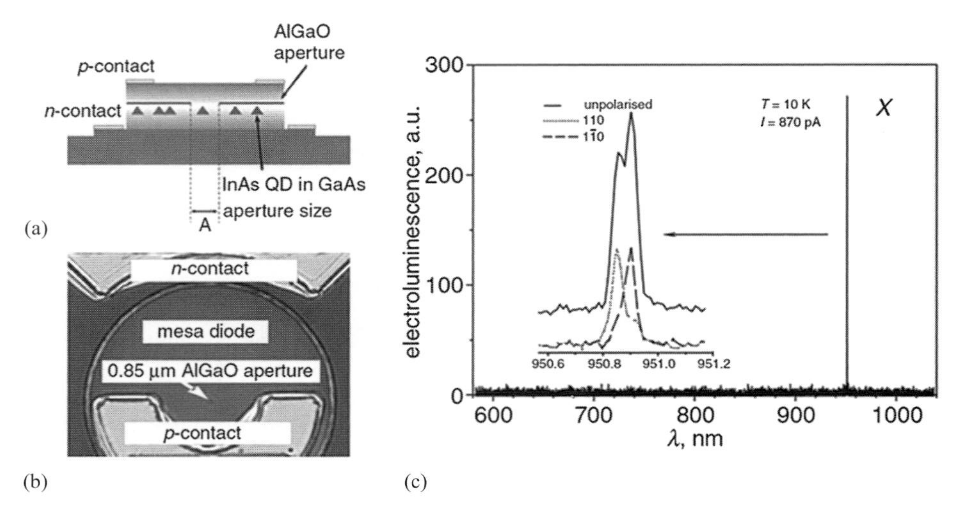

Figure 7.57 A single-quantum-dot-based single-photon source. (a) A schematic view of the device. The oxide aperture confines the current path in the mesa structure to provide for the excitation of a single quantum dot. (b) A microscopic image of the device. (c) The emission spectrum of the single-quantum-dot device. Reprinted with permission from Figures 1 and 2 of A. Lochmann, *et al.*, "Electrically driven single quantum dot polarised single photon emitter," *Electronics Letters*, 2006, **42** (13), pp. 774–775.

provided emission in a single-cavity mode, which increases the luminescence intensity by approximately a factor of 20 because of the larger spontaneous emission rate into the mode and larger out-coupling efficiency. The emission intensity increases with the current and saturates for currents ≥ 1 nA. The 1 nA current corresponds to an electron flux through the quantum dot equal to about five particles per nanosecond. Since the typical lifetime of excitation in the direct-bandgap InAs quantum dots is 1 ns, one can estimate the emission efficiency of this device as $\approx 20\%$.

The single-photon nature of the emission was proved by a photon-correlation measurement. Modeling of the perfect single-photon device with 1 GHz repetition rate (the full line in Fig. 7.57(c)) showed excellent agreement with the measured correlations. Thus, these measurements and modeling proved single-photon emission and the high-speed capability of the single-photon device on the basis of a single quantum dot. In general, emitting devices addressing a single quantum dot enable single-photon generation, open a new research field, and are suitable for quantum communication applications.

In Fig. 7.58, data on the threshold current densities of quantum dot lasers based on GaAs, Ge, SiGe, and Si are collected. As can be seen from this figure, the progress in quantum dot laser technology is dramatic in terms of the range of materials, the decreasing threshold current densities, the demonstration of III-nitride quantum dot lasers, and the demonstration of quantum dot lasers on group IV substrates.

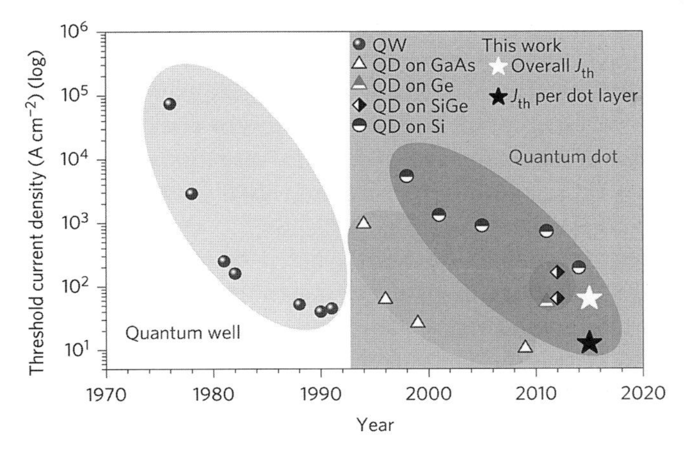

Figure 7.58 The evolution of the threshold current density of quantum dot lasers on GaAs, Ge, SiGe, and Si. The black star represents the case where the performance is normalized with respect to that of a single-quantum-dot layer. The white star indicates the threshold value achieved by the authors, who are the originators of this figure. Reprinted by permission from Fig. 1(a) in Springer Nature: Macmillan Publishers Ltd., *Nature Photonics*, "Electrically pumped continuous-wave III–V quantum dot lasers on silicon," S. Chen, *et al.*, Copyright 2016.

In concluding this section, we summarize that the transition from lasers based on quantum wells to lasers based on quantum wires and quantum dots reduces the threshold current (Fig. 7.58). Further progress in the field of quantum wire and quantum dot lasers will require considerable improvement in the technology of these structures. Smaller sizes, better interfaces, improved uniformity of the quantum wires, cavities with small optical losses, and improved fabrication of arrays of wires and dots are all necessary for achieving superior performance from this type of low-threshold laser.

7.6 Closing Remarks to Chapter 7

In this chapter, we have applied the results obtained in Chapters 4 and 5 to optical devices operating in the near-infrared, visible, and ultraviolet range that are based on interband phototransitions in quantum heterostructures. It has been shown that these heterostructures create the possibility of major improvements in key elements of optoelectronic technologies – the sources of spontaneous and stimulated emission, light-emitting diodes, and laser diodes.

We started with semiconductor injection lasers based on the homostructure p–n junction and then we considered the development of advanced heterojunction lasers. We traced the steps critical to the improvement of injection lasers: from a double heterojunction to single- and multiple-quantum-well lasers, and quantum wire and quantum dot lasers. Each of these steps leads to additional confinement of injected carriers and, as a consequence, to a considerable decrease in the threshold pumping current. It is of importance that the confinement both of carriers and of light occurs in these heterostructures. This increases significantly the gain of the confined optical modes and lowers the injection currents necessary for laser oscillations. It makes possible the design of microlasers excited by low currents as well as the design of arrays of such devices. Quantum heterostructure lasers operate from the infrared to the ultraviolet spectral region.

The short-wavelength lasers are mostly based on group-III-nitride quantum heterostructures. For these wide-bandgap materials, producing the high-quality structures and the p-doping necessary for light-emitting devices has presented serious problems. We showed that recent progress in the fabrication of high-quality single-crystal GaN, and of ternary and quaternary alloys such as AlGaN, InGaN, and AlGaInN, and the success of p-type doping technologies will facilitate the realization of efficient short-wavelength-emitting devices.

7.7 Control Questions

1. Show that the absorption rate, as discussed in Section 7.2.1, is proportional to $\Delta\mathcal{F} \equiv \mathcal{F}(E_v(\vec{k})) - \mathcal{F}(E_c(\vec{k}))$.
2. Show that, at low temperatures, as discussed in Section 7.2.1, $\Delta\mathcal{F} = +1$ if $\hbar\omega < E_{F_n} - E_{F_p}$, and $\Delta\mathcal{F} = -1$ if $\hbar\omega > E_{F_n} - E_{F_p}$.
3. Explain how the threshold current can be reduced by using a multiple-quantum-well structure instead of a typical double heterostructure with a single quantum well.
4. Explain how, in some cases, lasers using lattice-mismatched heterostructures can operate in spectral regions different from those accessible to lasers based on lattice-matched structures and can provide lower threshold currents and higher amplification.

8 Devices Based on Intraband Phototransitions in Quantum Structures and Silicon Optoelectronics

8.1 Introduction

In the previous chapter, we considered optoelectronic devices, in particular, light-emitting devices and lasers, which use interband phototransitions. Their action is based on phototransitions involving both the conduction band (electrons) and the valence band (holes). Another type of phototransition – intraband absorption/ emission of far-infrared light – was studied in Section 5.3. For this type of phototransition the free carriers absorb or emit photons while remaining in the same energy band i.e., either in the conduction band or in the valence band.

In an ideal bulk crystal, intraband phototransitions are forbidden by the energy and momentum conservation laws; however, they can be induced by phonons, impurities, and crystal imperfections, which ensure that the conservation laws are obeyed. In contrast to the case in bulk materials, intraband phototransitions are allowed in semiconductor heterostructures, as discussed in Section 5.3. Indeed, these phototransions occur between quantized levels in quantum heterostructures. These energy levels are artificially created by fabricating materials into structures of nanometric thickness (quantum wells, quantum wires, and quantum dots). Importantly, the energy of emitted/absorbed photons and the rate of phototransitions can be changed through variation of the structural parameters. This flexibility and the control of the optical properties portend new operating principles for optoelectronic devices.

In this chapter we consider various optoelectronic devices which exploit intraband phototransitions. These include unipolar cascade lasers operating in the mid-infrared and terahertz ranges, quantum well photodetectors, quantum dot photodetectors, devices based on two-dimensional crystals, etc. In view of the great importance of the integration of optical systems with silicon-based optoelectronics and microelectronics, we analyze silicon optoelectronics, which includes silicon-based waveguides, light modulators, etc.

8.2 Unipolar Intersubband Quantum-Cascade Lasers

Since intraband phototransitions are drastically different from interband transitions, laser action associated with these transitions should differ in a fundamental way from that studied for the laser schemes considered previously in this text. First of all, intraband-transition lasers employ only one type of carrier, i.e., such a laser is *a unipolar device*. Second, such laser action is based on electron transitions between confined states arising from the quantization in semiconductor heterostructures. In order to create a population inversion between two confined states, one needs to provide (i) electron injection into a higher lasing state and (ii) depletion of a lower lasing state. For this purpose, a vertical scheme of electron transport has been proposed. This scheme is illustrated in Fig. 8.1(a). The proposed heterostructure is a superlattice with a complex design for each period. Each of the periods consists of four AlInAs barriers, forming three GaInAs quantum wells and a digitally graded AlInGaAs region, which is doped. Under zero-bias conditions, the overall band diagram appears like a sawtooth structure. Under an applied electric field, the band diagram takes on a staircase structure, as shown in Fig. 8.1(a). The barriers form *three coupled quantum wells* with three quasi-bound levels. These three levels are

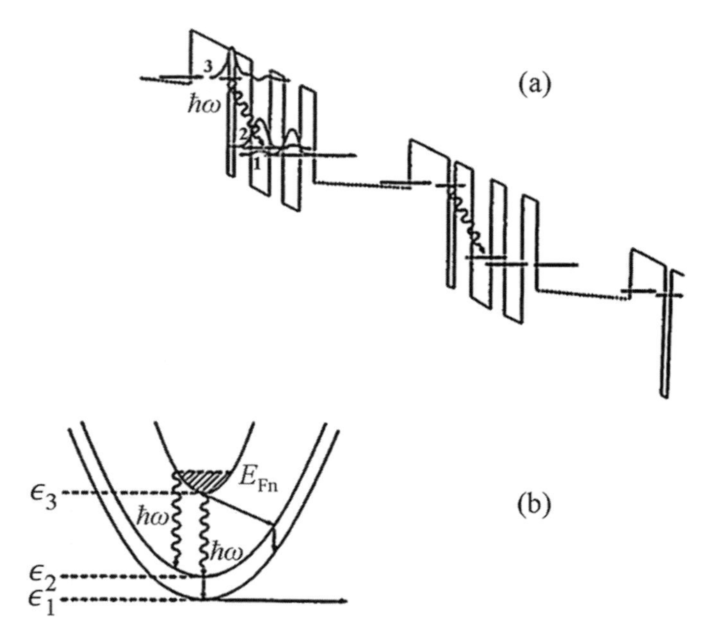

Figure 8.1 (a) Two periods of the 25-stage staircase coupled-well region of a quantum-cascade laser under operating conditions. The laser phototransitions are indicated by wavy arrows. They occur between levels (subbands) 3 and 2 with a photon energy of 295 meV. Level 2 depopulates through level 1 and subsequent tunneling. The energy separation between levels 2 and 1 is about 30 meV. (b) Energy dispersion for subbands 1, 2, 3, phototransitions and interband scattering processes (straight arrows). Figure 1 in F. Capasso, *et al.*, *Science* **264**, 553 (1994). Reprinted with permission from AAAS.

marked in Fig. 8.1(a) by 1, 2, and 3. Each of the confined states originates from one of the wells.

The structure is chosen so that there is a considerable overlap between the wave functions of the upper state 3 and of the intermediate state 2 as a result of tunneling processes. The same is valid for the wave functions of states 2 and 1. Under a voltage bias, the potential in the doped regions is almost flat, as shown in Fig. 8.1(a). The electrons are injected from the doped regions through the barrier in the confined state 3 of the first quantum well. From this state, they relax primarily to state 2. There are two processes of relaxation: phonon emission and photon emission. In Fig. 8.1(b), these processes are shown for electrons with different values of the in-plane wavevector, $\vec{k}$. The three indicated subbands, $\epsilon_{1,2,3}(\vec{k})$, correspond to the three confined states. The straight arrows represent intersubband phonon relaxation. The third confined state, 1, is selected to provide depletion of the state 2 as fast as possible. Thus, in this manner we have a three-level scheme where the upper level is pumped by the direct injection of electrons from the doped region. The second level is depleted due to strong coupling with the lowest level 1. From level 1, electrons escape to the next doped region. Then the processes are repeated in each subsequent period of the superlattice. One can say that the carriers make transitions down through such a *cascade structure*.

We can suppose that the intrasubband relaxation processes are faster than the intersubband processes. Then, one can introduce distribution functions for the three subbands: $\mathcal{F}_3(\vec{k}), \mathcal{F}_2(\vec{k})$, and $\mathcal{F}_1(\vec{k})$. We assume that quasi-Fermi distributions hold for each of the confined states. In this case, we need to define only the numbers of electrons in these states: n_3, n_2, and n_1. The criterion for population inversion between levels 2 and 3 is

$$n_3 > n_2.$$

We can write simple balance equations for n_3 and n_2:

$$\frac{dn_3}{dt} = -\frac{1}{e}j - \frac{n_3}{\tau_{32}}, \tag{8.1}$$

$$\frac{dn_2}{dt} = \frac{n_3}{\tau_{32}} - \frac{n_2}{\tau_{21}}, \tag{8.2}$$

where j is the density of the injection current, and τ_{32} and τ_{21} are the relaxation times between the states 3 and 2, 2 and 1, respectively. In Eq. (8.2), we have neglected the inverse $1 \to 2$ process since state 1 can be regarded as almost empty as a result of fast electron escape to the doped region. For steady-state conditions, we obtain the concentrations

$$n_3 = -\frac{1}{e}j\tau_{32} \quad \text{and} \quad n_2 = n_3\frac{\tau_{21}}{\tau_{32}},$$

and a population inversion,

$$\Delta n \equiv n_3 - n_2 = -\frac{1}{e}j\tau_{32}\left(1 - \frac{\tau_{21}}{\tau_{32}}\right). \tag{8.3}$$

Thus, to create a population inversion, one should design the laser so that

$$\tau_{21} < \tau_{32}. \tag{8.4}$$

In order to fabricate such unipolar laser structures with vertical electron transport, very precise and sophisticated semiconductor technology is necessary. In Fig. 8.2, a schematic cross-section of the unipolar n-type $Al_{0.48}In_{0.52}As/Ga_{0.47}$ laser structure is shown in detail. The structure is grown upon an n^+-InP substrate which serves as a contact. Then a waveguide cladding of a 5000 Å-thick AlGaAs doped layer follows. Additional cladding layers are designed at the bottom and the top of the structure to increase the refractive-index step between the core and the cladding, in order to enhance the optical confinement. The waveguide core is grown upon the waveguide cladding layer. It consists of AlGaln, AlGaInAs, and GaInAs layers,

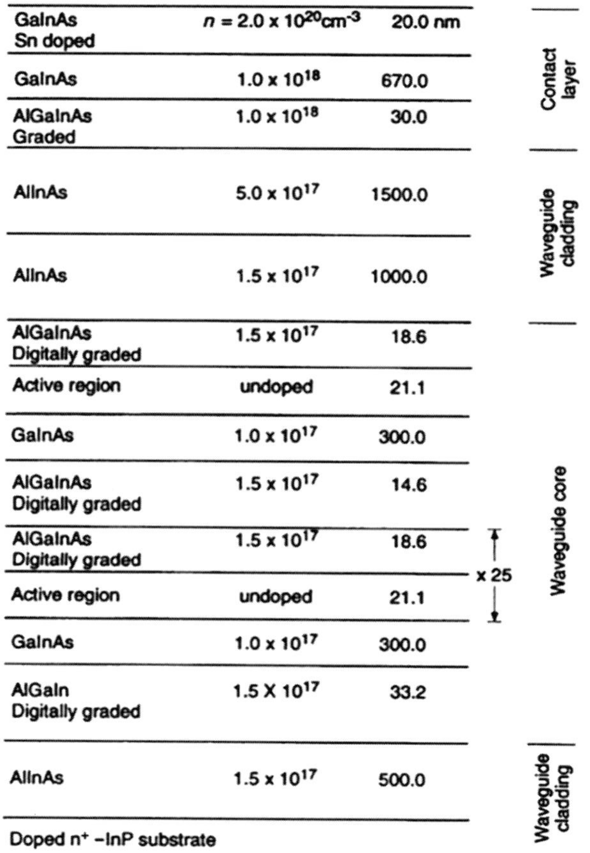

Layer	Doping	Thickness	Repeat	Region
GaInAs Sn doped	$n = 2.0 \times 10^{20}$ cm^{-3}	20.0 nm		Contact layer
GaInAs	1.0×10^{18}	670.0		Contact layer
AlGaInAs Graded	1.0×10^{18}	30.0		Contact layer
AlInAs	5.0×10^{17}	1500.0		Waveguide cladding
AlInAs	1.5×10^{17}	1000.0		Waveguide cladding
AlGaInAs Digitally graded	1.5×10^{17}	18.6		Waveguide core
Active region	undoped	21.1		Waveguide core
GaInAs	1.0×10^{17}	300.0		Waveguide core
AlGaInAs Digitally graded	1.5×10^{17}	14.6		Waveguide core
AlGaInAs Digitally graded	1.5×10^{17}	18.6	x 25	Waveguide core
Active region	undoped	21.1	x 25	Waveguide core
GaInAs	1.0×10^{17}	300.0		Waveguide core
AlGaIn Digitally graded	1.5×10^{17}	33.2		Waveguide core
AlInAs	1.5×10^{17}	500.0		Waveguide cladding
Doped n^+ -InP substrate				

Figure 8.2 A schematic cross-section of the cascade laser structure. The whole structure consists of 500 layers. From Fig. 2 in F. Capasso, *et al.*, *Science* **264**, 553 (1994). Reprinted with permission from AAAS.

which provide the optical confinement and optical amplification. For the structure presented in Fig. 8.2, the active region and the doped layer are repeated 25 times. The active region contains the barriers and quantum wells discussed previously. The waveguide provides a confinement factor for a single-mode waveguide of about $\Gamma \approx 0.46$; see Sections 5.2 and 7.2 as well as Eqs. (5.13) and (7.17). For this structure, the necessary cascade regime is reached at an electric field of about 10^5 V/cm. The characteristic times in Eqs. (8.1) and (8.2) are $\tau_{32} \approx 4.3$ ps and $\tau_{21} \approx 0.6$ ps, and the time of escape from level 1 is $\tau_{1,esc} < 0.5$ ps. That is, the criterion for population inversion as given by Eq. (8.4) is fulfilled. For this structure, it follows that spontaneous radiative processes are negligible in comparison with phonon relaxation: $\tau_R \approx 10$ ns, $\tau_{32}/\tau_R \approx 4.3 \times 10^{-4}$. For a particular device with an optical path of about 700 μm and mirror reflectivity $r_1 = r_2 = 0.27$, the laser output-current characteristics are shown in Fig. 8.3 for different temperatures. The insets of Fig. 8.3 show the current–voltage characteristics and the temperature dependence of the laser threshold current. The laser threshold current can be approximated by $I_{th} = C \times \exp(T/112)$, where the constant A is about 900 mA and T is measured in kelvin units. Using typical values of the electric current, one can estimate the values of the population inversion as given by Eq. (8.3): $\Delta n \approx 10^{11}$ cm^{-2} for each period of the cascade structure. From Fig. 8.3, it follows that the output power reaches tens of milliwatts.

The emission energy is in the range 275 meV to 310 meV. Spectra of the laser output for different currents at $T = 80$ K are presented in Fig. 8.4. For this case,

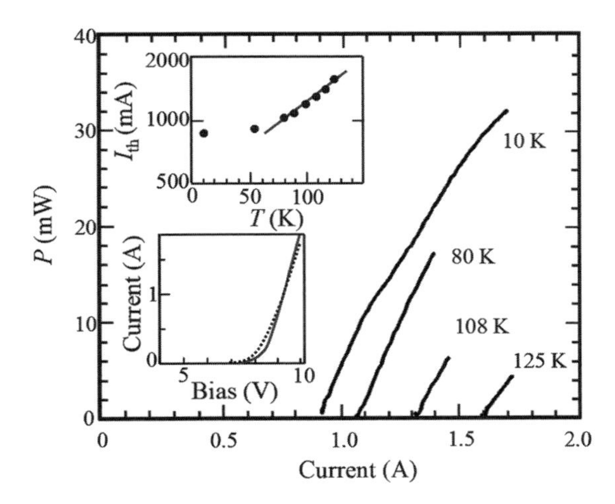

Figure 8.3 The measured optical power P from a single facet of the quantum-cascade laser with the structure presented in Fig. 8.2 and an optical cavity length of 1.2 μm. The results are given for different temperatures. Insets show the dependence of the laser threshold current as a function of temperature and the current–voltage characteristics of the device. From Fig. 2 in *Proceedings of 22nd Int. Conf. on the Physics of Semiconductors*, vol. 3, F. Capasso, *et al.*, ed. D. J. Lockwood, pp. 1536–1640, Copyright ©1995 World Scientific Publishing.

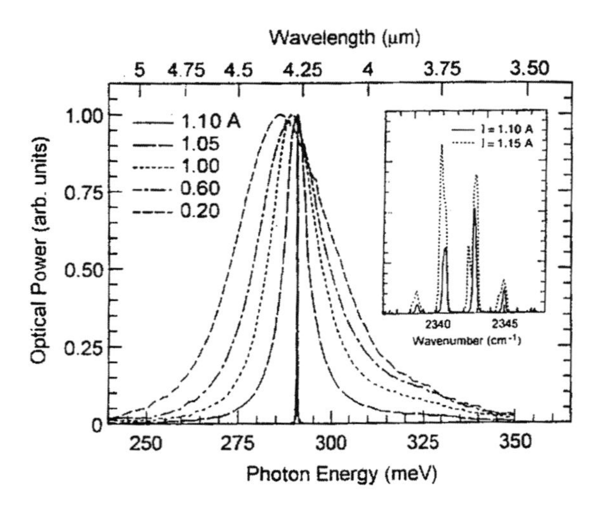

Figure 8.4 The spectra for the emission of the quantum-cascade laser presented in Figs. 8.2 and 8.3 at $T = 80$ K. Results are given for different pumping currents. The inset shows high-resolution spectra for the cascade laser with a shorter cavity (750 μm) at two pumping currents. From Fig. 3 in *Proceedings of 22nd Int. Conf. on the Physics of Semiconductors*, vol. 3, F. Capasso, *et al.*, ed. D. J. Lockwood, pp. 1536–1640, Copyright ©1995 World Scientific Publishing.

according to Fig. 8.3, the threshold current is about 1.06 A. Figure 8.4 clearly demonstrates that there is a sharp narrowing of the emission spectra above the laser threshold: the spectra reduce to a sharp peak at $I = 1.1$ A $> I_{th}$. The inset shows details of the laser emission spectra for particular values of the pumping currents which are greater than the threshold current; in particular, the inset depicts equally spaced longitudinal modes and a higher-order transverse mode for the higher current of 1.15 A. In accordance with the selection rules for intersubband phototransitions given in Table 5.1, the laser emission is polarized normal to the layers.

8.3 Terahertz Cascade Lasers

The quantum-cascade lasers discussed previously generate the coherent mid-infrared emission of photons with energies of hundreds of meV. As soon as the fabrication technologies reached the sub-meV-level for the control of the energy subbands in multilayered structures, the way was opened to quantum-cascade lasers in the terahertz range (photon energy range from 4 meV to 30 meV). Remember that at room temperature the thermal energy of the electrons is about 26 meV, i.e., it is in the same energy range; this hinders achieving the population inversion of electronic subbands at room temperature. Indeed, the first quantum-cascade laser provided stimulated emission at 4.4 THz with a maximum operating temperature of 50 K. Further improvement of quantum-cascade structures has led to an increase

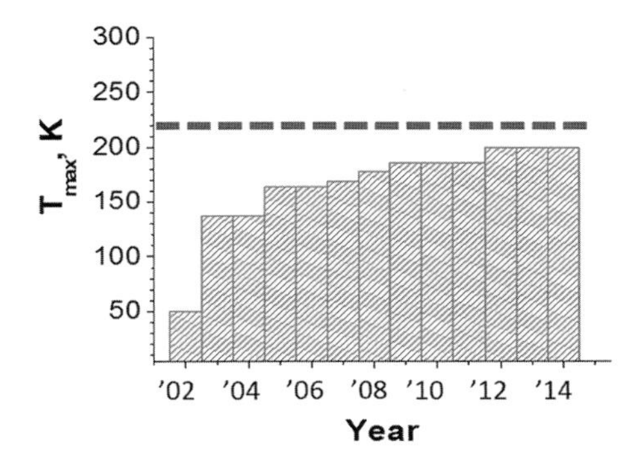

Figure 8.5 Maximum operating temperatures in pulsed mode achieved by terahertz quantum-cascade lasers from the initial demonstration in 2002 until mid 2015. From Fig. 2(b) in M. A. Belkin and F. Capasso, "New frontiers in quantum cascade lasers: high performance room temperature terahertz sources," *Physica Scripta* **90**, 118002 (2015). ©IOP Publishing. Reproduced with permission. All rights reserved.

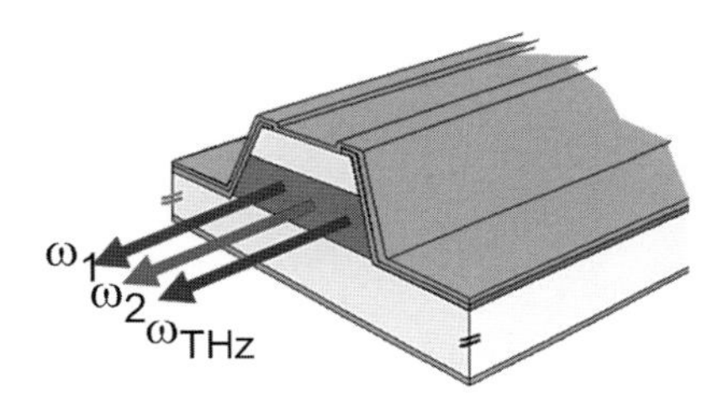

Figure 8.6 A schematic diagram of a terahertz difference-frequency-generation quantum-cascade laser. From Fig. 4(a) in M. A. Belkin and F. Capasso, "New frontiers in quantum cascade lasers: high performance room temperature terahertz sources", *Physica Scripta* **90**, 118002 (2015). ©IOP Publishing. Reproduced with permission. All rights reserved.

in the operating temperature of the terahertz emitting laser. However, these devices work at low temperatures (less than 200 K), as shown in the diagram in Fig. 8.5. Most recently, low-temperature terahertz lasing in the 1.2 THz to 5 THz range was demonstrated with a continuous-wave output power in excess of 100 mW and peak powers in excess of 1 W.

An alternative approach to terahertz generation is based on intracavity *difference-frequency generation*. In Fig. 8.6 we illustrate the concept: the current-pumped quantum-cascade device generates light consisting of two mid-infrared modes with similar frequencies, ω_1 and ω_2. Inside the laser cavity the electric-field amplitudes of these modes are high, which produces nonlinear optical effects (the second-order nonlinear effect), as described in Section 6.4.1. In particular, a nonlinear polarization $\mathcal{P}_{NL}$ arises at the difference frequency $\omega_{THz} = \omega_1 - \omega_2$, which, in turn, generates an electromagnetic wave at this difference frequency (see Eq. (6.15)). Since the frequencies of the two mid-infrared modes are close, the generated wave is in the terahertz range.

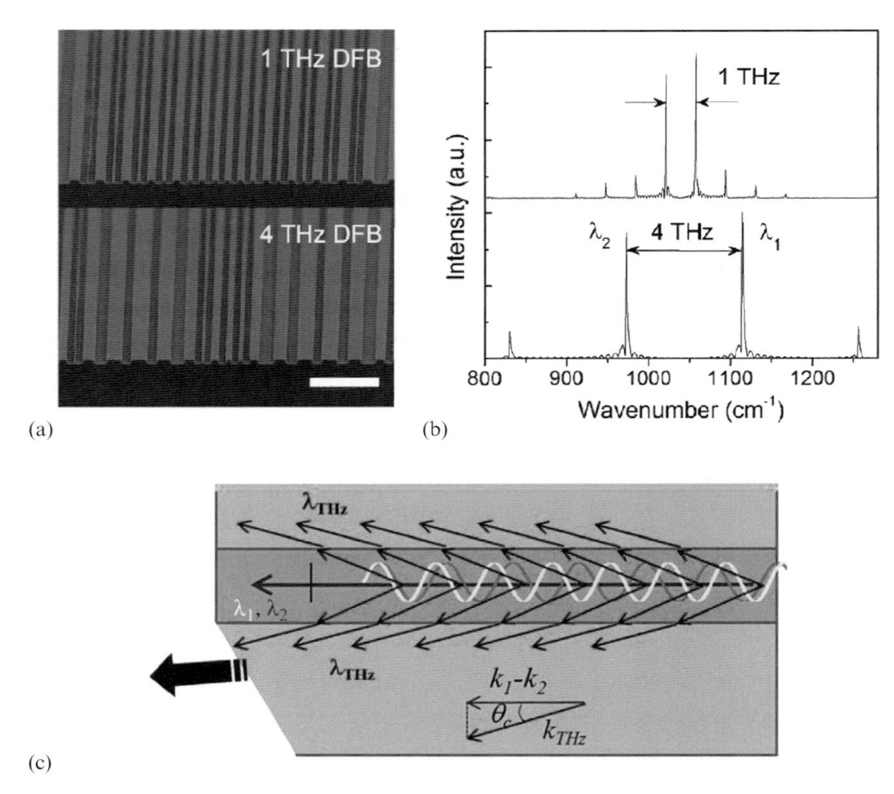

Figure 8.7 (a) Two examples of composite-grating cavities. (b) Mid-infrared spectra of the electromagnetic waves emitted into the cavities shown. (c) A schematic diagram showing Čerenkov phase matching in a terahertz-emitting quantum-cascade laser. The central black arrow indicates the propagation direction of the mid-infrared modes. After Figs. 3 and 4(a) in Q. Lu and M. Razeghi, *Photonics* **3**, 42 (2016). Licensed under CC BY 4.0.

It is important that the approach being discussed does not involve formation of the population inversion on terahertz transitions and that the mid-infrared lasing is reliably achieved at room temperature. Thus such a terahertz device can operate at temperatures up to room temperature. From the point of view of fabrication, this nonlinear optical device is very similar to traditional quantum-cascade lasers.

To excite two mid-infrared modes with similar frequencies, one can use a composite distributed-feedback grating, as illustrated in Fig. 8.7(a). The grating facilitates concentration of the infrared power in the two necessary wavelength components and, simultaneously, provides sufficient coupling strength between the modes and modal gain of these two components of special importance.

Figure 8.7(a) shows two electron-microscope pictures of composite distributed-feed-back gratings, each designed for two mid-infrared dominant modes with a given frequency difference of the mid-infrared modes. Figure 8.7(b) shows the mid-infrared spectra of the electromagnetic waves emitted into the composite cavities.

For the two cavities of Fig. 8.7(a), the differences of frequencies of the main modes are ≈1 THz and 4 THz. Such a grating design offers sufficient coupling strength for cavities of 2 mm to 3 mm length with high-reflection coatings.

It is worth mentioning the following effect, which should be taken into account in designing a terahertz cascade laser. In the active region of the quantum-cascade laser, the refractive index at terahertz frequencies, n_{THz}, is higher than that in the mid-infrared range, n_{MIR}. Accordingly, the mid-infrared waves propagate faster than does the difference-frequency terahertz wave. For effective generation of terahertz radiation, a phase-matching scheme should be used. For example, the so-called Čerenkov configuration is illustrated in Fig. 8.7(c). For this design the terahertz wave travels at an angle $\theta_{\check{C}} = \cos^{-1}[n_{\mathrm{MIR}}/n_{\mathrm{THz}}]$ with respect to the mid-infrared waves. Here $\theta_{\check{C}}$ is the Čerenkov angle. The Čerenkov configuration provides phase matching of the nonlinear polarization wave and the terahertz wave, and effective conversion of the energy of the mid-infrared waves to the terahertz wave.

In Fig. 8.8 we present the terahertz power as a function of the current in a quantum-cascade laser with intracavity difference-frequency generation. This terahertz quantum-cascade device operates at room temperature with a peak power of nearly 2 mW. The capabilities of quantum-cascade lasers have advanced dramatically in the last decade and a half. Multi-wavelength operation has been advanced, operating temperatures have risen, and output power levels have increased. The rapid progress in the peak power of terahertz cascade lasers is illustrated in Fig. 8.9.

In conclusion, the unipolar cascade laser is drastically different from lasers based on interband phototransitions. The properties of intersubband phototransitions and, consequently, the properties of the unipolar laser are determined to a large degree by quantum confinement; accordingly, this novel laser can be tailored for operation in the spectral region from the mid-infrared range to sub-millimeter waves.

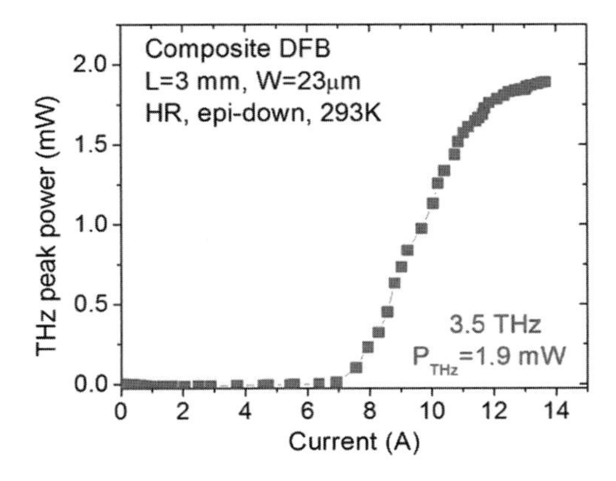

Figure 8.8 The 3.5 THz peak power versus electric current in a terahertz quantum-cascade laser. After Fig. 5(b) in Q. Lu and M. Razeghi, *Photonics* 3, 42 (2016). Licensed under CC BY 4.0.

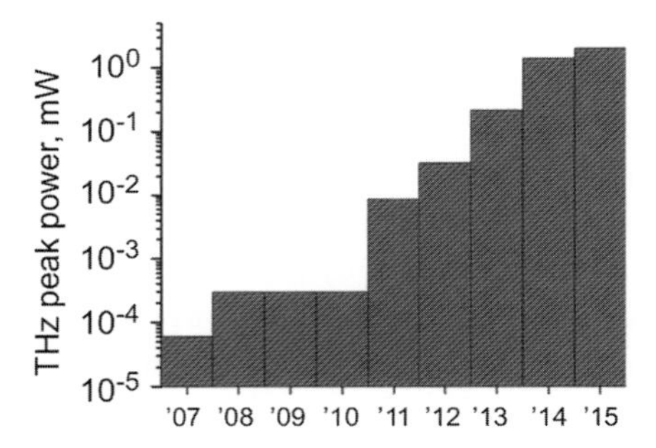

Figure 8.9 Progress in the maximum peak power output in pulse mode at room temperature achieved by terahertz difference-frequency-generated quantum-cascade lasers. The data are for room temperature, except for 2007, which is at 80 K. From Fig. 9 in A. Belkin and F. Capasso, "New frontiers in quantum cascade lasers: high performance room temperature terahertz sources," *Physica Scripta* **90**, 118002 (2015). ©IOP Publishing. Reproduced with permission. All rights reserved.

8.4 Photodetectors Based on Intraband Phototransitions

In general, the detection of light in semiconductor materials exploits changes in conductive processes. In bulk materials, there exist three main processes activated by light. The *intrinsic photoconductivity* involves the excitation of electrons from the valence band to the conduction band. The conductivity of the sample increases because additional concentrations of free electrons and holes are generated. This process occurs when the energy of the incident photons exceeds the bandgap; i.e., the process has a long-wavelength cut-off as determined by Eq. (4.57). The second photoconductive mechanism in bulk materials is the *extrinsic photoconductivity* which occurs in doped semiconductors. At low temperatures, most of the impurities are not ionized (i.e., neutral) and photoconductivity results from the excitation of an electron (hole) bound to a neutral donor (acceptor) into the conduction (valence) band. Such a process occurs at long wavelengths corresponding to the ionization energy of impurities. This type of photoconductivity is used for the detection of infrared light signals. The third process which affects the conductivity involves the absorption of light by free carriers. This absorption process heats up electrons, and detectors using this type of photoconductivity are referred to as *free-electron-absorbing bolometers*. These bolometers can work at very long wavelengths for the detection of far-infrared and terahertz signals. However, in bulk materials the intraband absorption is too weak and does not lead to the effective detection of light.

Though quantum heterostructures can be employed in photodetectors operating in different spectral regions, the most interesting applications of these structures are in the infrared frequency range, where it is hard to achieve high sensitivity, spectral tuning, and other desirable properties. According to Section 5.3, in quantum heterostructures with electron confinement, infrared absorption corresponds to intraband (i.e., intersubband) processes. These processes can be much more pronounced than those in bulk semiconductors and they can be controlled through the structure design. In a later discussion, we will show that intraband phototransitions can be used for the effective detection of long-wavelength light.

8.4.1 Photoconductive Detectors

We start with a brief review of the principles of operation of photodetectors and their basic characteristics. The simplest type of photodetector – a photoconductive device – is illustrated in Fig. 8.10. This photodetector is presented schematically in Fig. 8.10(a): a photosensitive sample of length L and cross-sectional area S is under a voltage bias Φ. When light of an appropriate wavelength illuminates the device, the electric current increases. The changes in the current induced by the light can be detected by a circuit, as illustrated in Fig. 8.10(b). The circuit includes a load resistor, R_L, and a capacitance, C. The latter is necessary if only an ac signal is to be detected. An important characteristic of photoconductive detectors is *the gain* of the device. The gain indicates how many electrons can be collected for each absorbed photon. Let g_l be the rate of photogeneration of free carriers; more specifically, let g_l be the number of carriers generated by photon absorption in a unit volume per unit time. If the lifetime of the free carriers is τ, the concentration of these carriers is $n_l = \tau g_l$. The total current, I, consists of two contributions. The first is the dark current, I_d, and the second is the photocurrent, I_l:

$$I = I_d + I_l, \quad I_d = e\mu n_d FS, \quad I_l = e\mu n_l FS,$$

where μ is the mobility, n_d is the carrier concentration contributing to the current in the absence of illumination, and F is the electric field. If we define the transit time of the electrons in the device by $t_{tr} = L/\mu F$, we can express the photocurrent in terms of τ and t_{tr}:

$$I_l = e\frac{\tau}{t_{tr}} g_l LS. \tag{8.5}$$

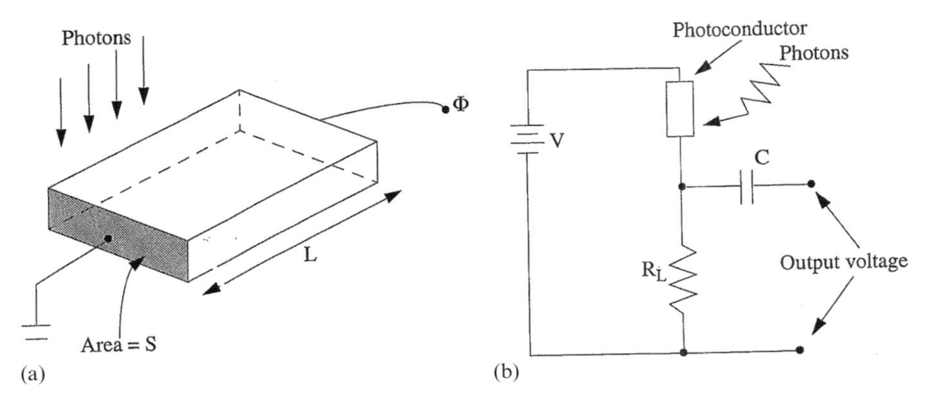

Figure 8.10 A schematic representation of a photoconductive detector: (a) the photoconductor geometry and (b) a typical photodetector circuit.

This is the current generated in the circuit by light. We can introduce the "primary photocurrent" $I_{l,p} = eg_l LS$, which is the photocurrent in the case when each generated carrier transfers one electron charge, e, from one contact to the other contact. Now we can define the gain of the detector as

$$G = \frac{I_l}{I_{l,p}} = \frac{\tau}{t_{tr}}. \tag{8.6}$$

Values of gain larger than one, $G \geq 1$, arise if a photogenerated electron can go around the circuit several times during its lifetime. It is easy to see that the gain increases with the applied electric field.

Another important figure of merit of a photodetector is the so-called *responsivity*, r, which is defined as

$$r = \frac{I_l}{P_l},$$

where P_l is the light power illuminating the detector. Since the rate of photogeneration of carriers is $g_l = \eta P_l/(\hbar\omega SL)$, where η is *the internal absorption efficiency* of the detector, the responsivity has the form

$$r = e\frac{\eta G}{\hbar\omega}. \tag{8.7}$$

Thus, the responsivity is determined by the internal efficiency and by the gain of the device.

Both figures of merit, the gain and the responsivity, characterize the photocurrent and the efficiency of the device. Despite the existence of a dark current, the photocurrent can certainly be distinguished from the dc dark current by using, for example, a time modulation of the photo signal with a frequency f and by detecting the electric current in a frequency band Δf around f.

However, there are always fluctuations of the dark current which constitute the so-called *current noise*. Specifically, there are small time-dependent deviations of the instantaneous current from its mean value, $\delta I(t) = I(t) - I$. These fluctuations can be caused by different random processes: shot noise, Nyquist noise, generation–recombination noise, etc. These fluctuations make a contribution to the electrical signal in any frequency range, including the one where the measurements are done. As a result, the fluctuations *limit* significantly the intensity of a detectable signal. Let the Fourier transform of the current fluctuations be ΔI_f. We can introduce the spectral density of current fluctuations, ΔI_f^2, through the following equation:

$$\langle \Delta I_f \Delta I_{f'} \rangle = \Delta I_f^2 \, \delta(f - f'), \tag{8.8}$$

where the angle brackets, $\langle \cdot \rangle$, denote a statistical average and $\delta(f - f')$ is the δ-function. It is easy to check that the average current fluctuations, $\langle (\Delta I(t))^2 \rangle$, are expressed via the spectral density, ΔI_f^2. Now we can characterize the current noise by the parameter

$$I_{ns} = \sqrt{\Delta I_f^2 \, \Delta f}. \tag{8.9}$$

The *signal-to-noise ratio*,

$$\mathcal{N}_{(\mathrm{sgn/ns})} = I_1^2 / I_{\mathrm{ns}}^2,$$

determines the detection limit of the device.

Often, one uses *the noise-equivalent optical power*, i.e., the power of the optical signal at which the photocurrent is equal to I_{ns}:

$$P_{\mathrm{NEP}} = \frac{I_{\mathrm{ns}}}{eG\eta}\hbar\omega. \tag{8.10}$$

In this section, we have defined three quantities, namely the gain G, the responsivity r, and the noise equivalent power P_{NEP}, which characterize the utility of any photodetector in converting an optical signal into an electrical signal.

8.4.2 Intraband Phototransitions and Electron Transport in Multiple-Quantum-Well Structures

Let us turn to an analysis of intraband absorption, electron transport, and current noise in multiple-quantum-well photodetectors.

In Section 5.3, we studied intraband phototransitions in a quantum well. We found that light absorption in a single well is relatively small. Since the main application of intraband phototransitions is the detection of weak infrared light signals, one needs to obtain as large an absorption as possible. For this purpose, multiple-quantum-well structures are employed. It is important to stress that an increase in the number of quantum wells leads to an increase of the electron capture rate that is proportional to the number of quantum wells. As a result, the lifetime of photogenerated carriers decreases as the number of wells increases. There are two different cases. In the first case, the barriers which separate the wells are assumed to be thick enough to prevent the formation of minibands due to electron tunneling; see Section 3.7. Intersubband absorption can activate an electric current through this system when a voltage bias is applied. Absorption by such a structure is presented in Fig. 5.12. For this particular case, the maximum absorption coefficient reaches $2000\,\mathrm{cm}^{-1}$. The conductivity is due to either the tunneling between higher states in different wells or thermally activated overbarrier transport. If the barriers of the heterostructure are of sufficient thickness, the modification of the electron spectrum in the energy region above the barrier is negligible and it can be considered as a continuum of the extended states studied in Section 5.3.3. The absorption coefficient for such a case, presented in Fig. 5.14, is about $1000\,\mathrm{cm}^{-1}$ to $1400\,\mathrm{cm}^{-1}$. In Fig. 8.11, the optical density of states for an electron energy spectrum is shown as a function of the photon energy for an ideal $Al_{0.25}Ga_{0.75}As$ multiple-quantum-well structure when the minibands are formed from the quantum states of the wells that are above the barrier. The well width is 40 Å and the period is $d = 240$ Å. The three peaks in the density of states are due to the formation of three minibands above the barriers. For such a structure, the absorption coefficient, α, can reach magnitudes of $200\,\mathrm{cm}^{-1}$ to $400\,\mathrm{cm}^{-1}$. We discussed the formation of minibands in Chapter 3

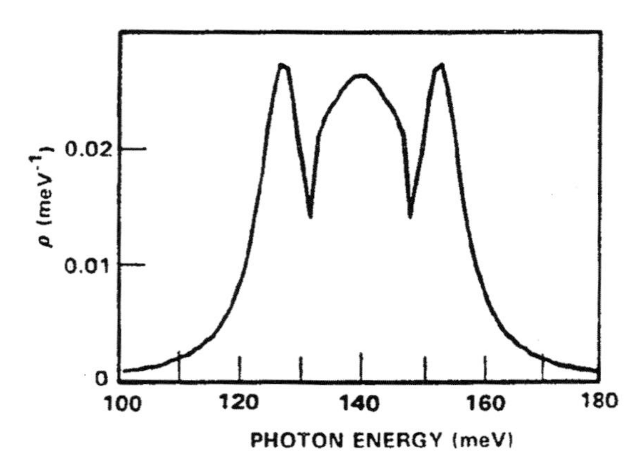

Figure 8.11 Calculated optical density of states of an $Al_{0.25}Ga_{0.75}As$ multiple-quantum-well structure with 40 Å wells and 200 Å barriers. Copyright 1993 from Fig. 3.5(a) of *Semiconductor Interfaces, Microstructures, and Devices. Properties and Applications* by Z. C. Feng. Reproduced by permission of Taylor and Francis Group, LLC, a division of Informa plc.

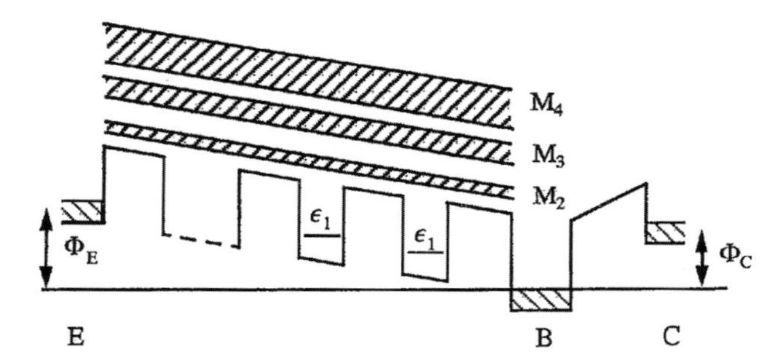

Figure 8.12 The band diagram of an electron energy spectrometer: E, B, and C denote the emitter, base, and collector, respectively. A single energy level is supposed to be in the wells; minibands are formed above the energies of the potential barriers, and Φ_E and Φ_C are the emitter and the collector biases. 1990 IEEE. Reprinted, with permission, from K. K. Choi, *et al.*, "InGaAs base infrared hot-electron transistors," in *International Electron Devices Meeting* 1991 [Technical Digest], pp. 809–812.

(see Figs. 3.26 and 3.27) and in Chapter 6 (see Fig. 6.9). Three minibands above the barriers are also shown in Fig. 8.12. Thus, different designs of multiple-quantum-well structures result in different absorption spectra and different magnitudes of absorption.

Consider next the important factors determining the detection of photo signals, i.e., the dark current and photocurrent through the structure. For photodetection, the wells are usually doped, thus there are always carriers at least in the lowest subband. If a voltage is applied to a multiple-quantum-well structure, a dark current flows through the structure even in the absence of illumination. The dark current is caused by different mechanisms: (i) sequential tunneling from one well to a neighboring well, which may be assisted by phonon absorption or emission; (ii) excitation to the second subband and subsequent tunneling; and (iii) thermoionization to the continuum and subsequent "free" propagation for energies above the barrier potential. According to the analysis, only the coherent tunneling is nearly independent of the temperature of the system. The other previously mentioned mechanisms in multiple-quantum-well structures with thick barriers are sharply activated with increasing temperature. For example, the thermionic current is an exponential function of the temperature.

The identification of the different mechanisms contributing to the dark electron transport can be accomplished straightforwardly by using the hot-electron spectroscopy method. For this purpose, a multiple-quantum-well photodetector structure should serve as the emitter part of the hot-electron spectrometer. Then, an additional doped layer should be grown to create a base. This base is separated from the collector by a barrier. The energy diagram of such a spectrometer is shown in Fig. 8.12. The diagram corresponds to the case where the wells have a single bound level, ϵ_1. The energy spectrum of the extended states can consist of several minibands, $M_2, M_3, \ldots$ The voltage drop on the multiple-quantum-well part of the device equals Φ_E, and the base–collector bias is Φ_C. By varying the emitter–base bias, Φ_E, and base–collector bias, Φ_C, one can separate electrons with different energies coming into the base from the multiple-quantum-well part of the structure. The electron distribution function $\mathcal{F}(E)$ can be found in the form of $\mathcal{F}(E_0 - e\Phi_E)$, where E_0 is a constant. That is, the electrons with small energies are detected at high voltage, Φ_E, while decreasing the voltage facilitates the detection of electrons with higher energy. Experimental results obtained by this method are shown in Fig. 8.13. The parameters of the multiple-quantum-well structure are those identified in the caption to Fig. 8.11. The structure consists of 50 periods of GaAs wells and $Al_{0.25}Ga_{0.75}As$ barriers. A thin, 300 Å pseudomorphic $In_{0.25}Ga_{0.85}As$ layer is grown on top of the multiple-quantum-well structure as a base layer. A 2000 Å layer of $Al_{0.25}Ga_{0.75}As$ separates the base from the GaAs collector. All the layers except the barriers are doped to $n_D = 1.2 \times 10^{18}\,\text{cm}^{-3}$. The distribution of electrons entering the base layer, $\mathcal{F}(\Phi_E)$, is presented as a function of the emitter–base bias, Φ_E. The position of the single subband, ϵ_1, is indicated by the arrow. Another arrow corresponds to the energy ϵ_2, which has a value of 225 meV. The results are given for different temperatures. It can clearly be seen that at low temperatures the distribution function is centered around ϵ_1, i.e., the major mechanism of electron transport is the tunneling through the lowest energy level, ϵ_1. With increasing temperature, the high-energy tail grows. Electrons with energies between ϵ_1 and ϵ_2 are associated with thermally assisted tunneling. Such thermally assisted tunneling prevails at temperatures around 80 K. Electrons coming into the base with energies close to ϵ_2 propagate as a result of thermionic processes. These processes become dominant at temperatures above 90 K to 100 K. Figure 8.14 shows the total dark (emitter) current and the collector current as functions of the emitter–base voltage at $T = 77$ K.

Consider another example of a multiple-quantum-well structure, with GaAs well layer thicknesses of 70 Å and $Al_{0.36}Ga_{0.64}As$ barrier widths of 140 Å. The doping

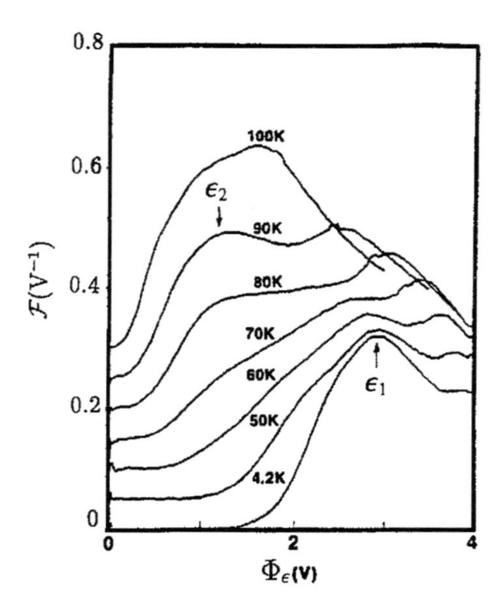

Figure 8.13 Results of the measurement of the electron distribution for the multiple-quantum-well structure as a function of the bias at different temperatures. The thicknesses of the wells and barriers are 40 Å and 200 Å, respectively; ϵ_1 indicates the position of the lowest subband, and ϵ_2 corresponds to the energy 225 meV. The origins of the curves are shifted upward for clarity. Copyright 1993 from Fig. 4.7 of *Semiconductor Interfaces, Microstructures, and Devices. Properties and Applications* by Z. C. Feng. Reproduced by permission of Taylor and Francis Group, LLC, a division of Informa plc.

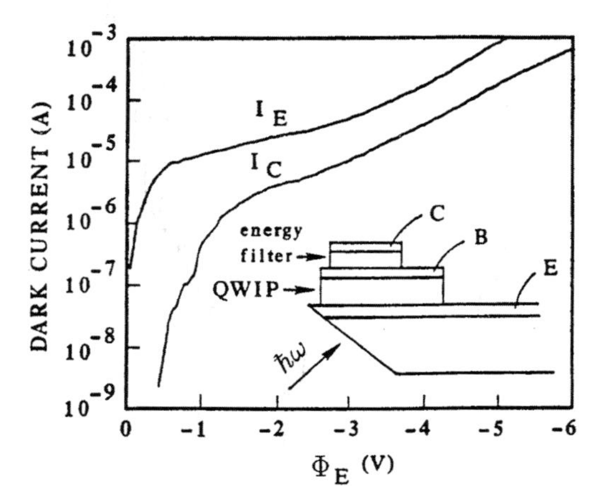

Figure 8.14 The emitter, I_E, and collector, I_C, dark currents for the device with the multiple-quantum-well structure at $T = 77$ K. The inset depicts the detector and the coupler with light. Copyright 1993 from Fig. 5.2 of *Semiconductor Interfaces, Microstructures, and Devices. Properties and Applications* by Z. C. Feng. Reproduced by permission of Taylor and Francis Group, LLC, a division of Informa plc.

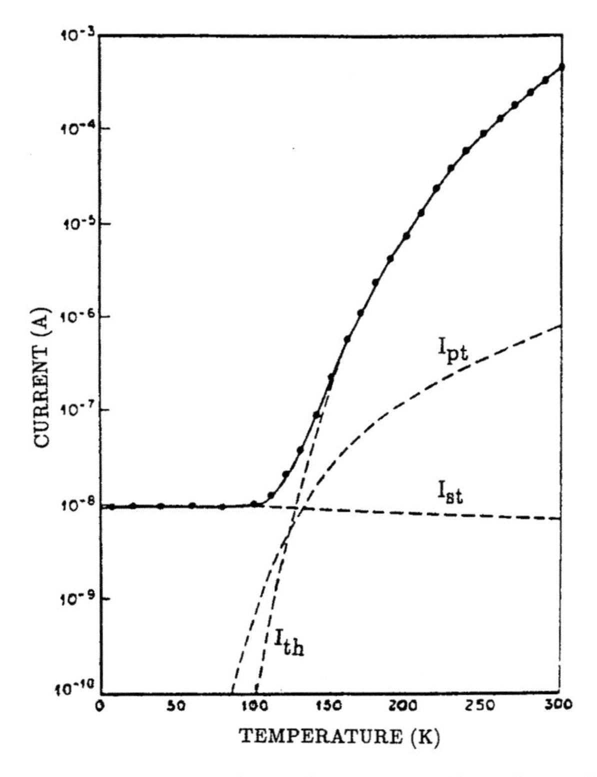

Figure 8.15 The experimental temperature dependence of the total current through a multiple-quantum-well structure at a low bias (shown by the circles). The dashed curves show the contributions to the current from different partial mechanisms: sequential tunneling (I_{st}), phonon-assisted tunneling (I_{pt}), and thermionic emission (I_{th}). Copyright 1993 from Fig. 4.2 of *Semiconductor Interfaces, Microstructures, and Devices. Properties and Applications* by Z. C. Feng. Reproduced by permission of Taylor and Francis Group, LLC, a division of Informa plc.

concentration of the wells is $1.4 \times 10^{18}\,\text{cm}^{-3}$. The barriers prevent large tunneling currents, so the broadening of the lowest energy level in the wells, $\Delta\epsilon_1$, is about 7 meV. The applied voltage is such that the voltage drop over one period of the structure is about $\Phi_p = 1$ meV. In Fig. 8.15, the relative contributions from sequential tunneling, I_{st}, phonon-assisted tunneling, I_{pt}, and thermionic emission, I_{th}, to the dark current are presented as functions of the temperature. From these results a general conclusion can be drawn: at low temperatures the tunneling contribution to the dark current dominates. This process is possible until the energy-level width exceeds the voltage drop, $\Delta\epsilon_1 > \Phi_p$. Phonon-assisted tunneling contributes over a relatively narrow temperature region, while the thermionic current dominates at temperatures above 100 K. Now, as we know the key characteristics of the dark current, let us compare it with the photocurrent. Let us consider a multiple-quantum-well structure with the parameters defined in Fig. 8.13. The geometry and

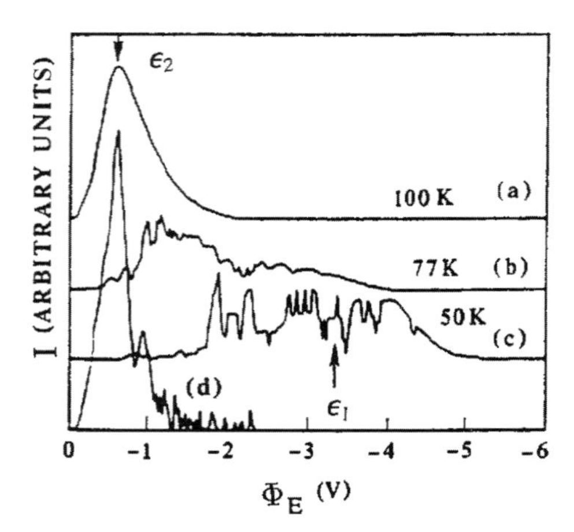

Figure 8.16 Hot-electron spectra at different temperatures. Curves (a), (b), and (c) correspond to dark (thermo-stimulated) currents at $T = 100$, 77, and 50 K. Curve (d) corresponds to the spectra under illumination at $T = 90$ K. The curves are shifted upward for clarity. Copyright 1993 from Fig. 5.3 of *Semiconductor Interfaces, Microstructures, and Devices. Properties and Applications*, by Z. C. Feng. Reproduced by permission of Taylor and Francis Group, LLC, a division of Informa plc.

the principal components of the photodetector are shown in the inset of Fig. 8.14: the coupler to the light, the emitter, E, the quantum well infrared photodetective (QWIP) region, the base, B, the barrier for filtering electrons with different energies, and the collector, C. The light is coupled into the device through a 45° angle to provide the interaction of the light with the quantum wells (see Table 5.1) (just as a reminder, we note that the electrons interact with light with polarization perpendicular to the quantum well layers, i.e., with the light propagating along the layers). This three-terminal device facilitates the investigation both of the dark current and of the photocurrent, as well as allowing control of the dark current. The thermo-stimulated collector current is shown in Fig. 8.16 as a function of the emitter–base voltage, Φ_E, at three different temperatures from 50 K to 100 K. The photocurrent is presented at $T = 90$ K. As one can see, this current corresponds to electron propagation at energies greater than the barrier potential. All the currents are in arbitrary units, but on the same scale. We can conclude that the photocurrent dominates over the dark current at temperatures below 100 K, and that the voltage bias on the photodetective part of the structure is about 0.5 V.

The measurement of the energy distribution of photo-excited electrons leads to conclusions which allow one to formulate a model of electron transport in multiple-quantum-well infrared photodetectors. In Fig. 8.16, the arrows marked by ϵ_1 and ϵ_2 have the same meanings as in Fig. 8.13. Thus, we deduce that the photocurrent corresponds to electron propagation at energies greater than the barrier potential. In particular, Fig. 8.16 clearly indicates that there are no ballistic electrons. The energy width of the photogenerated electrons is about several tens of meV, which is comparable to the energy corresponding to the device temperature of 90 K. That is, the distribution of photocarriers is consistent with that of the usual electron drift over barriers, despite the large applied electric fields. Note that, for a superlattice with N periods, the electric field in the detector can be estimated as $F = |\Phi_E|/(Nd) \approx 5 \times 10^3$ V/cm to 10^4 V/cm. These conclusions allow us to develop a simple model of the processes taking place in such devices.

In conclusion, intraband absorption in multiple-quantum-well structures allows one to design fundamentally new devices to detect weak infrared signals. They can be designed in selected spectral regions to yield greater sensitivity than is available from existing infrared technology based on doped bulk-like materials.

8.5 Silicon Photonics

Silicon and silicon-based structures are crucially important because of the dominant role of these materials in microelectronics and nanoelectronics. Over approximately the last 15 years there has been great progress in the effort to use silicon-based components for optical signaling between Si-based chips. Achieving such optics-based components for optical switching on a large scale offers the paradigm-changing use of Si-based electronic chips integrated functionally through optical interconnects. Such systems not only allow increased functionality but also address the growing problem of heat dissipation associated with Si-based chips – through the separation of optically interconnected chips – with ever increasing transistor counts. As discussed previously in Chapter 4, bulk Si is an indirect-bandgap material, and its ability to emit light in band-to-band radiative decay depends on phonon-assisted processes. Thus, the use of Si-based components for optical signaling between Si-based chips requires novel approaches. In this brief introduction to Si photonics it will become evident that insights based on an examination of the real and imaginary parts of the index of refraction, as discussed earlier in this text, have been used to advance the field of Si photonics. The recent progress in Si-based photonics is based, in large part, on the ability to fabricate nanoscale structures that facilitate the control of light. An illustrative example will be presented in the last part of this section.

Before introducing approaches where we avail ourselves of insights based on an examination of the real and imaginary parts of the index of refraction, we will return to a brief discussion on the indirect bandgap of bulk silicon. Figure 8.17 provides an illuminating comparison of a direct-bandgap material, InP, with indirect-bandgap bulk silicon. As shown in Fig. 8.17, for InP the minimum in the conduction band is at the same wavevector as the maximum in the valence band; this facilitates the emission of a photon – through direct recombination – while conserving energy and momentum since the photon momentum is very small on the wavevector scale for the Brillouin zone, as discussed in Chapters 4 and 5. In contrast, for bulk silicon the minimum in the conduction band edge is at a different wavevector (which is proportional to momentum since $p = \hbar k$) than that of the valence band maximum, so direct recombination is not possible given the small photon momentum, $p = \hbar\omega/c$, on the

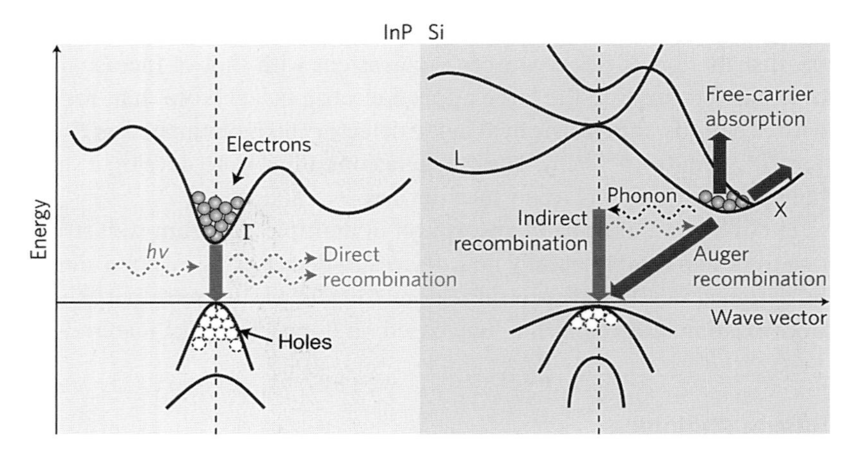

Figure 8.17 Comparison of the direct-band nature of InP with the indirect bandgap of bulk silicon. Reprinted by permission from Fig. 1 in Springer Nature: Macmillan Publishers Ltd., Nature Photonics, "Recent progress in lasers on silicon," Nature Photonics, D. Liang and J. E. Bowers, Copyright 2010.

scale of the wavevector of the Brillouin zone. As indicated in the band diagram for bulk silicon, indirect recombination is possible when higher-order (and therefore less likely) phonon-assisted processes occur; recall our discussion in Chapters 4 and 5. The Auger process indicated in Fig. 8.17 also makes recombination possible, since Auger scattering involves the scattering of two electrons whereby one loses energy and refills a hole, and the other electron gains energy and is excited to a higher energy in the conduction band. There have been successes in using nanoscale silicon for optical applications, such as those discussed in Chapter 5, where silicon nanocrystals, i.e., quantum dots, have been observed to be luminescent as a result of the altered band structure, on the nanoscale, which manifests a direct-band nature.

Many research approaches have been explored to realize Si photonic structures, and there have been major successes. As examples, we mention Si and Ge electro-optic modulators, where free carriers lead to a modulation of the real and the imaginary parts of the index of refraction, the optical emission of Er-doped SiO_2, Si-based hetero-structures containing layers such as SiGeSn, SiGe, and GeSn, and silicon-based plasmonic components with metal-containing structures of the type discussed in Section 6.5. These have been studied in the framework of the long-predicted sub-wavelength characteristics leading to light localization and enhanced emission due to plasmonic effects, and the integration of direct-bandgap optical elements directly on silicon substrates has been explored. The underlying physics for some of these approaches has already been discussed in our treatments of the role of the refractive index in optical devices as well as in our discussion on plasmonics.

In closing this section, we give a brief introduction to the fabrication techniques that have advanced silicon photonics. Low-loss silicon waveguides have been fabricated as essential elements of on-chip optical networks. In the ingenious approach

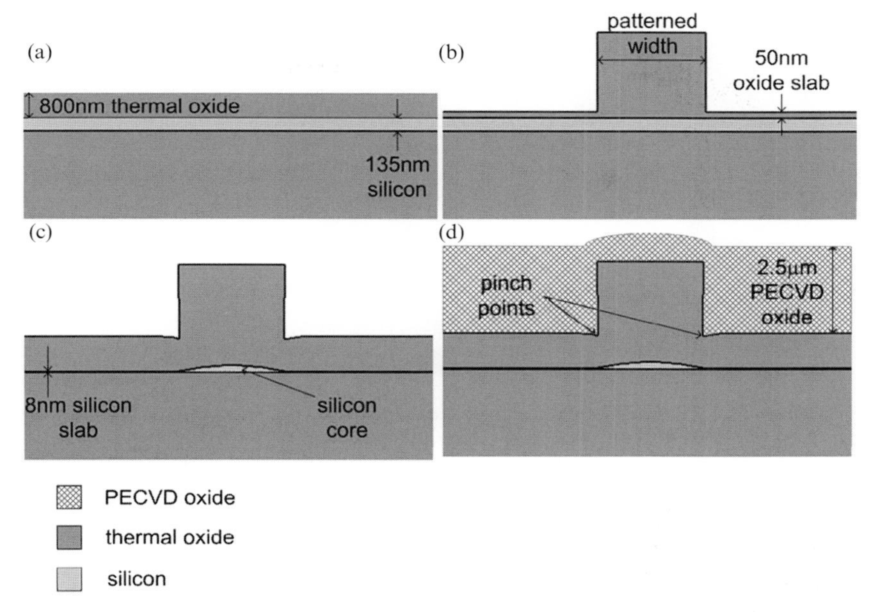

Figure 8.18 The thermal oxidation process for etchless-fabrication low-loss silicon waveguides starts with (a) an 800 nm-thick thermal oxide layer grown on a silicon-on-insulator wafer with a 3 μm buried oxide. This is followed by (b) the patterning of waveguides with electron-beam-based lithography and etching of the oxide, prior to (c) defining the waveguide core using thermal oxidation. Finally, (d) the waveguide is clad with a 2.5 μm-thick overlayer of SiO_2 deposited by plasma-enhanced chemical vapor deposition (PECVD). Adapted with permission from Fig. 1 in J. Cardenas, *et al.*, *Optics Express* **17**, 4752 (2009), Optical Society of America.

illustrated in Fig. 8.18, thermal oxidation – an etchless process that avoids the defects caused by etching – has been used to fabricate low-loss waveguides with smooth Si–SiO_2 interfaces that manifest reduced scattering losses near the sidewalls. Waveguides fabricated using this thermal oxidation technique have exhibited propagation losses of 0.3 dB/cm at 1.55 μm.

Waveguides fabricated with this etchless oxidation technique were found to support different modes, depending on the waveguide width. For example, a waveguide – patterned at 1 μm width – was observed to support only the fundamental quasi-TE (x-polarized) mode, and no quasi-TM (y-polarized) modes; see Fig. 8.19. The z-direction is out of the page, and the y-direction is from the bottom to the top of the figure. However, a 1.5 μm-wide waveguide supported higher-order modes.

It is evident from this discussion that state-of-the-art fabrication techniques are a primary part of the success of modern approaches for realizing optical devices for silicon photonic systems.

The replacement of metal interconnects in electronic devices and systems with optical interconnects has been a longstanding goal, since metal interconnects

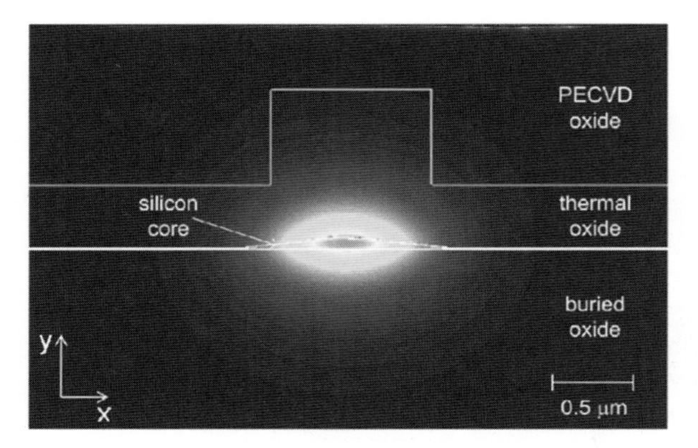

Figure 8.19 The transverse electric mode profile for a 1 μm-wide Si etchless waveguide with a cladding profile. This waveguide – patterned at 1 μm width – was observed to support only the fundamental quasi-TE (x-polarized) mode, and no quasi-TM (y-polarized) modes. The z-direction is out of the page, and the y-direction is from the bottom to the top of the figure. However, a 1.5 μm-wide waveguide supported higher-order modes. Reprinted with permission from Fig. 2 in J. Cardenas, *et al.*, *Optics Express* **17**, 4752 (2009), Optical Society of America.

represent a limiting factor and since optical interconnects offer the possibility of low power dissipation and broad-bandwidth operation. To facilitate the advent of optical interconnects (between chips or within chips), electro-optic modulators compatible with Si are essential. Herein, we discuss an experimental demonstration of a high-speed silicon optical modulator based on a ring resonator that translates phase variations into intensity variations; see Fig. 8.20. This modulator is based on a structure that confines light, as discussed above, for the case of a low-loss Si-based waveguide. The operating principle of this light-confining structure is based on its ability to enhance the sensitivity of light to small changes in the refractive index of the silicon and, therefore, in the transmission response. In the reported electro-optic modulator, high confinement facilitated the design of an electro-optical modulator of size a few micrometers. In this design, the effective refractive index of the ring resonator is electrically modulated through the injection of charge carriers using a p–i–n junction that is an integral part of the modulator design. The consequent modulation of the effective index of the ring waveguide changes the resonance wavelength of the ring resonator and leads to a strong modulation of the transmitted signal. The p–i–n ring resonator was fabricated on a silicon-on-insulator substrate, similar to the Si waveguide discussed above, with a 3 mm-thick buried oxide layer. The waveguide has a width of 450 nm and a height of 250 nm. The electron–hole pair density in the resonator increases with the forward bias on the p–i–n junction. This increase in the electron–hole pair density leads to a "blue-shift" of the resonance and changes the transmission of the device. Figure 8.20(b) shows the transmission spectra of the modulator near resonance at different values of the voltage bias at the p–i–n junction. In the inset, the relative transmission at

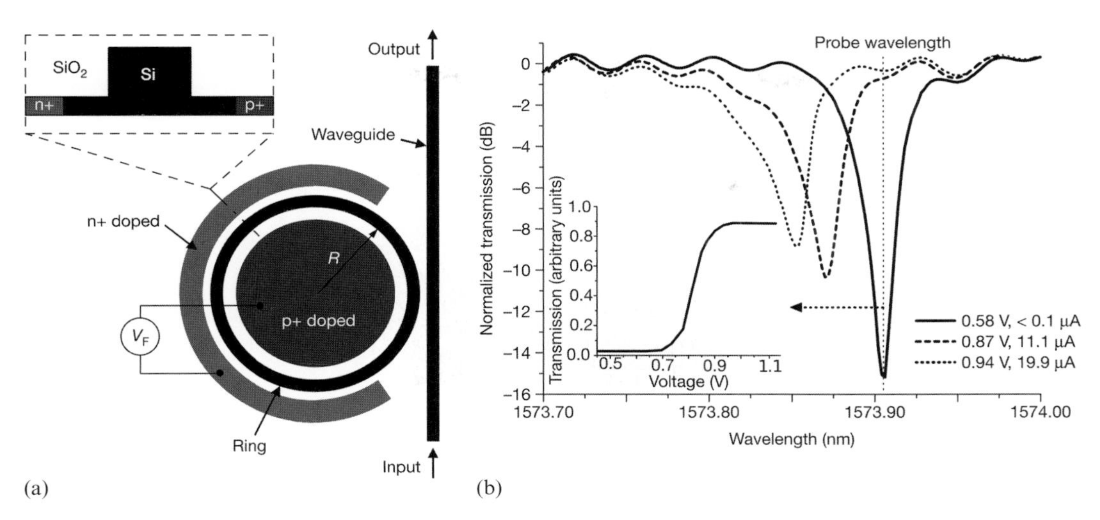

Figure 8.20 (a) An Si-based optical modulator based on a ring-resonator structure that uses phase variations to produce intensity modulation. The radius of the ring resonator is $R = 12$ mm. The voltage V_R is applied across the p–i–n junction. The inset shows the cross-section of the ring resonator; this should be compared with the cross-sections presented in Figs. 8.18 and 8.19. (b) The main panel shows the transmission spectra of the ring resonator at different bias voltages. The inset shows the transmission of the modulator for light of wavelength 1573.9 nm. Adapted by permission from Figs. 1 and 3 of Springer, *Nature*, "Micrometre-scale silicon electro-optic modulator," Q. Xu, *et al.*, Copyright 2005.

$\lambda = 1574$ nm is shown. One can see that the modulation practically reaches $\approx 100\%$. The modulation frequency is above 12 GHz.

This modulator is three orders of magnitude smaller than previously demonstrated ones, and could be one of the most critical components in optoelectronic integration.

8.6 Perspectives of Optoelectronic Devices Based on Two-Dimensional Crystals

In Chapters 2 and 3 we analyzed the basic properties of two-dimensional crystals; in Chapter 5 we presented the optical properties of these materials and provided some indications of possible device applications. The study of two-dimensional materials is in its early stages and their potential device applications have not been fully defined. In this section we briefly discuss several examples of optoelectronic applications of two-dimensional materials.

We begin with a consideration of graphene, a single sheet of carbon atoms in a hexagonal lattice, which attracts interest because of its exceptional electrical and optical properties (see Section 5.4). Indeed, direct-bandgap graphene couples strongly to light, which results in small, but easy observable, absorption (about 2.3%) which is almost constant across the infrared and visible ranges (the details

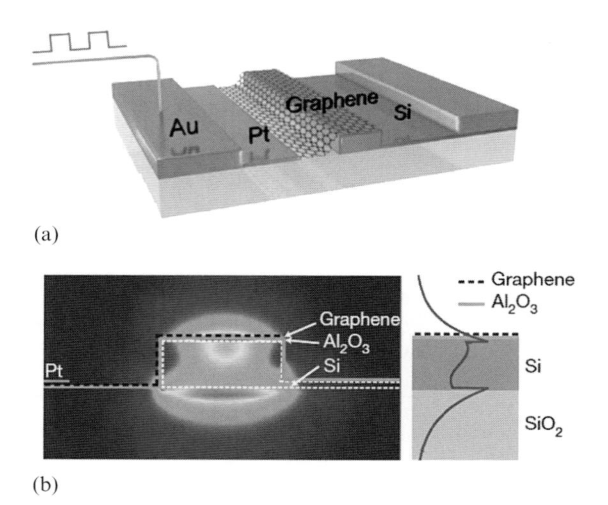

Figure 8.21 A graphene-based waveguide-integrated optical modulator. (a) A three-dimensional schematic illustration of the device. A graphene sheet is present on top of a silicon bus waveguide. (b) Left, cross-section of the device. Right, a cross-section through the center of the waveguide; the curve is the electric-field magnitude of the mode. Adapted by permission from Fig. 1(a) and (b) of Springer, *Nature*, "A graphene-based broadband optical modulator," M. Liu, *et al.*, Copyright 2011.

are presented in Section 5.4.1). In addition, due to the unique electronic structure, the essential electro-absorption phenomenon was observed for graphene. This implies that graphene has the potential to be used for a *broadband optical electro-absorption modulator*, when it is integrated with an optical waveguide. Because of the very high carrier mobility, which can exceed 2×10^5 cm^2/V s at room temperature, and fast relaxation processes (the generation and relaxation of the carriers occur on picosecond timescales), the graphene modulator promises high-speed operation.

Figure 8.21 illustrates a graphene-based optical modulator integrated with a silicon waveguide (for silicon waveguides, see the previous section). The graphene sheet is separated from the waveguide by a 7 nm-thick Al_2O_3 dielectric and the source is a platinum contact. The 250 nm-thick Si waveguide is doped and connected to the electrode through a thin layer of silicon. The waveguide has a length of 40 mm and carries a single optical mode. The mode was designed to maximize the electric field at the interface between the waveguide and the graphene.

For such a device design, a bias voltage, V, applied to the contacts induces the electro-absorption effect in graphene. In Fig. 8.22, the dependence of the transmission per unit length of the waveguide is shown as a function of V. One can see three different regimes of device operation, for which the processes of absorption are illustrated in the respective insets.

Thus, this example demonstrates a broadband high-speed electro-absorption modulator based on monolayer graphene. The modulation of the guided light at

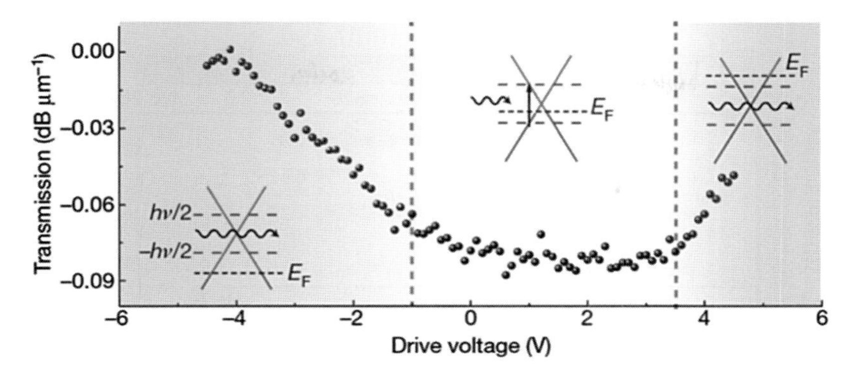

Figure 8.22 Electro-optical response of the device at different drive voltages. The insets illustrate the band structures and phototransitions for three different regimes of device operation. Adapted by permission from Fig. 2 of Springer, *Nature*, "A graphene-based broadband optical modulator," M. Liu, *et al.*, Copyright 2011.

frequencies over 1 GHz is achieved for a broad operation spectrum, in particular, for wavelengths near 1.35 μm to 1.6 μm, which are of interest in telecommunications.

The exciting properties and discovery of graphene take on added significance because of the emergence of two-dimensional crystals with *nonzero bandgaps*, which can be used to form heterostructures, in particular those based on van der Waals interactions. As discussed in Chapters 2, 4, and 5, these nonzero bandgaps have opened the way to exploiting these materials for new optical and optoelectronic applications. In Fig. 8.23, we present a comparison of the bandgap values for different two-dimensional and bulk semiconductor materials. Among these two-dimensional materials with nonzero bandgaps are: silicene, germanene, and tinene; black phosphorus; monochalcogenides, dichalcogenides, and trichalcogenides; and boron nitride. As illustrated in Fig. 8.23, two-dimensional materials have bandgaps spanning the range from nearly 0 eV to about 6 eV, with black phosphorus having a bandgap near 0.5 eV, Mo-based and W-based dichalcogenides spanning the 1 eV to 2 eV region, and BN having a bandgap of about 6 eV. Together, these two-dimensional materials have the potential for optical applications spanning a wide range of energies. Moreover, these two-dimensional materials can be used to form heterostructures based on weak van der Waals interactions without the need for strict lattice matching, as for conventional heterostructures, as explained in Chapter 2. Many of the two-dimensional materials have very large exciton binding energies, making excitonic effects important even at room temperature, with a consequent effect on the absorption and emission spectra of these materials. When considering their potential device applications, it is worth noting again that the study of these two-dimensional materials is still in an early stage, but that the optical properties of these structures signify widespread optical and optoelectronic applications. Examples of these applications include photovoltaic energy harvesting and photocatalysis

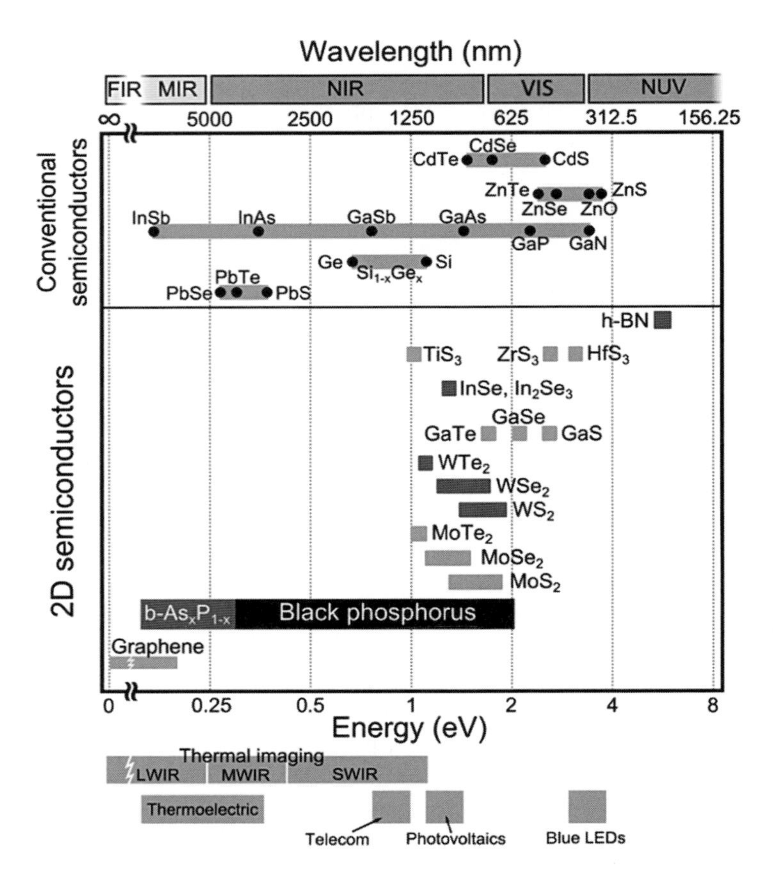

Figure 8.23 A comparison of the bandgap values for different two-dimensional semiconductor materials. The bandgap values for conventional semiconductors are also included for comparison. The horizontal bars spanning a range of bandgap values indicate that the bandgap can be tuned over that range by changing the number of layers, straining, or alloying. In conventional semiconductors, the bar indicates that the bandgap can be continuously tuned by alloying the semiconductors (e.g. $Si_{1-x}Ge_x$ or $In_{1-x}Ga_xAs$). The ranges of bandgap values required for certain applications are given at the bottom of the figure to illustrate potential applications of the different semiconductors. Adapted with permission from figure 2 in A. Castellanos-Gomez, *J. Phys. Chem. Lett.* **6**, 4280 (2015). Copyright 2015 American Chemical Society.

applications (bandgaps of 1.2 eV to 1.6 eV are optimal), fiber-optic telecommunications (the corresponding photon energies are 0.8 eV to 1 eV), thermal imaging (requiring semiconductors with bandgaps of 0.1 eV to 1 eV), and thermoelectric power generation (bandgaps of 0.2 eV to 0.3 eV are required). These potential applications are indicated at the bottom of Fig. 8.23. From Fig. 8.23 it follows that the bandgaps spanned by black phosphorus nanolayers make this two-dimensional material suitable for numerous applications, namely thermal imaging and thermoelectric, telecommunication, and photovoltaic applications, and also make it a prospective replacement for conventional narrow-gap semiconductors in applications requiring thin, flexible, and quasi-transparent material. In Chapters 2 and 3, we discussed the anisotropy of black phosphorus and indicated that this anisotropy results in a photoresponsivity that makes it a candidate for a broadband photodetector made of a layered black phosphorus structure where the photoresponse depends on the angle of polarization relative to the crystal x-axis and y-axis of the black phosphorus (the crystalline x- and y- axes of black phosphorus are shown in Fig. 3.15(a)).

Recent studies have suggested that thin layers of black phosphorus may function as a broadband saturable absorber for ultrashort-pulse generation. Specifically, these studies have examined mechanically exfoliated layers of black phosphorus approximately 300 nm thick that were transferred onto a fiber core (note that multilayer black phosphorus was briefly discussed in Section 3.3.3). Experiments with pulsed excitation at 1560 nm wavelength have demonstrated that black phosphorus acts as a saturable absorber that is polarization-sensitive. The fabricated device was used to generate optical pulses of bandwidth 10.2 nm and duration 272 fs. It is thought that the short pulse length is due to nonlinear effects in the black phosphorus. The power-dependent transmittance of a black phosphorus saturable absorber is shown in Fig. 8.24. In these studies, a fiber connector with a deposited black phosphorus layer was connected with a clean fiber via a fiber adapter, thus forming a fiber-integrated device that acts as a saturable absorber. The measurements were performed with the use of a polarized light pulse: the direction of the

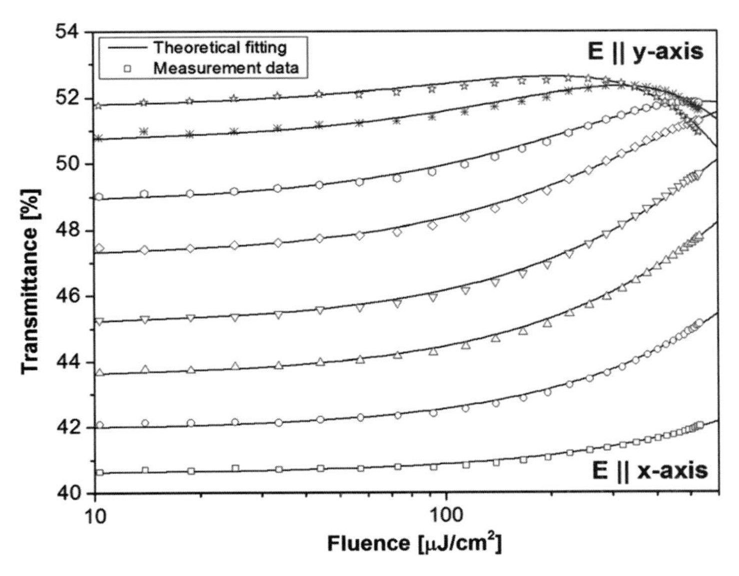

Figure 8.24 The measured, and theoretically fitted, nonlinear transmittance of a 300 nm-thick exfoliated black phosphorus layer. The power-dependent transmittance was measured for polarization azimuths changing from $E\,||\,x$-axis to $E\,||\,y$-axis. Reprinted from Fig. 2 of J. Sotor, *et al.*, *Appl. Phys. Lett.*, **107**, 051108 (2015), with the permission of AIP Publishing.

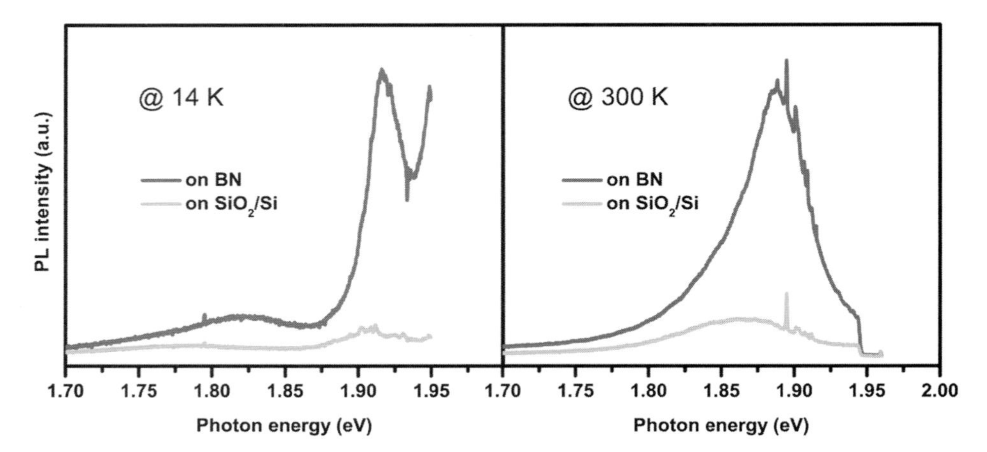

Figure 8.25 Photoluminescence spectra of monolayer MoS_2 on boron nitride (BN) and on SiO_2/Si. Reprinted by permission from Fig. 2 of Springer, *Nature Nanotechnology*, "Control of valley polarization in monolayer MoS_2 by optical helicity," Kin Fai Mak, *et al.*, ©2012.

electric field of the light wave, $\vec{E}$, was varied from the x-axis to the y-axis. One can see that the observed nonlinear transmission is strongly dependent on the light polarization.

The two-dimensional transition-metal dicalchogenide crystals considered in Chapters 2 and 5 (see Table 5.2) are also very promising for optoelectronic aplications. Indeed, in these direct-bandgap materials the probabilities of phototransitions depend strongly on the specific polarization of the incident radiation, which facilitates selective valley population under illumination by polarized light. According to Chapter 2, strong excitonic effects are present in the two-dimensional transition-metal dicalchogenides; for MoS_2 there are so-called A and B exciton states near the K and K$'$ points. Specifically, as discussed in Chapter 2, strong exciton resonances are associated with transitions between two valence band states (split by spin interactions) and the conduction band. In the experiments reported in Fig. 8.25, only excitons near the K point are created by left-handed circularly polarized radiation in resonance with the A exciton, because of the optical selection rule.

In Fig. 8.25, the photoluminescence spectra of monolayer MoS_2 on boron nitride (BN) and on SiO_2/Si for the case of excitation by 1.96 eV (633 nm) left-handed circularly polarized radiation at 14 K (left) and at 300 K (right) are shown. The photoluminescence persists over a wide range of temperatures; it is about an order of magnitude higher for samples on BN than it is for those on SiO_2/Si. In the study reported in Fig. 8.25 it was found that interband transitions at the two valleys, K and K$'$, are allowed for optical excitations of opposite helicity incident along the c-axis, that is, left-circularly polarized and right-circularly polarized radiation, respectively. It was also found that these selection rules carry over to the excitons, because the selection rules for valley pumping are nearly exact over a large region around the two valleys. The ability to selectively confine carriers to selected states

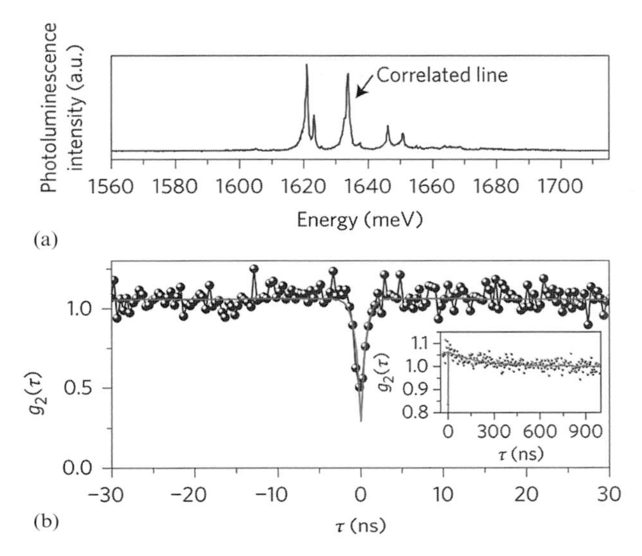

Figure 8.26 The narrow-line photoluminescence centers are interpreted as sources of single-photon emission. (a) The narrow-line emission collected from a microspot located at the edge of a thick WSe_2 flake. (b) The photon coincidence correlation, $g_2(t)$, where t is the time interval between the coincidence counts for one (designated as the "correlated line") of the narrow emission lines in (a). The effect of photon antibunching (with a characteristic time of 600 ps) is evident in the correlation function. The inset shows a weak bunching effect on a longer timescale (200 ns), likely caused by temporal jittering of the center of the line, an effect often seen for single-photon emitters. Reprinted by permission from Fig. 2(a) and (b) of Springer, *Nature Nanotechnology*, "Single photon emitters in exfoliated WSe_2 structures," M. Koperski, *et al.*, ©2015.

and valleys by the use of optical helicity signifies the advent of *switching devices* of a type referred to as "valleytronic devices"; see also Figs. 5.22–5.24.

Another example of the application of two-dimensional transition-metal dichalcogenide crystals is related to WSe_2 (tungsten diselenide; for the parameters of this crystal see Table 5.2). The study cited in Fig. 8.26 reports a comprehensive optical micro-spectroscopy study of thin layers of WSe_2. At the edges of WSe_2 flakes that have been transferred onto Si/SiO_2 substrates, these authors found that, at low temperatures, sharp emission lines with approximate linewidths of 100 meV occur. These narrow emission lines are interpreted as evidence of photon antibunching, which is known to be associated with single-photon emitters. In this study, the characteristic coincidence time in the autocorrelation function – an upper bound for the lifetime of the emitting state – is estimated to be ~600 ps for the line analyzed in Fig. 8.26; this time was observed to be as large as a few nanoseconds for other lines investigated in the study. The emitters reported in this study were tentatively recognized as nanoflakes of the WSe_2 monolayer, which was located at the edges of the flakes. These results suggest that, with advances in the control of edge imperfections, thin films of WSe_2 may find applications in the optoelectronics of single-photon emitters.

These examples indicate the potential for a variety of optical devices based on two-dimensional materials. The examples, coupled with the range of available bandgaps as well as the strong exciton effects, indicate that, even though the study of these two-dimensional materials is still in an early stage, these structures promise widespread optical and optoelectronic applications.

8.7 Adaptive Photodetectors

Currently, infrared photodetectors are used widely in imaging applications. As discussed above, quantum well infrared photodetectors have the advantage that their detection frequency can be relatively easily modified by changing either the thickness or material of the quantum wells or the host material in which the quantum wells are embedded. The parameters of quantum well structures are well controlled on the large lateral scale, so they can be used in imaging, but, once a structure has been designed and grown, the response frequency of the photodetector is fixed. The development and implementation of sensors with adaptable parameters would provide efficient use of sensing resources and in this way enhance real-time detection and simultaneously decrease the cost of the sensing system. Reconfigurable nanomaterials are expected to demonstrate wide spectral tunability, an adjustable dynamic range for the detection of high- and low-intensity radiation, possibilities to operate under high-level background radiation, scalability for use in large sensor arrays for imaging applications, light-weight and low-power consumption, and manageable trade-off characteristics, such as operating time and sensitivity.

The adaptivity of a detector implies possibilities for changing parameters over wide ranges. A long operating time increases the responsivity and improves the noise characteristics. By sacrificing the sensitivity, one can significantly increase the operating speed, i.e., decrease the acquisition time.

In Section 3.9 we demonstrated that selective doping is one of the most effective and reliable ways to manage nanomaterial properties by controlling the electric charge of quantum dots and quantum wells. Here we will discuss how those nanomaterials can be used in adaptive detectors.

8.7.1 Voltage Control of Frequency Response in Asymmetrically Doped Double-Quantum-Well Based Photodetectors

Double-quantum-well structures are good examples of nanomaterials whose optoelectronic characteristics depend strongly on the charge distribution in nanoblocks, as discussed in Section 3.9.2 (see Fig. 3.42). The interplay of the external electric field with the doping-induced field in such structures can substantially shift the response frequency of a photodetector. An example of a double-quantum-well-structure photodetector is shown in Fig. 8.27. The photodetector was grown on a semi-insulating GaAs wafer to permit back-side illumination. The growth sequence started with a 500 nm undoped GaAs buffer layer, followed by an 800 nm

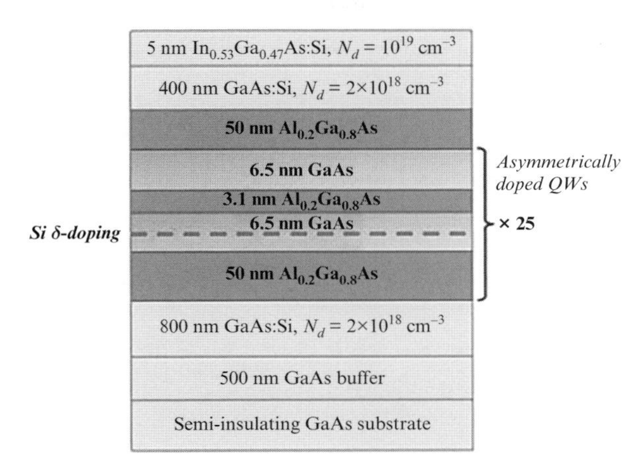

Figure 8.27 A schematic representation of the epitaxial layer structure of a double-quantum-well infrared photodetector with asymmetrically doped detection units. After Fig. 2 in J. K. Choi, *et al.*, *Jpn. J. Appl. Phys.* **51**, 074004 (2012). Copyright 2012 The Japan Society of Applied Physics.

heavily doped ($N_d = 2 \times 10^{18}$ cm^{-3}) GaAs contact layer, 25 stages of the detection unit, a 400 nm layer of GaAs doped to 2×10^{18} cm^{-3} as a contact layer, and finally a 5 nm InGaAs layer doped to 10^{19} cm^{-3}. One unit of the structure is composed of a 6.5 nm GaAs layer doped with Si, with doping sheet density 5×10^{11} cm^{-2}, and a 6.5 nm undoped GaAs layer, separated by a 3.1 nm-thick $Al_{0.2}Ga_{0.8}As$ layer. The units are separated from each other by 50 nm $Al_{0.2}Ga_{0.8}As$ barriers (see Fig. 8.27).

The spectral characteristics of the photoresponse of this detector are presented in Fig. 8.28. The relative intensities and the widths of the peaks depend on the bias voltage. Moreover, the spectrum depends strongly on the polarity of the bias voltage, due to the asymmetric doping, and the photoresponse can be varied in the range between about 7.5 μm and 11 μm (see Fig. 8.28(d)). The peaks can be attributed to transitions from the ground states to excited states that are effectively coupled to the extended states. For convenience we have duplicated Fig. 3.42 here and have marked ground states as E_1 and excited states as E_2. Both levels are split into two, and the wave functions of the lower double-split levels are located mostly in the right well (levels E_1 and E_2), while the wave functions of the higher double-split levels are located mostly in the left well (levels E_1' and E_2'). Electron repopulation between the left and right wells leads to a change in the frequency of the photoresponse. The photoresponses are very distinct for +5 V and −5 V and, for the intermediate bias, both spectral peaks are observed with small overlap. Table 8.1 shows the energy differences between the levels as well as the relative population of the wells. Switching from −5 V to +5 V increases the population of the left well from 1% to 24% of the total electron concentration as the split between the ground levels E_1 and E_1' decreases from 30 meV to 15 meV.

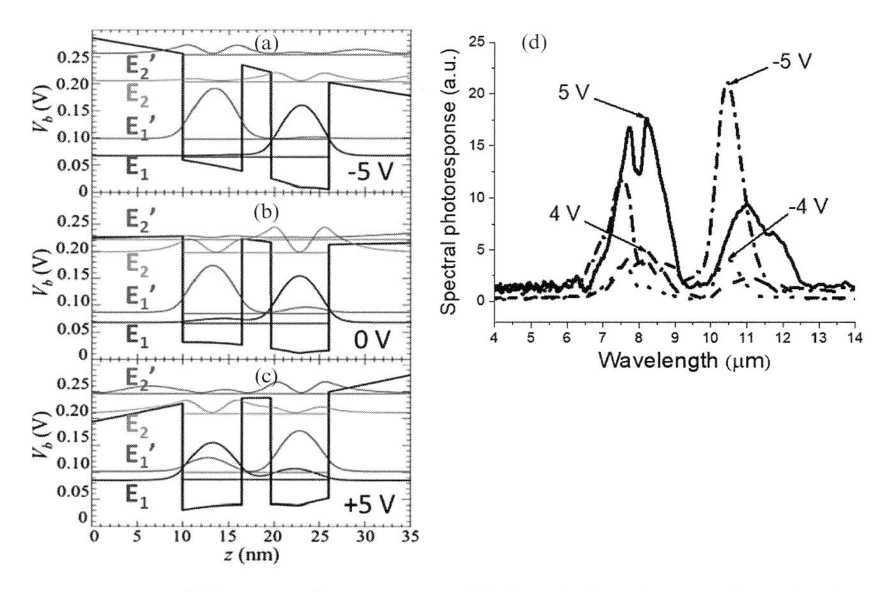

Figure 8.28 Band diagrams of a quantum well infrared photodetector shown in Fig. 8.27 under biases −5 V (a), 0 V (b), and +5 V (c), and (d) the spectral dependence of the photoresponse at $T = 80$ K under positive biases 4 V, 5 V and negative biases −4 V, −5 V. Part (d) after Fig. 2 in X. Zhang, *et al.*, "Nanoscale engineering of photoelectron processes in quantum-well and quantum-dot structures for sensing and energy conversion," *J. Phys. Conf. Ser.*, **906**, 1, 012026 (2017). Licensed under CC BY 3.0. ©IOP Publishing. All rights reserved.

Table 8.1 Calculated energies of electron transitions and relative populations of electrons in double quantum wells with asymmetric doping

	$E_1 \to E_2$	$E_1 \to E_2'$	$E_1' \to E_2$	$E_1' \to E_2'$
−5 V	135 meV 9.2 μm	188 meV 6.6 μm	105 meV 11.8 μm	158 meV 7.8 μm
Relative population	99%		1%	
5 V	125 meV 9.9 μm	165 meV 7.5 μm	110 meV 11.3 μm	150 meV 8.3 μm
Relative population	76%		24%	

This table was prepared by Xiang Zhang (PhD student of Dr. Mitin) during the work on the paper cited in Fig. 8.28, but it was not included in the paper due to the page limit.

It is important to emphasize that an asymmetrically doped double-quantum-well-based photodetector is a two-color photodetector, and it replaces two photodetectors employing a conventional technology. The photoresponse at each wavelength depends on the temperature of the emitter of radiation and is proportional to the intensity of the emitted radiation, while the ratio of the photoresponses at the two distinct wavelengths depends only on the temperature of the emitter. So, the photodetector discussed here allows the remote measurement of temperature.

8.7.2 Control of Quantum-Dot-Based Detector Performance by Selective Doping

Photodetectors based on quantum dots operate like photodetectors based on quantum wells: i.e., electrons from the quantized levels in quantum dots are photoexcited into the conductive continuum of the matrix. There are at least two distinctive advantages of quantum-dot-based photodetectors: (i) they are sensitive to normally incident light, so no grating or faceting is required as in quantum well photodetectors that are sensitive only to light polarization perpendicular to the planes of quantum wells; and (ii) in quantum well photodetectors the photoconductive electrons always cross quantum wells and, while passing through the quantum wells, they can be captured. In contrast, in quantum dot photodetectors the electrons can pass through the quantum dot layers while passing between the dots; see Section 3.9.3 and Figs. 3.43 and 3.44.

As with quantum wells, photon-induced transitions cover wide spectral ranges from the terahertz range to infrared, and can be effectively controlled by the size and form of the quantum dots as well as by the material of the dots and the matrix in which those dots are embedded. Quantum dot structures are essentially three-dimensional nanomaterials, where the processes of photon absorption in the quantum dots are spatially separated from photocarrier transfer along the conducting channels formed between dots by the nanoscale potential barriers created by the charge on the dots; see Section 3.9.3. Therefore quantum dot nanomaterials provide high optoelectronic gain and more flexibility in managing critical kinetic and transport characteristics such as the photocarrier lifetime and photocarrier mobility. To optimize the potential of quantum dot detectors, the photocarrier lifetime should be adjusted to values required by the intended applications. A long lifetime cannot provide effective sensing in a dynamic background, but at the same time too short a lifetime decreases the sensitivity of the detector. In this subsection we discuss how doping of the barrier material (the matrix) and dots changes the responsivity of photodetectors due to the change in the lifetime of photocarriers. Also, a substantial dependence of the lifetime on the applied electric field (see Fig. 3.43) allows switching between high sensitivity (a long lifetime in the low-electric-field limit) and short acquisition time (reduced lifetime) in a high applied electric field.

Quantum dot detectors are unipolar devices, and for this reason unipolar doping either in the quantum dot layer (see an example in Fig. 8.29(a)) or in the barrier layer (see an example in Fig. 8.29(b)) is employed, mainly to supply electrons to quantum dots. If all dopants intended for doping in the quantum dot layer are located in the quantum dots, the quantum dots will be neutral. During quantum dot growth, a thin wetting layer is formed between the matrix and the dots, as shown schematically in Fig. 8.29. A substantial number of dopants will stay in the wetting layers, and they supply electrons to the quantum dots. As a result, the quantum dots always have negative charge due to unintentional doping outside the dots, even when doping is introduced during the growth of the quantum dots.

The charge on the dots determines the repulsive potential of the dots and, as the charge increases, the preferential electron paths are shifted around the dots and

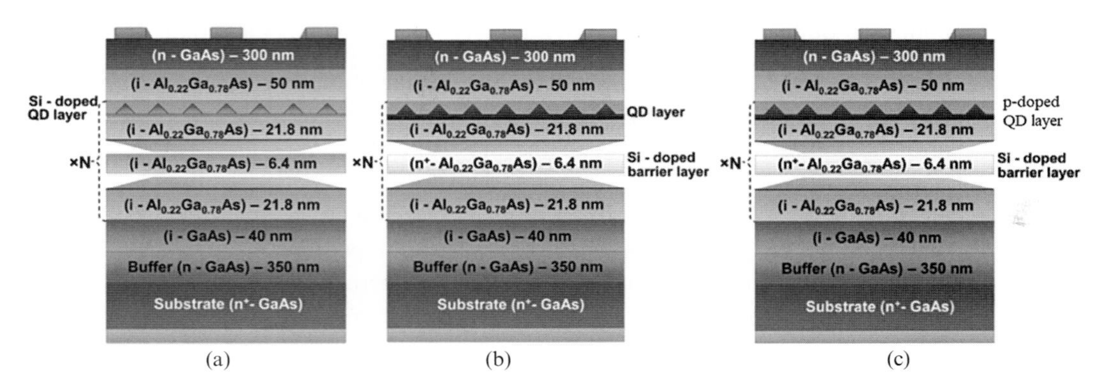

Figure 8.29 Examples of quantum dot photodetector structures in a matrix with (a) doping by donors in the quantum dot layer; (b) doping by donors in the barrier layer; and (c) doping by acceptors in the quantum dot layer and by donors in the barrier layer (bipolar doping). Parts (a) and (b) reprinted from Fig. 1 in K. Sablon, *et al.*, "Effective harvesting, detection, and conversion of IR radiation due to quantum dots with built-in charge," *Nanoscale Res. Lett.* **6**, 584 (2011). Licensed under CC BY 2.0. ©Sablon, *et al.*; licensee Springer.

Table 8.2 Doping of quantum dot infrared photodetectors presented in Figs. 8.29(a), 8.29(b), and 8.30(a)

Device	B44	B45	B52	B53
Doping	Intra-dot	Inter-dot	Intra-dot	Inter-dot
Donor concentration (10^{11} cm^{-2})	2.7	2.7	5.4	5.4
Number of electrons in dot, n	2.7	2.8	4.7	6.1
Built-in dot charge	1.8	2.8	3.45	6.1

Reprinted from Table 1 in K. Sablon, *et al.*, "Effective harvesting, detection, and conversion of IR radiation due to quantum dots with built-in charge," *Nanoscale Res. Lett.* **6**, 584 (2011). Licensed under CC BY 2.0. ©Sablon *et al.*; licensee Springer.

their lifetime increases (see Fig. 3.44). Figure 8.30(a) depicts the responsivity of the photodetectors as a function of the built-in dot charge, measured by the number of electrons per dot. Details of the sample doping are presented in Table 8.2. Figure 8.30(a) clearly demonstrates that an increase of the repulsive potential of the dots leads to a substantial increase of the photoresponse. On the basis of the average number of electrons per dot, shown in Table 8.2, and on the size of the dots, the repulsive potential for the device B53 reaches about 130 meV, which is substantially higher than the thermal energy, $\sim$6.7 meV, at 80 K. So, the exponential factor of Eq. (3.120) leads to a substantial increase of the electrons' lifetime.

In spite of considerable control of the photodetector performance, unipolar doping does not solve all problems. First, to increase the photocarrier lifetime the potential barrier around a quantum dot should be three to four times larger than $k_B T$. Therefore, at liquid-nitrogen temperatures the nanoscale barriers should be at least 25 meV. The corresponding doping needs to be above two to three electrons per dot. Stronger unipolar doping changes the spectral characteristics of the detector

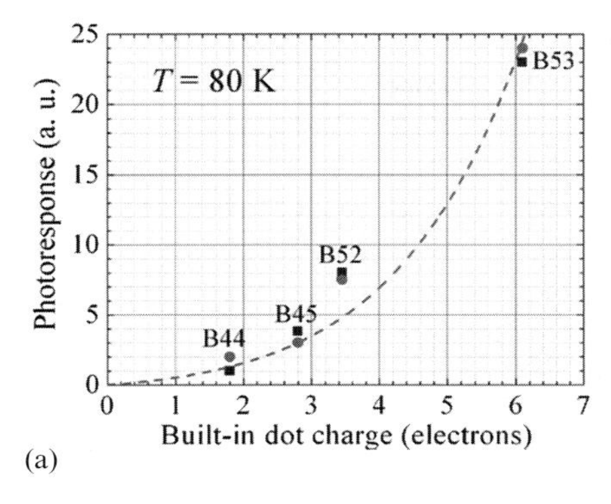

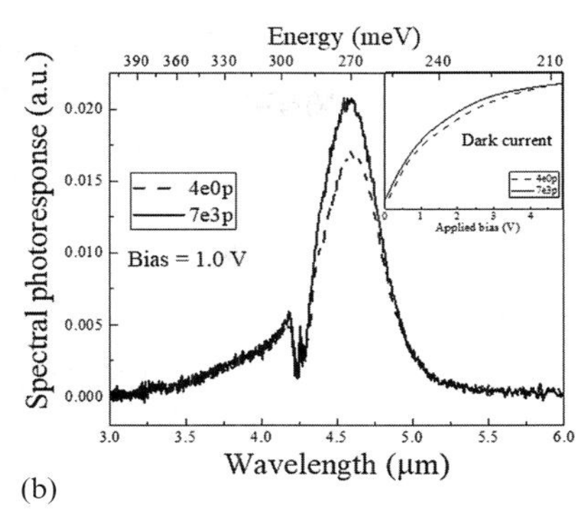

Figure 8.30 (a) The dependence of the relative photocurrent on the built-in dot charge for the structures shown in Figs. 8.29(a) and (b) and the average concentration of electrons for each sample shown in Table 8.2. Experimental results are shown by squares and the numerical simulation by circles; the dashed line is the theoretical dependence for the case of inter-dot doping. (b) A comparison of the spectral photoresponse of two photodetectors shown in Figs. 8.29(b) and (c) with approximately the same average number of electrons (about four) per dot, with monopolar doping (dashed curve) and bipolar doping (solid curve). The inset shows that the dark currents in these two samples are practically the same. Part (a) reprinted by permission from Fig. 6 in K. Sablon, *et al.*, "Effective harvesting, detection, and conversion of IR radiation due to quantum dots with built-in charge," *Nanoscale Res. Lett.* **6**, 584 (2011). Licensed under CC BY 2.0. ©Sablon *et al.*; licensee Springer. Part (b) reprinted from Fig. 4a in X. Zhang, *et al.*, "Nanoscale engineering of photoelectron processes in quantum-well and quantum-dot structures for sensing and energy conversion," *J. Phys. Conf. Ser.* **906**, 1, 012026 (2017). Licensed under CC BY 3.0. ©IOP Publishing. All rights reserved.

and substantially increases the noise current, as a result of fluctuations in the dark current.

The photocarrier lifetime and noise current are two key optoelectronic parameters of the photodetector. The detector response time is directly given by the photocarrier lifetime, and the noise bandwidth is given by the inverse lifetime. So, a long lifetime is really desirable, but a high concentration of electrons increases the dark current and noise. So, independent control and tuning of the lifetime (i.e., the height of the potential barrier) and the numbers of free electrons in the dots (i.e., the dark current and noise) would provide adequate adjustment of the detector to application requirements and substantially increase the commercial potential of quantum dot photodetectors. The quantum dot nanomaterials with bipolar doping shown in Fig. 8.29(c) provide a long and controllable photocarrier lifetime and, therefore, an enhanced photoresponse together with independently manageable dark and noise currents. A quantum dot structure with bipolar doping combines p-doping of the quantum dot layers with an acceptor concentration per dot n_a and n-doping of the inter-dot space with dopant concentration per dot n_d. The electrons from donors populate the acceptors that are in quantum dots and the remaining electrons occupy the quantum levels of the dots.

As a result, the charge of a quantum dot is equal to the number of donors per dot, n_d, but the number of free carriers in a dot is equal to $n_d - n_a$. In the limit $n_d = n_a \gg 1$, carriers in the quantum dots are practically absent, while the dots are strongly charged and surrounded by huge potential barriers (see Fig. 3.43).

So, bipolar doping results in a significant decrease in the thermally generated carrier density that is determined by $n_d - n_a$, and an exponential increase of the photoelectron lifetime that is determined by n_d. Figure 8.30(b) compares the spectral photoresponse of two detectors with about the same concentrations of free electrons per dot. The first detector is doped by donors in the barrier layer (see Fig. 8.29(b)) with a donor concentration that supplies on average four electrons per dot; compare this with samples B45 and B53 in Table 8.2. The second detector is doped by donors in the barrier to supply on average seven electrons per dot, and three of those electrons go onto the acceptors that are placed in the quantum dot layer, so the number of free electrons per dot is the same as in the first sample. As shown in Fig. 8.30(b), the second sample has a photoresponse about 25% higher and about the same dark current. Figure 8.31 shows schematically the placement of dopants and their levels (left panel) and the charge distribution due to electron transfer from donors into the levels of acceptors and quantum dots. Note how the relocation of dopants from the quantum dots into the wetting layer for doping of the quantum dot layers leads to opposite results in the cases of monopolar and bipolar doping. In the first case the effect is positive as those dopants lead to the creation of a repulsive potential of the quantum dot; see the last row in Table 8.1, which displays the number of electrons in the dots that are not compensated by donors that move to the wetting layer. In the second case the effect is negative as the electron charge in the dots decreases because some electrons would be located on acceptors that are in the wetting layer rather than located in the dots. So, progress in growth technology would further improve the performance of quantum dot photodetectors.

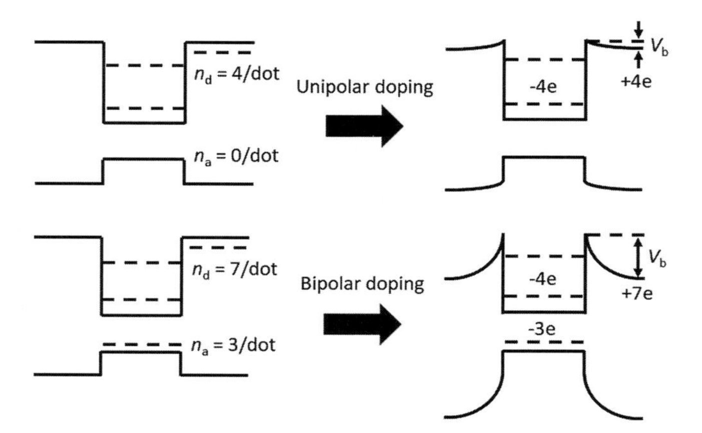

Figure 8.31 Schematic diagrams of monopolar and bipolar doping (left) and the positioning of charges on donors, acceptors, and on the quantum levels of dots (right).

8.7.3 Terahertz and Infrared Detectors Based on Graphene

It was shown in Chapter 5 (see Figs. 5.15 and 5.16) that graphene has zero bandgap and its absorption coefficient does not depend on frequency. This appears to indicate that the frequency-independent absorption coefficient is a huge advantage insofar as graphene can serve as a detector in any frequency from the visible to the terahertz range. However, it is also a disadvantage because a detector is designed predominantly for a particular frequency, i.e., detectors should have selectivity. The selectivity problem can be resolved, for example, by using filters that would allow only a specific frequency to pass through. Such an approach is used in bolometric detectors, as a bolometer registers the incoming energy and has no frequency selectivity. Filters introduce additional complexity to the system. Moreover, a lack of nanometer-sized passing filters excludes the usage of nonselective detectors in arrays that should include detectors of different frequencies. Here we give a couple of examples of how to make graphene-based detectors selective and adaptable.

The first example is a selective and adaptable detector that is based on the double-graphene-layer structure that was discussed in Section 3.9.1 (see Figs. 3.40(b) and 3.41(a)). The resonant tunneling from the n layer to the p layer is shown in Fig. 3.41(a) for the case when the Dirac points are at resonance. Figure 8.32(a) shows the off-resonance condition when the Dirac point of the p layer is higher on the energy scale than the Dirac point of the n layer. As we stated when discussing the current–voltage characteristics shown in Fig. 3.41(b), horizontal tunneling transitions (shown by the dotted line "a" in Fig. 8.32(a)) are forbidden, as the state in the p layer corresponds to a different momentum. Photon-assisted tunneling helps to satisfy both energy and momentum conservation. These transitions are

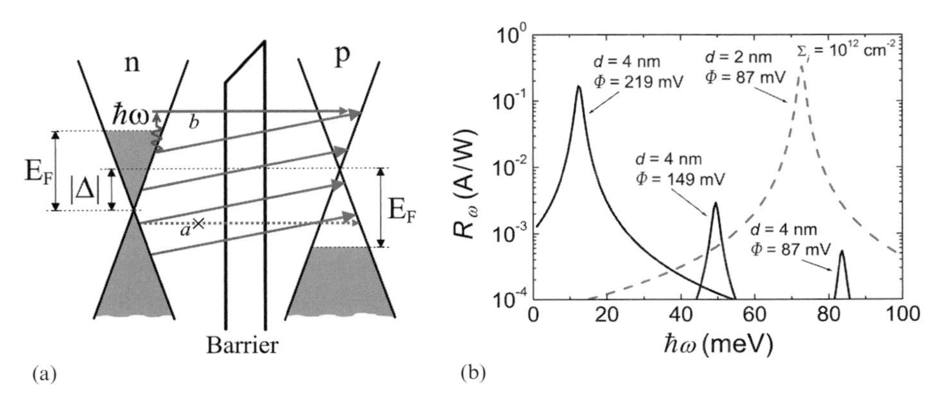

Figure 8.32 (a) The energy-band diagram of a double-graphene-layer structure with n doping on the left and p doping on the right. (b) The dependence of the responsivity of a photodetector based on a vertical p–n junction double-graphene-layer structure on the photon energy corresponding to the inter-graphene-layer barrier thickness at different applied voltages (solid lines and dashed line). Adapted from Figs. 2b and 3 of V. Ryzhii, *et al.*, *Appl. Phys. Lett.* **104**, 163505 (2014), with the permission of AIP Publishing.

schematically shown by the four inclined lines and, to demonstrate that the transitions are photon-assisted, the absorption of photon energy is shown by a wavy line and tunneling is shown by line "b" in Fig. 8.32(a). As stated earlier, photons as well as electrons in graphene have linear dispersion relations, but the speed of light is two orders of magnitude higher than the electron velocity in graphene. So, the wavevector of a photon is very small compared with the wavevector of an electron, and thus the photon energy, $\hbar\omega$, is practically equal to the energy gap between the Dirac points in the n and p layers, which is denoted by Δ in Fig. 8.32(a). The energy gap, Δ, depends on the voltage Φ applied between the n and p graphene layers; see Fig. 3.40(b). Figure 8.32(b) depicts the dependence of the responsivity R_ω (R_ω is the ratio of the photocurrent to the absorbed radiation power, see Eqs. (8.6) and (8.7)) on the frequency for different voltages, Φ, applied between the n and p layers. The doping levels, Σ, in the n and p layers are the same and equal to $\approx 10^{12}\,\mathrm{cm}^{-2}$. The three curves for a sample with barrier thickness $d = 4\,\mathrm{nm}$ have fairly large peaks, associated with tunneling between graphene layers accompanied by absorption of the incident photons. The peak values of the responsivity are high, and the position of the peak depends strongly on the applied voltage. These features provide strong evidence for the selectivity and adaptivity of these double-graphene-layer photodetectors. Moreover, the detectors can operate at room temperature and cover the terahertz frequency range. The dashed curve indicates that the reduction of the barrier width leads to a substantial increase in the current since the tunneling probability depends exponentially on the barrier width.

The second example is a vertical transistor with a graphene layer serving as the base of the transistor, as shown in Fig. 8.33(a). In the case where the graphene base is filled by electrons (Fig. 8.33(c)), the device can be referred to as an N–n–N heterostructure hot-electron transistor; if the graphene base is filled by holes, it is referred to as a heterostructure bipolar transistor (Fig. 8.33(b)). The graphene base can be undoped and its filling can be controlled by the voltage Φ_B applied between the emitter and the graphene base; compare the cases shown in Figs. 8.33(b) and (c). Also, the variation of the voltage results in a substantial variation of the collector current J_C that is determined by the number of electrons that leave the emitter and reach the collector. Figures 8.33(b) and (c) show schematically the path of those electrons as well as the small fraction of electrons that are captured into the graphene base. As the graphene base is one monolayer thick, the probability of capture into the graphene base is very low and the gain of such a device is very high. In addition to the dc voltages, a small ac voltage $\delta\Phi_\omega$ is applied between the left terminal and the right terminal of the graphene base, chosen as the x-direction. The voltage $\delta\Phi_\omega$ is supplied by a terahertz antenna that is not shown in Fig. 8.33(a). If the frequency of the ac signal coincides with the frequency of plasma oscillations in the graphene base, the plasma oscillations in the graphene base are excited, and the local potential, $\delta\phi_\omega$, in the graphene base will depend on x, that is, $\delta\phi_\omega = \delta\phi_\omega(x)$. As the tunneling from the n-type emitter through the top of the barrier between the emitter and the graphene base depends on the local potential difference between the emitter and the graphene base, the collector current will acquire an additional ac

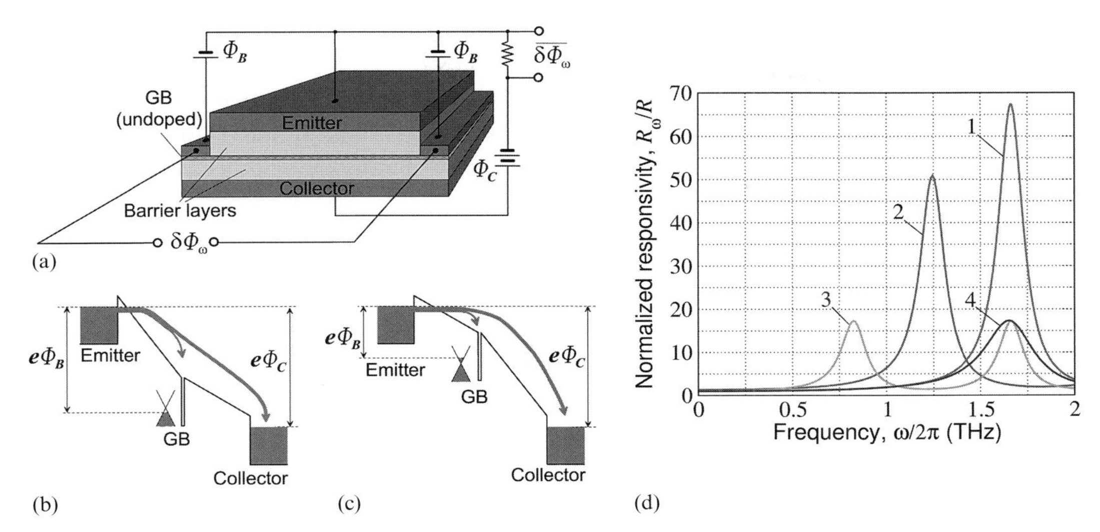

Figure 8.33 (a) A hot-electron transistor with a graphene base and the potential profile in the direction perpendicular to the graphene base when the graphene base is filled by (b) holes and (c) electrons. The arrows show the paths of electrons across the barrier layers and the capture of a small proportion of the electrons into the graphene base. (d) The dependence of the responsivity R_ω on the frequency for the device shown in (a) with different parameters: curve 1, $\tau = 1$ ps, $s = 5 \times 10^8$ cm/s, $L = 1.5$ μm; curve 2, $\tau = 1$ ps, $s = 5 \times 10^8$ cm/s, $L = 2$ μm; curve 3, $\tau = 1$ ps, $s = 2.5 \times 10^8$ cm/s, $L = 1.5$ μm; and curve 4, $\tau = 0.5$ ps, $s = 5 \times 10^8$ cm/s, $L = 1.5$ μm; These parameters correspond to plasma oscillation frequencies in the graphene base equal to 5/3, 5/4, 5/6, and 5/3 THz, respectively. Adapted from Figs. 1 and 3 of V. Ryzhii, *et al.*, *Appl. Phys. Lett.* **118**, 204501 (2015), with the permission of AIP Publishing.

component that depends on $\delta\phi_\omega(x)$. This is a high-frequency current and it is very difficult to measure, but, in addition, there is a dc component, which is proportional to $|\delta\phi_\omega(x)|^2$. Since $|\delta\phi_\omega(x)|^2$ is proportional to the plasma excitation signal $|\delta\Phi_\omega|^2$, this device can measure the intensity of terahertz radiation. As was discussed earlier, the frequency of plasma oscillations in graphene depends on the lateral size $2L$ (in our case $2L$ is the distance between the left and right contacts of the graphene base, Fig. 8.33(a)) and on the plasma wave velocity s, which increases with increasing carrier density, which in turn is controlled by Φ_B. The quality factor of the plasma resonance is limited primarily by the carrier momentum relaxation time associated with the scattering on impurities, various imperfections, and phonons.

Figure 8.33(d) depicts the dependence of the responsivity on the terahertz frequency in devices for the four sets of parameters that we have just discussed. The pronounced peaks in the responsivity correspond to plasma excitations in the graphene base. Curve 3 has a second peak at double the frequency of the first. Reduction of the momentum relaxation time reduces the amplitude of the plasma oscillations, which leads to a decrease of the photoresponse; compare curves 1 and 4, which correspond to the same plasma frequency but different quality factors. The device has excellent selectivity and adaptivity.

8.8 Closing Remarks to Chapter 8

In this chapter we studied applications of quantum heterostructures to optoelectronic devices utilizing intraband phototransitions. It was shown that these heterostructures make it possible to realize essentially new optoelectronic devices working from the mid-infrared to the terahertz spectral range. These devices include sources of spontaneous and stimulated long-wavelength emission, as well as infrared, far-infrared, and terahertz photodetectors.

One of the important classes of devices using intraband phototransitions consists of the quantum-cascade lasers. These laser sources are based on unipolar injection in quantum heterostructures. The properties of these unipolar lasers are determined to a large degree by quantum confinement. We showed that quantum-cascade lasers can be tailored for operation from the mid-infrared to the terahertz regions. The terahertz quantum-cascade lasers work at low temperatures (below 200 K). To overcome these low-temperature limitations of their operation, intracavity difference-frequency generation has been proposed for long-wavelength quantum-cascade lasers. This approach facilitates achieving the operation of terahertz quantum-cascade devices at room temperature.

We studied photodetectors based on intraband phototransitions in multiple-quantum-well structures. We found that this technology provides a new approach for the detection of weak infrared signals. Multiple-quantum-well photodetectors can be designed to operate in the infrared spectral region with great sensitivity. We presented photodetectors that are based on quantum dot structures. They have considerable potential for further development and will compete with conventional extrinsic infrared photodetectors.

We also presented treatments of silicon photonics by providing discussions of key and recent contributions to these fields. Silicon photonics is important for the realization of optics-based components for optical switching on a large scale, and offers the paradigm-changing use of silicon-based electronic chips integrated functionally through optical interconnects.

Finally, we indicated some perspectives regarding potential applications of two-dimensional materials for optoelectronic technologies.

8.9 Control Questions

1. In bulk semiconductor materials, intraband phototransitions are forbidden unless a "third body" (a defect, a phonon, an inhomogeneity, etc.) participates in the process. Explain why, in heterostructures, intraband phototransitions (namely intersubband ones) are allowed and can be used in different applications.
2. A quantum-cascade laser can be interpreted in terms of the following three-step laser scheme: (i) pumping of the upper working level; (ii) phototransition between the upper and lower working levels; and (iii) depopulation of the lower

working level. Specify the relation between the pumping and depopulating rates in quantum-cascade lasers.

3. A population inversion between two electronic states can be achieved in relatively simple four-barrier heterostructures, as shown in Fig. 8.1. Explain the necessity for the use of multiple-period (cascade) structures for the practical realization of monopolar lasers.
4. For electromagnetic waves of frequencies 1 THz and 5 THz calculate the energies of quanta corresponding to these frequencies. Compare your results with average electron energies at ambient temperatures of $T = 77$ K and 300 K. By using this comparison, explain why terahertz frequency generation is difficult to achieve with quantum-cascade lasers at elevated temperatures (see Fig. 8.5).
5. A quantum-cascade laser with a composite cavity can produce two mid-infrared modes of similar frequencies, $|\omega_1 - \omega_2| = \omega_{THz}$, such that their difference corresponds to the terahertz frequency range. Intracavity nonlinear optical effects provide a nonlinear polarization wave of terahertz frequency, which results in a terahertz emitted wave (see Fig. 8.6). Assume that one of the mid-infrared modes has wavelength $\lambda_{0,1} = 10\,\mu m$ ($\omega_{0,1} = 2\pi c/\lambda_{0,1}$, $\omega_{0,2} = 2\pi c/\lambda_{0,2}$) (see Section 4.2.1). Determine the possible wavelengths of the second mid-infrared mode, $\lambda_{0,2}$, necessary to generate the 5 THz emitted wave. (Answer: $\lambda_{0,2} = 8.57\,\mu m$ or $12\,\mu m$).
6. In a photoconductive detector (see Fig. 8.10), the photocurrent, I_l, and the photodetector gain, G, are the most important characteristics (see Eqs. (8.5) and (8.6)). Present qualitative explanations for why both characteristics increase with the voltage applied to the detector. Determine how I_l and G depend on the geometric parameters of the detector at a given applied voltage.
7. In Section 8.4.2, multiple-quantum-well-based photodetectors were discussed, and it was concluded that such detectors can be designed to achieve high sensitivity in a given infrared spectral region (providing thereby a "one-color" detector). Explain how it is possible to develop a multi-color detector that is based on multiple-quantum-well structures.
8. Explain why it is possible to achieve the confinement of infrared light and waveguiding in Si/SiO_2 systems.
9. In the Si-based optical modulator shown in Fig. 8.21(a), the control of light is brought about by varying the voltage applied to the p–i–n junction. Explain the mechanisms for the modulation of the phase and intensity of the light in this ring-resonator structure.

Appendix A Basic Statements and Formulae of Quantum Physics

In this appendix we present key concepts and formulae of quantum physics which are necessary for an understanding and analysis of the properties of quantum heterostructures.

We start with a reminder of the de Broglie wavelength, which is one of the basic quantum characteristics of a particle. An electron in a semiconductor can be characterized by the effective mass m^*, which is usually less than the free-electron mass m. The de Broglie wavelength of an electron is

$$\lambda = \frac{h}{p} = \frac{h}{\sqrt{2m^* E}} = \lambda_0 \sqrt{\frac{m}{m^*}}, \tag{A.1}$$

where h is the Planck constant, p and E are the electron momentum and energy, respectively, and λ_0 is the de Broglie wavelength of an electron in vacuum. From Eq. (A.1) it follows that the de Broglie wavelength of an electron in a semiconductor is greater than that of a free electron. In Fig. A.1 the value λ is depicted as a function of m^*/m. Points 1–4 on the curve indicate wavelengths for electrons in bulk crystals of InSb, GaAs, GaN, and SiC, respectively. We have taken the effective masses, m^*/m, to be 0.014, 0.067, 0.172, and 0.41, respectively, for these materials. Absolute values are shown for electrons with thermal energy $E = k_B T$, where $T = 300$ K is the ambient temperature and k_B is the Boltzmann constant, $k_B = 1.381 \times 10^{-23}\ \mathrm{J\,K^{-1}}$. We see that the de Broglie wavelength of an electron in typical semiconductors with m^* in the range $(0.01–1)m$ is of the order of 73–730 Å. As the temperature decreases to 3 K, the de Broglie wavelength increases by one order of magnitude. Thus the de Broglie wavelength is comparable to the sizes of the semiconductor structures and devices fabricated by modern nanotechnology. This is one of the major reasons why quantum-mechanical phenomena must be considered in order to understand the physics of nanostructures.

Comparison of the electron de Broglie wavelength with the dimension sizes of a sample, or a device, allows us to draw qualitative conclusions about the quantization of the electron spectrum. Indeed, let the geometrical dimensions of a semiconductor sample be L_x, L_y, L_z. If the system is free of randomness and other scattering mechanisms are sufficiently weak, the electron motion is *quasi-ballistic*, and the only length that the geometrical dimensions need to be compared with is the electron de Broglie wavelength λ. Since only an integer number of half waves of the electrons can be put – i.e., fitted – into any finite system, instead of a continuous energy spectrum

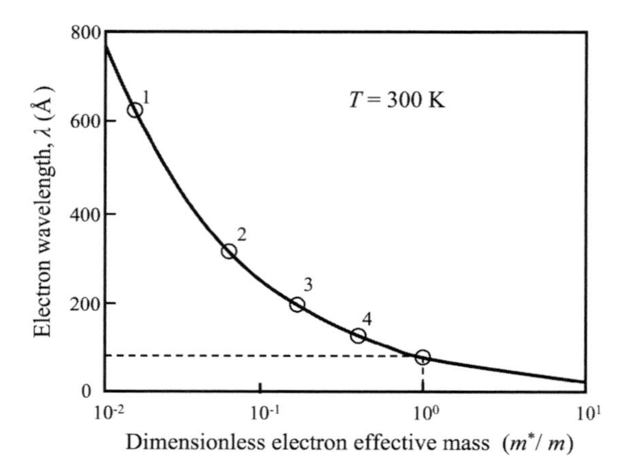

Figure A.1 Electron wavelength versus the electron effective mass for room temperature ($E = k_B T$, $T = 300$ K). Points 1–4 correspond to InSb ($m^*/m = 0.014$), GaAs (0.067), GaN (0.172), and SiC (0.41), respectively.

and a continuous number of electron states, one obtains a set of discrete electron states and energy levels, each of which is characterized by the corresponding number of half wavelengths. This is frequently referred to as the *quantization of electron motion*. Depending on the system dimensions one can distinguish different cases:

(a) the three-dimensional or bulk-like case, when the quantization of the electron spectrum is not important at all,

$$\lambda \ll L_x, L_y, L_z, \tag{A.2}$$

and an electron behaves like a free particle characterized by the effective mass m^*;

(b) the two-dimensional or quantum well case, when quantization of the electron motion occurs in one direction, while in the two other directions the electron motion is free,

$$\lambda \simeq L_x \ll L_y, L_z; \tag{A.3}$$

(c) the one-dimensional or quantum wire case, when the quantization occurs in two directions, so that the electron moves freely only along the wire,

$$L_x \simeq L_y \simeq \lambda \ll L_z; \tag{A.4}$$

(d) the zero-dimensional or quantum box (dot) case, when the quantization occurs in all three directions and the electron cannot move freely in any direction,

$$L_x \simeq L_y \simeq L_z \simeq \lambda. \tag{A.5}$$

The last three cases also illustrate *quantum size effects* in one, two, or three dimensions, respectively.

If at least one geometrical dimension of a device is comparable to the electron wavelength, a quantum-mechanical treatment of the problem is strictly required.

The Schrödinger Equation

A quantum-mechanical treatment is based on the statement that a particle is characterized by a *wave function*, $\psi(\vec{r})$, such that the value $|\psi(\vec{r})|^2 \, d\vec{r}$ gives the probability of finding a particle inside a small volume $d\vec{r}$ around the point $\vec{r}$. Thus the wave function, ψ, may be interpreted as the probability amplitude corresponding to the probability density for finding a particle at a particular point of space, $\vec{r}$.

The wave function, Ψ, of an electron or of an electron system satisfies the principal equation of quantum mechanics, the *Schrödinger equation*,

$$i\hbar \frac{\partial \Psi}{\partial t} - \mathcal{H}\Psi = 0, \tag{A.6}$$

where the operator $\mathcal{H}$ is the *Hamiltonian* of the system:

$$\mathcal{H} = -\frac{\hbar^2 \nabla^2}{2m} + \mathcal{U}(\vec{r}). \tag{A.7}$$

Here $\mathcal{U}(\vec{r})$ is the potential energy (note that $\mathcal{U}(\vec{r})$ might be a function of time as well) and the first term is the operator of the kinetic energy,

$$\nabla^2 = \frac{\partial^2}{\partial x^2} + \frac{\partial^2}{\partial y^2} + \frac{\partial^2}{\partial z^2}, \tag{A.8}$$

which is the divergence of the gradient, or the Laplacian operator. The form of Eq. (A.6) is that of a wave equation and its solutions are expected to be wave-like in nature. If $\mathcal{U}(\vec{r})$ is time-independent, one may separate the dependences on the time and spatial coordinates,

$$\Psi(\vec{r}, t) = e^{-iEt/\hbar}\psi(\vec{r}), \tag{A.9}$$

where ψ is a function only of the spatial coordinates. By substituting Eq. (A.9) into Eq. (A.6) one gets the time-independent Schrödinger equation:

$$\left[-\frac{\hbar^2 \nabla^2}{2m} + \mathcal{U}(\vec{r})\right] \psi(\vec{r}) = E\psi(\vec{r}). \tag{A.10}$$

In Eqs. (A.9) and (A.10) E is the energy of a particle, or, as we shall explain, of a system of particles. The major task of quantum mechanics is to solve the Schrödinger equation for wave functions corresponding to different systems (i.e., different potential energies $\mathcal{U}(\vec{r})$).

Very conditionally, applications of the Schrödinger equation to solid-state electronics and optoelectronics can be divided into two groups. The first is related to electron bound states in heterostructures (impurities, quantum wells, quantum wires, quantum dots, etc.). For these problems, the major goals are to find the discrete energies of bound states, to calculate the relaxation of excited states due to interactions with free electrons, phonons, and defects, and to understand the results of interactions with electromagnetic radiation. For the same group of problems we may define the stationary-state problems of a free electron, i.e., calculation of the electron energy spectra and electron interactions with crystal vibrations and defects as well as with other electrons and with electromagnetic waves. This group of problems requires solution of the stationary Schrödinger equation (A.10) and, in general, the application of perturbation theory.

The second group of problems concerns the dynamics of free carriers in time-dependent strong fields, for example, electromagnetic fields. In general, this group of problems requires the solution of the time-dependent Schrödinger equation (A.6).

An exact value of the energy, E, characterizes the system only for the stationary-state case, when the potential energy and the Hamiltonian do not depend on time. If they depend on time, the Schrödinger equation becomes time-dependent. Sometimes the time-dependent part of the Hamiltonian constitutes just a small addition to the stationary Hamiltonian. In this very important case, the small addition can be treated as a perturbation and the change in the wave function due to the perturbation may be described in terms of transitions between the stationary (unperturbed) states. Perturbation theory describes such electron transitions due to the perturbation of the stationary Hamiltonian by lattice vibrations, other electrons, crystal defects, and interactions with light. The latter interaction is the focus of this book.

As we shall see, Eq. (A.10) may be cast into the form of an eigenvalue equation. The energy, E, is its eigenvalue and the wave function, $\psi_E(\vec{r})$, is its eigenfunction. The eigenvalue E may run over discrete or continuum values, depending on the shape of the potential function, $\mathcal{U}(\vec{r})$, and the boundary conditions on the wave function.

In order to illustrate both possible types of solutions of the Schrödinger equation and energy states, as well as to clarify the tasks which arise, let us consider the one-dimensional problem for a system with potential energy $\mathcal{U}(\vec{r}) = \mathcal{U}(x)$ as shown in Fig. A.2. Here the vertical axis depicts the energy, E, and x represents only one space coordinate. The potential has a minimum at $x = 0$, vanishes at $x \to -\infty$, and saturates to a finite value $\mathcal{U}_\infty$ at $x \to \infty$. This potential is the most general form of a potential well.

At this point a short qualitative discussion serves to emphasize that the boundary conditions define the type of solution. These solutions will be obtained and discussed in more detail in due course. Among the possible solutions of the Schrödinger equation (A.10) with a chosen potential there can exist solutions with negative energy $E = E_i < 0$. The wave function corresponding to energy $E = E_1 < 0$ is sketched in Fig. A.2 as curve 1. One of the peculiarities of solutions with negative energy is that the spatial region with classically allowed motion, where the kinetic energy $p^2/2m = E - \mathcal{U}(x) > 0$, is certainly restricted. In the classically forbidden regions where $\mathcal{U}(x) > E$, the ψ function decays for $|x| \to \infty$. The states of particles like those just described are so-called bound states, and they are characterized by a discrete energy spectrum. Consider next the possible solutions for the energy range $0 \leq E \leq \mathcal{U}_\infty$ as shown by the line E_2 in Fig. A.2. These solutions exist for any values of E; they are finite at $x \to -\infty$ and penetrate slightly into the barrier region $\mathcal{U}(x) > E$, as shown by curve 2 in Fig. A.2. In the limit $x \to -\infty$ these solutions may be represented as a sum of two waves traveling in opposite directions: one

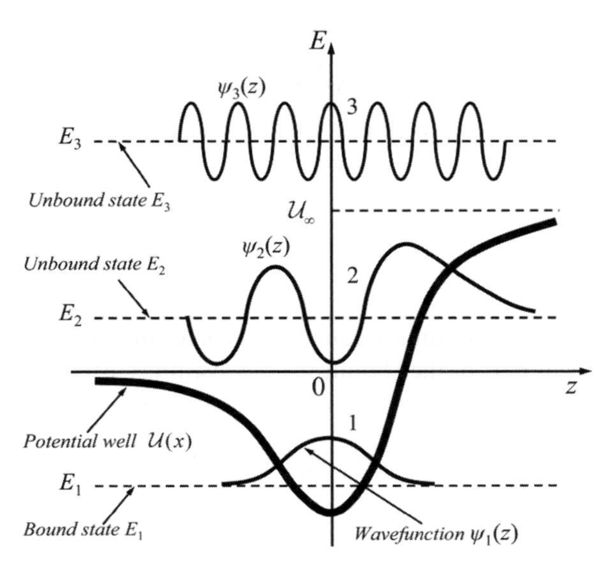

Figure A.2 A one-dimensional potential well $\mathcal{U}(x)$, and three types of solutions (1, 2, 3) of the Schrödinger equation corresponding to the bound state (energy E_1) and unbound states (energies E_2 and E_3).

wave is incident on the barrier, and the other is reflected from the barrier. For each energy in the range $0 \leq E \leq \mathcal{U}_\infty$, there is only one solution satisfying the physical requirements. For any energy $E > U_\infty$ there exist two independent solutions; see curve 3 corresponding to the energy E_3. One solution can be chosen in the form of a wave propagating from left to right. At $x \to \infty$ it has only one component, namely the wave overcoming the barrier, and at $x \to -\infty$ there is a superposition of both incident and reflected waves. It is emphasized that reflection of the wave from the barrier even when its energy exceeds the height of the barrier arises only in quantum physics. The reflected wave function can be chosen in the form of waves propagating from right to left. Examples of continuum energy spectra and non-bound states are given by considering cases with energies $E > \mathcal{U}_\infty$ and wave functions which are finite far away from the potential barrier.

The previous considerations demonstrate the importance of boundary conditions for the Schrödinger problem: if the decay of wave functions (i.e., $\Psi \to 0$ at $x \to \pm\infty$) away from the potential well is required, the discrete energies and the bound states can be found and they are of principal interest. If the boundary conditions correspond to an incident particle with $E > \mathcal{U}(\infty)$, continuous energy spectra will be obtained. To conclude the discussion of boundary conditions, let us mention that we often use potential energies with discontinuities. In this case at the point of discontinuity of the potential we require on physical grounds both *continuity of the wave function and continuity of the derivative of the wave function with respect to coordinates*.

Since the Schrödinger equation is linear, it is clear that, if a function, Ψ, is a solution of the equation, then any function of the form of a constant times Ψ is also a solution of the same equation. To eliminate this ambiguity, we have to take into account the probabilistic character of the wave function. Indeed, if a physical system is enclosed in a finite volume, the actual probability of finding a particle in this volume must equal 1, i.e.,

$$\int |\Psi(\vec{r},t)|^2 \, d\vec{r} = 1. \tag{A.11}$$

Equation (A.11) is known as the normalization condition. It provides the condition needed to determine the constant multiplier of the wave function for the case of a system of finite extent. Note that wave functions of different energy states are orthogonal to each other,

$$\int \Psi_{E_i} \Psi_{E_j} \, d\vec{r} = \delta_{ij}. \tag{A.12}$$

Because of the probabilistic character of the description of quantum-mechanical systems we have to clarify how to determine the average values of quantities which characterize such systems. The simplest case is the calculation of the average coordinate of a particle. Indeed, the absolute square of the normalized wave function gives the actual probability per unit volume of finding a particle at a particular point in space. Hence, the average value of a particular coordinate, say x, is given by

$$\langle x \rangle = \int x |\psi|^2 \, d\vec{r} = \int \psi^* x \psi \, d\vec{r}. \tag{A.13}$$

Thus the integral over space gives the mean, or *expectation* value, of coordinate x. The last expression, Eq. (A.13), represents a particular example of a general form for the calculation of the expectation value of any observable a:

$$\langle a \rangle = \langle \mathcal{A} \rangle = \int \psi^* \mathcal{A} \psi \, d\vec{r}, \tag{A.14}$$

where the operator $\mathcal{A}$ is associated with the observable a. For example, if $\mathcal{A} = \mathcal{H}$ then averaging as in Eq. (A.14) gives the expectation value of the particle energy. For $\mathcal{A} = -i\hbar \nabla$, we obtain the average particle momentum, etc.

This summary of general quantum-mechanics statements is far from complete. More detailed information on this subject can be found in textbooks on quantum mechanics cited in the Further Reading.

The Schrödinger Equation for Envelope Functions

Now we shall give a brief description of the quantum properties of electrons in crystals. In a crystal an electron moves in the field created by the positively charged ions (or neutral atoms) and all the other electrons. This field is frequently referred to as the *crystalline potential*. For an ideal crystal the crystalline potential is periodic, with the period of the crystalline lattice. Let $\vec{a}_i$, $i = 1, 2, 3$, be three basic vectors of the Bravais lattice, which define the three primitive translations. The periodicity of the crystalline potential $W(\vec{r})$ implies that

$$W\left(\vec{r}+\sum_i n_i\vec{a}_i\right)=W(\vec{r}), \tag{A.15}$$

where $\vec{r}$ is an arbitrary point of the crystal, and n_i are any integers. The one-particle wave function should satisfy the Schrödinger equation

$$H_{\rm cr}\psi(\vec{r})=\left[-\frac{\hbar^2}{2m}\nabla^2+W(\vec{r})\right]\psi(\vec{r})=E\psi(\vec{r}), \tag{A.16}$$

where $H_{\rm cr}$ is the crystalline Hamiltonian and m is the *free-electron mass*. The translation invariance of the crystalline potential of Eq. (A.15) provides solutions of the Schrödinger equation (A.16) in the *Bloch form*:

$$\psi(\vec{r})=e^{i\vec{k}\vec{r}}u_{\vec{k}}(\vec{r}), \tag{A.17}$$

where $\vec{k}$ is a constant vector and $u_{\vec{k}}(\vec{r})$ is the so-called *Bloch function*, which is a periodic function (see also subsection 2.4.1 in Chapter 2):

$$u_{\vec{k}}(\vec{r}+\vec{a}')=u_{\vec{k}}(\vec{r}), \quad \vec{a}'=\sum_i n_i\vec{a}_i.$$

Therefore, a stationary one-particle wave function in a crystalline potential has the form of a plane wave modulated by a Bloch function having the lattice periodicity. The wavevector $\vec{k}$ is called the *wavevector of the electron in the crystal*. This wavevector is one of the quantum numbers of electron states in crystals. The allowed quasi-continuum values of $\vec{k}$ form the so-called *first Brillouin zone* of the crystal. It is important that the symmetry of the Brillouin zone in $\vec{k}$-space is determined by the crystal symmetry.

Let the one-particle energy corresponding to the wavevector $\vec{k}$ be $E=E(\vec{k})$. If the wavevector changes within the Brillouin zone, one gets a continuum energy band, i.e., an *electron energy band*. At fixed $\vec{k}$, the Schrödinger equation (A.16) has a number of solutions in the Bloch form:

$$\psi_{\alpha,\vec{k}}(\vec{r})=\frac{1}{\sqrt{\mathcal{V}}}e^{i\vec{k}\vec{r}}u_{\alpha,\vec{k}}, \tag{A.18}$$

where α enumerates the energy bands. Here we normalize the wave function $\psi_{\alpha,\vec{k}}$ for the crystal volume $\mathcal{V}$. For the chosen normalization condition, one can easily obtain

$$\frac{1}{\mathcal{V}_0}\int_{\mathcal{V}_0}|u_{\alpha,\vec{k}}|^2\,d\vec{r}=1, \tag{A.19}$$

where the integral is calculated over the volume of the primitive cell, $\mathcal{V}_0$. The latter formula allows one to estimate the order of magnitude of $u_{\alpha,\vec{k}}$: $|u_{\alpha,\vec{k}}|\approx 1$. Note that for Bloch functions with different α and $\vec{k}$ the following orthogonality conditions are valid:

$$\frac{1}{\mathcal{V}_0}\int_{V_0} u^*_{\alpha,\vec{k}} u_{\alpha',\vec{k}'}\, d\vec{r} = \delta_{\alpha\alpha'}\delta_{\vec{k}\vec{k}'}. \tag{A.20}$$

The Bloch function $u_{\alpha,\vec{k}}(\vec{r})$ satisfies the following equation:

$$\left[-\frac{\hbar^2}{2m}\nabla^2 - i\frac{\hbar^2}{m}(\vec{k}\nabla) + W(\vec{r})\right] u_{\alpha,\vec{k}} = \left(E_\alpha(\vec{k}) - \frac{\hbar^2 k^2}{2m}\right) u_{\alpha,\vec{k}}. \tag{A.21}$$

Owing to the periodicity of the crystal, the Bloch function can be calculated within a single primitive cell.

To find $u_{\alpha,\vec{k}}$ and $E_\alpha(\vec{k})$ one should know the crystalline potential $W(\vec{r})$ in detail. However, most aspects of the electron properties in a crystal require knowledge of these quantities in a small range of the $\vec{k}$-space around the energy band extrema. Let an extremum of $E_\alpha(\vec{k})$ be at the point $\vec{k}=0$. Then, for small $\vec{k}$ and under the condition of a nondegenerate band, we can find the quadratic dispersion for the electron energy in the vicinity of the extremum of $E_\alpha(\vec{k})$ at $\vec{k}=0$. For the electron energy in the conduction band we get the conventional form of the kinetic energy:

$$E_{\rm c} = E_{\rm c}(0) + \frac{\hbar^2\vec{k}^2}{2m^*_{\rm e}}, \tag{A.22}$$

where $m^*_{\rm e}$ is the effective mass for electrons. For the valence band with a maximum of $E_\alpha(\vec{k})$, we obtain

$$E_{\rm v} = E_{\rm v}(0) - \frac{\hbar^2\vec{k}^2}{2m^*_{\rm h}}, \tag{A.23}$$

where $m^*_{\rm h}$ is the effective mass for holes, and again a simple nondegenerate parabolic dependence on $\vec{k}$ is supposed for $E_{\rm v}(\vec{k})$. Note that the effective masses $m^*_{\rm e}$ and $m^*_{\rm h}$ are both positive.

It is worth emphasizing that we have obtained results of great importance in solid-state physics: electron motion in a crystalline potential can be described by a quasi-continuum wavevector $\vec{k}$. One can introduce a quasi-momentum of the electron, $\vec{p}=\hbar\vec{k}$. That is, an electron in a crystal can be described like a free electron in the vacuum, but with an effective mass instead of the free-electron mass m. The effective mass differs from the free-electron mass, m. Notably, it can be anisotropic, as discussed in Chapter 2 for several particular crystals.

If an external potential $U(\vec{r})$ is applied to the crystal, an electron moves in a very complicated potential field composed of the crystalline potential $W(\vec{r})$ and $U(\vec{r})$:

$$W(\vec{r}) + U(\vec{r}). \tag{A.24}$$

For the case of a potential $U(\vec{r})$ that varies slowly with spatial coordinate, the effective-mass approximation can be applied to describe the quantum-mechanical behavior of an electron in the potential $U(\vec{r})$. Indeed, in such a potential the electron moves while remaining within a certain energy band, say α, i.e., no interband transitions occur. The solution of the total Schrödinger equation

$$[H_{\rm cr} + U(\vec{r})]\,\psi(\vec{r}) = E\psi(\vec{r}) \quad \text{(A.25)}$$

can be approximated as

$$\psi(\vec{r}) \approx F_\alpha(\vec{r})u_{\alpha,0}(\vec{r}), \quad \text{(A.26)}$$

where $F_\alpha(\vec{r})$ is a function varying slowly on scales of the order of the lattice constant. For small $\vec{k}$ in this band the energy is $E_\alpha(\vec{k}) = E(0) + \hbar^2 k^2/2m^*_\alpha$, so instead of Eq. (A.25) one can obtain the following equation for the so-called *envelope function* $F_\alpha(\vec{r})$:

$$\left[-\frac{\hbar^2}{2m^*_\alpha}\nabla^2 + U(\vec{r})\right] F_\alpha(\vec{r}) = (E - E(0))\,F_\alpha(\vec{r}). \quad \text{(A.27)}$$

The latter equation has the form of the Schrödinger equation for an electron with an effective mass m^*_α moving in the external potential $U(\vec{r})$. Thus, in a slowly varying external potential the electron can be described in term of a slow modulation on the Bloch function $u_{\alpha,0}$. For Bloch functions normalized according to Eq. (A.19) the envelope function satisfies a similar normalization condition, $\int d\vec{r}\,|F_\alpha(\vec{r})|^2 = 1$.

The effective-mass approximation can also be used for the analysis of semiconductor structures with abrupt junctions, as shown in Fig. 3.1, if the characteristic sizes of quantum wells, quantum wires, and quantum dots considerably exceed the lattice constant. In such cases, the spatial dependence of the energy band can be treated as a potential (the "heterostructure potential"). For smooth portions of this potential one can apply the effective-mass expression in Eq. (A.27), which should be supplemented by appropriate boundary conditions at the abrupt junctions.

In the main text we study an *idealized picture* of electron states for four types of quantum structures: quantum wells, quantum wires, quantum dots, and superlattices. For each of these structures our goals are to study the quantum-mechanical features and to reveal the pure effects of quantization. We consider exciton effects and electron states on defects in these idealized quantum structures.

By idealized we mean that the following set of simplifications applies. We describe an electron by an isotropic effective mass m^*, which is independent of both the position and the energy of the electron. The effective mass m^* characterizes the motion of an electron in the periodic potential of a crystal and differs from the free-electron mass. In general, m^* changes when the electron passes through a heterojunction, but in our idealized approach we neglect this effect. The second simplification is related to the profile of the potential energy in a heterostructure. The idealized approach is based on a step-like potential profile, which can be analyzed easily.

Besides the discontinuities of energy bands, there are other factors which control the profile of the potential and other properties of electrons. In real devices these factors include the space-dependent composition of compounds, the dependence of doping on spatial coordinates, the electrostatic potential, the different effective masses in layers composing a heterostructure, possible mechanical strain of layers, etc. These factors can be taken into account by detailed analysis of real heterostructures. However, they do not change the general features of the electron properties of quantum heterostructures.

Spin, Many-Particle Systems, and Electron Statistics

Though any semiconductor material consists of vast numbers of electrons, ions, atoms, etc., the major properties of semiconductor devices are generally determined by the electron subsystem. For many-electron systems, quantum physics brings additional features which are absent in the classical description. These features are associated with the fact that elementary particles, including electrons, are identical and the impossibility, in principle, of specifying the coordinates and tracing a given electron.

So far we have studied how an electron's wave function evolves in space as revealed by quantum mechanics. In addition to the parameters which characterize this spatial evolution, quantum-mechanical particles possess an additional "internal" degree of freedom, which is referred to as *spin*. Though one can compare spin with classical rotation, in fact spin is a strictly quantum-mechanical quantity and differs substantially from its classical analog. The principal quantitative characteristic of spin is a dimensionless quantity called the *spin number*. It is well established experimentally that an electron has a spin number equal to $1/2$. If one fixes an axis in space, the projection s of the electron spin on this axis can be either $+1/2$ or $-1/2$. Thus, a complete description of an electron state requires at least two quantum numbers: one corresponding to the spatial wave function, say n, and another corresponding to the spin quantum number, s.

In the absence of a magnetic field, one often may neglect the weak interaction between the electron spin and electron translational (or orbital) motion, which is known as the spin–orbit interaction. In this case, the electron spin does not affect the electron's spatial properties. Hence, an electron occupying any energy level may have two possible spin orientations. This case corresponds to a two-fold degeneracy for each energy level. A weak spin–orbit interaction is rather characteristic for electrons in the conduction band, while for a carrier in the valence band this interaction can be essential to understand the electronic and optical properties; see the analysis of the III–V compounds in Section 4.4.3.

Though for many actual cases the electron spin is not important in altering the energy spectra, or the spatial dependence of wave functions, etc., there is one crucially important consequence of the fact that the electron spin number is a half-integer. The point is that particles with half-integer spin numbers obey the *Pauli exclusion principle*: *any quantum state* $\{n, s\}$ *can be occupied by only a single particle*. In other words, two electrons in a system cannot be simultaneously in the same quantum state. Stated another way, two electrons may be in the same energy state if their spin quantum numbers are different; that is, if one spin quantum number is $+1/2$ the other must be $-1/2$. This restriction leads to a new, nonclassical statistics

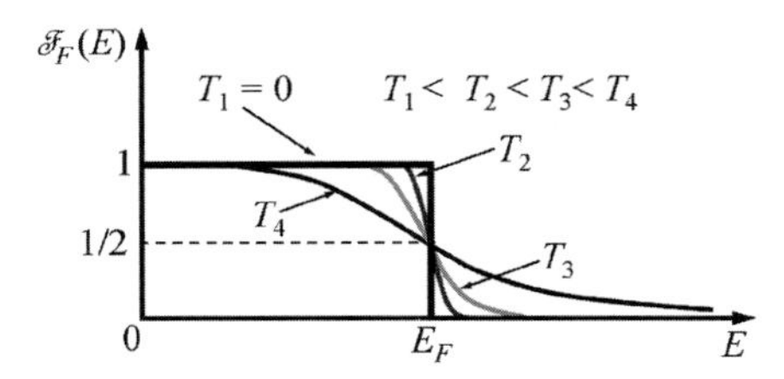

Figure A.3 Fermi distribution function for different crystal temperatures.

of electrons known as *Fermi statistics*. Under equilibrium the occupation of the energy levels is described by the *Fermi distribution function*

$$\mathcal{F}_{\mathrm{F}}(E_{n,s}) = \frac{1}{1 + e^{(E_{n,s} - E_{\mathrm{F}})/k_{\mathrm{B}}T}}, \tag{A.28}$$

where $k_{\mathrm{B}}T$ is Boltzmann's constant times the temperature of the system, E_{F} is the so-called *Fermi energy* or *Fermi level*, and $E_{n,s}$ is the energy of the quantum state characterized by two quantum numbers, n and s. The Fermi function $\mathcal{F}_{\mathrm{F}}(E)$ is shown in Fig. A.3 for different temperatures. The Fermi energy depends on the electron concentration, and may be found from the following normalization condition:

$$\sum_{n,s} \mathcal{F}_{\mathrm{F}}(E_{n,s}) = N. \tag{A.29}$$

Here N is the total number of the electrons in the system.

If the positions of the energy levels are independent of spin, $E_{n,s} \equiv E_n$, the occupation of these levels is simply equal to $2\mathcal{F}_{\mathrm{F}}(E_n)$, and the normalization condition may be written as

$$2\sum_{n} \mathcal{F}_{\mathrm{F}}(E_n) = N. \tag{A.30}$$

One can see that in accordance with the Pauli principle the occupation of any energy state, $\{n, s\}$, defined by Eq. (A.28), is always less than or equal to 1.

At high temperatures the Fermi distribution is close to the Boltzmann distribution,

$$\mathcal{F}_{\mathrm{F}}(E) \approx e^{(E_{\mathrm{F}} - E)/k_{\mathrm{B}}T}. \tag{A.31}$$

The corresponding curves are shown schematically in Fig. A.3 for different values of T. The evolution of $\mathcal{F}_{\mathrm{F}}(E)$ with temperature is illustrated by curves corresponding to temperatures $T = T_1, T_2, T_3, T_4$, where $T_1 = 0$. In the limit of low temperatures, $T \to 0$, the function $\mathcal{F}_{\mathrm{F}}$ is transformed into a step function: $\mathcal{F}_{\mathrm{F}}(E) = 1$ for the energy levels below the Fermi energy E_{F}, since all levels with $E < E_{\mathrm{F}}$ are occupied and $\mathcal{F}_{\mathrm{F}}(E) = 0$ for energies above E_{F}, since these levels are empty. In this limit, the electron subsystem is frequently referred to as a highly degenerate electron gas.

For a *three-dimensional electron gas*, the index n may be replaced by the wavevector $\vec{k}$ and the summation over $\vec{k}$ in the expression of Eq. (A.30) may be converted to an integration,

$$\sum_{\vec{k}} = \frac{\mathcal{V}}{(2\pi)^3} \int d\vec{k}, \tag{A.32}$$

where $\mathcal{V}$ is the volume of the system under consideration. Performing the integration in the spherical coordinate system and using the parabolic dispersion relation as in Eq. (A.22), one obtains from Eq. (A.30) the following expression for the Fermi energy of the degenerate electrons:

$$E_{\mathrm{F}} = (3\pi^2)^{2/3} \frac{\hbar^2 n_{3\mathrm{D}}^{2/3}}{2m^*}, \quad T \to 0, \tag{A.33}$$

with m^* the effective mass of the electrons. The Fermi energy increases as the two-thirds power of the electron concentration $n_{3\mathrm{D}} = N/\mathcal{V}$ (N is the total number of the electrons in the volume $\mathcal{V}$). Since the Fermi function contains the exponential factor, the low-temperature limit corresponds to

$$E_{\mathrm{F}} \gg k_{\mathrm{B}} T. \tag{A.34}$$

In metals and heavily doped semiconductors the electron gas remains degenerate up to room temperature.

Similarly, for *two-dimensional electrons* in the low-temperature limit defined by Eq. (A.34), we can find the Fermi energy in the form

$$E_{\mathrm{F}} = \frac{\pi \hbar^2}{m^*} n_{2\mathrm{D}}, \quad T \to 0, \tag{A.35}$$

i.e., it increases as the first power of the electron concentration $n_{2\mathrm{D}}$, where $n_{2\mathrm{D}} = N/S$ is the concentration per unit area.

For *one-dimensional electrons,* the Fermi energy is given by

$$E_{\mathrm{F}} = \frac{\pi^2 \hbar^2}{8m^*} n_{1\mathrm{D}}^2, \quad T \to 0, \tag{A.36}$$

where $n_{1\mathrm{D}} = N/L$ is the concentration per unit length of the one-dimensional electron gas.

All three of the formulae (A.33), (A.35), and (A.36) may be unified by introducing the Fermi wavevector k_{F}, i.e., the wavevector of electrons with the Fermi energy:

$$E_{\mathrm{F}} = \frac{\hbar^2 k_{\mathrm{F}}^2}{2m^*}. \tag{A.37}$$

Comparison of these three formulae gives k_{F} for all three dimensionalities:

$$k_{\mathrm{F,3D}} = (3\pi^2 n_{3\mathrm{D}})^{1/3}; \quad k_{\mathrm{F,2D}} = (2\pi n_{2\mathrm{D}})^{1/2}; \quad k_{\mathrm{F,1D}} = \frac{1}{2}\pi n_{1\mathrm{D}}. \tag{A.38}$$

From these results one can conclude that on lowering the dimensionality of the electron gas the effects of degeneracy become more pronounced. Another important conclusion is that for a light mass, m^*, the effects of degeneracy are always stronger; cf. Eqs. (A.33)–(A.36).

The concept of Fermi statistics is one of the fundamental ideas of modern solid-state physics, electronics, and optoelectronics.

Control Questions

1. If you know the exact position of a particle that obeys quantum-mechanical laws (i.e., $\Delta x = 0$), what is the uncertainty in its momentum, Δp?
2. Using a plane wave description of a free particle, what is the probability of finding a freely moving particle in any point of space?
3. What is the square magnitude of the particle's wave function?
4. What can we know from a particle's wave function?
5. The probability of finding a single electron during an experiment at point x is equal to 30%. What would be observed in a single experiment?
6. Which of the following entities are linear operators: $x, x\partial/\partial x, \partial^2/\partial^2 x$?
7. What does the stationary Schrödinger equation allow one to find?

Appendix B Tables of Units

Table B.1 SI base units

Quantity	Name	Symbol
Length	meter	m
Mass	kilogram	kg
Time	second	s
Electric current	ampere	A
Temperature	kelvin	K
Amount of substance	mole	mol

Table B.2 Conversion of SI units to Gaussian units

Quantity	SI unit	Gaussian unit
Length	1 m	10^2 cm
Mass	1 kg	10^3 g
Force	1 N	10^5 dyne = 10^5 g cm s^{-2}
Energy	1 J	10^7 erg = 10^7 g cm^2 s^{-2}
	1 eV	1.602×10^{-19} J = 1.602×10^{-12} erg

Table B.3 SI derived units

Quantity	Name	Symbol	Equivalent
Plane angle	radian	rad	m/m = 1
Solid angle	steradian	sr	$m^2/m^2 = 1$
Speed, velocity			$m\,s^{-1}$
Acceleration			$m\,s^{-2}$
Angular velocity			$rad\,s^{-1}$
Angular acceleration			$rad\,s^{-2}$
Frequency	hertz	Hz	s^{-1}
Force	newton	N	$kg\,m\,s^{-2}$
Pressure, stress	pascal	Pa	$N\,m^{-2}$
Work, energy, heat	joule	J	N m, $kg\,m^2\,s^{-2}$
Impulse, momentum			N s, $kg\,m\,s^{-1}$
Power	watt	W	$J\,s^{-1}$
Electric charge	coulomb	C	A s
Electric potential, emf	volt	V	$J\,C^{-1}$, $W\,A^{-1}$
Resistance	ohm	Ω	$V\,A^{-1}$
Conductance	siemens	S	$A\,V^{-1}$, Ω^{-1}
Magnetic flux	weber	Wb	V s
Inductance	henry	H	$Wb\,A^{-1}$
Capacitance	farad	F	$C\,V^{-1}$
Electric field strength			$V\,m^{-1}$, $N\,C^{-1}$
Magnetic flux density	tesla	T	$Wb\,m^{-2}$, $N\,A^{-1}\,m^{-1}$
Electric displacement			$C\,m^{-2}$
Magnetic field strength			$A\,m^{-1}$
Celsius temperature	degree Celsius	°C	
Luminous flux	lumen	lm	cd sr
Illuminance	lux	lx	$lm\,m^{-2}$
Radioactivity	becquerel	Bq	s^{-1}
Catalytic activity	katal	kat	$mol\,s^{-1}$

Table B.4 Physical constants

Constant	Symbol	Value	Units
Speed of light in vacuum	c	$2.9979 \times 10^8 \approx 3 \times 10^8$	$m\,s^{-1}$
Elementary charge	e	1.602×10^{-19}	C
Electron mass	m_e	9.11×10^{-31}	kg
Electron charge-to-mass ratio	e/m_e	1.76×10^{11}	$C\,kg^{-1}$
Proton mass	m_p	1.67×10^{-27}	kg
Boltzmann constant	k_B	1.38×10^{-23}	$J\,K^{-1}$
Gravitation constant	G	6.67×10^{-11}	$m^3\,kg^{-1}\,s^{-2}$
Standard acceleration of gravity	g	9.807	$m\,s^{-2}$
Permittivity of free space	ϵ_0	$8.854 \times 10^{-12} \approx 10^{-9}/(36\pi)$	$F\,m^{-1}$
Permeability of free space	μ_0	$4\pi \times 10^{-7}$	$H\,m^{-1}$
Planck's constant	h	6.6256×10^{-34}	J s
Impedance of free space	$\eta_0 = \sqrt{\mu_0/\epsilon_0}$	$376.73 \approx 120\pi$	Ω
Avogadro constant	N_A	6.022×10^{23}	mol^{-1}

Table B.5 Standard prefixes used with SI units

Prefix	Abbreviation	Meaning	Prefix	Abbreviation	Meaning
atto-	a-	10^{-18}	deka-	da-	10^{1}
femto-	f-	10^{-15}	hecto-	h-	10^{2}
pico-	p-	10^{-12}	kilo-	k-	10^{3}
nano-	n-	10^{-9}	mega-	M-	10^{6}
micro-	μ-	10^{-6}	giga-	G-	10^{9}
milli-	m-	10^{-3}	tera-	T-	10^{12}
centi-	c-	10^{-2}	peta-	P-	10^{15}
deci-	d-	10^{-1}	exa-	E-	10^{18}

Table B.6 Conversion of SI units to Gaussian units

Quantity	SI unit	Gaussian unit
Length	1 m	10^2 cm
Mass	1 kg	10^3 g
Force	1 N	10^5 dyne = 10^5 g cm s^{-2}
Energy	1 J	10^7 erg = 10^7 g cm^2 s^{-2}
	1 eV	1.602×10^{-19} J = 1.602×10^{-12} erg

Appendix C List of Pertinent Symbols*

a_0, a_i	lattice constants
c	velocity of light
d, L	geometrical sizes of samples
$\vec{D}$	dipole moment
E, $E(\vec{k})$	energy of electrons in a crystal
E_g	bandgap
E_F	Fermi energy, Fermi level
$\mathcal{E}$, $\mathcal{D}$	electric field and dielectric displacement
f, ω	frequency and angular frequency of wave
$\mathcal{F}$, $\mathcal{F}_F$	distribution functions
h, $\hbar = h/2\pi$	Planck's constant and reduced Planck constant
H, $\mathcal{H}$	Hamiltonian
$\mathcal{H}$, $\mathcal{B}$	magnetic field strength and magnetic field
$\mathcal{I}$, I	intensity of light wave
j, J	electron current density
$\vec{k}$, $\vec{p}$	wavevector and momentum of electron
k_B	Boltzmann's constant
m^*	effective mass of an electron in a solid
n	electron concentration
$n(\omega)$, n_{ref}	refractive index
R_{abs}, $R_{st.em}$	rates of absorption and emission processes
$\mathcal{V}_0$, $\mathcal{V}$	volume of primitive cell, volume of crystal
$u_{\alpha,\vec{k}}(\vec{r})$	Bloch functions
$V(\vec{r})$, $U(\vec{r})$	potential energy
$W(\vec{r})$	crystalline potential
W	density of electromagnetic energy
x, y, z, $\vec{r}$	coordinates, vector of coordinates
α	absorption/emission coefficients
ϵ, ϵ_i	energy of electron subbands
ϵ_0	electric permittivity of vacuum
κ, $\kappa(\omega)$	dielectric permittivity of medium

λ, $\vec{q}$	wavelength and wavevector of wave
$\rho(E)$	density of states
ϕ	phase, electrostatic potential
Φ	potential, applied voltage
$\psi(\vec{r})$, $\Psi(\vec{r})$, $\chi(z)$	wave functions

*Others symbols are self-evident and relate to the specific analysis

Further Reading

Further Reading for Chapter 1

The trends in optics and optoelectronics of quantum heterostructures and two-dimensional crystals, and the use of these nanostructures and devices, were analyzed in Chapter 1.

Progress Made on the Fabrication of Scaled-Down, Microelectronic, and Optoelectronic Devices, Quantum Structures, and Their General Properties and Perspectives

V. V. Mitin, V. A. Kochelap, and M. A. Stroscio, *Quantum Heterostructures for Microelectronics and Optoelectronics* (Cambridge University Press, New York, 1999).

J. Zhang, Z. L. Wang, J. Liu, S. Chen, and G.-Y. Liu, *Self-Assembled Nanostructures* (Kluwer Academic/Plenum Publishers, Dordrecht, 2003).

R. Soref, "Mid-infrared photonics in silicon and germanium," *Nat. Photon.*, **4**, 495–497 (2010).

D. Bimberg and U. W. Pohl, "Quantum dots: promises and accomplishments," *Mater. Today*, **14**, 388–397 (2011).

B. Rogers, J. Adams, and S. Pennathur, *Nanotechnology: The Whole Story* (CRC Press, Boca Raton, 2013).

Z. M. Wang, *MoS_2 Materials, Physics, and Devices* (Springer International Publishing, New York, 2014).

Technologies and General Trends Related to Two-Dimensional Crystals (in Particular to Graphene)

S. Shafraniuk, *Graphene: Fundamentals, Devices, and Applications* (Pan Stanford Publishing, Singapore, 2015).

A. K. Geim, *Random Walk to Graphene*, Nobel Lecture – 2010, www.nobelprize.org/nobel_prizes/physics/laureates/2010/geim-lecture.pdf.

K. S. Novoselov, *Graphene: Materials in the Flatland*, Nobel Lecture – 2010, www.nobelprize.org/nobel_prizes/physics/laureates/2010/novoselov-lecture.pdf.

T. Enoki and T. Ando, *Physics and Chemistry of Graphene: Graphene to Nanographene* (Pan Stanford Publishing, Singapore, 2013).

J. H. Warner, F. Schäffel, A. Bachmatiuk, and M. H. Rümmeli, *Graphene: Fundamentals and Emergent Applications* (Elsevier Science, Waltham, 2012).

A. R. Ashrafi, F. Cataldo, A. Iranmanesh, and O. Ori, eds., *Topological Modelling of Nanostructures and Extended Systems* (Springer Netherlands, Amsterdam, 2013).

M. J. O'Connell, ed., *Carbon Nanotubes: Properties and Applications* (Taylor and Francis, Basingstoke, 2006).

The Fascinating Story of the Invention of Short-Wavelength Light Emitters and Lasers

S. Nakamura, *Background Story of the Invention of Efficient Blue InGaN Light Emitting Diodes*, Nobel Lecture – 2014, www.nobelprize.org/nobel_prizes/physics/laureates/2014/nakamura-lecture.pdf.

General Facts of Light–Matter Interaction, Including Formation of Plasmons and Plasmon-Polaritons, and Effects on Sub-Wavelength Scales

S. A. Maier, *Plasmonics. Fundamentals and Applications* (Springer Science+Business Media LLC, Berlin, 2007).

V. M. Shalaev and S. Kawata, eds., *Nanophotonics with Surface Plasmons* (Elsevier, Amsterdam, 2007).

J. Weiner and F. Nunes, *Light–Matter Interaction: Physics and Engineering at the Nanoscale* (Oxford University Press, Oxford, 2017).

The Rapid Development of Sub-Wavelength Optics and the Emerging Field of Nanophotonics

R. Yan, D. Gargas, and P. Yang, "Nanowire photonics," *Nat. Photon.*, **3**, 569–576 (2009).

N. Xi and K. W. C. Lai, *Nano-Optoelectronic Sensors and Devices – Nanophotonics from Design to Manufacturing* (Elsevier, Amsterdam, 2012).

M. Ohtsu, ed., *Progress in Nanophotonics 2* (Springer, Berlin, 2013).

V. A. Soife, ed., *Diffractive Nanophotonics* (CRC Press, Boca Raton, 2014).

O. Minin and I. Minin, *Diffractive Optics and Nanophotonics: Resolution below the Diffraction Limit* (Springer, Berlin, 2016).

Further Reading for Chapter 2

Different materials for optoelectronic applications were considered in Chapter 2. There, the classification of dielectrics, semiconductors, and metals and general properties of the crystal lattices and crystal electrons were discussed, including basic types of semiconductor heterostructures, as well as two-dimensional monolayer crystals (graphene, silicene, black phosphorus, monochalcogenides, boron nitrides, etc.).

Elements of Solid-State Theory, Including Crystal Symmetry, Electron Energy Spectra, Classification of Solid-State Materials Important for Optoelectronics

C. Kittel, *Quantum Theory of Solids* (Wiley, New York, 1963).

I. Ipatova and V. Mitin, *Introduction to Solid-State Electronics* (Addison-Wesley, New York, 1996).

History of Semiconductor Heterostructures and Clear Motivation for Their Implementation

H. Kroemer, "Nobel Lecture: Quasielectric fields and band offsets: teaching electrons new tricks," *Rev. Mod. Phys.* **73**, 783–793 (2001).

Semiconductor Alloys and Heterojunctions, and Methods for Their Fabrication, Doping, and Processing

J. Singh, *Physics of Semiconductors and Their Heterostructures* (McGraw-Hill, New York, 1993).

V. V. Mitin, V. A. Kochelap, and M. A. Stroscio, *Quantum Heterostructures for Microelectronics and Optoelectronics* (Cambridge University Press, New York, 1999).

Calculations and Discussions of Electron Energy Spectra and Wave Functions in Various Quantum Semiconductor Structures, Including Si-Based Structures, III–V-Compound Structures, etc.

T. Ando, A. B. Fowler, and F. Stern, "Electronic properties of two-dimensional systems," *Rev. Mod. Phys.* **54**, 437–672 (1982).

G. Bastard, *Wave Mechanics Applied to Semiconductor Heterostructures* (Halsted, New York, 1988).

C. Weisbuch and B. Vinter, *Quantum Semiconductor Structures* (Academic, San Diego, 1991).

Results on Especially Important Wide-Bandgap Semiconductors

F. A. Ponce and D. P. Bour, "Nitride-based semiconductors for blue and green light emitting devices," *Nature*, **386**, 351 (1997).

H. Morkoc, *Nitride Semiconductors and Devices* (Springer, Berlin, 1999).

Carbon Nanotubes and Carbon Buckyballs

M. S. Dresselhaus, G. Dresselhaus, and P. C. Eklund, *Science of Fullerenes and Carbon Nanotubes* (Academic, San Diego, 1996).

P. Moriarty, "Nanostructural materials," *Rep. Prog. Phys.* **64**, 297–381 (2001).

Further Reading for Chapter 3

In Chapter 3 we presented the basic electronic properties of quantum heterostructures and two-dimensional crystals.

Models for the Potential Energy of Carriers Used to Describe the Principal Quantum Effects Occurring in Quantum Wells, Quantum Wires, and Quantum Dots (One-, Two-, and Three-Dimensional Models) as well as in Superlattices

L. I. Schiff, *Quantum Mechanics* (McGraw-Hill, New York, 1968).

R. P. Feynman, R. B. Leighton, and M. Sands, *The Feynman Lectures on Physics* (Addison-Wesley, New York, 1964), Vol. 3.

The Critically Important Effective-Mass Approximation, Discussions of Quantum Heterostructures with Complex Energy-Band Structures as well as Detailed Analysis of Particular Examples of the Heterostructures Based on the SiO_2/Si System and III–V Compounds

T. Ando, A. B. Fowler, and F. Stern, "Electronic properties of two-dimensional systems," *Rev. Mod. Phys.* **54**, 437–672 (1982).

G. Bastard, *Wave Mechanics Applied to Semiconductor Heterostructures* (Halsted, New York, 1988).

C. Weisbuch and B. Vinter, *Quantum Semiconductor Structures* (Academic, San Diego, 1991).

V. Mitin, D. Sementsov, and N. Vagidov, *Quantum Mechanics of Nanostructures* (Cambridge University Press, Cambridge, 2010).

Detailed Analysis of Novel Two-Dimensional Crystals, Including Graphene, Bigraphene, Transition-Metal Dichalcogenides, Black Phosphorus, etc.

A. H. Castro Neto, F. Guinea, N. M. R. Peres, K. S. Novoselov, and A. K. Geim, "The electronic properties of graphene," *Rev. Mod. Phys.* **81**, 109–162 (2009).

S. Das Sarma, S. Adam, E. H. Hwang, and Enrico Ross, "Electronic transport in two-dimensional graphene," *Rev. Mod. Phys*. **83**, 407–470 (2011).

K. F. Mak, C. Lee, J. Hone, J. Shan, and T. F. Heinz, "Atomically thin MoS_2: a new direct-gap semiconductor," *Phys. Rev. Lett.* **105**, 136805 (2010).

Z. Y. Zhu, Y. C. Cheng, and U. Schwingensch, "Giant spin–orbit-induced spin splitting in two-dimensional transition-metal dichalcogenide semiconductors," *Phys. Rev.* **B 84**, 153402 (2011).

D. Xiao, G.-B. Liu, W. Feng, X. Xu, and W. Yao, "Coupled spin and valley physics in monolayers of MoS_2 and other group-VI dichalcogenides," *Phys. Rev. Lett.* **108**, 196802 (2012).

A. Castellanos-Gomez, "Black phosphorus: narrow gap, wide applications," *J. Phys. Chem. Lett.* **6**, 4280–4291 (2015).

Further Reading for Chapter 4

Interactions of electromagnetic radiation with semiconductors including the processes of emission, absorption, and scattering of light, as well as various nonlinear optical phenomena, etc. were studied in Chapter 4.

The Optics of Solid-State, Including Semiconductor, Materials

C. Kittel, *Introduction to Solid State Physics* (Wiley, New York, 1986).

C. Kittel, *Quantum Theory of Solids* (Wiley, New York, 1963).

D. K. Ferry, *Semiconductors* (Macmillan, New York, 1991).

I. Ipatova and V. Mitin, *Introduction to Solid-State Electronics* (Addison-Wesley, New York, 1996).

J. I. Pancove, *Optical Processes in Semiconductors* (Prentice-Hall, New Jersey, 1971).

R. J. Elliott and A. F. Gibson, *An Introduction to Solid State Physics* (Harper & Row, New York, 1974).

J. D. Jackson, *Classical Electrodynamics*, 2nd edn (Wiley, New York, 1975).

The Elements of the Optics of Different Quantum Structures

V. V. Mitin, V. A. Kochelap, and M. A. Stroscio, *Quantum Heterostructures for Microelectronics and Optoelectronics* (Cambridge University Press, New York, 1999).

B. E. A. Saleh and M. C. Teich, *Fundamentals of Photonics*, 2nd edn (Wiley, New York, 2007).

E. Hecht, *Optics*, 4th edn (Pearson–Addison Wesley, San Francisco, 2002).

Wide-Bandgap Semiconductors

F. A. Ponce and D. P. Bour, "Nitride-based semiconductors for blue and green light emitting devices," *Nature* **386**, 351–359 (1997).

H. Morkoc, *Nitride Semiconductors and Devices* (Springer, Berlin, 1999).

S. Nakamura, *Background Story of the Invention of Efficient Blue InGaN Light Emitting Diodes*, Nobel Lecture – 2014, www.nobelprize.org/nobel_prizes/physics/laureates/2014/nakamura-lecture.pdf.

Further Reading for Chapter 5

The optical properties of quantum structures were presented in Chapter 5. There we studied the optical spectra, stimulated emission, and other optical characteristics of quantum structures, including the basic features of the optics of one- and few-monolayer crystals.

The Optics of Solid-State Structures, Including Semiconductor Quantum Structures

C. Kittel, *Introduction to Solid State Physics* (Wiley, New York, 1986).

C. Kittel, *Quantum Theory of Solids* (Wiley, New York, 1963).

D. K. Ferry, *Semiconductors* (Macmillan, New York, 1991).

I. Ipatova and V. Mitin, *Introduction to Solid-State Electronics* (Addison-Wesley, New York, 1996).

J. I. Pancove, *Optical Processes in Semiconductors* (Prentice-Hall, New Jersey, 1971).

R. J. Elliott and A. F. Gibson, *An Introduction to Solid State Physics* (Harper & Row, New York, 1974).

B. E. A. Saleh and M. C. Teich, *Fundamentals of Photonics*, 2nd edn (Wiley, New York, 2007).

E. Hecht, *Optics*, 4th edn (Pearson–Addison Wesley, San Francisco, 2002).

J. D. Jackson, *Classical Electrodynamics*, 2nd edn (Wiley, New York, 1975).

The Optics of Heterostructures Based on Compound Semiconductors

G. Bastard, *Wave Mechanics Applied to Semiconductor Heterostructures* (Halsted, New York, 1988).

L. Banyai and S. W. Koch, *Semiconductor Quantum Dots* (World Scientific, Singapore, 1993), Vol. 2.

C. Weisbuch and B. Vinter, *Quantum Semiconductor Structures* (Academic, San Diego, 1991).

P. Bhattacharya, *Semiconductor Optoelectronic Devices*, 2nd edn (Pearson, London, 1997).

D. Bimberg, M. Grundmann, and N. N. Ledentsov, *Quantum Dot Heterostructures* (Wiley, Chichester, 1999).

P. Harrison, *Quantum Wells, Wires and Dots* (Wiley, Chichester, 2000).

P. S. Zory, Jr., *Quantum Well Lasers* (Academic, Boston, 1993).

W. P. Risk, T. R. Gosnell, and A. V. Nurmikko, *Compact Blue–Green Lasers* (Cambridge University Press, Cambridge, 2003).

J. I. Pankove and T. D, Moustakas, *Gallium Nitride (GaN)* (Academic, New York, 1998), Vols. I and II.

Intraband Phototransitions

G. Bastard, *Wave Mechanics Applied to Semiconductor Heterostructures* (Halsted, New York, 1988).

M. O. Manasreh, ed., *Semiconductor Quantum Wells and Superlattices for Long-Wavelength Infrared Detectors* (Artech House, Boston, 1993).

Optical Properties of Two-Dimensional Crystals

A. H. Castro Neto, F. Guinea, N. M. R. Peres, K. S. Novoselov, and A. K. Geim, "The electronic properties of graphene," *Rev. Mod. Phys.* **81**, 109–162 (2009).

M. Orlita and M. Potemski, "Dirac electronic states in graphene systems: optical spectroscopy studies," *Semicond. Sci. Technol.* **25**, 063001 (2010).

Z. Y. Zhu, Y. C. Cheng, and U. Schwingensch, "Giant spin–orbit-induced spin splitting in two-dimensional transition-metal dichalcogenide semiconductors," *Phys. Rev.* **B 84**, 153402 (2011).

D. Xiao, G.-B. Liu, W. Feng, X. Xu, and W. Yao, "Coupled spin and valley physics in monolayers of MoS_2 and other group-VI dichalcogenides," *Phys. Rev. Lett.* **108**, 196802 (2012).

A. Castellanos-Gomez, "Black phosphorus: narrow gap, wide applications," *J. Phys. Chem. Lett.* **6**, 4280–4291 (2015).

Further Reading for Chapter 6

Electro-optical effects and nonlinear optical effects in quantum heterostructures, including quantum wells, double and multiple-quantum-well structures, and superlattices were considered in Chapter 6. We showed that these effects have great potential for the optoelectronic applications of using a field to control light frequency and intensity, and so realizing the tunable generation of emission of microwave, terahertz, infrared and visible radiation.

Electro-Optics and Nonlinear Optics

N. Bloembergen, *Nonlinear Optics* (Benjamin, New York, 1965).

R. Loundon, *The Quantum Theory of Light* (Clarendon Press, Oxford, 1973).

A. Yariv, *Quantum Electronics* (Wiley, New York, 1991).

B. E. A. Saleh and M. C. Teich, *Fundamentals of Photonics*, 2nd edn (Wiley, New York, 2007).

V. Mitin and D. Sementsov, *Introduction to Applied Electromagnetics and Optics* (CRC Press, Boca Raton, 2016).

D. A. B. Miller, D. C. Chemla, and S. Schmitt-Rink, "Electric field dependence of optical properties of semiconductor quantum wells," in *Optical Nonlinearities and Instabilities in Semiconductors*, H. Haug, ed. (Academic, Boston, 1988), p. 325.

D. S. Chemla, D. A. B. Miller, and S. Schmitt-Rink, "Nonlinear optical properties of semiconductor quantum wells," in *Optical Nonlinearities and Instabilities in Semiconductors*, H. Haug, ed. (Academic, Boston, 1988), p. 83.

Plasmonics and Sub-Wavelength Effects

S. A. Maier, *Plasmonics. Fundamentals and Applications* (Springer Science+Business Media LLC, Berlin, 2007).

W. L. Barnes, A. Dereux, and T. W. Ebbesen, "Surface plasmon subwavelength optics," *Nature* **424**, 824–830 (2003).

A. A. Maradudin, J. R. Sambles, and W. L. Barnes, eds., *Handbook of Surface Science in Modern Plasmonics* (Elsevier, Amsterdam, 2014), Vol. 4.

Further Reading for Chapter 7

In Chapter 7, we analyzed applications of quantum heterostructures to devices emitting near-infrared, visible, and ultraviolet radiation. These semiconductor devices exploit the phototransitions between the valence and conduction bands, i.e., the interband phototransitions studied in Chapters 4 and 5.

Heterostructure Emitters, Lasers, Lasers Based on Quantum Wells, Wires, and Dots

A. Yariv, *Quantum Electronics* (Wiley, New York, 1991).

B. E. A. Saleh and M. C. Teich, *Fundamentals of Photonics*, 2nd edn (Wiley, New York, 2007).

Y. Suematsu, K. Iga, and S. Arai, "Advanced semiconductor lasers," *Proc. IEEE* **80**, 383–397 (1992).

E. Kapon, "Quantum-wire lasers", *Proc. IEEE* **80**, 398–410 (1992).

P. S. Zory, Jr., *Quantum-Well Lasers* (Academic, Boston, 1993).

D. Bimberg, M. Grundmann, and N. N. Ledentsov, *Quantum-Dot Heterostructures* (Wiley, Chichester, 1999).

D. Bimberg and U. W. Pohl, "Quantum-dots: promises and accomplishments," *Mater. Today* **14**, 388–397 (2011).

Short-Wavelength Light-Emitting Diodes and Lasers

H. Morkoc, *Nitride Semiconductors and Devices* (Springer, Berlin, 1999).

W. P. Risk, T. R. Gosnell, and A. V. Nurmikko, *Compact Blue–Green Lasers* (Cambridge University Press, Cambridge, 2003).

H. Hirayama, "Quaternary InAlGaN-based high-efficiency ultraviolet light-emitting diodes," *J. Appl. Phys.* **97**, 091101 (2005).

Y. Taniyasu, M. Kasu, and T. Makimoto, "An aluminum nitride light-emitting diode with a wavelength of 210 nanometres," *Nature* **441**, 325–328 (2006).

S. Nakamura, *Background Story of the Invention of Efficient Blue InGaN Light Emitting Diodes*, Nobel Lecture – 2014, www.nobelprize.org/nobel_prizes/physics/laureates/2014/nakamura-lecture.pdf.

M. Kneiss and J. Rass, eds., *III-Nitride Ultraviolet Emitters* (Springer, Cham, 2016).

Further Reading for Chapter 8

Various optoelectronic devices which exploit intraband phototransitions were analyzed in Chapter 8. These include unipolar cascade lasers operating in the mid-infrared and terahertz ranges, quantum well photodetectors, quantum dot photodetectors, devices based on two-dimensional crystals, etc. Moreover, because of the great importance of the integration of optical systems with silicon, we analyzed silicon optoelectronics.

Quantum-Cascade Lasers

J. Faist, F. Capasso, D. L. Sivko, C. Sirtori, A. L. Hutchinson, and A. Y. Cho, "Quantum cascade laser," *Science* **264**, 553–556 (1994).

Q. Lu and M. Razeghi, "Recent advances in room temperature, high-power terahertz quantum cascade laser sources based on difference-frequency generation," *Photonics* **3**, 42 (2016).

M. A. Belkin and F. Capasso, "New frontiers in quantum cascade lasers: high performance room temperature terahertz sources", *Physica Scripta* **90**, 118002 (2015).

Z. M. Wang, ed., *Quantum Dot Devices* (Springer, New York, 2012).

Heterostructure Applications in Infrared Photodetectors

M. O. Manashreh, ed., *Semiconductor Quantum Wells and Superlattices for Long-Wavelength Infrared Detectors* (Artech House, Boston, 1993).

K. L. Wang and R. P. G. Karunasiri, "Infrared detectors using SiGe/Si quantum well structures," in *Semiconductor Interfaces, Microstructures, and Devices: Properties and Applications*, Z. C. Feng, ed. (Institute of Physics Publishing, Bristol, 1993).

W. W. Chow, S. W. Koch, and M. Sargent, *Semiconductor Laser Physics* (Springer, Berlin, 1994).

Silicon Optoelectronics

Q. Xu, B. Schmidt, S. Pradhan, and M. Lipson, "Micrometre-scale silicon electro-optic modulator," *Nature* **435**, 325–327 (2005).

J. Cardenas, C. B. Poitras, J. T. Robinson, K. Preston, L. Chen, and M. Lipson, "Low loss etchless silicon photonic waveguides," *Opt. Express* **17**, 4752–4757 (2009).

Perspectives of Two-Dimensional Crystals for Optoelectronic Applications

K. F. Mak, K. He, J. Shan, and T. F. Heinz, "Control of valley polarization in monolayer MoS_2 by optical helicity," *Nature Nanotech.* **7**, 494–498 (2012).

M. Koperski, *et al.*, "Single photon emitters in exfoliated WSe_2 structures," *Nature Nanotech.* **10**, 503–506 (2015).

A. Castellanos-Gomez, "Black phosphorus: narrow gap, wide applications," *J. Phys. Chem. Lett.* **6**, 428–491 (2015).

T. Cao, *et al.*, "Valley-selective circular dichroism of monolayer molybdenum disulphide," *Nat. Commun.* **3**, 887–889 (2012).

Background Materials

Basic Principles of Quantum Physics

L. Schiff, *Quantum Mechanics* (McGraw-Hill, New York, 1968).

D. Saxon, *Elementary Quantum Mechanics* (Holden-Day, San Francisco, 1968).

R. P. Feynman, R. B. Leighton, and M. Sands, *Lectures on Physics* (Addison-Wesley, New York, 1964), Vol. 3.

V. Mitin, D. Sementsov, and N. Vagidov, *Quantum Mechanics of Nanostructures* (Cambridge University Press, Cambridge, 2010).

Basic Solid-State Physics

C. Kittel, *Introduction to Solid State Physics* (Wiley, New York, 1986).

J. M. Ziman, *Principles of the Theory of Solids* (Cambridge University Press, London, 1972).

I. Ipatova and V. Mitin, *Introduction to Solid-State Electronics* (Addison-Wesley, New York, 1996).

V. V. Mitin, V. A. Kochelap, and M. A. Stroscio, *Quantum Heterostructures for Microelectronics and Optoelectronics* (Cambridge University Press, New York, 1999).

Index